Precision agriculture for sustainability

It is widely recognised that agriculture is a significant contributor to global warming and climate change. Agriculture needs to reduce its environmental impact and adapt to current climate change whilst still feeding a growing population, i.e. become more 'climate-smart'. Burleigh Dodds Science Publishing is playing its part in achieving this by bringing together key research on making the production of the world's most important crops and livestock products more sustainable.

Based on extensive research, our publications specifically target the challenge of climate-smart agriculture. In this way we are using 'smart publishing' to help achieve climate-smart agriculture.

Burleigh Dodds Science Publishing is an independent and innovative publisher delivering high quality customer-focused agricultural science content in both print and online formats for the academic and research communities. Our aim is to build a foundation of knowledge on which researchers can build to meet the challenge of climate-smart agriculture.

For more information about Burleigh Dodds Science Publishing simply call us on +44 (0) 1223 839365, email info@bdspublishing.com or alternatively please visit our website at www.bdspublishing.com.

Related titles:

Managing soil health for sustainable agriculture Volume 1: Fundamentals
Print (ISBN 978-1-78676-188-0) Online (ISBN 978-1-78676-190-3; 978-1-78676-191-0)

Managing soil health for sustainable agriculture Volume 2: Monitoring and management
Print (ISBN 978-1-78676-192-7) Online (ISBN 978-1-78676-194-1; 978-1-78676-195-8)

Integrated weed management for sustainable agriculture
Print (ISBN 978-1-78676-164-4) Online (ISBN 978-1-78676-166-8; 978-1-78676-167-5)

Water management for sustainable agriculture
Print (ISBN 978-1-78676-176-7) Online (ISBN 978-1-78676-178-1; 978-1-78676-179-8)

Chapters are available individually from our online bookshop: https://shop.bdspublishing.com

climate•SMART•publishing
IN AGRICULTURAL SCIENCE

BURLEIGH DODDS SERIES IN AGRICULTURAL SCIENCE

NUMBER 52

Precision agriculture for sustainability

Edited by Dr John Stafford, Silsoe Solutions, UK

burleigh dodds
SCIENCE PUBLISHING

Published by Burleigh Dodds Science Publishing Limited
82 High Street, Sawston, Cambridge CB22 3HJ, UK
www.bdspublishing.com

Burleigh Dodds Science Publishing, 1518 Walnut Street, Suite 900, Philadelphia, PA 19102-3406, USA

First published 2019 by Burleigh Dodds Science Publishing Limited

Library of Congress Control Number: 2018961023

British Library Cataloguing in Publication Data
A catalogue record for this book is available from the British Library

ISBN 978-1-78676-204-7 (Print)
ISBN 978-1-78676-207-8 (PDF)
ISBN 978-1-78676-206-1 (ePub)
ISSN 2059-6936 (print)
ISSN 2059-6944 (online)

DOI 10.19103/AS.2017.0032

Typeset by Deanta Global Publishing Services, Chennai, India
Printed by Lightning Source

Contents

Series list

Title	Series number
Achieving sustainable cultivation of maize - Vol 1 From improved varieties to local applications Edited by: Dr Dave Watson, CGIAR Maize Research Program Manager, CIMMYT, Mexico	001
Achieving sustainable cultivation of maize - Vol 2 Cultivation techniques, pest and disease control Edited by: Dr Dave Watson, CGIAR Maize Research Program Manager, CIMMYT, Mexico	002
Achieving sustainable cultivation of rice - Vol 1 Breeding for higher yield and quality Edited by: Prof. Takuji Sasaki, Tokyo University of Agriculture, Japan	003
Achieving sustainable cultivation of rice - Vol 2 Cultivation, pest and disease management Edited by: Prof. Takuji Sasaki, Tokyo University of Agriculture, Japan	004
Achieving sustainable cultivation of wheat - Vol 1 Breeding, quality traits, pests and diseases Edited by: Prof. Peter Langridge, The University of Adelaide, Australia	005
Achieving sustainable cultivation of wheat - Vol 2 Cultivation techniques Edited by: Prof. Peter Langridge, The University of Adelaide, Australia	006
Achieving sustainable cultivation of tomatoes Edited by: Dr Autar Mattoo, USDA-ARS, USA & Prof. Avtar Handa, Purdue University, USA	007
Achieving sustainable production of milk - Vol 1 Milk composition, genetics and breeding Edited by: Dr Nico van Belzen, International Dairy Federation (IDF), Belgium	008
Achieving sustainable production of milk - Vol 2 Safety, quality and sustainability Edited by: Dr Nico van Belzen, International Dairy Federation (IDF), Belgium	009
Achieving sustainable production of milk - Vol 3 Dairy herd management and welfare Edited by: Prof. John Webster, University of Bristol, UK	010
Ensuring safety and quality in the production of beef - Vol 1 Safety Edited by: Prof. Gary Acuff, Texas A&M University, USA & Prof. James Dickson, Iowa State University, USA	011
Ensuring safety and quality in the production of beef - Vol 2 Quality Edited by: Prof. Michael Dikeman, Kansas State University, USA	012
Achieving sustainable production of poultry meat - Vol 1 Safety, quality and sustainability Edited by: Prof. Steven C. Ricke, University of Arkansas, USA	013
Achieving sustainable production of poultry meat - Vol 2 Breeding and nutrition Edited by: Prof. Todd Applegate, University of Georgia, USA	014
Achieving sustainable production of poultry meat - Vol 3 Health and welfare Edited by: Prof. Todd Applegate, University of Georgia, USA	015

Introduction

Precision agriculture is based on the ability to identify within-field spatial variability and to use this information for more targeted crop management. By using resources more efficiently, precision agriculture can make agriculture more productive, sustainable and reduce its environmental impact. This volume reviews the key advances in precision agriculture technology and its range of applications. It offers a comprehensive review of key technologies in precision agriculture, from proximal and remote sensing to decision support systems and variable rate technologies The volume surveys key applications of precision agriculture from controlled traffic farming to site-specific nutrient and water management, and includes discussion of the economics of precision agriculture.

Part 1 Information gathering and processing

The chapters in the first part of the volume look at monitoring techniques, including proximal soil and crop sensors and remote sensing technologies. Today's farmers are increasingly reliant on sensors in their farming operations. The theme of Chapter 1 is the use of proximal crop sensor technologies to evaluate a crop during the growing season. After briefly discussing crop properties, the chapter reviews the evolution of crop sensors and then examines three issues in current sensor development: the limitations of current crop canopy sensors; sensing for stresses other than nitrogen; and sensor fusion and high-throughput phenotyping. The chapter includes three detailed case studies showing crop sensing in practice.

Complementing the previous chapter's focus on proximal crop sensor technologies, Chapter 2 looks at proximal soil surveying and monitoring techniques. Understanding soil fertility is fundamental to agricultural production. However, soils can show great variation both spatially and temporally. The chapter outlines the key challenges faced in monitoring such variation and the solutions that have been developed to aid in optimizing crop production. The chapter focuses on proximal soil sensing where a sensor is placed at a very short distance from the soil surface and contact sensors where the sensor is in contact with or buried in the soil. These types of sensors provide information which is particularly useful in developing high resolution soil maps for use in precision agriculture. After discussing sampling methods, the chapter reviews a wide array of sensors, focusing on how such sensors may be integrated into agricultural management.

Moving from proximal to remote sensing systems, Chapter 3 considers airborne and satellite remote sensors for precision agriculture. An overview of airborne imaging systems and multispectral systems based on industrial and consumer-grade cameras is offered, before going on to consider airborne hyperspectral sensors. Examples of the application of airborne imaging systems are reviewed. The chapter offers an overview of high-resolution satellite sensors and the variety of systems available, as well as examples of the application of satellite sensors.

Chapter 4 continues the theme of remote sensing systems by considering the use of unmanned aerial vehicles (UAVs) in precision agriculture. The rapid development of small unmanned aerial vehicles (UAVs) in recent years has triggered considerable interest in their application for precision agriculture. Decreasing cost, ultra-high spatial resolution, and the increased flexibility of image acquisition have made UAS remote sensing an ideal tool for identifying field patterns in close-to-real-time. The chapter provides an overview of UAV platforms and sensors, and flight planning and imagery acquisition, before moving on to consider stitching and ortho-rectification in UAV image processing. The chapter reviews the applications of UAV imagery and methods of image analysis, and includes a detailed case study of practical applications.

The subject of Chapter 5 is the key challenges and methods in identifying management zones. Management zones are sub-regions in a field with relatively homogeneous characteristics which allow targeted treatment (e.g. with fertilizers or pesticides). The chapter examines methods to delineate management zones, including empirical, geostatistical and clustering methods. The chapter offers three detailed case studies, focusing on the delineation of management zones in wine grapes, table grapes and olive oil tree plantations which then allow the use of variable-rate and other delivery systems. The section's final chapter, Chapter 6, looks at modelling and decision support systems (DSSs) in precision agriculture. The chapter examines the key issues associated with deploying DSS in precision agriculture before moving on to consider the human and social aspects of DSS. The chapter provides a number of detailed case studies of the application of DSS, with an emphasis on nitrogen management. Finally, the chapter considers research options for decision support systems to improve productivity in a precision agriculture framework.

Part 2 Delivery systems

The focus of the second part of the volume is the delivery systems required in precision agriculture. Chapter 7 considers variable rate application (VRA) technologies using GNSS (global navigation satellite systems) for accurate location. The application of precision agriculture can be viewed as cyclical in nature, involving data collection, development of management plans, implementing those plans in the field and evaluation of their effectiveness. The chapter examines the characteristics of variable-rate application (VRA) control systems (both liquid and dry), before going on to consider the implementation of both liquid and dry VRA systems. The chapter reviews a number of detailed VRA case studies and summarises the current status of research in this field.

Chapter 8 moves on to examine the deployment of spray technologies in precision agriculture. The application of plant protection products plays a key role in the production of most crops. Equipment designed to operate with field crops are examined (although many of the principles discussed also relate to other application systems). The chapter describes features of field crop sprayers for precision agriculture, including control of delivered dose, spatial resolution, matching physical characteristics of sprays to target requirements, and minimising drift and exposure of systems outside the treatment area. Two detailed case studies are included which focus on designing and developing a system for spot treatment of volunteer potatoes and implementing a patch spraying system for applying herbicides to field crops.

Complementing the themes of the preceding chapters, Chapter 9 looks at the use of intelligent machinery for precision agriculture. The innovative core of modern precision agriculture is the use of agricultural machinery to perform precise, responsive field operations in large-scale mechanized precision crop production. Integration of autonomous functions into such machinery endows it with capabilities which could be described as 'intelligent'. These include perception, reasoning and control of operations in the field with the flexibility to achieve predetermined operational goals without human supervision. The chapter reviews the current state-of-the-art in intelligent agricultural machinery and discusses automated guidance systems based on GNSS (global navigation satellite systems), as well as path planning, automated actuation systems and implement controls. Finally, the chapter discusses the prospects for creating increasingly autonomous systems and considers the potential implications of such technology.

The concluding chapter of the section, Chapter 10, looks at controlled traffic farming (CTF) in precision agriculture. In the past few decades, there has been a continuous drive towards the development and adoption of larger, and more powerful, agricultural machinery. Larger machinery is often related with timeliness, higher work rates and lower labour requirements, which has led to significant improvements both in efficiency and productivity but with the serious drawback of damage to soil structure. The chapter defines and outlines the requirements of controlled traffic farming systems based on GNSS (global navigation satellite systems), examines the sustainability of controlled traffic farming and looks at the feasibility of coupling controlled traffic farming with precision agriculture.

Part 3 Applications

The third and final section of the volume examines applications for precision agriculture technologies, beginning with a consideration of precision tillage systems in Chapter 11. The commercial development of information-intensive precision agriculture technologies has given rise to a number of applications aimed at improving agricultural production through site-specific management, including precision tillage. The chapter describes developments in on-the-go soil strength sensing and the state of the art in sensor technology as applied to precision tillage. The chapter examines depth-prescribed tillage to control subsoil compaction, techniques of soil sensing to enable precision tillage, and how to extract information from soil sensors. Finally, the chapter looks at the challenge of implementing depth control and tractor/implement guidance technology.

Chapter 12 moves from precision tillage to consider variable seeding systems for precision agriculture. Variable-rate seeding (VRS) is the practice of adjusting the seeding rate or amount across a field in order to maximise yield and profit. This method of precision agriculture relies on site-specific data such as soil type, landscape position and yield maps. In the last decade, adoption of this method has dramatically increased as new planter technologies have emerged. The chapter summarises current practices in VRS, reviewing the available technology and discussing the key strategies in its implementation. It also discusses the economic value of VRS and the conditions in which it could produce the greatest economic return.

The theme of Chapter 13 is site-specific nutrient/fertiliser management systems. The chapter describes the processes that inform site-specific nutrient management, including grid sampling, construction of management zones, in-season crop sensing and grain quality sensing. The chapter considers regional perspectives on nutrient/fertiliser management systems, including case studies taken from both the sub-humid US Midwest and semi-arid US West.

Continuing with the theme of site-specific systems, Chapter 14 considers site-specific irrigation systems. Conventional irrigation management is based on uniform application of water across a field, which may result in both over- and under-irrigation. The chapter focuses on site-specific variable rate irrigation (SS-VRI) using center pivot and lateral move irrigation systems, which are mainly used for row-crop irrigation. Via detailed case studies, the chapter examines site-specific data acquisition and mining approaches, soil mapping, zone delineation, site-specific yield estimation and irrigation scheduling. The chapter also looks ahead to new forms of irrigation that may become available in the future for fixed sprinkler, drip and surface irrigation systems, as well as potential applications in nursery and urban irrigation settings.

The theme of Chapter 15 is precision crop protection systems. The spatial distribution of plant pests within crops is often heterogeneous particularly during growth periods. Within-field heterogeneity implies that plants at some sites may need pest control while plants at other sites do not. Automated detection, identification and quantification of diseases and invertebrate pests on a small scale are the prerequisites for a site-specific pest management strategy. The chapter reviews how sensor technologies in combination with informatics and modern application technologies may enable pest control where and when it is actually needed. The chapter summarizes and discusses the potential of aerial and proximal remote sensing using various types of sensors for applications ranging from quarantine, to the production of speciality crops and crop loss assessment.

Complementing the preceding chapter's focus on crop protection, Chapter 16 looks specifically at precision weed management systems. Weed populations in arable fields are often spatially and temporally heterogeneous, and high-density weed patches require intensive weed management. In areas with no or few weeds, input for weed control can be reduced. Manual and GIS-based weed mapping approaches and sensor-based weed identification technologies and their application to create weed maps are described. The chapter also discusses the adoption of patch sprayers based on two alternative approaches; application maps and real-time sensor information. It is shown that site-specific weed control in cereals, oil-seed rape, sugar beet and maize using GNSS-controlled patch sprayers can reduce herbicide input by 20–60%. Looking ahead, the development of precision farming technologies for mechanical weeding, including automatic steering systems for intra-row hoeing and robotic inter-row weeding are considered.

Chapter 17 examines precision livestock farming and pasture management systems. Animal management seeks to first measure the variability of animal production, disease or welfare and then prescribe targeted management strategies and treatments. The chapter reviews a range of technologies currently applied or in development which facilitate this individual animal management across the spectrum of extensive grazing production systems using GNSS (global navigation satellite systems). The challenges facing the livestock industry and the potential solutions offered by precision management, focussing particularly on precision pasture management, are examined. The chapter includes detailed case studies on using a basic farm GIS to determine sustainable long term stocking rates, and integrating satellite multispectral imaging with plant growth modelling to manage livestock rotations.

The final chapter of the volume, Chapter 18, focuses on the economics of precision agriculture. Topics reviewed include data on use of precision agriculture in the production of agronomic crops, whether economic research has correctly predicted adoption of precision agriculture technologies by farmers. Several precision agriculture technologies that recent economic research has identified as having profit and adoption potential are discussed. The chapter examines the adoption of precision agriculture technology, the economics of precision agriculture adoption, and seeks to predict future trends based on recent studies of precision agriculture profitability.

Part 1

Information gathering and processing

Chapter 1

Proximal crop sensing

Richard B. Ferguson, University of Nebraska-Lincoln, USA

1 Introduction

1.1 The importance of sensors

The use of sensors to evaluate crop status is, in one sense, ancient; farmers have manually evaluated plants and crop development from the beginning of crop cultivation. They could observe if individual plants appeared to be growing normally, and could begin to estimate yield potential when fruiting structures developed as plants matured. They learned to associate certain symptoms with specific stresses and could differentiate water deficit from nutrient deficiency and foliar disease from insect infestation. Farmers today are no different from their ancestors in this regard. However, one challenge for today's crop producer is scale. While a farmer a thousand years ago might manage a fraction of a hectare manually and observe his crops on a daily basis, farmers today may manage hundreds or even thousands of hectares, enabled by mechanization, herbicides, irrigation and other tools of modern agriculture. Consequently it is not possible for a farmer today to be physically in contact with crops during the growing season in the same way as their ancestors.

Today, farmers are increasingly reliant on sensors in their farming operations. All the equipment used in crop production – tractors, planters, combines and so forth – have a multitude of sensors to help optimize machine operation. Sensors are also becoming increasingly useful in monitoring crop and soil conditions. This chapter addresses the use of proximal sensors to evaluate a crop during the growing season. The term 'proximal' indicates the sensor is in close proximity to the crop. Thus, its use is differentiated from other sensor applications such as remote sensing (using aerial or satellite sensor platforms)

http://dx.doi.org/10.19103/AS.2017.0032.02

or soil sensing (using proximal or *in situ* sensors to evaluate soil properties, rather than crop properties). The term 'proximal' is fuzzy, as developments in unmanned aerial systems (UASs) have brought into the proximal realm what had been historically classified as remote sensing technologies. In this chapter, proximal crop sensing will refer to any sensor used to evaluate the properties of a crop, ranging from physical contact with the crop canopy to a few metres above the canopy.

1.2 Crop properties of interest

There are a variety of crop properties for which proximal sensing may be useful. From the perspective of the crop producer, properties that influence yield, or in some cases crop quality, and issues that can be detected and managed if necessary during the growing season are of greatest interest. Biomass accumulation, crops' water status, nutrient deficiency (particularly nitrogen), disease onset and weed and insect infestation are all factors that crop growers may wish to monitor throughout the season. For a number of these properties there are specific periods during the growing season where monitoring is particularly critical, depending on the crop and region in which it is grown. Proximal sensing, as opposed to remote sensing, has a potential advantage in that sensors on satellite and aerial platforms will be more influenced by weather and clouds, potentially limiting the utility of remote sensing at critical times in the growing season.

2 The evolution of crop sensors

2.1 Contact or *in situ* sensors

Crop sensors typically infer specific crop properties, rather than directly measuring them. There are some exceptions to this generalization, mostly sensors directly attached to or placed among plants. Sap flow sensors, for example, can be attached directly to a plant stem to estimate transpiration (Steinberg et al., 1989). Leaf area index (LAI) can be measured with ground-based sensors as well as with destructive methods (Bréda, 2003). However, measurement of sap flow and LAI with static instruments is time-consuming and does not provide spatial information, at least without intensive effort. A plant contact sensor which is a relatively simple biomass sensor developed for use with cereals is the Crop-meter (Ehlert and Dammer, 2006). This sensor, based on the pendulum principle, is mounted on a vehicle driven through the crop. The greater the biomass of the standing crop, the more the Crop-meter is deflected. For small-grain cereals, crop biomass at later growth stages can be a good predictor of grain yield potential, and thus, can be used for site-specific management of nitrogen fertilizer, growth regulators and fungicide in real time based on Crop-meter input.

2.2 Ranging sensors

Another sensor approach to estimate biomass and or crop height is range-finding, or distance measurement, using acoustic, laser or radar sensors. A review article by Dworak et al. (2011) cites a range of approaches using acoustic and electromagnetic (EM) wave ranging to assess crop stand, estimate biomass or evaluate canopy characteristics. Lee and Ehsani (2009), for example, demonstrated use of a laser scanner to quantify geometric characteristics of citrus trees, which can be useful for yield prediction, water consumption estimation, health monitoring and long-term productivity monitoring. Shiratsuchi (2011) demonstrated that integration of acoustic ranging sensors to estimate crop canopy height with multispectral and thermal sensors was useful for the detection of water stress in maize.

2.3 Electromagnetic (EM) sensors

The majority of crop canopy sensors in use or development today utilize the EM spectrum to characterize crop properties (Fig. 1). Such sensors can be passive – using reflected, scattered or emitted energy – or active, using an internal pulsed or modulated energy source and measuring reflectance or fluorescence from that source. Radar, for example, is an active sensor. Most crop canopy sensors use reflectance, or in some cases transmission, in visible and near-infrared wavelengths to infer plant properties. The EM spectrum of interest ranges between the visible (400–700 nm), near-infrared (700–1300 nm), mid-infrared (1300–2500 nm) and thermal infrared regions (typically the principal region of interest is 8000–14 000 nm). The primary factors influencing light absorption/reflectance in the visible portion of the spectrum are plant pigments, such as chlorophyll, carotenoids and anthocyanins. Cell structure, water content and canopy architecture primarily influence canopy reflectance in the near-infrared region, and water content in leaves influences reflectance in the mid-infrared region (Fig. 1).

In many cases, absolute reflectance in a given wavelength is not very informative about plant stress. Reflectance is related to various plant attributes, such as plant pigment concentrations, cellular structure, leaf water content or canopy architecture. The concept of a vegetation index (VI) allows the inference of specific plant properties and sources of stress. A VI uses the relationship of reflectance in two or more spectral regions, combined in an equation, to infer plant properties. More than 150 VIs have been described, with the most

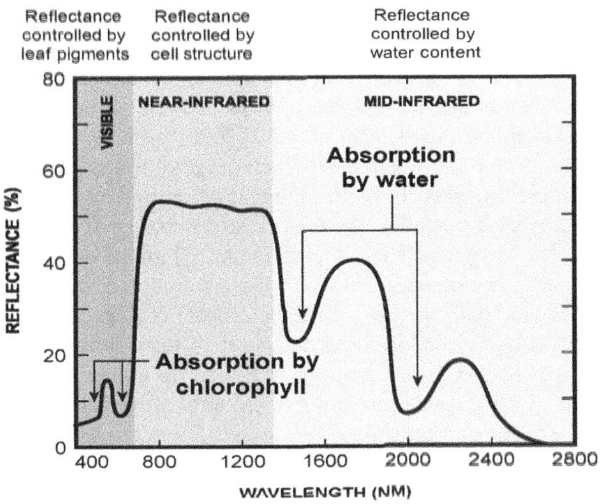

Figure 1 Primary factors influencing plant canopy reflectance in visible-to-mid-infrared wavelengths.

widely known being the normalized difference vegetation index (NDVI) (Rouse et al., 1974). The NDVI uses the equation below, resulting in values ranging from −1.0 to 1.0.

$$NDVI = (NIR - RED)/(NIR + RED) \quad (1)$$

Wallihan (1973) described the development of a portable reflectance meter to estimate chlorophyll content of plant leaves. This can be thought of as a precursor to the chlorophyll meter, which is an instrument commonly used in plant research today. The most commonly used chlorophyll meter available today is the Konica Minolta SPAD 502 Plus (Konica Minolta, 2017) which measures light transmittance through a leaf surface at wavelengths of 650 and 940 nm when a light source and detector clamp are placed around a plant leaf. Researchers have documented the performance of this meter or preceding models in estimating leaf chlorophyll concentration (Markwell et al., 1995), and, given the high correlation between leaf chlorophyll and leaf nitrogen content, the capacity of the chlorophyll meter to detect N stress and predict the need for N fertilization (Piekielek and Fox, 1992; Follett et al., 1992; Blackmer and Schepers, 1995; Bundy and Andraski, 2004; Scharf et al., 2006). The SPAD-502 meter was adapted by Yara International ASA (Oslo, Norway) (N-Tester) with calibrations for N management of various crops. Ortuzar-Iragorri et al. (2005) found that normalization of N-Tester measurements with a non-N-limiting reference enabled the prediction of plant N concentration for soft red winter wheat. Varvel et al. (2007) described the concept of normalizing chlorophyll meter readings within a field for N fertilizer management of maize, using a non-limiting fertilizer N rate as a reference. This concept, termed a sufficiency index (SI), recognized that absolute chlorophyll meter readings may vary with cultivar, soil N supply and spatial distribution, within a field. In this concept, chlorophyll meter information is calibrated to a specific cultivar and field. Such calibration can also normalize for stresses other than N, such as sulphur, that may influence chlorophyll levels within leaves. Varvel et al. (2007) developed a fertilizer N recommendation algorithm for maize based on the SI concept.

Another type of handheld sensor uses principles of EM transmission as well as fluorescence to estimate leaf pigments and, consequently, nutrient stress or other abiotic stresses. The Dualex Scientific™ (Force-A, 2017) clips to a leaf and uses light-emitting diodes (LEDs) in ultraviolet (UV), visible and NIR wavelengths to estimate chlorophyll and polyphenol content. Polyphenols have been shown to be associated with plant stress factors such as N availability (Samborski et al., 2009). Tremblay et al. (2007) found that use of the Dualex in combination with a SPAD chlorophyll meter to calculate a ratio was more sensitive in detecting corn N status than using either instrument alone. Force-A company also manufactures portable handheld fluorometers measuring at UV and visible wavelengths (Multiplex Research™ and Multiplex 330™) intended for the assessment of abiotic stresses, in particular, disease detection (Sankaran and Ehsani, 2012).

While a contact sensor such as the chlorophyll meter may provide useful information about a given plant, it becomes laborious to use it to represent a group of plants or a field. Researchers have found the chlorophyll meter to be a useful proximal sensing tool in small-plot research, where readings from many plants within a plot can be averaged to represent effects of a treatment on chlorophyll levels. There are extension publications on the use of a chlorophyll meter for crop management (Shapiro et al., 2013); yet, the labour challenges associated with chlorophyll meter use preclude widespread adoption by crop producers for N fertilizer management.

2.4 Mobile EM sensors

Passive sensor systems

Among the first mobile EM sensors intended for commercial, field-scale evaluation of crop N status was the Yara N-Sensor™ (Yara International ASA, 2017). The N-Sensor™ is a passive, tractor-mounted spectrometer system consisting of two spectrometers, with one to scan the crop canopy to the side of the vehicle, and the other to measure ambient light to correct the reflected signal in real time in selected wavelengths between 450 and 900 nm (Zebarth et al., 2003; Samborski et al., 2009). Reflectance is used to calculate NDVI or other VIs of interest. The tractor-mounted system then uses an algorithm to calculate an optimal N fertilizer rate for the scanned region and relay this information to a controller to vary the application rate of N fertilizer on the go. The N-Sensor™ has been used primarily for spatial N management of small grains such as wheat (Berntsen et al., 2006; Jørgensen and Jørgensen, 2007), but has also been used for corn at early growth stages (Tremblay et al., 2009) and potato (Zebarth et al., 2003).

Active sensor systems

While a passive system can be potentially corrected for ambient light as with the N-Sensor™, there are still challenges with clouds, sun angle and early/late time of day use. To precisely calculate reflectance, passive sensors need to measure incident light, to know the characteristics of the incoming light resource and to measure reflectance from the crop canopy at the same time (Raun et al., 2001). Consequently, active sensor systems which use an internal modulated light source to address the limitations of passive sensor systems have been developed (de Souza et al., 2010). Holland et al. (2012) described the radiometric principles of proximal active optical sensors, in particular, the inverse-square law of optics and how that influences operation and use of active sensors. Raun et al. (2002) first described the use of an active crop canopy sensor, measuring reflectance at 671 ± 6 and 780 ± 6 nm, to control the rate of fertilizer N applied to winter wheat. Sensor-based in-season N application improved nitrogen use efficiency (NUE) by >15% compared to traditional practices with uniform N rates. This system, developed at Oklahoma State University, became the GreenSeeker system (Trimble Inc., Sunnyvale CA, USA; Trimble, 2017a), which was one of the first commercially available active crop canopy sensors for N fertilizer management. The system relies on an in-field reference which is considered to be non-limiting for N in order to calibrate the sensor to specific cultivar and field conditions. The system produces NDVI values; in-field reference NDVI is divided by NDVI from the target area to generate a response index (RI – essentially the inverse of the SI described by Varvel et al., 2007). The RI is then used with an in-season estimate of yield (INSEY – Lukina et al., 2001) in a variety of crop- and locale-specific algorithms to generate an on-the-go fertilizer N rate. The GreenSeeker system can be used for N management of a variety of crops, but is mostly used for N management with corn at V6–V14 growth stages (Abendroth et al., 2011) and winter wheat at Feekes growth stages 4–6 (Feekes, 1941; Teal et al., 2006; Raun et al., 2002). Recently Trimble has developed a handheld version of this sensor for scouting purposes with an integrated power supply, a data logger and a GPS.

Another set of active crop canopy sensors is the Crop Circle suite of sensors (Holland Scientific, Lincoln, NE, USA). The initial sensor developed by Holland Scientific was the

ACS-210, a two-band active sensor measuring reflectance at wavebands of 590 ± 5.5 and 880 ± 10 nm from an internal, modulated LED light source. Solari et al. (2010) described the development of an N rate recommendation algorithm for this sensor for corn at V11 and V15 growth stages, which was based on prior relationships of chlorophyll meter readings to N rate by Varvel et al. (2007). The ACS-210 is no longer commercially available and has been replaced by Holland Scientific with several sensors primarily for research purposes: the ACS-430, a three-band sensor (670, 730 and 780 nm); the ACS-470, a three-band sensor with interchangeable filters (user-selected between 420 and 800 nm) and the RapidScan CS-45, a handheld unit with the same optics as the ACS-430 and an integrated power supply, a data logger and a GPS.

The Holland Scientific set of sensors are intended primarily for research use. For commercial implementation, the technology is incorporated into the OptRx® crop sensor system (Ag Leader, Inc., Ames, IA, USA). The OptRx® system uses the same three-band sensor system as the Holland Scientific ACS-430, measuring reflectance at 670, 730 and 780-nm wavebands. One of the advantages of the OptRx® and other sensors using a red-edge band (730 nm) in place of the red band in the NDVI equation is the greater sensitivity of reflectance in the red-edge region to canopy chlorophyll and thus N content with high biomass plant canopies. Reflectance saturation occurs in the red region with high biomass crops such as corn as growth advances beyond LAI of ~ 2–3 (Thenkabail et al., 2000; Gitelson et al., 2005). The OptRx® system requires in-field calibration as does the GreenSeeker system and can be used with a managed high N reference area. However, the system is designed primarily for use with a universal N recommendation algorithm approach which can be adjusted by the user (Holland and Schepers, 2010) and using a virtual reference approach (Holland and Schepers, 2013). The virtual reference approach uses the top 5% of VI readings in an area of the field considered to represent the range of N status of the crop for the field and uses this value in the SI calculation. This approach precludes the need for the grower to establish a high N reference strip in the field.

Another active crop canopy sensor system available today is the Yara™ N-Sensor ALS (active light source), first offered commercially in 2005. A similar system is also marketed as the Topcon CropSpec™ crop canopy sensor in some countries. The N-Sensor ALS system uses a xenon flash lamp light source rather than an LED as in GreenSeeker and Crop Circle/OptRx systems. Other operating characteristics of the Yara™ N-Sensor ALS are similar to those of the passive sensor system N-Sensor™. The CropSpec™ system uses a laser light source, measuring reflectance in 730–740 and 800–810-nm bands, with other operating characteristics similar to those of the N-Sensor™ and N-Sensor™ ALS.

3 Current issues in sensor development

3.1 Crop canopy sensor limitations

The use of EM-based crop canopy sensors, particularly active sensors, has tremendous potential to improve the accuracy of N management through both refined rate and application timing. However, there are factors which can be sources of error and influence the accuracy of N fertilizer recommendations based on canopy sensor information. Samborski et al. (2009) provided an excellent review of potential sources of error for EM canopy sensors. These include sensor operating characteristics, such as sensor wavelengths and resulting VIs, sensor height and footprint/sensitivity issues; seasonal variation, related to algorithm calibration for specific growth stages; genotypic effects and most significantly, other stresses, such as water, nutrient, disease and insect stresses. The general assumption for use of proximal crop canopy sensors to manage N fertilization is that other sources of stress are absent or minimal or, even if present, will be equally present in reference areas as in the target areas, and thus, effects of other stresses will cancel out – assuming there are no interactions between N stress and other stresses. Water stress is perhaps of greatest concern, as water stress can be common, even in irrigated environments, and could confound sensor-based N management at critical times of the growing season. Researchers have investigated sensor N management when water and N stress are confounded (Shiratsuchi et al., 2011; Bronson et al., 2017). There is evidence that some VIs can correctly differentiate N stress in the presence of water stress; typically these are indices that include three wavebands in the equation: the Canopy Chlorophyll Content Index (CCCI, Barnes et al., 2000), the DATT index (Datt, 1999) and the Meris Terrestrial Chlorophyll Index (MTCI, Dash and Curran, 2004). None of these indices are currently available with commercial active crop canopy sensor systems.

3.2 Proximal sensing for stresses other than nitrogen

As is evident from the literature cited earlier, the majority of research and resulting commercial application of proximal crop sensing is for N fertilizer management for grain crops. There is substantial potential for adaptation of proximal crop sensing techniques for detecting other sources of stress, with several areas of ongoing research. However, there has been little commercial adoption of such sensor technologies to date. One technology which is commercially available for real-time detection and management of weeds is the WeedSeeker Spot Spray System (Trimble, 2017b). The WeedSeeker uses similar active canopy sensor technology to that found in the GreenSeeker canopy sensor, but adapted to detect weeds and then activate a herbicide application on-the-go. Sui et al. (2008) documented that the WeedSeeker system could accurately detect and manage weed infestations in cotton. Slaughter et al. (2007) provided a comprehensive review of autonomous weed detection and control research, and the use of various machine vision systems with associated weed recognition algorithms along with canopy reflectance for on-the-go weed control.

Disease and insect detection and intervention is another area in which proximal canopy sensing is being utilized. Franke and Menz (2007) have shown that early detection of disease onset in wheat can be accomplished with high-resolution multispectral imagery. Feng et al. (2016) found that reflectance in visible to red-edge bands (580–710 nm) could be used to monitor for powdery mildew infestation in mid-to-late growth stages of wheat. Hillnhütter et al. (2011) explored a range of VIs developed from hyperspectral imaging for their ability to detect the onset of damage to sugar beet from nematodes and fungal pathogens and noted that there was potential for field use of sensor technology to detect such stresses. Bravo et al. (2004) demonstrated that multispectral fluorescence images were useful for the detection of foliar disease in winter wheat. The primary proximal sensors commercially available today for disease detection are the suite of portable fluorometers from Force-A (Force-A, 2017). Sankaran and Ehsani (2012) found that the ratio of yellow fluorescence (UV excitation) to simple fluorescence (green excitation) was useful in the detection of Huanglongbing disease in citrus canopies.

Studies have used remote sensing technologies for detecting Russian wheat aphids in winter wheat (Mirik et al., 2012) and spider mites in corn (Fitzgerald et al., 2004). Liu et al. (2017) found that UV reflectance, used by birds along with visible wavelengths, could be useful in sensor detection of invertebrates on green leaves. Such sensor approaches have been little utilized with proximal sensors. Puig et al. (2015) did demonstrate that UAV-based RGB sensors could be used to detect white grub infestation in sorghum.

3.3 Sensor fusion and high-throughput phenotyping

An area of current research is the use of multiple proximal sensors to provide greater insight into plant characteristics than any single sensor – a concept termed 'sensor fusion'. Adamchuk et al. (2011) described the benefits of sensor fusion for precision agriculture, though their article primarily addressed sensor fusion for soil properties. They noted that the crop is an excellent bioindicator of variable growing conditions. They commented that '... a combination of conceptually different sensing techniques and integrating the subsequent data holds promise for providing more accurate property estimates, leading to more robust management and increased adoptability of sensor-based crop management'. A commercially available multi-sensor system for soil assessment is the Veris Technologies set of sensors which integrate soil apparent electrical conductivity (EC_a), soil pH and soil organic matter assessment in one platform (Veris Technologies, 2017). Data from multi-sensor systems can be integrated to generate variable-rate application maps for lime, fertilizer and seeding density. One of the earlier efforts to integrate multiple sensor information to assess crop properties was that of Shiratsuchi et al. (2009), who used a prototype system with a multispectral optical sensor for canopy reflectance, a thermal sensor for canopy temperature and an ultrasonic distance sensor for canopy height to assess crop N status. Vories et al. (2014) documented use of a similar system including multispectral canopy reflectance, an ultrasonic sensor for crop height and a thermal sensor for canopy temperature to assess water and N status of cotton. Weis et al. (2013) documented an effort to integrate four sensors [light detection and ranging (LIDAR), passive spectrometer, ultrasonic ranging and WeedSeeker™ active multispectral sensor] to regulate herbicide dosage in a study. Long and McCallum (2015) discussed the use of an on-combine set of sensors (mass flow yield monitor, in-line near-infrared spectrometer and LIDAR) to evaluate environmental stress in wheat at harvest. Sensor information documented crop yield, grain protein content and straw yield simultaneously. Lamb et al. (2014) used an airborne, high-intensity, LED-active multispectral sensor in conjunction with a passive thermal infrared sensor to map a cotton field at a height of 50 m above the canopy. They compared simple ratio (SR) VI maps from the active sensor to those developed from RapidEye satellite imagery (with the active sensor footprint and RapidEye pixel of similar dimensions), and found that the active sensor map closely resembled the RapidEye map. Combining information from the multispectral active and passive thermal sensors could accurately differentiate areas with full-to-partial plant cover from those with partial cover to bare areas.

One of the most active areas of investigation using sensor fusion is that of high-throughput phenotyping, primarily for accelerated identification of genetic traits for yield and stress tolerance for breeding purposes. This concept can involve applications that are lab- and greenhouse-oriented as well as field investigations. Busmeyer et al. (2013) described a multi-sensor platform for field phenotyping, which incorporated a colour camera, time-of-flight cameras, LIDAR, a hyperspectral imaging sensor and light curtain imaging systems. In this system, all sensors operated on a shrouded carrier to prevent exposure to ambient light. Consequently the height of plants over which the sensor could operate was limited; the authors documented repeatability of measurement results over triticale (*Triticosecale* Wittmack L.). The system could estimate values for plant moisture content, lodging, tiller density and biomass yield. Araus and Cairns (2014) provided an overview of high-throughput phenotyping platforms (HTTPs) and their capacity to inform genetic selection. In many cases, HTTPs utilize multispectral active crop canopy sensors, ultrasonic or LIDAR sensors for crop height and thermal sensors for canopy temperature. Wang et al. (2016) discussed the challenges with proper geolocation of data collected by HTTPs; with small plots normally used in breeding field trials, it is critical to accurately associate sensor data with precise location information. A relatively new integrated sensor designed specifically for phenotyping is the Crop Circle Phenom (Holland Scientific, 2017), which includes sensors for upwelling and downwelling photosynthetically active radiation, canopy temperature, air temperature, relative humidity, atmospheric pressure and canopy reflectance at wavelengths of 670, 730 and 780 nm.

4 Case studies

4.1 Case study 1: chlorophyll meter data – seasonal and management influences

This study illustrates the use of a simple crop canopy sensor, the chlorophyll meter, in understanding soil/fertilizer dynamics and resulting crop response during the growing season. This information is from a three-year study conducted on cooperating farmer's fields in Merrick County, Nebraska, USA. More detailed information from this study is available in Maharjan et al. (2016) and Peng et al. (2015). This particular data set is from the 2010 growing season. For clarity of illustration, only a subset of treatments is shown with this example. The cooperator planted corn on 15 April 2010, and nitrogen fertilizer was broadcast on 14 May 2010, shortly after crop emergence. To minimize the risk of ammonia volatilization in this study, fertilizers were lightly incorporated with a rotary tiller immediately after broadcast application. The study investigated several factors influencing crop yield and N fertilizer use efficiency, including N fertilizer source and rate. The most commonly used N fertilizer source in Nebraska is urea ammonium nitrate (UAN) solution – typically containing either 28% or 32% N. This fertilizer is favoured by farmers for a variety of reasons, including versatility in application methods and timing. At application, this fertilizer contains approximately 50% urea-N, 25% ammonium-N and 25% nitrate-N. Consequently, depending on the method of application and soil/weather conditions, N from UAN solution may be susceptible to loss to ammonia volatilization, leaching or denitrification. The soil at the study site in 2010 was a loamy sand (mixed, superactive, mesic oxyaquic Haplustolls), and consequently the greatest risk for N loss was via nitrate leaching. The other fertilizer evaluated was a polymer-coated urea (PCU) product, ESN® (Environmentally Smart Nitrogen, Agrium, Inc., Calgary, Canada). The controlled-release of PCU provides gradual release of urea-N to the soil and crop with the intent of minimizing N loss through ammonia volatilization, nitrate leaching or denitrification and matching N supply to crop demand.

Figure 2 illustrates daily rainfall during the first month after fertilizer application. There were light precipitation events during the first two weeks after fertilization. A 17-mm rainfall event occurred on day 16, followed by 20 mm on day 20, 21 mm on day 25, 57 mm on day 27, 55 mm on day 29, intermixed with several other lighter rainfall events. By three weeks following fertilization, basically all urea in the UAN solution would have hydrolysed and nitrified to nitrate-N (Peng et al., 2015). Consequently, much of the N contained in UAN was highly susceptible to leaching on this soil with these conditions.

Chlorophyll (SPAD) meter readings were collected with a Minolta SPAD-502 meter in the middle of the uppermost fully expanded leaf until approximately the V14 growth stage (Abendroth et al. 2011), then from the ear leaf thereafter. Approximately 30 readings were collected from each plot, with mean values reported. Figure 3 illustrates chlorophyll meter data collected beginning at the V6 growth stage on 22 June 2010 – slightly more than two months after fertilizer application. It is at this growth stage or slightly thereafter that rapid uptake of N occurs with corn, between V8 and VT growth stages (Abendroth et al., 2011). Consequently, there were no significant differences in canopy chlorophyll status as influenced by N fertilizer source or rate at the V6 growth stage; only the unfertilized check was substantially lower in chlorophyll content at V6. This pattern was similar six days (28 June) later at V8. However, eight days later at V10, there began to be some separation in chlorophyll content among treatments – especially with UAN at the lower application rate of 168 kg N ha^{-1}. More separation among N sources and rates continued at subsequent growth stages, with chlorophyll content with PCU at either 168 or 280 kg N ha^{-1} substantially above that for UAN at either rate. These data indicate fairly early in the growing season – at the V10 growth stage – that N availability to the crop was an issue with UAN. Chlorophyll levels with UAN tended to peak around the V12 growth stage, while chlorophyll levels with PCU continued to increase to VT and R2 growth stages. The gradual decline in canopy chlorophyll content during later reproductive growth stages is typical for corn, as nutrients increasingly are transferred to developing grain in the ear.

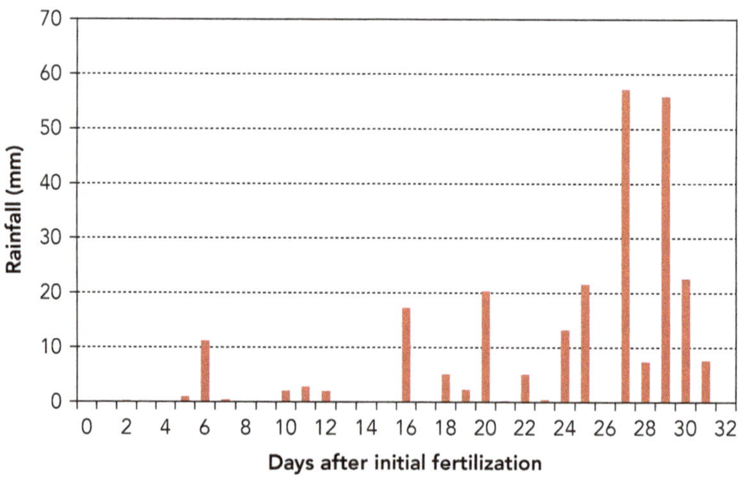

Figure 2 Daily precipitation for the first month after fertilization in 2010, Merrick County, Nebraska, USA.

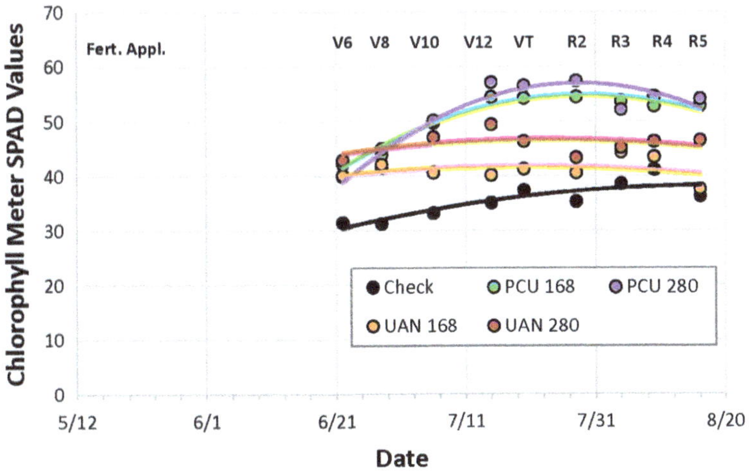

Figure 3 Chlorophyll meter readings as influenced by date/growth stage of corn during the 2010 growing season, fertilizer N source [urea ammonium nitrate (UAN) solution or polymer-coated urea (PCU)] and application rate (168 or 280 kg N ha^{-1}). Field location is in Merrick County, Nebraska, USA.

Figure 4 illustrates the impact of N fertilizer source and rate on grain yield in this study. At both 168 and 280 kg N ha^{-1} rates, grain yield with PCU was approximately 4 Mg ha^{-1} greater than for UAN at the same rates. These results confirm that substantial N loss from UAN reduced yield potential compared to PCU. Chlorophyll meter readings measured as early as the V10 growth stage indicated N deficiency with UAN treatments relative to PCU. Analyses by Peng et al. (2015) and Maharjan et al. (2016) indicate that hydrolysis of urea contained in UAN was mostly complete within a week after fertilization, followed by nitrification of fertilizer-derived ammonium to nitrate within the following couple of weeks. Consequently, it is most likely that rainfall during the period 16–30 days after fertilization leached much of the fertilizer N from UAN below the root zone on this coarse-textured site. Urea-N in PCU was gradually released through the growing season, and mostly protected against early season nitrate leaching.

This data set illustrates two attributes of proximal crop canopy sensing. For a research study, crop canopy sensing helps document interactions of treatments with soil and weather conditions over time. For a crop producer, proximal crop canopy sensing can provide invaluable management information during the growing season. If UAN fertilization at planting had been the standard management approach for a grower, canopy sensor information would have let the producer be informed of concerns about N supply at the V10 growth stage – early enough to intervene with supplemental fertilization to protect yield potential. One key issue here is that absolute chlorophyll meter readings are of little use – these can vary with hybrid and field conditions other than N supply. Consequently an in-field, non-limiting N reference, as suggested by Varvel et al. (2007), is necessary to accurately interpret chlorophyll meter information. A second key issue, as already noted, is that use of a handheld chlorophyll meter is impractical for management at the field scale, and other sensing approaches are advised.

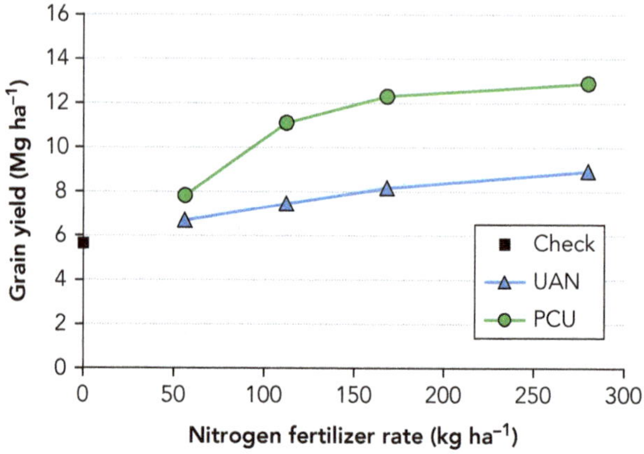

Figure 4 Grain yield as influenced by N source and rate, 2010, Merrick County, Nebraska, USA.

4.2 Case study 2: use of an active crop canopy sensor for in-season N management and improved environmental stewardship

Corn producers in Nebraska, and generally in the United States, have steadily increased the efficiency with which they use N fertilizer. One parameter for measuring NUE which is relatively easily measured is partial factor productivity for N (PFP_N), which is the mass of grain produced per mass of fertilizer N applied. In Nebraska, PFP_N for corn has increased from around 35 kg grain kg N^{-1} in 1968 to around 65 kg grain kg N^{-1} in 2012 (Ferguson, 2015). This has been achieved through multiple factors, including improved genetics, changes in cropping systems, as well as more efficient fertilizer management practices. However, there has been little change in PFP_N since 2000, suggesting that current fertilizer management practices may be limiting further increases in efficiency, and that new approaches to N management are needed to continue increasing NUE. While active crop canopy sensors have been on the market in the United States for over a decade, there has been limited adoption of this technology for N fertilizer management to date. As with any new technology, there can be a range of factors influencing producer decisions to adopt, including cost of the technology, impacts on profit, unfamiliarity with the technology or uncertainty about its application to specific production systems. In order to address some of these issues relative to active canopy sensor N management for corn, a project was started in 2015 in Nebraska, USA, as part of an on-farm research initiative. The project, called Project SENSE (Sensors for Efficient Nitrogen Use and Stewardship of the Environment), is a collaborative effort of the University of Nebraska-Lincoln, the Nebraska corn board and five natural resource districts (NRDs) in Nebraska (NRDs are governmental agencies with boundaries organized along watersheds charged with management of soil and water resources in their district). The objective of Project SENSE is to encourage improved NUE through in-season fertilization. One of the tools promoted by the project is the use of active crop canopy sensor-based in-season application. The project gives cooperating producers the opportunity to see an active canopy sensor-equipped high clearance applicator operate on their fields and to compare their standard N management approaches to sensor-based N management in a scientific manner. Project SENSE is a component of the Nebraska On-Farm Research Network, an organization of growers who conduct scientifically valid field-scale research on their fields and share results in annual conferences and on the project website (Nebraska On-Farm Research Network, 2017).

Approximately 20 producers participate in Project SENSE each year. A typical site compares the grower's standard N management to sensor-based N management in field - length strips, usually with six replications of each treatment. The widths of treatment strips are based on the cooperator's fertilizer application and harvest equipment and typically are 8–16 rows (6–12 m) wide. Growers use their choice of N fertilizer application rate and timing; most use split application timing, with some fertilizer applied at or prior to planting and the remainder side-dressed early in the growing season. Practically all cooperators use uniform N application rates. A base rate of 70–80 kg N ha^{-1} was applied to sensor-based treatment strips by the cooperator at or prior to planting. For Nebraska soil and weather conditions, this amount of fertilizer N, coupled with soil-supplied N, allows the crop to reach the V8–V9 growth stage without yield-limiting N stress. The project uses the AgLeader OptRx® active crop canopy sensor system, including the standard N recommendation algorithm approach included with the sensors. This includes a virtual reference approach (Holland and Schepers, 2013). To determine the reference, an area considered to represent the range of crop N status for the field is driven over by the high clearance applicator (Fig. 5) and sensed to build a virtual reference data set. Once the virtual reference is established, sensor-based treatments were applied. In-season application occurred between V8 and V14 growth stages, with a target of application

Figure 5 Project SENSE high clearance applicator (Hagie DTS-10) equipped with OptRx® active crop canopy sensors. Typically an applicator of this width would use 2–3 OptRx sensors; this applicator is equipped with 7 sensors, to allow flexibility with treatment strip width and to collect supplemental reflectance data for research purposes.

as close as possible to V9 growth stage. Otherwise management was at the grower's discretion, including fertilizers other than N, hybrid and seed density, use of pesticides and irrigation. The cooperator harvested each treatment strip with a yield-mapping combine. Results were then analysed statistically to determine treatment effects on fertilizer use, grain yield, NUE and partial profit.

Figure 6a–b illustrates results from one 2016 site. Figure 6a is a map of Normalized Difference Red Edge Vegetation Index (NDRE, Barnes et al., 2000) reflectance values for sensor-based treatment strips; the grower treatment strips are randomized among the sensor-based treatment strips. This illustrates a range in crop N status across the field, primarily influenced by soil variation. Figure 6b shows the resulting side-dress N rate calculation based on canopy reflectance relative to reference values within the study area; side-dress N application rates ranged from 34 kg N ha^{-1} (the minimum rate applied) to 204 kg N ha^{-1}. Figure 6c shows resulting grain yield from all treatment strips, including the cooperator's uniform rate treatment strips. Figure 6d shows the mean results comparing the grower treatment to sensor-based treatment. The average fertilizer N rate for the sensor-based treatment was 40 kg N ha^{-1} less than the grower rate, while grain yield was statistically the same, at around 9.4 Mg ha^{-1}. In this case, profit increased with sensor-based treatment due to the use of less fertilizer N, while yield remained the same. Figure 7a–d illustrates the same information (NDRE, side-dress N application rate, grain yield and treatment mean total N applied and grain yield) for a second site in 2016. At this site, total overall N applied was basically the same between the grower and sensor-based treatments – around 170 kg N ha^{-1}. However, primarily due to more timely application, the sensor-based treatments produced on average 2.5 Mg ha^{-1} greater yield than the grower treatment.

Overall results from the project for 2015 and 2016 are shown in Table 1. In both years, sensor-based N application significantly reduced the total amount of N applied compared to grower practices by 44 kg N ha^{-1} in 2015 and 38 kg N ha^{-1} in 2016. This resulted in slightly reduced average grain yields compared to grower standard practices in both years, by 0.31 Mg ha^{-1} in 2015 and 0.19 Mg ha^{-1} in 2016. However, fertilizer NUE (PFP_N) increased substantially with sensor-based application. Grower treatments produced an average of 65 kg grain kg N^{-1} in 2015 and 63 kg grain kg N^{-1} in 2016. Sensor-based PFP_N was 80 kg grain kg N^{-1} in 2015 and 75 kg grain kg N^{-1} in 2016. Perhaps most importantly to the growers, marginal net return was increased both years with sensor-based management – by \$19.14 ha^{-1} in 2015 and \$17.39 ha^{-1} in 2016. Depending on the type of sensor and applicator control system a grower would purchase, it would typically require use over 218–437 ha over four years to pay for the sensor system purchase at current equipment prices of \$15 000–\$30 000, and commodity prices of \$0.12 kg^{-1} corn and \$1.00 kg^{-1} N.

While overall increased profit is of concern to the farmer, reduced environmental impacts from fertilizer use are of concern to society. Grower sites in Project SENSE are located in areas of Nebraska with elevated groundwater nitrate-N levels, resulting from past overuse of N fertilizers coupled with excessive irrigation resulting in nitrate leaching. Typically, groundwater nitrate-N concentrations in these areas can range from 10 to 30 mg L^{-1} or more (irrigation water nitrate-N credits were accounted for in calculating sensor-based N applications at these sites). The reduction of 30–40 kg N ha^{-1} or more at these sites, with little impact on grain yield and an increase in profit, with the use of sensor-based N application will have a positive impact on groundwater quality over time.

(a)

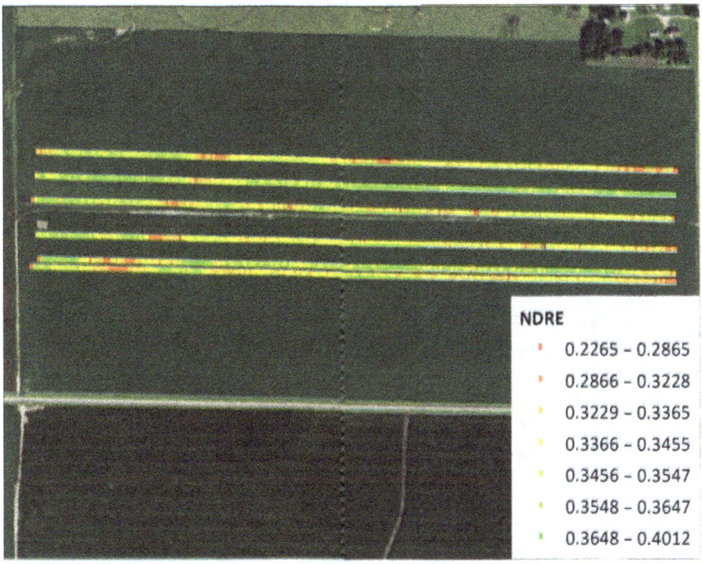

(b)

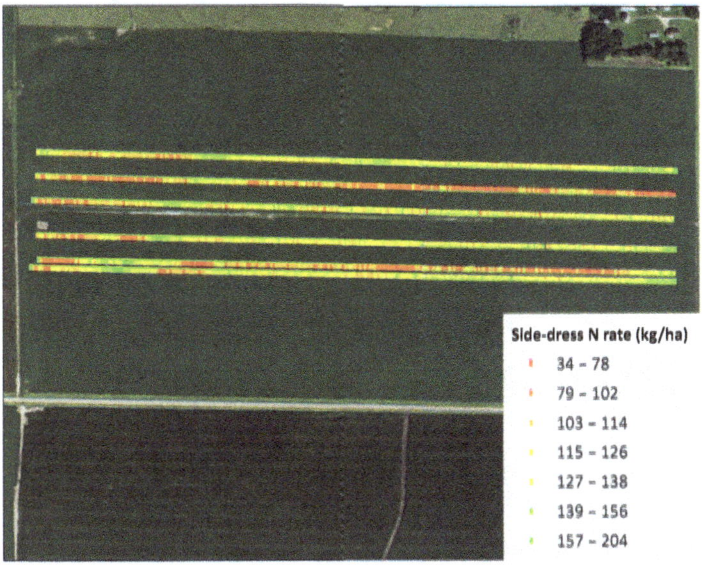

Figure 6 (a) Normalized Difference Red Edge (NDRE) for Project SENSE site 16BK, 2016. NDRE is shown only for crop canopy sensor-based treatment strips. (b) Side-dress fertilizer N rate for Project SENSE site 16BK, 2016. Nitrogen rate is shown only for crop canopy sensor-based treatment strips. (c) Grain yield for Project SENSE site 16BK, 2016. Yield is shown for both sensor-based treatment strips and grower treatment strips. (d) Total mean fertilizer N applied (red markers) and mean grain yield (bars) for the grower and sensor-based treatment strips, Project SENSE site 16BK, 2016.

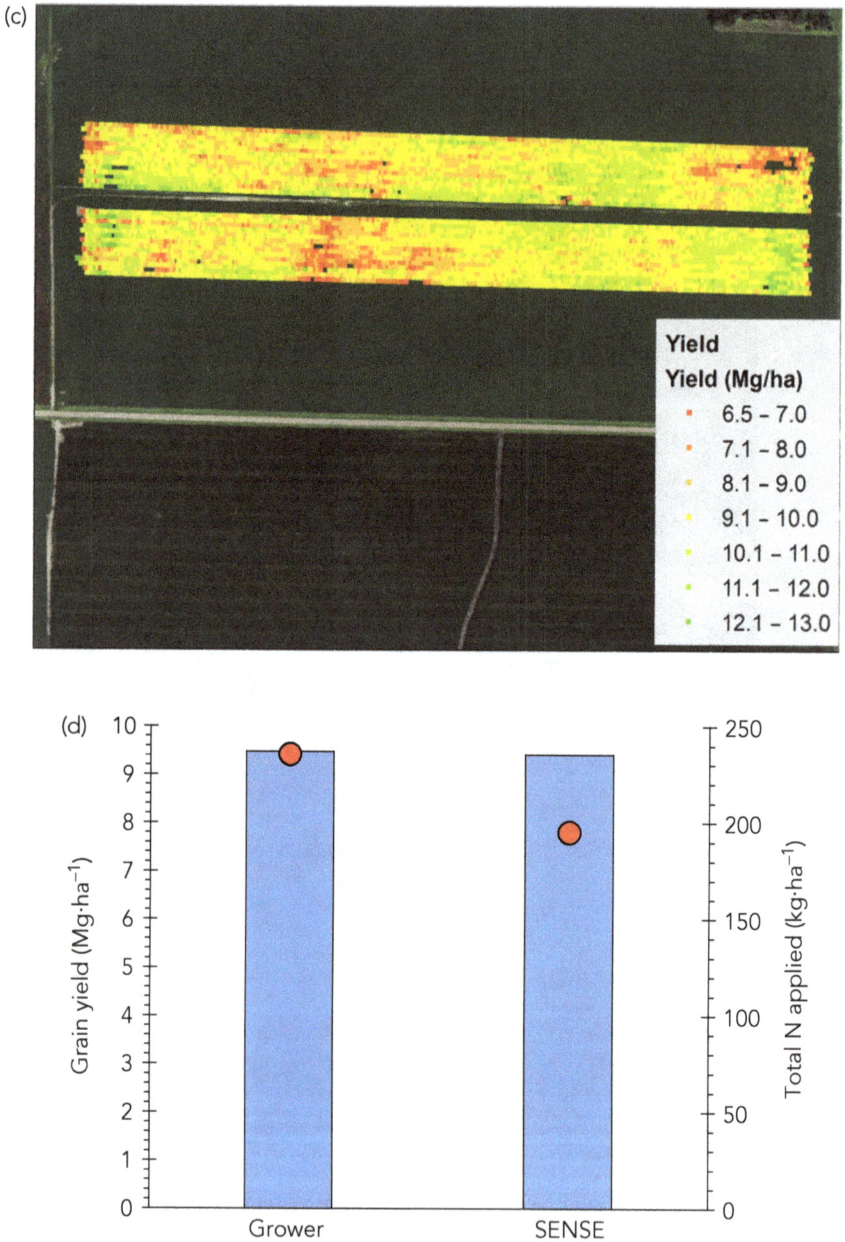

Figure 6 (*Continued*)

(a)

(b)

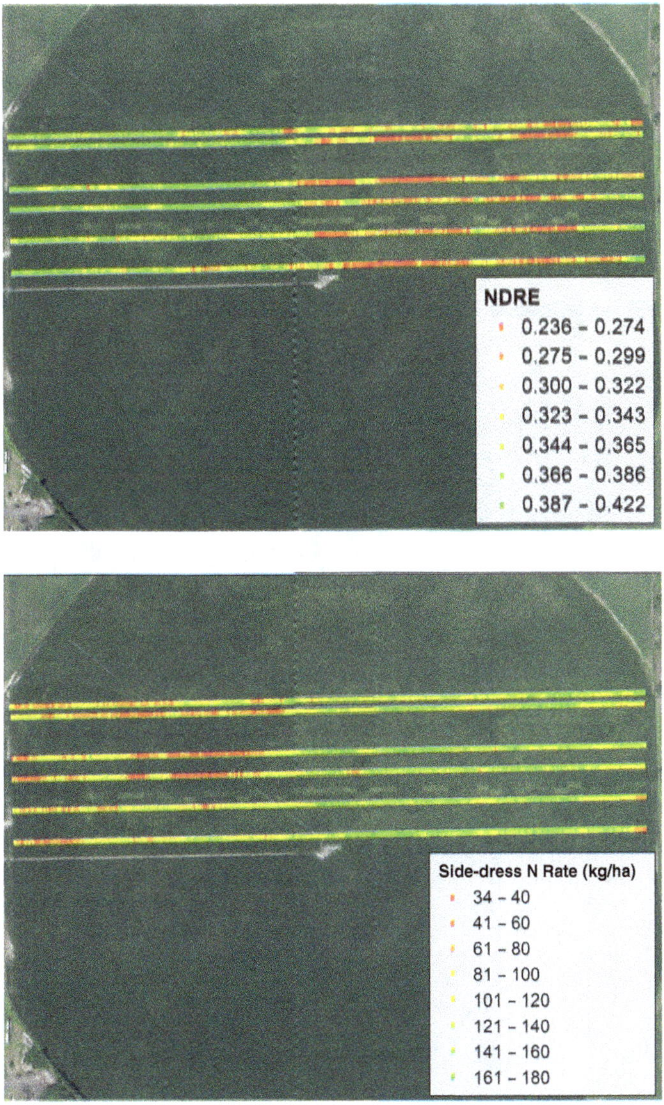

Figure 7 (a) Normalized Difference Red Edge (NDRE) for Project SENSE site 16KF, 2016. NDRE is shown only for crop canopy sensor-based treatment strips. (b) Side-dress fertilizer N rate for Project SENSE site 16KF, 2016. Nitrogen rate is shown only for crop canopy sensor-based treatment strips. (c) Grain yield for Project SENSE site 16KF, 2016. Yield is shown for both sensor-based treatment strips and grower treatment strips. (d) Total mean fertilizer N applied (red markers) and mean grain yield (bars) for the grower and sensor-based treatment strips, Project SENSE site 16KF, 2016.

(c)

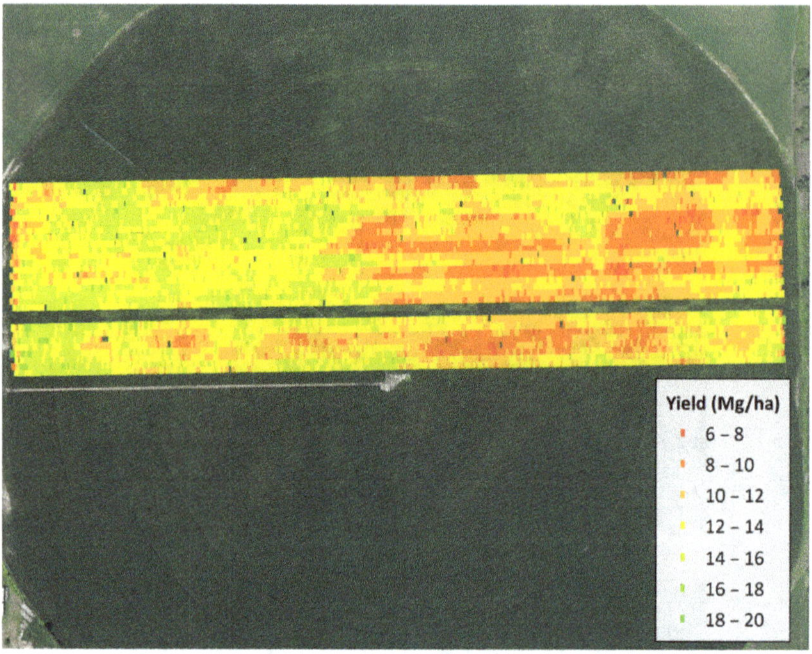

(d)

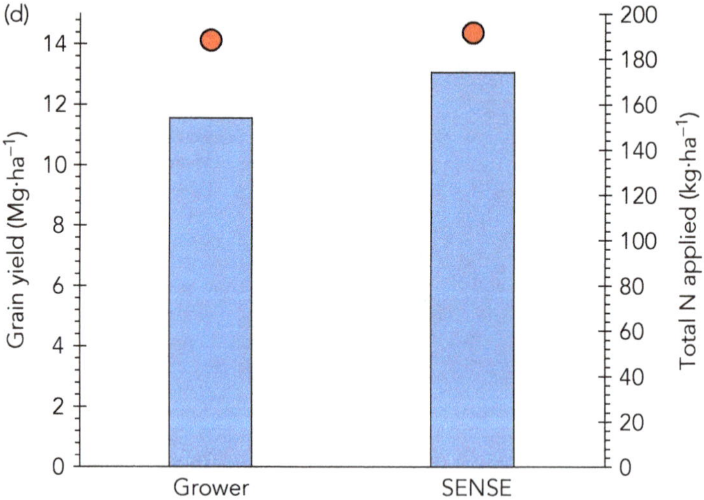

Figure 7 (*Continued*)

Table 1 Mean fertilizer N application, grain yield, NUE and profit, Project SENSE sites (n = 17 for 2015, n = 20 for 2016). Marginal net return based on: 2015 N price of \$1.44 kg N^{-1} and corn price of \$0.14 kg^{-1}; 2016 N price of \$1.00 kg^{-1} and corn price of \$0.12 kg^{-1}

	2015		2016	
	Grower treatment	Sensor-based treatment	Grower treatment	Sensor-based treatment
Total N applied (kg ha^{-1})	218	194	212	174
Grain yield (Mg ha^{-1})	14.25	13.94	12.68	12.49
PFP_N (kg grain kg N^{-1})	65	80	63	75
Marginal net return (\$ ha^{-1})	1733.44	1752.59	1310.19	1327.58

4.3 Case study 3: use of an unmanned, aerial system–mounted (UAS-mounted) active crop canopy sensor

As previously mentioned, the rapid growth in availability of UAS with associated airframes and sensor packages is blurring the traditional boundary between proximal and remote sensing. UAS can operate in close proximity to the crop canopy, allowing spatial and temporal resolution well beyond that achievable with remote sensing systems based on aircraft or satellites. Optical sensing systems used with UAS are typically passive sensors, which restrict their utility based on time of day and weather conditions. There are several advantages to the use of active sensors with a UAS; they can be used regardless of sun conditions, day or night; they are not imaging sensors, so images do not need to be processed, geo-rectified or stitched; data are thus immediately available for use, with geographic position information, without extensive processing. However, light sources used with active canopy sensors are relatively of low power, and sensors may have limited sensitivity beyond a few metres above the crop canopy. GreenSeeker and OptRx sensors, for example, typically are designed for use within a couple of metres or less from the crop canopy. Previous examples of an airborne active sensor are provided by Lamb et al. (2009) who used a standard CropCircle™ active sensor mounted to an aircraft and flown at a height of 3–5 m above a sorghum canopy, and Lamb et al. (2011) who used an active sensor with a high-powered LED system (Raptor™, Holland Scientific) to map NDVI at heights of 15–45 m above a wheat canopy. In both studies, data collected with airborne active sensors showed good correlation with data collected from a passive multispectral aerial sensor in the first study, or with a ground-based active sensor in the second study.

A project conducted in Nebraska beginning in 2014 evaluated the potential for use of active canopy sensors with UAS. The project used a UAS-mounted Holland Scientific RapidScan CS-45 (Fig. 8), modified such that it would log reflectance information and position every second, rather than averaging values over a period of time. The RapidScan

Figure 8 Manual operation of a UAS with RapidScan CS-45 active sensor.

CS-45 sensor is intended for handheld use at a distance of 0.3–3 m above the crop canopy. Krienke et al. (2017) evaluated the sensitivity of the UAS-mounted RapidScan to height above the crop canopy and investigated the relationships between fertilizer N rate, NDRE and corn grain yield. The study found that up to a height of 1.5 m above a turfgrass canopy (selected due to its lack of complexity and uniform depth of canopy relative to a maize canopy), sensor readings were relatively similar. However, above 1.5 m, the slope of NDRE relative to fertilizer N rate was different from that below 1.5 m (Fig. 9). For this particular sensor they concluded that use at heights of 1.5 m or less above the canopy

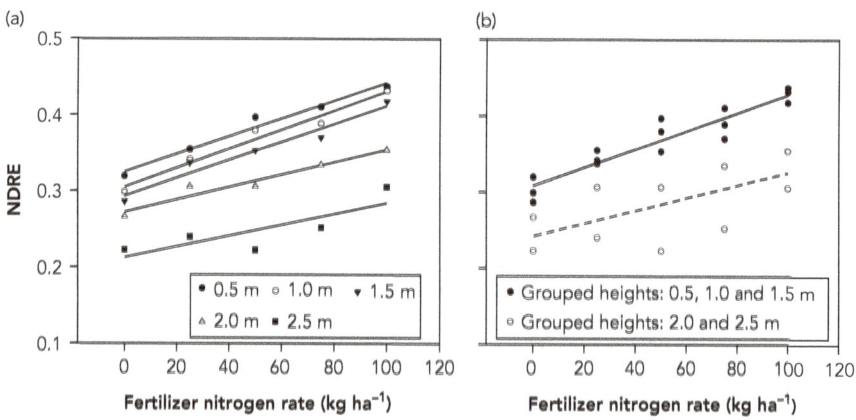

Figure 9 Relationship of UAS-mounted RapidScan CS-45 at various heights above the crop canopy (0.5, 1.0, 1.5, 2.0 and 2.5 m) to fertilizer N rate and NDRE (a) and grouped heights to fertilizer N rate and NDRE (b).

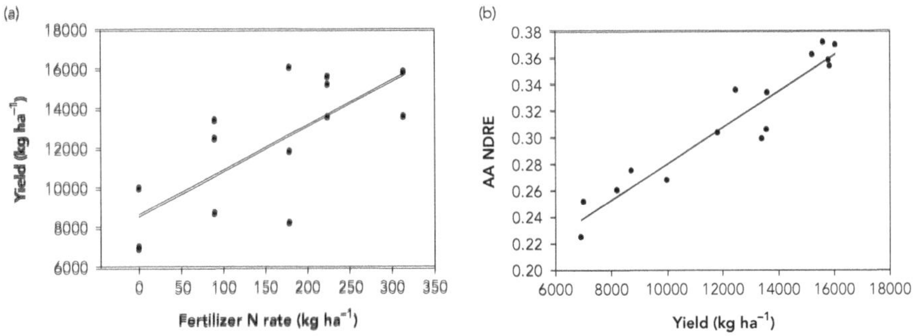

Linear Regression Model: $y = b_0 + b_{1x}$	Adj r^2
8630.42 + 22.66*N Rate	0.55

Linear regression model: $y = b_0 + b_{1x}$	Adj r^2
0.14 + 0.00001*Yield	0.90

Figure 10 Relationship of fertilizer N rate and corn grain yield (a) and grain yield and NDRE (b).

would produce reliable estimates of crop N status. In a separate trial, they evaluated the relationship of active sensor NDRE above a corn study with varying fertilizer N rates. The study had significant scatter in the relationship between fertilizer N rate and grain yield (Fig. 10a), due to small-scale variation in soil N supply even within small plots. However, the relationship between NDRE from the UAS-mounted active sensor at the V11 growth stage and final grain yield was quite strong (Fig. 10b).

This study illustrates the potential for active sensor use with a UAS. The sensor used in the above study was not designed with a light source intended for UAS operation. Maintaining a constant height of 1.0–1.5 m as in this study manually is not practical with current elevation control systems available for UAS. However, a more intense light source would allow greater distance between the UAS sensor and the crop canopy while still obtaining accurate canopy reflectance information. Also, ranging sensors, either ultrasonic or LIDAR, could be incorporated into the flight control system of the UAS to maintain an accurate height above the crop canopy. Such a UAS system could operate at any time of the day – or night, if regulations allowed, at a height of a few metres above the canopy in airspace with little concern for air traffic, other than perhaps agricultural spray planes. The extended times of data collection could also allow a UAS equipped with an active sensor to operate at dawn or dusk, when wind speeds in many areas are typically lower, allowing for more stable UAS operation and improved data accuracy.

5 Conclusion: sustainability and environmental implications

The concept of proximal crop sensing is directly related to improved sustainability and reduced environmental impact of producing crops. One of the primary concepts of precision agriculture is site-specific management – adjusting crop inputs, or even the crop cultivar or species, based on the capacity of the underlying soil resource. Both proximal and remote crop sensing enable such spatial management, but proximal sensing adds the capacity to better manage temporal variation as well, due to inherent temporal constraints with remote sensing via aircraft or satellite platforms. Both remote and proximal sensors allow growers to monitor the crop during the growing season and adjust management according to such information. The potential for more judicious use of resources such as soil and water, and crop inputs such as fertilizer, pesticides or tillage seems evident when growers can detect and manage emerging issues in a very timely and site-specific manner, rather than use of blanket, prophylactic amendments which may not be necessary and have the potential for environmental degradation when used in excess.

6 Future trends for research

As indicated earlier, most proximal crop sensors today are either passive sensors using optical reflectance from the canopy using the sun as a light source or active sensors using internal, modulated light sources. There appears to be significant potential for research and development of active canopy sensors for use with UAS, for rapid, more timely spatial surveys of crops and to expand properties of interest beyond N, to water, other nutrient stresses, as well as disease and insect stress. There is significant potential to further explore multispectral fluorescence as an indicator of specific stresses. The mid-infrared and thermal EM regions have not yet been widely studied as indicators of stress, other than for water stress, primarily due to the lack of lower cost sensors in these wavelength ranges. Finally, there seems to be significant potential for sensors other than EM sensors, for example, pheromone or spore detectors for the onset of insect or disease or other chemo-detectors in general. These seem especially suited to UAS use, if detection of a stress indicator can be easily coupled with geographic position within a field, allowing rapid mapping of field for stress indicators.

7 Where to look for further information

Further information on proximal crop sensing can be found through conferences sponsored by the International Society of Precision Agriculture (ISPA; www.ispa.org) and from research journals, such as Precision Agriculture (springer.com/journal/11119).

8 References

Abendroth, L. J., R. W. Elmore, M. J. Boyer and S. K. Marlay (2011), 'Corn growth and development', PMR 1009. Ames, IA, USA, Iowa State University.

Adamchuk, V., R. Viscarra Rossel, K. Sudduth and P. Lammers (2011), 'Sensor fusion for precision agriculture', In: C. Thomas (Ed.), *Sensor Fusion – Foundation and Applications.* InTech Inc., Vienna, Austria. DOI: 10:5772/680. https://www.intechopen.com/books/sensor-fusion-foundation-and-applications/.

Ag Leader (2017), OptRx® Crop Sensors. http://www.agleader.com/products/directcommand/optrx-crop-sensors/ (Accessed 20 April 2017).

Araus, J. L. and J. E. Cairns (2014), 'Field high-throughput phenotyping: The new crop breeding frontier', *Trends in Plant Science* 19: 52–61. http://dx.doi.org/10.1016/j.tplants.2013.09.008.

Barnes, E. M., T. R. Clarke, S. E. Richards, P. D., Colaizzi, J. Haberland and M. Kostrzewski (2000), 'Coincident detection of crop water stress, nitrogen status and canopy density using ground-based multi-spectral data [CD-ROM]', In: P. C. Robert, R. H Rust and W. E. Larson (Eds), *Proceedings 5th International Conference on Precision Agriculture*. Madison, WI, USA, ASA, CSSA, and SSSA.

Berntsen, J., A. Thomas, K. Schelde, O. M. Hanse, L. Knudsen, N. Broge, H. Hougaard and R. Hørfarter (2006), 'Algorithms for sensor-based redistribution of nitrogen fertilizer in winter wheat', *Precision Agriculture* 7: 65–83.

Blackmer, T. M. and J. S. Schepers (1995), 'Use of a chlorophyll meter to monitor nitrogen status and schedule fertigation for corn', *Journal of Production Agriculture* 8: 56–60.

Bravo, C., D. Moshou, R. Oberti, J. West and A. McCartney, L. Bodria and H. Ramon (2004), 'Foliar disease detection in the field using optical sensor fusion', *Agricultural Engineering International: The CIGR Journal of Scientific Research and Development* VI: 1–13, manuscript FP 04 008.

Bréda, N. J. J. (2003), 'Ground-based measurements of leaf area index: A review of methods, instruments and current controversies', *Journal of Experimental Botany* 54: 2403–17.

Bronson, K. F., J. W. White, M. M. Conley, D. J. Hunsaker, K. R. Thorp, A. N. French, B. E. Mackey and K. H. Holland (2017), 'Active optical sensors in irrigated durum wheat: Nitrogen and water effects', *Agronomy Journal* 109: 1–12.

Bundy, L. G. and T. W. Andraski (2004), 'Diagnostic tests for site-specific nitrogen recommendations for winter wheat', *Agronomy Journal* 96: 608–14.

Busemeyer, L., D. Mentrup, K. Möller, E. Wunder, K. Alheit, V. Hahn, H. P. Maurer, J. C. Reif, T. Würschum, J. Müller, F. Rahe and A. Ruckelshausen (2013), 'BreedVision – a multi-sensor platform for non-destructive field-based phenotyping in plant breeding', *Sensors* 13: 2830–47. doi:10.3390/s130302830.

Dash, J. and P. J. Curran (2004), 'The MERIS terrestrial chlorophyll index', *International Journal of Remote Sensing* 25: 5403–13. doi:10.1080/0143116042000274015.

Datt, B. (1999), 'A new reflectance index for remote sensing of chlorophyll content in higher plants: Tests using Eucalyptus leaves', *Journal of Plant Physiology* 154: 30–6. doi:10.1016/S0176-1617(99)80314-9.

De Souza, E. G., P. C. Scharf and K. A. Sudduth (2010), 'Sun position and cloud effects on reflectance and vegetation indices of corn', *Agronomy Journal* 102: 734–44.

Dworak, V., J. Selbeck and D. Ehlert (2011), 'Ranging sensors for vehicle-based measurement of crop stand and orchard parameters: A review', *Transactions of the American Society of Agricultural and Biological Engineers* 54(4): 1497–510.

Ehlert, D. and K. H. Dammer (2006), 'Widescale testing of the Crop-meter for site-specific farming', *Precision Agriculture* 7: 101–15.

Feekes, W. (1941), 'De tarwe en haar mileu [Wheat and its environment]', *Verslagen van de Technische Tarwe Commissie* (English summary) 17: 523–888.

Feng, W., W. Shen, L. He, J. Duan, B. Guo, Y. Li, C. Wang and T. Guo (2016), 'Improved remote sensing detection of wheat powdery mildew using dual-green vegetation indices', *Precision Agriculture* 17: 608–827. doi: 10.1007/s11119-016-9440-2.

Ferguson, R. B. (2015), 'Groundwater quality and nitrogen use efficiency in Nebraska's Central Platte River Valley', *Journal of Environmental Quality* 44: 449–59.

Fitzgerald, G., S. Maas and W. Detar (2004), 'Spider mite detection and canopy component mapping in cotton using hyperspectral imagery and spectral mixture analysis', *Precision Agriculture* 5: 275–89.

Follett, R. H., R. F. Follett and A. D. Halvorson (1992), 'Use of a chlorophyll meter to evaluate the nitrogen status of dryland winter wheat', *Communications in Soil and Plant Analysis* 23: 687–97.

Force-A (2017), Dualex Scientific™, Multiplex Research™, Multiplex 330™. http://www.force-a.com/en/ (Accessed 17 April 2017).

Franke, J. and G. Menz (2007), 'Multi-temporal wheat disease detection by multi-spectral remote sensing', *Precision Agriculture* 8: 161–72.

Gitelson, A. A., A. Viña, V. Ciganda, D. C. Rundquist and T. J. Arkebauer (2005), 'Remote estimation of canopy chlorophyll content in crops', *Geophysical Research Letters* 32, L08403, doi:10.1029/2005GL022688.

Hillnhütter, C., A. Mahlein, R. Sikora and E. Oerke (2011), 'Use of imaging spectroscopy to discriminate symptoms caused by *Heterodera schachtii* and *Rhizoctonia solani* on sugar beet', *Precision Agriculture* 13: 17–32.

Holland, K. H. and J. S. Schepers (2010), 'Derivation of a variable rate nitrogen application model for in-season fertilization of corn', *Agronomy Journal* 102: 1415–24.

Holland, K. H., D. W. Lamb and J. S. Schepers (2012), 'Radiometry of proximal active optical sensors (AOS) for agricultural sensing', *IEEE Journal of Selected Topics in Applied Earth Observations and Remote Sensing* 5: 1793–801.

Holland, K. H. and J. S. Schepers (2013), 'Use of a virtual-reference concept to interpret active crop canopy sensor data', *Precision Agriculture* 14: 71–85.

Holland Scientific (2017), Crop Circle Phenom. http://hollandscientific.com/portfolio/crop-circle-phenom/ (Accessed 8 May 2017).

Jørgensen, J. R. and R. N. Jørgensen (2007), 'Uniformity of wheat yield and quality using sensor assisted application of nitrogen', *Precision Agriculture* 8: 63–73.

Konica Minolta (2017), Chlorophyll meter SPAD-502Plus. https://www.konicaminolta.com/instruments/download/catalog/color/pdf/spad502plus_catalog_eng.pdf (Accessed 14 April 2017).

Krienke, B., R. Ferguson and B. Maharjan (2017), 'Using an unmanned aerial vehicle to evaluate nitrogen variability and height effect with an active crop canopy sensor', *Precision Agriculture* 18:900–15.

Lamb, D., M. Trotter and D. Schneider (2009), 'Ultra low-level airborne (ULLA) sensing of crop canopy reflectance; A case study using a CropCircle™ sensor', *Computers and Electronics in Agriculture* 69: 86–91.

Lamb, D., D. Schneider, M. Trotter, M. Schafer and I. Yule (2011), 'Extended-altitude aerial mapping of crop NDVI using an active optical sensor; A case study using a Raptor™ sensor over wheat', *Computers and Electronics in Agriculture* 77: 69–73.

Lamb, D., D. Schneider and J. Stanley (2014), 'Combination active optical and passive thermal infrared sensor for low-level airborne crop sensing', *Precision Agriculture* 15: 523–31.

Lee, K. H. and R. Ehsani. (2009), 'A laser scanner based measurement system for quantification of citrus tree geometric characteristics', *Applied Engineering in Agriculture* 25(5): 777–88.

Liu, H., S. Lee and J. Chahl (2017), 'An evaluation of the contribution of ultraviolet in fused multispectral images for invertebrate detection on green leaves', *Precision Agriculture* 18: 667–83.

Long, D. and J. McCallum (2015), 'On-combine, multi-sensor data collection for post-harvest assessment of environmental stress in wheat', *Precision Agriculture* 16: 492–504.

Lukina, E. V., K. W. Freeman, K. J. Wynn, W. E. Thomason, R. W. Mullen, M. L. Stone, J. B. Solie and W. R. Raun (2001), 'Nitrogen fertilization optimization algorithm based on in-season estimates of yield and plant nitrogen uptake', *Journal of Plant Nutrition* 24: 885–98.

Maharjan, B., R. B. Ferguson and G. P. Slater (2016), 'Polymer-coated urea improved corn response compared to urea-ammonium nitrate when applied on a coarse-textured soil', *Agronomy Journal* 108: 509–18.

Markwell, J., J. C. Osterman and J. L. Mitchell (1995), 'Calibration of the Minolta SPAD-502 leaf chlorophyll meter', *Photosynthesis Research* 46: 467–72.

Mirik, M., R. Ansley, G. Michels Jr. and N. Elliott (2012), 'Spectral vegetation indices selected for quantifying Russian wheat aphid (*Diuraphis noxia*) feeding damage in wheat (*Triticum aestivum* L.)', *Precision Agriculture* 13: 501–16.

Nebraska On-Farm Research Network (2017), http://cropwatch.unl.edu/on-farm-research (Accessed 30 April 2017).

Ortuzar-Iragorri, M. A., A. Alonso, A. Castellón, G. Besga, J. M. Estavillo and A. Aizpurua (2005), 'N-Tester use in soft winter wheat: Evaluation of nitrogen status and grain yield prediction', *Agronomy Journal* 97: 1380–9.

Peng, X., B. Maharjan, C. Yu, A. Su, V. Jin and R. B. Ferguson (2015), 'A laboratory evaluation of ammonia volatilization and nitrate leaching following nitrogen fertilizer application on a coarse-textured soil', *Agronomy Journal* 107: 871–9.

Piekielek, W. P. and R. H. Fox (1992), 'Use of a chlorophyll meter to predict sidedress nitrogen requirements for maize', *Agronomy Journal* 84: 59–65.

Puig, E., F. Gonzalez, G. Hamilton and P. Grundy (2015), 'Assessment of crop insect damage using unmanned aerial systems: A machine learning approach', In: T. Weber, M. J. McPhee and R. S. Anderssen (Eds), *MODSIM2015, 21st International Congress on Modelling and Simulation.* Modelling and Simulation Society of Australia and New Zealand, Gold Coast, Australia, December 2015, pp. 1420–6. ISBN: 978-0-9872143-5-5. http://www.mssanz.org.au/modsim2015/F12/puig.pdf (Accessed 3 May 2017).

Raun, W. R., J. B. Solie, G. V. Johnson, M. L. Stone, E. V. Lukina, W. E. Thomason and J. S. Schepers (2001), 'In-season prediction of potential grain yield in winter wheat using canopy reflectance', *Agronomy Journal* 93: 131–8.

Raun, W. R., J. B. Solie, G. V. Johnson, M. L. Stone, R. W. Mullen, K. W. Freeman, W. E. Thomason and E. V. Lukina (2002), 'Improving nitrogen use efficiency in cereal grain production with optical sensing and variable rate application', *Agronomy Journal* 94: 815–20.

Rouse Jr., J. W., R. H. Haas, J. A. Schell and D. W. Deering (1974), 'Monitoring vegetation systems in the Great Plains with ERTS', In: *Third Earth Resources Technology Satellite-1 Symposium-Volume I: Technical Presentations. NASA SP-351*, Washington, DC, 10–24 December 1973. Washington, DC, NASA, pp. 309–17.

Samborski, S. M., N. Tremblay and E. Fallon (2009), 'Strategies to make use of plant sensor-based diagnostic information for nitrogen recommendations', *Agronomy Journal* 101: 800–16.

Sankaran, S. and R. Ehsani (2012), 'Detection of Huanglongbing Disease in citrus using fluorescence spectroscopy', *Transactions of the American Society of Agricultural and Biological Engineers* 55(1): 313–20.

Scharf, P. C., S. M. Brouder and R. G. Hoeft (2006), 'Chlorophyll meter readings can predict nitrogen need and yield response of corn in the North-Central USA', *Agronomy Journal* 98: 655–85.

Shapiro, C. A., D. D. Francis, R. B. Ferguson, G. W. Hergert, T. M. Shaver and C. S. Wortmann (2013), 'Using a chlorophyll meter to improve N management', NebGuide G1632. University of Nebraska-Lincoln, Lincoln, Nebraska, USA. http://extensionpublications.unl.edu/assets/pdf/g1632.pdf (Accessed 14 April 2017).

Shiratsuchi, L. S., R. B. Ferguson, V. I. Adamchuk, J. F. Shanahan and G. P. Slater (2009), 'Integration of ultrasonic and active canopy sensors to estimate the in-season nitrogen content for corn', In: *Proceedings of the 39th North Central Extension-Industry Soil Fertility Conference.* Norcross Georgia, USA, International Plant Nutrition Institute, pp. 182–8.

Shiratsuchi, L. S. (2011), 'Integration of plant-based canopy sensors for site-specific nitrogen management', *Theses, Dissertations and Student Research in Agronomy and Horticulture*, University of Nebraska-Lincoln Digital Commons. http://digitalcommons.unl.edu/agronhortdiss/36 (Accessed 14 April 2017).

Shiratsuchi, L., R. Ferguson, J. Shanahan, V. Adamchuk, D. Rundquist, D. Marx and G. Slater (2011), 'Water and nitrogen effects on active canopy sensor vegetation indices', *Agronomy Journal* 103: 1815–26.

Slaughter, D. C., D. K. Giles and D. Downey (2007), 'Autonomous robotic weed control systems: A review', *Computers and Electronics in Agriculture* 61: 63–78.

Solari, F., J. F. Shanahan, R. B. Ferguson and V. I. Adamchuk (2010), 'An active sensor algorithm for corn nitrogen recommendations based on a chlorophyll meter algorithm', *Agronomy Journal* 102: 1090–8.

Steinberg, S., C. H.M. Van Bavel, and M. J. McFarland (1989), 'A gauge to measure mass flow rate of sap in stem and trunks of woody plants', *Journal of the American Society of Horticultural Science* 114: 466–72.

Sui, R., J. A. Thomasson, J. Hanks and J. Wooten (2008), 'Ground-based sensing system for weed mapping in cotton', *Computers and Electronics in Agriculture* 60: 31–8.

Teal, R. K., B. Tubana, K. Girma, K. W. Freeman, D. B. Arnall, O. Walsh and W. R. Raun (2006), 'In-season prediction of corn grain yield potential using normalized difference vegetation index', *Agronomy Journal* 98: 1488–94.

Thenkabail, P. S., R. B. Smith and E. De Pauw (2000), 'Hyperspectral vegetation indices and their relationships with agricultural crop characteristics', *Remote Sensing of the Environment* 71: 158–82.

Tremblay, N., Z. Wang and C. Bélec (2007), 'Evaluation of the Dualex for the assessment of corn nitrogen status', *Journal of Plant Nutrition* 30(9): 1355–69.

Tremblay, N., Z. Wang, B. Ma, C. Belec and P. Vigneault (2009), 'A comparison of crop data measured by two commercial sensors for variable-rate nitrogen application', *Precision Agriculture* 10: 145–61.

Trimble (2017a), https://agriculture.trimble.com/precision-ag/products/greenseeker/ (Accessed 19 April 2017).

Trimble (2017b), https://agriculture.trimble.com/precision-ag/products/weedseeker/ (Accessed 29 April 2017).

Varvel, G. E., W. W. Wilhelm, J. F. Shanahan and J. S. Schepers (2007), 'An algorithm for corn nitrogen recommendations using a chlorophyll meter based sufficiency index', *Agronomy Journal* 99: 701–6.

Veris Technologies (2017), http://www.veristech.com/ (Accessed 5 May 2017).

Vories, E., A. Jones, J. Sudduth, S. Drummond and N. Benson (2014), 'Sensing nitrogen requirements for irrigated and rainfed cotton', *Applied Engineering in Agriculture* 30: 707–16.

Wallihan, E. F. (1973), 'Portable reflectance meter for estimating chlorophyll concentration in leaves', *Agronomy Journal* 65: 659–62.

Wang, X., K. Thorp, J. White, A. French and J. Poland (2016), 'Approaches for geospatial processing of field-based high-throughput plant phenomics data from ground vehicle platforms', *Transactions of the American Society of Agricultural and Biological Engineers* 59: 1053–67.

Weis, M., D. Andujar, G. G. Peteinatos and R. Gerhards (2013), 'Improving the determination of plant characteristics by fusion of four different sensors', In: J. V. Stafford (Ed.), *Precision Agriculture'13, Proceedings of the 9th European Conference on Precision Agriculture*. The Netherlands, Wageningen Academic Publishers, pp. 63–9.

Yara International ASA (2017), http://yara.com/ (Accessed 17 April 2017).

Zebarth, B. J., H. Rees, N. Tremblay, P. Fournier and B. Leblon (2003), 'Mapping spatial variation in potato nitrogen status using the "N-Sensor"', *Acta Horticulturae* 627: 267–73.

Chapter 2

Proximal soil surveying and monitoring techniques

R. Gebbers, Leibniz Institute for Agricultural Engineering and Bioeconomy (ATB), Germany

1 Introduction

Soil is a fundamental resource in agriculture. The contribution of soil to crop yield is summarized as soil fertility or soil productivity (White, 2006). The impact of soil fertility on crop production can be specified by the physiological, phytosanitary and technological functions of soil (Kundler, 1989):

- The physiological function comprises the supply of nutrients, water, air and heat to the crop roots.

http://dx.doi.org/10.19103/AS.2017.0032.01

Table 1 Soil properties relevant for plant production as indicated by the notion of soil fertility

Physical	Chemical	Biological
• Particle size: gravel, sand, loam, clay • Water: water content, water potential • Structure: bulk density, porosity • Thermal properties: thermal diffusivity, heat capacity, heat conductivity, specific heat • Relief: slope, exposure	• Mineralogy: quartz, clay type • Organic matter: total content, fractions (labile, stable) • Nutrients: total content, plant available (e.g. N, P, K) • Acidity: pH • Redox potential: O_2 • Toxic substances	• Mineralization, fixation • Beneficial organisms • Bioturbation by animals and plants

- The phytosanitary function encompasses the influence of soil-borne pests and pollutants on crop health.
- The technological function refers to soil-related influences on cultivation methods (e.g. tillage) and management conditions (e.g. trafficability).

Agricultural research has identified several physical, chemical and biological soil properties that are relevant for plant production and can be used to characterize soil fertility (Table 1).

Knowledge about soil fertility parameters is relevant for optimizing plant production. With respect to manageability, soil fertility properties can be divided into two main groups:

- soil properties that can be influenced with reasonable efforts and within reasonable time, for example nutrient content, pH, bulk density
- soil properties that can be regarded as almost unchangeable such as soil texture and mineralogy

Modifiable soil properties can be managed directly (e.g. by fertilization, irrigation) in order to optimize them and crop yields.

2 Key challenges in monitoring soils

Due to the complex nature of soil, assessment of soil fertility parameters is difficult. Soil components are present in solid, liquid and gaseous phases. The solid phase (soil matrix) consists of mineral as well as of living and dead organic matter. These components are not uniformly mixed but form structures like aggregates and pores due to endogenous and exogenous processes (weathering, erosion, biological turbation by animals and plants). Soil properties vary on the micro scale due to the granularity of the material. Macro-variability is caused by soil-forming factors including parent material, time, climate, relief and human activities (Jenny, 1980). Variation extends to the field scale and beyond (regional and global scale). Beside spatial variation, several relevant soil properties show high temporal dynamics, for example nitrate, ammonium, sulphate and water content. Within-field variability of soils is one of the main drivers of precision agriculture. It can be quantified by geostatistical analysis (e.g. variography).

Scientists and agronomists investigate spatial and temporal soil characteristics in various ways. These investigations may include:

- observation by the human sensory system
- the collection of soil samples and subsequent analysis in the laboratory

- or the use of sensors that are in direct contact with the soil (*in situ*) or collect data remotely (*ex situ*, e.g. cameras on airplanes)

Observations that primarily address the spatial distribution of soils are often called *soil surveys* while the term *soil monitoring* is often used when observing temporal changes (De Gruiter et al., 2006).

The traditional approach to the investigation of soil fertility parameters is based on the collection of samples in the field and the consecutive preparation and analysis of the samples in a laboratory. This is particularly true for chemical parameters and soil texture. However, these methods are slow, labour-intensive and often expensive. Capturing the spatio-temporal variability of soils using these traditional methods in a way that makes precision agriculture viable is often not practical. The feasibility of precision agriculture depends on obtaining spatial and time-resolved information about relevant soil factors in a quick and cost-efficient manner (Stafford, 1988).

One approach to tackle this challenge is the use of sensor-based technologies, which collect data directly in the field. However, even though many sensors are available for detecting physical, chemical and biological properties in industrial applications, there are few sensor systems that have been proven to be useful for soil investigations for agricultural purposes. This is mainly because soil is a difficult medium for sensors:

- it is a complex mixture of many substances in three phases which makes it hard to separate the sensor stimuli
- the mineral components and the density of the matrix cause mechanical stress and wear for sensors that are inserted into the soil
- remote sensors usually receive signals only from the soil's surface since soil attenuates/absorbs energy from most parts of the electromagnetic spectrum.

As a result, remote sensing from space- or airborne platforms has limited success in soil mapping for precision agriculture.

3 Soil sampling methods

Key elements of sampling design are shown in Box 1.

Box 1 The elements of spatial sampling design

Sampling designs for soil monitoring need to take account of the following elements (De Gruiter et al., 2006; Legendre and Legendre, 2012):

- Sample support: The size of the elementary sampling unit. It is the surface or volume of matter supporting the measurements.
- Sampling interval: The average distance between neighbouring sampling units.
- Extent: The total length, area or volume included in the study. In the case of precision agriculture, the maximum extent is typically given by the field border.
- Sampling pattern: The geometric arrangement of sampling points, for example systematic (regular grids) or random locations and equal or variable distances.

Traditionally, soil monitoring in agriculture is carried out by collecting soil samples in the field with augers/borers and analysing these samples in the lab, typically using wet chemical methods. For a given field border (fixed extent) the sampling interval in this traditional soil monitoring approach is a function of the number of samples (sampling units). The number of samples in turn is usually controlled by economic considerations (the financial resources and the expected benefit) and sometimes by official regulations, for example by national laws that dictate a minimum number of samples per hectare and field. Usually, the sampling intervals that can be afforded when using traditional methods are too wide to capture spatial variability (Fig. 1a). To overcome these limitations, several sampling strategies have been developed as shown in Figs. 1b and c.

3.1 Grid-based sampling

Grid-based sampling (Fig. 1a) is the traditional approach to assess soil fertility. The field is divided into regular grid cells (sub-field) according to the number of samples or according to a given cell size. Samples are taken within these grid cells. Due to the variability at the centimetre and metre scale, it is common practice to collect sub-samples and bulk them in order to form the elementary sample to be analysed. The distance between the sub-samples differs between the area composite sampling and point composite sampling approaches. In the area composite sampling approach, sub-samples should be coming from the whole extent of the grid cell, for example by collecting them along a diagonal or z-shaped line. In the centre-point approach, sub-samples are collected within a distance of a few metres around the centre of the cell. Sampling is replicated over time according to best management practices or national regulations. For example, in Germany (VDLUFA, 1997) it is recommended to repeat sampling for pH, P, K and Mg after one crop rotation or after 4–6 years (AHDB, 2017).

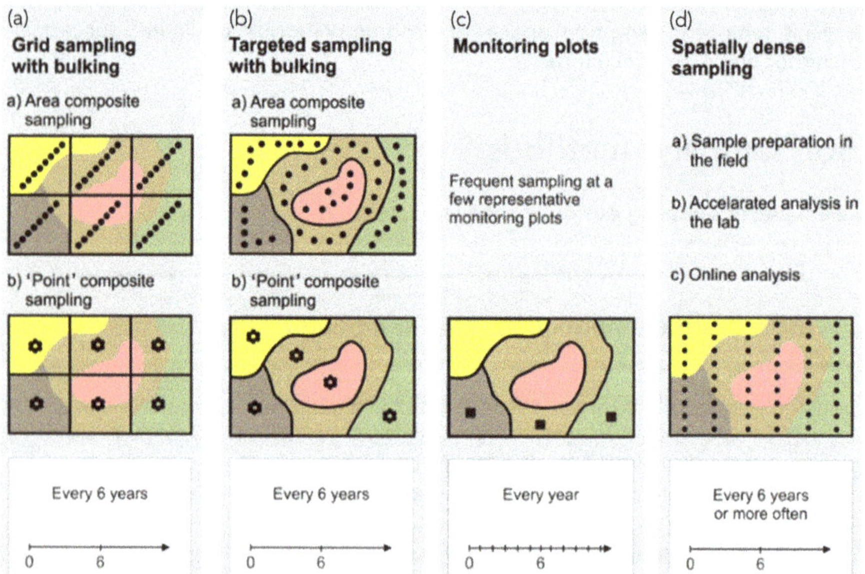

Figure 1 Soil sampling strategies in agriculture.

3.2 Targeted sampling

Targeted sampling (Fig. 1b) is based on ancillary data like pedogenetic soil maps, elevation models, remote sensing images, yield maps or maps from proximal soil sensors. These data should be available at high resolution and ideally at moderate or no cost (e.g. publically available soil maps or satellite data). It is assumed that these ancillary data are related to the soil variables under consideration. If this assumption holds true, following pre-existing patterns from ancillary data can (1) help to assess the whole range of soil property values and (2) can be used to create maps by using the ancillary data as auxiliary variables (e.g. by regression or regression kriging). So targeted sampling can make mapping more efficient and/or more precise with respect to spatial resolution. As with grid-based sampling, two approaches of forming composite samples exist, the grid cell and the centre-point approach. A problem with targeted sampling is that often just one ancillary map or type of data is used and that this ancillary data is not necessarily correlated with all soil fertility parameters in the same way. As a result, targeted sampling can be efficient for one variable but might be ineffective for another.

3.3 Monitoring plots

The monitoring plots approach (Fig. 1c) intends to achieve a high temporal resolution. It is assumed that a few locations within the field can be identified that represent the most important soil conditions. At these locations, intensive sampling and measuring over time is carried out, for example by permanently installing stationary soil sensors. The drawback of this approach is that, with these few monitoring plots, there is a high risk of missing out spatial variation. Relying on monitoring plots requires excellent knowledge about the soil properties present in the field. Detailed soil mapping will always be a prerequisite for establishing a soil monitoring plots approach.

3.4 Spatially dense sampling

Spatially dense sampling (Fig. 1d) is preferred if detailed soil maps for precision agriculture are required. In this case, many samples are collected in the field or the soil is directly analysed in the field (*in situ*). The elementary samples are taken within sampling intervals that should be smaller than the range of the spatial auto-correlation of the respective variable (typically less than 30 m). However, dense sampling is only affordable if costs for gathering and analysing an elementary sample are much lower than in the traditional approach. To lower the costs, the following approaches have been taken (Fig. 2):

- automation of sampling and packaging in the field (Lütticken, 2000, SpeedProbe system by Bodenprobetechnik Nietfeld GmbH, Quakenbrück, Germany)
- analysing samples at the field in mobile labs using robust and fast instruments (e.g. 'lab-in-a-box' by SoilCares, Wageningen, The Netherlands)
- assessment of soil *in situ* by mobile sensors, also called sensor-based sampling (De Gruijter et al., 2010; Viscarra Rossel et al., 2011). The degree of automation varies from portable handheld sensor systems to autonomous robotic sensor platforms.

The *in situ* analysis of soil by stationary and mobile sensors is also referred to as proximal soil sensing. This approach will be discussed in more detail later in this chapter.

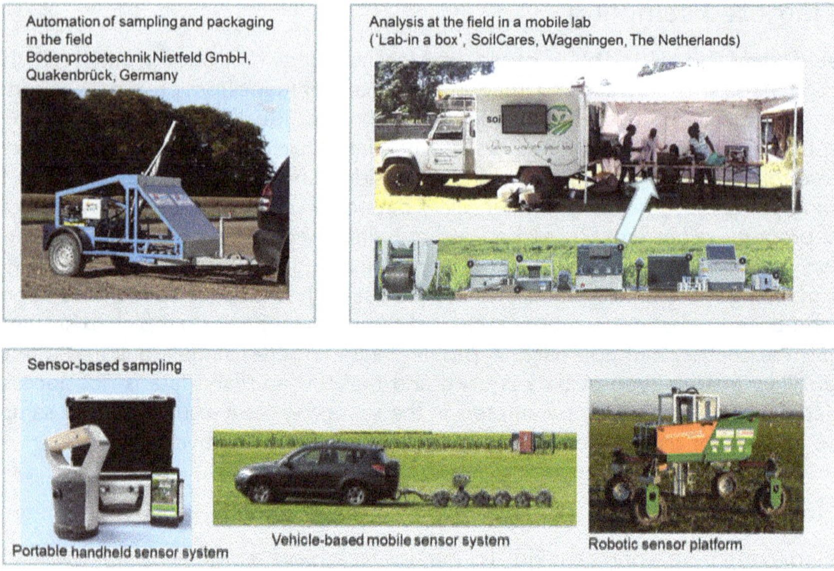

Figure 2 Approaches to spatially dense soil sampling.

4 Proximal soil sensing: definitions and principles

The nature of soil requires sensor placement in the soil or at a very short distance from the surface. This approach is called proximal soil sensing and has become a key means of developing high-resolution soil maps for precision agriculture. Proximal soil sensing is defined as 'the development and application of field-based sensors from close by (say within 2 m), or within the soil body' (Viscarra Rossel et al., 2011). Proximal sensing systems have to provide direct measurements of relevant soil properties. A sensor responds to physical or chemical stimuli (such as heat, light, electrochemical potential) and converts them typically into digital signals (usually into varying voltage which is then digitized) which can then be analysed. A key requirement is a known dependency between the stimuli measured and the soil property under investigation. Systems then require data analysis methods to extract the relevant information from the sensor data. Finally, the output from the sensor systems has to be incorporated into agronomic decision making in a suitable way. Sensor system design also needs to take account of requirements such as robustness, safety, economic efficiency and ease of handling.

Simple handheld or stationary proximal soil sensing systems consist of:

- one or more soil sensors
- a unit for data recording, processing (e.g. for linking sensor readings to spatial co-ordinates and a time stamp from a global navigation satellite system (GNSS) receiver), calibration and/or transmission to a central unit

More complex mobile automated proximal soil sensing systems for soil mapping additionally include: a platform for carrying the sensors, sometimes with a sampling unit, sensor heads and/or equipment for sample preparation, as well as a GNSS. Table 2 identifies a range of characteristics describing key elements and requirements for proximal soil sensor systems.

Table 2 Characterization criteria of proximal soil sensor systems for precision agriculture

Mobility	*Stationary sensors* are installed at the same position permanently or over a longer period (several days). Typically, they are used for the assessment of temporal variations (monitoring). Common examples are soil moisture and temperature sensors that are buried in the soil. *Kinematic (on-the-go) sensors* are dragged over the field while continuously collecting data. These systems are regularly used for soil mapping (surveying) in precision agriculture. Electrical conductivity sensors are the most popular examples. *Stop-and-go sensor systems* are mobile but require remaining stationary over a short period during the measurement. An example is the cone penetrometer, which is pushed vertically into the soil.
Temporal responsiveness	Temporal responsiveness refers to the time for measurement, sensor data evaluation (post-processing), sensor-based decision making and application. Sensor systems, which can provide data immediately and do not require lengthy post-processing, can be used in *real-time (online) applications*. For example, stationary soil moisture sensors can be used for real-time irrigation management. If sensor response is slow and/or time consuming, post-processing of data is necessary; the resulting data must be used in *map-based (offline) applications*. This is currently the case for most mobile soil sensor systems since their data need to be cleaned and calibrated.
Proximity	*In situ* sensors are in direct contact with the soil (e.g. tensiometers) while *ex situ* systems can measure from a distance (e.g. electromagnetic induction sensors) or require sample extraction (e.g. for ion-selective electrodes).
Sample conditioning	*Non-invasive* sensors do not require soil conditioning or sample preparation. Electromagnetic induction sensors are used without soil preparation. *Invasive* sensors require some sort of sample preparation to obtain reasonable measurements. The preparation may include flattening of the soil surface, opening or excavating the soil.
Source of energy for stimuli generation	*Passive sensor systems* rely on ambient sources of energy (e.g. sunlight, electrochemical potential, natural radioactive decay) to obtain the physical or chemical stimuli from the sample. *Active sensor systems* provide their own source of energy like light, electrical power or pulling force.
Selectivity	Selectivity indicates the extent to which a sensor can determine a particular analyte without interference from other components of the sample (Bănică, 2012). For example pH electrodes have a relatively high selectivity compared to geo-electrical sensors since apparent electrical conductivity, as measured by geo-electrical sensors, is affected by several relevant soil properties like water content, texture, salinity, temperature and bulk density. High selectivity is preferred. However, there is often a trade-off between the selectivity of a sensor on the one hand and its costs, robustness or responsiveness on the other hand.
Measurement method	Several measurement methods exist in proximal soil sensing. They are often classified by the primary parameters (measurand). The most common methods used for precision agriculture research and practice are based on the primary detection of thermal, electromagnetic (conductivity, resistivity, capacitance), electrochemical, optical, radioactive radiation and mechanical properties.
Sensor combination	*Single sensor systems* have been used most commonly. However, to address more relevant soil properties and/or to better discriminate soil properties affecting sensor readings, it can be advantageous to combine several sensors. When different sensors are mounted on the same platform and measure at the same time they form a *multi-sensor platform*; this is also called *sensor fusion*. The deployment of several sensors, whether they have been used at the same time on the same platform or not, generates the challenge of *sensor data fusion*.
Power consumption and source of energy	Securing power supply is often highly relevant for stationary sensors that are installed in remote areas. Thus, lowering power consumption and energy harvesting are important design aspects.

5 Electrical soil sensors

Sensors that use electric circuits to measure the capability of soils to conduct or accumulate electrical charge, become polarized or support the formation of magnetic fields are categorized as electrical sensors (Lesmes and Friedman, 2005). These electromagnetic properties of soils are assessed in a frequency range from 0 Hz (direct current) to about 300 GHz (radar). Electrical sensors are the most common proximal soil sensors for soil mapping. Many of the stationary soil sensors that are used in practice – in particular those for assessing soil water content – are electrical sensors.

All electric methods relevant for agriculture are active. As active methods, they modify the electrical and/or electromagnetic field of the adjacent soil. The displacement of the electrical charges results in a change in the electrical potential which is detected by the sensor. The physical properties of soils that are most relevant for the modification of applied electromagnetic fields include (Knödel et al., 2005; Santamarina et al., 2009):

- electrical conductivity
- dielectric permittivity (dielectric constant)
- magnetic permeability

These are discussed below.

5.1 Electrical properties of soils

5.1.1 Electrical conductivity

Electrical conductivity depends on the availability and mobility of charges. Three conduction mechanisms are usually involved (Knödel et al., 2005):

1. Electrical (metallic) matrix conductivity is based on the movement of free electrons in solids. In soils, matrix conductivity is often negligible since most soil constituents have poor electrical conductivity and the concentration of conductive minerals (iron oxides like hematite, pyrites) is usually low.
2. Electrolytic conductivity is based on the mobility of ions dissolved in water. The degree of pore filling (water content), the pore volume, the ion concentration and the specific mobility of the hydrated ion determines the electrolytic conductivity of soils. The electrolytic conductivity is used as an indicator of soil salinity.
3. Boundary layer conductivity (surface conductivity) is based on the double layer which occurs on the interface between particles with surface charges (in particular clays) and the soil solution. Permanent and variable, pH-dependent charges can be distinguished. Permanent charges of soil minerals arise out of the substitution of elements within the crystal structure while pH-dependent charges arise through the reversible dissociation of protons from surface groups (White, 2006). The 'double layer' formed by the solid surface and the ionic solution is in fact an 'ionic-cloud' with a diffuse distribution of anions and cations in solution as described by the Stern layer model (White, 2006). The electrical charge is carried along the surfaces of the particles.

Pathways of electrical current in soils are shown in Fig. 3.

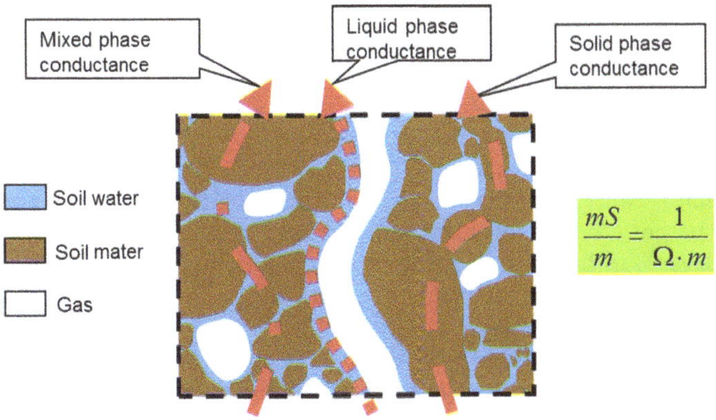

Figure 3 Pathways of electrical current in the soil. Adapted from Corwin and Lesch (2003).

Since all three mechanisms can contribute to the conduction of electrical charge in soils, this overall conduction is termed apparent (bulk) electrical conductivity (ECa in mS/m) or its reciprocal, the apparent (bulk) electrical resistivity (ERa in $\Omega \cdot$ m). The term ERa is more common in geophysics, where the method was developed, because ERa was measured first with resistivity instruments (see GRC below). However, in agriculture, ECa is preferred because ECa is often positively correlated with clay content and the ECa values in mS/m are in measurable range (usually below 100 mS/m for non-saline soils), which is easier to memorize by the user (Fig. 4).

In most soils, the electrolytic and the boundary layer conduction dominate. Therefore, the ECa is mainly associated with the following soil properties (Robinson et al., 2008):

- Water content
- Surface charges (amount and composition charged particles, in particular clay minerals)
- Concentration of electrolytes
- Hydraulic permeability
- Porosity/bulk density
- Temperature

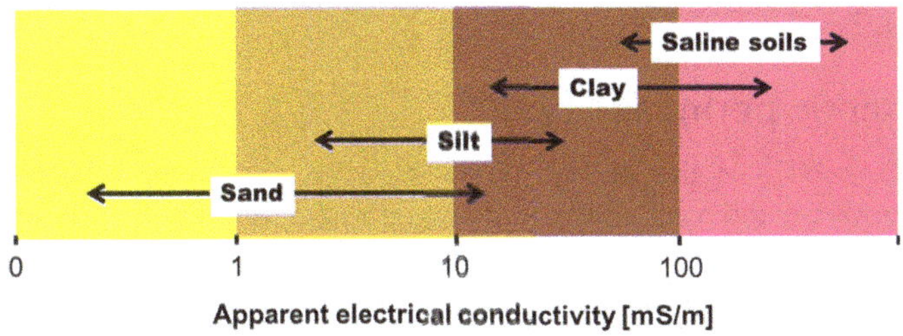

Figure 4 Ranges of apparent electrical conductivity depending on soil texture and salinity. Adapted from Veris Technologies: https://www.veristech.com/.

When mapping the spatial variation of ECa in non-saline soils, clay content (soil texture) is often the driver of these variations since it is a key factor that influences many other soil properties related to ECa.

In areas that are not affected by salinization of soils, the primary EC response is due to the soil texture variation within the field:

- Coarse sands, with limited particle contact, low moisture holding capacity and low surface charges are extremely poor conductors.
- Heavy clays, with high particle-to-particle contact, high moisture holding capacity and high surface charges are highly conductive.

In saline areas, the EC response can be affected by the presence of dissolved salts in the soil's pore water. EC signals from saline spots are typically much higher than the signals from heavy clays.

5.1.2 Dielectric permittivity (dielectric constant)

Dielectric permittivity is the ability of a medium to become polarized in response to an electrical field (Santamarina et al., 2009). The electric field displaces negative charges (electrons and anions) and positive charges (protons and cations) and it can polarize polar molecules (orientation polarization) and charged interfaces. Permittivity increases as frequency decreases and it is typically measured by soil sensors at frequencies between about 1 MHz and 1 GHz.

Measured at frequencies <1 GHz, the relative dielectric permittivities of most soil-forming minerals range between three and ten while dielectric permittivity of free water is about 80 (Santamarina et al., 2009). At frequencies above 200 MHz, the permittivity of soils is dominated by the orientation polarization of free water, which is used to estimate the volumetric water content (Santamarina et al., 2009). At lower frequencies, the displacement of hydrated ions and polarization of double layers become more relevant and thus the ionic concentration, the specific surface of the soil particles and their arrangement affect the electrical permittivity. As pointed out by Evett and Parkin (2005), none of the electromagnetic sensors used for *in situ* soil investigations (see below) can measure dielectric permittivity directly and they suggest using the term apparent complex permittivity to indicate that the values are derived from measured travel times or frequencies. Consequently, interfering soil properties have to be taken into account. They include ECa and related properties (electrolytic conductivity, ion exchange capacity), magnetic permeability, temperature and signal frequency (Evett and Parkin, 2005; Mohamed, 2008).

5.1.3 Magnetic permeability

Magnetic permeability is the ability of the medium to support the formation of a magnetic field. Water and most soil-forming minerals are non-ferromagnetic (Santamarina et al., 2009). However, soils can contain ferromagnetic compounds like iron oxides and superparamagnetic minerals from atmospheric deposition, human activities, parent rocks or pedogenesis (Dearing et al., 1996; Hannam et al., 2009; Mohamed, 2008). If they are present, the magnetic permeability of the soil mass is proportional to the volume fraction of the compounds. Instead of magnetic permeability, many instruments used for proximal soil sensing provide readings of magnetic susceptibility. The magnetic susceptibility is the ratio of the permeability of the medium to the permeability of free space minus one.

Magnetic properties can be measured by specific magnetometers or by the electromagnetic induction sensors that are commonly used for ECa measurements in agriculture. Allred et al. (2008b) describe the passive magnetometers that record the magnitude of the earth's local magnetic field. Magnetic susceptibility or permeability data measured by active sensors have mainly been used in studies on paedogenesis, archaeology and environmental pollution (Hannam et al., 2009). Only a few studies have been published that use magnetic permeability mapping of soil in an agricultural context (Barrios et al., 2017).

5.1.4 Frequency dependency of electrical measurements

With respect to electrical measurement methods, the relevance of the three electromagnetic properties mentioned above depends on the frequency of the applied electromagnetic field. The electrical conductivity is mainly relevant for lower frequencies methods while the dielectric constant and the magnetic permeability are more relevant for high frequency methods like time domain reflectometry (TDR) and radar. The higher the frequency, the more they are associated with electromagnetic wave propagation and with the dielectric permittivity. At frequencies <100 MHz polarization, mechanisms cause the permittivity response of soils to increase with decreasing frequencies while frequencies >100 GHz cause the permittivity to decrease as the relaxation frequency of water molecules is approached (Lesmes and Friedman, 2005). Under practical conditions, the impact of electrical conduction and dielectric permittivity cannot be distinguished completely. However, electrical sensors are classified based on frequency into two main groups:

- pure electrical methods (also called direct conductivity methods), operating at frequencies below 50 Hz, which only measure electrical potentials (voltage)
- electromagnetic methods, which measure electrical and/or magnetic fields

A further grouping is based on the energy transfer into the soil. This can be galvanic, capacitive or inductive (via a magnetic field).

5.2 Types of electrical soil sensor

This section reviews a range of types of electrical sensors, their principles of operation, applications, strengths and weaknesses.

5.2.1 Galvanic coupled electrical methods for soil electrical resistivity

Galvanic coupled electrical resistivity (GCR) methods use metal electrodes that are in direct (galvanic) contact with the soil to measure how strongly the soil resists the flow of electricity (Fig. 5). This can be accomplished by using just two electrodes and an ohmmeter. However, more accurate readings are obtained with the four-electrode approach, also called a Wheatstone bridge. This uses two metal electrodes to apply electrical current to the soil (injecting electrodes) while a second pair of metal electrodes, which are also inserted into the soil, measure the potential difference (measuring

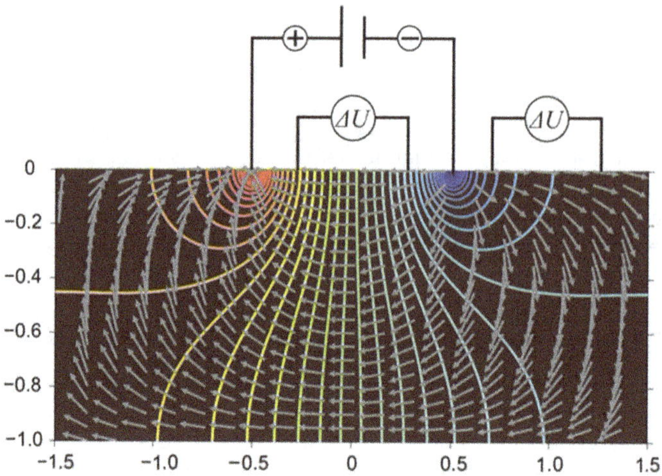

Figure 5 Measurement principle of the galvanic coupled resistivity method.

electrodes). The geometry of the electrode arrangement and the distance between the electrodes control the depth and the depth sensitivity. Based on the current, the voltage, the electrode spacing and arrangement, the apparent (bulk) resistivity (ERa) is calculated as Ω·m. Galvanic coupled electrical methods for determining the electrical resistivity work at low frequencies, usually below 50 Hz. Direct current (0 Hz) is not used in frequent measurements because the electrodes become polarized by migration of ions which can create unwanted effects. The application of higher frequencies is used to analyse polarization effects.

GCR is a relatively cost-efficient and robust method. Resistivity measurements at low frequency can be done with relatively cheap ohmmeters. Electrodes can be made of durable metal and their arrangement is very flexible, which allows the assessment of soil volumes in the cubic centimetre to cubic metre range. Power consumption is low and GCR is less susceptible to distortions by ambient electromagnetic sources than other electrical methods. The main disadvantages of GCR are the invasive nature of the galvanic coupling and the requirement of good contact. In frozen, dry or stony soils GCR methods often fail. In these cases they are outperformed by electromagnetic induction methods (EMI). Even if good galvanic contact exists, GCR methods are more sensitive to resistive soil structures than EMI. In particular, if the resistive soil (e.g. sand) is in the vicinity of the electrodes, the injected current will be prevented from flowing through a larger soil volume.

5.2.2 Electromagnetic induction

Electromagnetic induction (EMI) sensors have had a great impact on precision agriculture (Corwin, 2008; Heil and Schmidhalter, 2017). The electromagnetic induction method is based on two or more electric coils (solenoids) operating at frequencies from about 100 Hz to 60 kHz (0.4–40 kHz, Corwin, 2008). The transmitter coil creates the primary magnetic field in the soil. Conductive zones within the soil react by establishing a secondary field (eddy current) (Fig. 6). The superposition of the primary and secondary

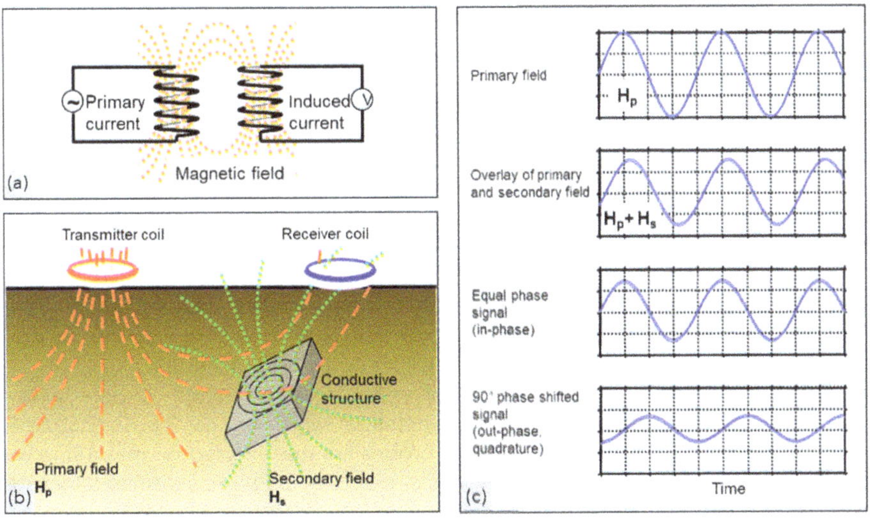

Figure 6 Measurement principle of the electromagnetic induction method.

magnetic field results in a bulk magnetic field which is sensed by a receiver coil. The electrical component, which is coupled to the magnetic field, is not measured. Depending on the soil's electrical conductivity the bulk magnetic field differs from the primary field with respect to amplitude, orientation and phase. All of these components can be measured. The phase difference between the primary and the bulk field is used to extract the secondary (induced) field.

The portion of the bulk field intensity (amplitude) which has the same phase as the primary field is called real part, in-phase or 0°-phase. The portion of the bulk field intensity which shows a phase shift of 90° is called imaginary part, out-of-phase, out-phase, quadrature-phase or 90°-phase. Based on the assumption of 'low induction numbers' (McNeil, 1980) the out-of-phase can be used to determine apparent electrical conductivity. The assumption holds only for conductivities up to 1000 mS/m. Beyond this value, for example in the vicinity of metal, many EMI instruments produce unrealistic readings that can even be negative. Electromagnetic systems prefer to look through an insulator (e.g. sand) to a conductor rather than through a conductor to an insulator (McNeil, 1980). Thus, EMI sensors tend to 'see' deeper if the soil next to the coils is more resistive (sandy, dry).

EMI instruments can provide readings of magnetic susceptibility. However, only a few have been using this parameter for soil characterization with respect to agriculture. Most of these studies are from Brazil where high concentrations of iron oxides are characteristic for strongly weathered tropical soils like Oxisols. In these soils, correlations of magnetic properties were found between sand, soluble iron, humus, available phosphorus (P) (Barrios et al., 2017), adsorbed phosphate (Camargo et al., 2016) and clay content, cation exchange capacity and P sorption capacity (Siqueira et al., 2010). In temperate climate Fluvisoils and Gley soils, Jordanova et al. (2013) used magnetic parameters and pH to predict the presence of nitrogen, carbon and sulphur.

5.2.3 Capacitive methods: Capacitance, frequency domain reflectometry and capacitively coupled resistivity

The methods described in this section use the capacitor principle to derive electromagnetic properties. A capacitor consists of two electrodes (plates) and a dielectric medium between them (Corwin, 2008). An electrical oscillator operating at frequencies from about 0.1 to 0.25 GHz (38 to 150 MHz) is connected to the electrodes and creates an electrical field which penetrates the dielectric medium. With capacitance probes and frequency domain reflectometry (FDR), the capacitor principle is used to analyse the soil which forms the dielectric medium. Due to the frequency range of 0.1–0.25 GHz, it is assumed that the applied electric field is mainly influenced by the dielectric permittivity of the soil. In capacitance sensors, the soil's dielectric permittivity is determined by measuring the charge time of a capacitor made with that medium (Muñoz-Carpena, 2005).

In FDR, the oscillator frequency is varied within a certain range to determine the resonant frequency (at which the amplitude is greatest), which is used as an indicator of the dielectric permittivity. Capacitance and FDR probes are mainly used for measuring water content. Compared to TDR probes (see below), they may be more sensitive to variations in clay content and temperature (Robinson et al., 2008). However, electrode geometry of FDR and capacitance probes is more flexible and they are cheaper because they work at lower frequencies where standard circuitry can be used. The capacitor principle is also used for continuous mobile mapping of ERa. The capacitively coupled resistivity (CCR) method draws upon the classical GCR approach, which uses a four-point electrode array, except that the galvanic electrodes are replaced by capacitive plates/antennae (Kuras et al., 2006).

5.2.4 Time domain reflectometry and ground penetrating radar

TDR and ground penetrating radar (GPR) rely on a similar principle in that they measure travel times and amplitudes of an electromagnetic pulse (Corwin, 2008). Due to the high frequencies used, the travel time should mainly be affected by the dielectric permittivity of the soil, which in turn should be determined by soil water content. TDR uses two (or more) buried electrodes to send the electromagnetic signal into the soil while GPR uses one transmitter and one or more receiver antennas placed on the surface. The metallic electrodes of TDR form transmission lines restricting the signal path while the surface antennae of GPR sense in any direction.

TDR does not operate at a specific frequency but measures the propagation velocity of a step voltage pulse with a bandwidth of around 20 kHz to 1.5 GHz (Noborio, 2001; Robinson et al., 2003; Bittelli et al., 2008). The first reflection from the end of the probe is evaluated by extracting the pulse rise time. Many TDR probes operate at frequencies above 0.5 GHz which should make them less susceptible to interference from the apparent electrical conductivity (Robinson et al., 2008).

The surface antenna of a GPR emits a wide band pulse (10 MHz to 4 GHz) around a 'centre frequency' of, for example, 50–500 MHz. The pulse is propagated into the soil and picked up by the receiver antenna. Within the soil, electromagnetic waves can be scattered when they impinge on interfaces of contrasting electric permittivity. Some of the scattered waves return to the receiver antenna as reflections. Their amplitudes are recorded by the receiver as a function of time. These are called 'traces'. Traces from different locations can be combined to form radargrams. Based on the known velocity of the waves, distances (depths) of layers of contrasting permittivities can be calculated from the transit time. The principle is analogous to ultrasonic distance measurement.

The evaluation of radar signals (radargrams) is quite challenging and requires expert knowledge. Details can be found in Daniels et al. (1988), Knight (2001) and Allred et al. (2008a–c). Depth of investigation is decreased with increasing bulk dielectric permittivity because of attenuation. Thus GPR is not suitable for detecting structures in wet or clay soils. However, variation in the bulk dielectric properties and the respective radar attenuation can often be attributed to variations in water content (Huisman et al., 2003; Lambot et al., 2004). Depth of investigation can be increased by using longer wavelengths but this comes at the cost of losing spatial resolution.

5.2.5 Other electrical measurement methods

Other, less common electrical measurement methods are available that are variants of TDR or FDR. These include:

- amplitude domain reflectometry (ADR)
- phase transition
- time domain transition (TDT)

They are briefly described in Muñoz-Carpena (2005). Global navigation satellite signal reflectometry (GNSS-R) is an approach which evaluates interference patters of direct GNSS signals and GNSS signals reflected from the soil (Rodriguez-Alvarez et al., 2009, 2011). As GNSS signals are electromagnetic waves of a working frequency of about 1.6 GHz, they are influenced by water. A special antenna is used to receive the coherently added signals. The observables are the signal level and angular patterns. They are related to the dielectric permittivity of the soil. Compared to the common stationary electrical soil moisture sensors GNSS-R has a much larger footprint of more than 1 m^2.

5.3 Stationary electrical sensors

Stationary sensors are well suited to monitor soil water dynamics. Advantages of electrical soil sensors for stationary soil water measurements over more traditional methods such as tensiometers include the higher range of observable tension (up to 1500 Pa), the ability to be left in place over a long period and lower costs in some cases. Disadvantages compared to tensiometers include slow response and interference by soil salinity (Durner et al., 2017). Detailed descriptions and a comparison of many commercial soil moisture sensors can be found in Charlesworth (2005) and Muñoz-Carpena (2005).

Although sometimes claimed otherwise, most of the stationary soil moisture sensors need careful calibration to provide absolute measures (Evett and Parkin, 2005; Cosh et al., 2016). Several problems arise from the small footprint and the sensitivity to small-scale variations caused for example by installation, bioturbation by animals or plants, crusting of the soil surface or frost heave. An example is given by Durner et al. (2017) from an experiment with 107 sensors installed on a homogenized 60 m^2 test site (14 types of water potential sensors, 15 types of water content sensor). They observed an increase of variation between the sensor readings after 2 months due to small-scale variations at the soil surface, for example crusting or growth of algae.

Another set of problems arise from the measurement principle itself since the influence of water on electromagnetic properties cannot be entirely distinguished from influences of salinity, texture, bulk density and other properties (Inoue et al., 2008). Recommendations for the application of stationary electrical sensors include standardized installation practices and procedures, use of more than one sensor at a plot, use of rain gauge records for detecting erroneous readings and troubleshooting (Cosh et al., 2016; Durner et al., 2017). Detailed descriptions and a comparison of many commercial soil moisture sensors can be found in Charlesworth (2005) and Muñoz-Carpena (2005), among others. Only a brief review is given here.

5.3.1 Electrical sensors for measuring soil water content

TDR and capacitance sensors are mainly used for stationary measurements. Their electrodes are placed directly in the soil. TDR probes use metal rods to guide the electromagnetic pulse. Capacitance sensors are more flexible in their design and can be made more cheaply. Based on the higher frequencies, TDR sensors should be more accurate but comparative studies report mixed results (Mittelbach et al., 2012; Evett and Parkin, 2005; Cosh et al., 2016; Durner et al., 2017). According to Durner et al. (2017), who tested 15 different sensor systems (capacitive, FDR, TDR), most of the soil moisture sensors follow the temporal dynamics reasonably well, but considerable differences can be observed with respect to absolute values. Compensation for temperature changes is particularly necessary for capacitance sensors.

5.3.2 Electrical sensors for measuring soil water potential indirectly via the equivalent porous medium approach

The approach to assess soil water potential indirectly via an equivalent porous medium is often referred to as the gypsum block method. It is based on a porous block (gypsum or fibreglass), with embedded electrodes, which is placed in the soil. Ideally, the soil water content will be in equilibrium with the water content in the block and the latter is reflected by electromagnetic properties measured by the sensor. The soil water potential is inferred from water retention curves which have to be calibrated for each soil. The simplest and cheapest implementation of the measurement principle uses a gypsum block with two wires inside connected to a resistivity meter. However, gypsum blocks need individual calibration and they decompose over time. Blocks made of fibreglass are more durable and can be manufactured in a more uniform way to obtain stable and transferable calibrations. Advantages of electrical soil sensors for stationary soil water measurements over tensiometers include the higher range of observable tension (up to 1500 Pa), the ability to be left in place over a long period and the low costs for some of them. The disadvantages compared to tensiometers include the slow response (Durner et al., 2017) and interference by salinity.

5.3.3 Multi-electrode arrays for electrical resistivity tomography

For research purposes, the GCR method is used with multi-electrode arrays to obtain depth profiles along a transect. Several electrodes (usually 20–100) are placed equidistantly along a line. With the help of dedicated software and cables, four of the electrodes are switched on at different locations and with different spacing. The variation of the electrode separation provides readings from different depths. By means of a numeric procedure called inversion, the surface measured bulk resistivities can be discretized laterally and vertically into a mesh of a large number of cells. The minimum spacing of the electrodes controls depth resolution. The method is known in geophysics as electrical resistivity tomography. This method is a good option if spatio-temporal dynamics of water content need to be monitored at high resolution (Pellerin and Wannemaker, 2005; Käthner et al., 2017).

5.4 Mobile electrical soil sensors

Mobile electrical sensors for continuous (on-the-go) mapping were the first sensors used for soil spatial variability in precision agriculture (Corwin and Lesch, 2003, 2005; Samouëlian et al., 2005). In particular, sensor systems based on GCR and electromagnetic induction (EMI) are still the most common soil mapping sensors in practical precision agriculture. They have been related to many soil properties including soil texture, soil moisture and bulk density. Allred et al. (2006) and Gebbers et al. (2009) compared GCR, EMI and capacitively coupled sensors for which more details will be given in the next sections. One important characteristic of these sensors is the depth of investigation (DOI). There is often a lack of clarity about the DOI of electrical measurements from the surface. Actually, there is no strict limit since the electrical field tends to spread in all directions. However, the density of the field decreases with distance from the injecting electrodes. By convention, DOI is often defined as the depth of 70% signal contribution. Table 3 provides an overview of methods used for mobile electromagnetic sensors.

5.4.1 Galvanic coupled resistivity and electromagnetic induction sensors

Mobile GCR sensors are based on the four-point method with at least four metallic wheels (rolling electrodes, coulters) acting as electrodes. The spacing and the spatial arrangement of the electrodes control the depth sensitivity. A common arrangement is the Wenner array where all electrodes are arranged along a straight line with equal spacing between them. The four-point arrangement can easily be extended by adding more pairs of measuring electrodes. In this way, a GCR system can capture readings from several depths. Typical GCR instruments (Veris 3100, ARP03) operate at frequencies of less than 500 Hz. The Veris 3100 (Veris Technologies, Salina (Kansas) USA (https://www.veristech.com)) works with two depths (Sudduth et al., 2003). The injected current is less than 100 mA at a frequency of about 150 Hz. The six rolling electrodes (smooth coulters) are arranged in a line forming a Wenner array. For shallow measurements, separation distance of the electrodes ('a') is about 24 cm and for deep measurements 'a' is 72 cm. The ARP 03 is from the company Geocarta in France (http://www.geocarta.net). ARP stands for Automatic Resistivity Profiling. The rolling electrodes (coulters with spikes) of the ARP system are arranged in a trapezoid pattern (Samouëlian et al., 2005). Current electrodes have a spacing of 1 m; potential electrodes have spacing of 0.5, 1 and 2 m. The frequency is about 225 Hz, voltage can be varied between 0 and 200 V depending on contact resistance of the site.

The more advanced GEOPHILUS system uses frequencies up to 1 kHz to analyse polarization effects (Lueck and Ruehlmann, 2012). The GEOPHILUS (Geophilus GmbH, Germany (http://www.geophilus.de/)) consists of six pairs of galvanic coupled electrodes

Table 3 Overview of methods of mobile electromagnetic sensors (Knödel et al, 2005, modified and extended)

Method	Frequency range	Transmitter, receiver	Coupling	Directly measured quantity (measurand)	Derived electromagnetic quantity
GCR	<50 Hz	Injecting electrodes, potential electrodes	Galvanic	Voltage (potential difference) and current input	Apparent electrical resistivity
Induced polarization, complex resistivity	10 Hz to 10 kHz	Injecting electrodes, potential electrodes	Galvanic and inductive	Voltage (potential difference), voltage input, phase shift between injected and received current	Chargeability, metal factor, frequency-dependent complex apparent specific resistivity
Capacitively coupled resistivity (CCR)			Capacitive		
Electromagnetic induction (EMI)	100 Hz to 60 kHz	Transmitter coil, inductor coil	Inductive	Magnetic field	Apparent electrical conductivity, normalized secondary magnetic field
Ground penetrating radar (GPR)	20 MHz to 1 GHz	Antennas	Capacitive	Electric field strength	Apparent electrical conductivity, dielectric constant, propagation velocity

arranged in an equatorial dipole–dipole array and a special resistivity meter which can also analyse spectral-induced polarization (SIP). For SIP measurements, the instrument records the complex resistivity (absolute value and phase) simultaneously at four frequencies within the range of 1 kHz to 1 MHz (preferred frequencies are 62.5, 125, 187.5 and 562.5 Hz). The first pair of electrodes, next to the vehicle, supplies the current while the voltage drop is measured simultaneously from the five other pairs. The electrode spacing within one dipole is 1 m while the (equatorial) spacing between the injecting dipole and measuring dipoles is 0.5, 1.0, 1.5, 2.0 and 2.5 m. The DOI (calculated according to Roy and Apparao, 1971) varies accordingly from 0.5 to 1.5 m.

The GRC approach has also been integrated into borehole probes. Borehole probes are inserted into the soil and can take readings along the depth profile. They can provide high-resolution cm-range information about soil layering. However, they have to be operated in a stop-and-go mode which makes them less suitable for dense mapping over large areas. In order to increase efficiency, it is sensible to integrate more sensors into the borehole probe. Probe-based GCR have been offered by Eijkelkamp (https://en.eijkelkamp.com) and Veris Technologies (https://www.veristech.com/).

Electromagnetic induction instruments have had a great impact in the field of mobile sensors (Corwin, 2008; Heil and Schmidhalter, 2017). In particular, the EM38 instrument from Geonics Ltd. (Mississauga, ON, Canada: http://www.geonics.com) has gained popularity in the precision agriculture community. It has been used in many studies, and several agricultural service providers offer mapping services based on the EM38. A recent review of EM38 applications in agriculture was given by Heil and Schmidhalter (2017). DOI characteristics are influenced by coil spacing, coil orientation and frequency. EMI instruments, such as EM38 and EM38DD with 1 m intercoil spacing (both Geonics Ltd., Canada), often provide readings of the electrical conductivity and the magnetic susceptibility alternately or simultaneously. The EM38-DD consists of two EM38 attached together in order to have simultaneous measurements in the vertical coplanar mode (VCP, vertical coils and horizontal dipole) and horizontal mode (HCP, horizontal coils and vertical dipole). To avoid interference, one of the paired instruments of the EM38-DD works at a slightly different frequency. The EM38-MK2 has three coils and works at a frequency of 14.5 kHz. The two receiver coils are separated by 0.5 and 1 m from the transmitter.

The DUALEM-1 (Dualem, Milton, Ontario, Canada: http://www.dualem.com) has 1 m separation between its transmitter and two receivers. Its depths of conductivity measurement are 0.5 and 1.5 m. In-phase measurements register magnetic susceptibility in parallel to the electrical conductivity. Unlike the EM instruments from Geonics, DUALEM requires no calibration to obtain proper ECa readings. The GEM-2 (Geophex Ltd., Raleigh, NC, USA: www.geophex.com) operates at frequencies from 30 Hz to 93 kHz. Changing the transmitter frequencies can be used for depth sounding. The Topsoil Mapper (Geoprospectors, Traiskirchen, Austria (www.geoprospectors.com/gb/products-services/agriculture/) is a relatively new sensor system which was launched in November 2015. With its four receiver coils, it is capable of obtaining layer conductivities by inversion (see below) of measurements from one position. At the typical mounting height of 35 cm, its DOI is 1.1 m. The Topsoil Mapper was the first EMI particularly designed for precision agriculture and offers a couple of interesting features. The sensor can be mounted directly in front of a tractor because the system suppresses interference from the metal parts. The manufacturer offers calibration models for estimating bulk density, soil texture and water content. The system can be extended to be used for real-time control of tillage depth and seed rate.

5.4.2 Capacitance and CCR sensors

Mobile sensors that use the capacitance principle and are galvanically coupled to the soil have been reviewed by Adamchuk et al. (2009). In 2018, Veris Technologies released a commercial version for moisture sensing. Mobile capacitively coupled sensors (CCR) have been commercially available for more than 10 years but they are rarely used for mobile soil mapping in agriculture. Their advantage over galvanic coupled systems is that they are suitable for situations when the electrical conductivity of the ground is very low and mechanical contact is difficult (e.g. hard, dry or frozen soil). However, in typical situations with more conductive soils, the voltage of the receiver dipole becomes too small to be measured reliably (Allred et al., 2008a–c). They use either coaxial cables (line antenna type) or flat metallic conductors as capacitor plates placed on the surface (Loke et al., 2013). One pair of these 'plates' is connected to an oscillator and forms a non-grounded electric dipole to generate current flow in the ground. A second similar dipole is used to measure the resulting potential distribution at the ground surface (Kuras et al., 2006; Loke et al., 2013). Since the capacitor plates are located outside of the medium they constitute an open capacitor and thus, the approach is also called 'electrostatic' method (Souffaché et al., 2010).

An example of a commercial CCR system is the OhmMapper (Geometrics Inc., San Jose, USA (http://www.geometrics.com/)) which implements capacitive coupling by line electrodes. Two pairs of insulated coaxial cable sections of 2.5 and 5 m worked as receiver and transmitter dipoles of 5 and 10 m length. Voltage difference between the injected and the received field is measured. The amount of injected current is communicated to the receiver by encoding a signal. Another encoded signal is used to keep the receiver in step with the transmitter. Operating frequency is 16.5 kHz and current is set automatically to adapt the transmitter current to the soil conditions (Geometrics). DOI is controlled by the length of the dipoles and the distance between them. Distances between the dipoles could be set arbitrarily.

5.4.3 Mobile TDR and GPR sensors

While GPR systems for continuous mapping are commercially available, only information on prototype mobile TDR systems can be found in the literature. Enlarged versions of the stationary TDR probes with longer (0.3–0.75 m) and stronger rods fitted with a hydraulic hammer and mounted on a mobile platform were designed for automated stop-and-go measurements by Long et al. (2002) and Thomsen et al. (2005). To enable continuous mobile measurements, the classical TDR probe geometry has to be modified and several aspects have to be considered such as contact, wear and heating due to friction (Wraith et al., 2005). One approach to achieving a continuous TDR mapping system is described by Inoue et al. (2001). GPR is common in geophysical exploration and archeology but only used for research purposes in agriculture. The main difficulty with GPR is the time-consuming data analysis.

5.5 Inversion of multi-depth measurements

In geophysics, inversion of multi-depth measurements is a numerical method that estimates electrical resistivity/conductivity of discrete layers from surface measurements. Several electrical surface methods can vary their DOI. With GCR, DOI is largely controlled by the electrode spacing; with EMI, coil spacing, coil orientation, frequency and height above the ground affect the DOI. This can be used to obtain information about soil layering. If the DOI is varied systematically and the surface readings are recorded, one obtains a depth response curve. This approach is known as vertical electrical sounding (VES). However, the VES curve contains redundant information because the surface measurement of greater depth encompasses the soil volume that was sensed by the measurement with a smaller DOI. To get rid of this overlap (redundancy), inversion is used. The basic idea of inversion is to calculate possible VES curves based on known physical laws describing the influence of the specific electrical conductivity (EC) of soil layers on the ECa measured above the soil surface. The formulae of the physical laws are called the forward model. The results of forward modelling are compared with the actual measurements and when they are different, estimations for input parameters are adjusted by means of numerical routines. Technical details can be found in Loke and Barker (1995), Hendrickx et al. (2002) and Pellerin and Wannemaker (2005). Alternatively, machine learning methods can be used (Pan et al., 2014). To obtain good inversion results it is important to have many surface measurements with different DOIs. A minimum of four to five DOIs is recommended. Only a few mobile commercial sensors offer this feature (e.g. Geophilus, DUALEM). Pan et al. (2014) describe an approach where just two pairs of electrodes are used which are automatically shifted away from each other while measuring the resistivity. Inversion is mainly limited due to the principle of equivalence which says that there are several possible combinations of layer conductivities creating the same surface response.

6 Optical soil sensors

Optical sensors use the reflection, absorption or scattering of light (photons) to analyse a sample. The optical portion of the electromagnetic spectrum covers roughly wavelengths from 10 nm to 1 mm (Grant, 2011). It is divided into the main regions of ultraviolet, visible and infrared which are further divided into subdivisions (Table 4). The definition of regions and subdivisions is not always consistent between different authors.

Optical sensor systems for soil analysis usually come with their own light sources which emit light, either over a broad range of the spectrum or at selected wavebands. If the primary photons of the light source hit the sample they are absorbed, reflected and/or transmitted. In the case of the typical *in situ* soil analysis, with little or no sample

Table 4 Division of the optical electromagnetic spectrum (Grant, 2011, modified)

Name of the waveband	Waveband limits	Interaction with matter
Ultraviolet (UV)	200 . . . 400 nm	Electron excitation
Visible (Vis)	360 . . . 780 nm	
Near infrared (NIR)	0.78 . . . 1.1 µm	Overtone and combined molecule vibration
Shortwave infrared (SWIR)	1.1 . . . 2.5 µm	
Mid-infrared (MIR), Midwave infrared (MWIR)	2.5 . . . 7 µm	Fundamental molecule vibration
Longwave infrared (LWIR)	7 . . . 15 µm	

preparation, only reflected light is measured. To be more precise, what is measured is the diffuse reflection which is composed of two portions: directly reflected and scattered light. Large portions of the reflected light do not return directly towards the light source but are scattered in many directions causing diffuse reflection. The returning photons are captured by detectors and converted to digital signals which are proportional to the number of photons. In spectral analysis, it is convenient to calculate the reflectance which is the ratio between the reflected and the primary light. Reflectance values are given in a range from 0% to 1% or 0% to 100%.

The principle mechanism of reflection and absorption depends on the energy of the photons. Absorption occurs when an atom or a molecule can be transferred between discrete levels of energy. This can only happen if the energy quantum of the light (corresponding to the wavelength) is equal to the difference between the two energy levels. The energy of photons decreases with increasing wavelength. Based on the wavelength and its energy two basic absorption mechanism can be distinguished:

- Ultraviolet (UV) and visible (Vis) radiation can be absorbed by promoting electrons from the outer orbits of atoms or molecules from their ground state to an excited state. Finally, the absorbed energy generates heat or splits molecules into smaller compounds.
- Infrared (IR) radiation can be absorbed if its frequency corresponds to the vibrational frequency of the molecule. In this case, the radiation increases the amplitude of the molecular vibration.

Optical sensors can be used in three ways:

- For *spectral analysis*, the light returning from the sample is divided into wavebands and the intensity at each waveband is measured. The spectral assessment of soil properties is currently the most widespread use of optical methods in soil analysis and will be discussed in more detail below.
- For *geometric (spatial) analysis*, light reflecting from an area of the sample is captured as an image of multiple pixels in one shot (camera) or by scanning the surface (laser scanning, line cameras). With stereovision (images from multiple viewpoints) and with laser distance measurements (e.g. time-of-flight (TOF) cameras) three-dimensional data can be generated (Shao et al., 2014). If outlines of objects are extracted and analysed, this is regarded as *shape* analysis. If spatial patterns within objects are analysed, this is called (image) *texture* analysis. The geometric measurements can be used to assess soil structure and soil texture (Sudarsan et al., 2018).
- *Multi- and hyperspectral imaging* is the combination of the previous approaches (Robles-Kelly and Huynh, 2013). *Multispectral* imagers provide readings from up to ten wavebands, compared to a typical consumer camera which as three wavebands (red, green, blue). *Hyperspectral* cameras have more than ten wavebands. Hyperspectral imaging has been used to characterize small scale soil profiles in the lab (Buddenbaum and Steffen, 2011; Jung and Vohland, 2014). Multi- and hyperspectral imaging has for long been the domain of remote sensing because of the high cost and the complexity of these cameras in the past. Cheaper and more robust systems are becoming available which may boost their use for closed range soil sensing in the future.

In proximal soil sensing, spectral analysis in the Vis-NIR range is currently the most important tool. It enables simultaneous assessment of various soil physical and chemical properties such as soil colour, organic matter, nitrogen, clay and sand content, iron oxides, pH and moisture content (Viscarra Rossel et al., 2005; Waiser et al., 2007; Christy, 2008; Mouazen et al., 2009; Schirrmann et al., 2013; Tekin et al., 2013). Important upcoming technologies are MIR, Raman and laser-induced breakdown spectroscopy (LIBS). All of these are discussed in the following sections.

6.1 Diffuse reflectance spectroscopy in the Vis-NIR range

Most spectrometers used for proximal soil sensing in the Vis to NIR range (300–2500 nm) are based on an optical detector array which includes one photo diode for each waveband to be detected (Fig. 7 and 8). These detector array spectrometers are robust and fast. Those working in the Vis to SWIR range are relatively cheap. In a typical detector array spectrometer, the light is collected by a sensor head and transferred to the spectrometer via an optical fibre (made of glass or plastic). The light is filtered by an entrance slit (typically 10–500 μm wide), which controls the angle of the light that enters the spectrometer. The entering light is reflected by a spherical mirror (collimator mirror) in parallel rays onto a dispersive element (typically a diffraction grating). The dispersive element splits the light into different wavebands, which propagate in slightly different directions. A focussing mirror is used to optimize the optical separation of the bands and focus them on the detector array. Spectral resolution, spectral range and sensitivity of a spectrometer are

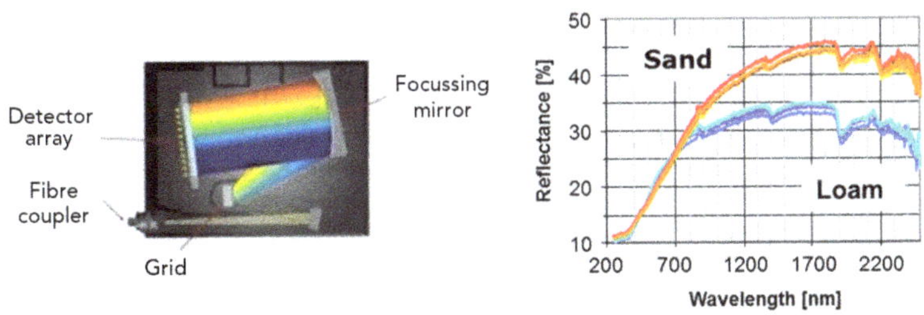

Figure 7 Principle of a detector array spectrometer and soil Vis-NIR soil spectra.

Figure 8 Example of a mobile Vis-NIR sensor system.

mainly determined by the slit width, the grating constant, the material of the detector array and the number of pixels of the detector array.

Over the last 20 years there has been a significant increase in scientific papers on Vis-NIR soil spectroscopy. Most studies have been conducted with prepared samples in the lab, with some researchers using handheld spectrometers in the field while two studies used vehicle-based sensor systems for continuous mapping. Vis-NIR soil spectroscopy has been thoroughly reviewed by Stenberg et al. (2010a,b). Compared to MIR and Raman spectra, Vis-NIR spectra look rather unspecific and calibration of Vis-NIR sensors requires a lot of data processing. As a result, much of the research in Vis-NIR soil spectroscopy has been devoted to (1) establishment of spectral libraries (Viscarra Rossel and Bouma, 2016), (2) pretreatment of spectra such as baseline removal, normalization, filtering and derivatives, (3) calibration algorithms like partial least squares regression, support vector machines (SVM), artificial neural networks, random forest and so on (Viscarra Rossel and Behrens, 2010a,b).

Continuous mapping of Vis-NIR spectra with a mobile sensor platform was pioneered by Professor Sakae Shibusawa and co-workers (Shibusawa et al., 2000). Other platforms for research were developed by the group led by Professor Abdul Mouazen at Cranfield University (UK) and the group led by Dr Gerhard Welp and Dr Stefan Pätzold at Bonn University (Germany). Commercial systems are available from Shibuya, Japan, and Veris Technologies, USA.

6.2 MIR

The absorption of the MIR radiation can be an excellent indicator of the molecular composition of a sample because within the spectrum there are clearly distinguishable wavebands that are related to the vibrations of known chemical properties (functional groups) (Fig. 9). Thus, interpretation of MIR spectra is easier than Vis-NIR spectra and very good calibration results have been found for many soil properties.

However, MIR spectroscopy is not common in proximal soil sensing because sensors are expensive and fragile. Additionally, some form of sample preparation is recommended to obtain the best results. In particular, surface roughness of the sample can have a strong impact on the measurement result. Thus, MIR in the lab is usually done with KBr discs. This involves grinding the sample to a fine powder, mixing it with KBr and producing the disc in a hydraulic press.

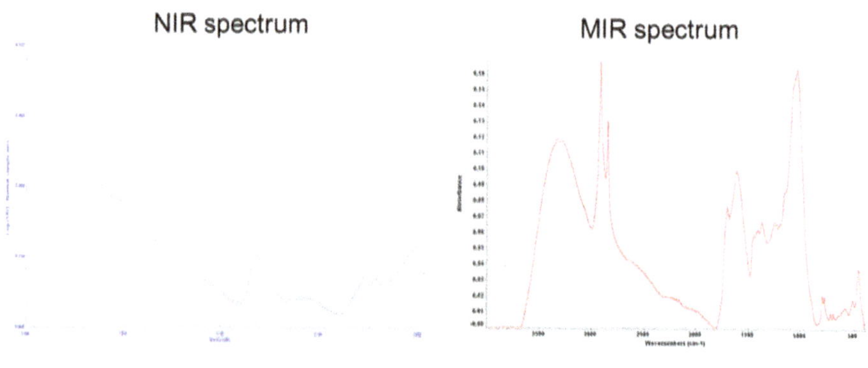

Figure 9 Comparison of NIR and MIR.

For measurement of (almost) undisturbed samples, two approaches are currently feasible: (1) attenuated total reflectance (ATR) uses a diamond or KRS-5 crystal which is pressed against the sample; (2) diffuse reflectance spectroscopy (DRIFTS) over a free optical path. ATR spectra are less affected by surface roughness and are easier to interpret because they are almost like laboratory MIR spectra of KBr discs. MIR DRIFTS can be noisy and absorption bands can be quite different (even inverted) compared to laboratory MIR spectra. However, DRIFTS MIR is very attractive because of avoidance of sample treatment and the speed of measurement. Ji et al. (2016) have reviewed results from diffuse reflectance MIR measurements *in situ*.

6.3 Raman spectroscopy

Raman spectroscopy is based on inelastic scattering of monochromatic light of a laser beam. The high energy induces in molecules an elevated 'virtual energy state'. Most of the molecules relax directly back to the original energy state, as a result of which light of the same energy as the exciting light is emitted (Rayleigh scattering) (Hof and Machán, 2014). A very small percentage of the excited molecules relax back to a vibrational excited state, hence they emit photons of smaller energy (lower frequency) than the exciting photons. The emitted photons are called Stokes lines and they are part of the Raman scattering (the other part is called anti-Stokes lines). Stokes lines are measured by a spectrometer after filtering out the light of the exciting beam and of higher frequencies. Raman intensities are recorded and plotted as a function of the Raman wavenumber, which is calculated from the difference between the scattered and the exciting light. For excitation, lasers within the visible and infrared range are used; typical wavelengths are 532, 660 and 785 nm. Shorter wavelength of the exiting light creates stronger Raman scattering. However, unwanted fluorescence (as a side effect) can increase as well. Because only a very small percentage of the molecules generate Stokes lines, Raman scattering is always of very low intensity and its detection requires very sensitive, high-quality spectrometers.

Raman spectroscopy is able to discriminate many complex molecules, including primary and secondary plant metabolites, like proteins, lipids, carbohydrates, flavonoids, alkaloids and so on (Schulz and Baranska, 2006). However, this kind of analysis only works well without the presence of a 'fluorescence background' which often masks other spectral features. Fluorescence in raw soil samples is created by organic substances and clay minerals. Thus, most of the (very limited) research on Raman spectroscopy with soils has been conducted after intensive sample preparation or by introducing enhanced Raman methods. Haskin et al. (1997) and Tomić et al. (2010) identified minerals, Stokes et al. (2003) and Martin et al. (2005) analysed soil organic matter, and Zheng et al. (2012) quantified phosphorus. Currently, shifted excitation Raman difference spectroscopy (SERDS) is under development that can eliminate fluorescence effects by using two lasers with very similar wavelengths (Maiwald et al., 2016). In its detection profile, Raman spectroscopy is very similar to mid-infrared (MIR) spectroscopy. Compared to MIR, Raman spectroscopy is less sensitive to matric (surface roughness) effects and to water content. If lasers for SERDS can be offered at low prices, Raman spectroscopy might become more cost-effective than MIR spectroscopy.

6.4 Laser-induced breakdown spectroscopy

LIBS assesses the elemental composition of a sample by analysing the optical spectra of a plasma created by a high-energy laser pulse (Anabitarte et al., 2012). The laser is focussed on a small spot, typically less than 1 mm. The laser pulse is in the nano- or femtosecond range. The created plasma consists of atomic and ionized matter. When the plasma cools down, the excited atoms and ions return to a lower energy level while emitting light with characteristic waves. The emission of the characteristic radiation takes only about 10 μs. LIBS spectra show many very sharp peaks and they require high-resolution spectrometers to capture these fine details. The spectral emission lines are unique for each element and usually several peaks are associated with the abundance of an element in the sample. LIBS is therefore an excellent qualitative method. For quantitative detection, the height of some peaks is evaluated. Under laboratory conditions, LIBS is able to detect relevant elements like K, P, Mg and Ca in plants and soils (Pouzar et al., 2009). However, the sample has to be dried since water strongly affects quantification. This currently limits the applicability of LIBS *in situ*. Compared to XRF, LIBS is better suited to detect light elements (Mg, Al, Si, Li) but accuracy is often not as high as with XRF. Since LIBS has a small footprint (sample support), several measurements of one soil sample have to be taken in order to even out the inherent micro variability.

6.5 Transmittance spectroscopy and dye photometry

Optical transmittance spectroscopy and dye photometry can be applied to liquid extracts of soil samples or leachates. The liquid extract is filled into a transparent measuring cell (cuvette) and the transmittance is measured. A dye can be added to increase the sensitivity. The dye is a chromophore molecule which reacts with the target substance. The colour of the solution changes according to the concentration of the target which is detected by the sensor at a given wavelength. Due to the necessary sample preparation, transmittance spectroscopy is mainly used in stationary systems, for example for groundwater monitoring or in labs at the field. In particular, UV transmittance spectroscopy is used for detection of nitrate (Nitraxa plus sc, by Hach Lange GmbH, Düsseldorf, Germany). Dye photometric test kits are available.

7 Soil temperature sensors

7.1 Thermal sensors for temperature

Thermal sensors for measuring absolute temperature or relative temperature (differences) are among the oldest sensor systems used in agriculture. Temperature is usually measured with electrical and infrared (IR) thermometers. While electrical temperature sensors like thermistors and thermocouples have to be in direct contact with the object of interest, IR thermometers can be used for stand-off measurements. Thermistors rely on the change of electrical resistance as affected by temperature while thermocouples use the thermoelectric effect. Electrical thermometers are relatively inexpensive. They are often integrated into sensor systems in order to compensate for temperature effects (e.g. in electrical sensor systems for measuring water content such as TDR and FDR). An IR thermometer analyses the thermal radiation emitted from an object. The instrument includes a lens to focus the radiation onto a detector, which is sensitive in a range of about 0.7–14 μm.

Temperature and the associated thermal properties are relevant for many physical, chemical and biological processes. Thermal properties relevant for heat transfer include volumetric heat capacity, thermal conductivity and thermal diffusivity. These can be used in dynamic soil-crop models (e.g. CANDY) for the simulation of mineralization, germination, growth and other processes.

7.2 Heat-pulse sensors for water content and other properties

Heat-pulse (heat dissipation) sensors are active sensors that analyse temperature rise after briefly heating up the sample. They include a heat generator and one or more thermocouples (Campbell et al., 1991; Bristow, 1998). Heat-pulse sensors were primarily developed to estimate water content or water potential. The approach relies on the fact that the specific heat of water is higher than that of other soil constituents. The temperature rise after a heat pulse is applied depends on the volumetric heat of the medium. The higher the water content, the lower the temperature rise.

Several commercial sensor systems use this approach for measuring soil water potential. In these applications the heat generator and thermocouple are encapsulated in a porous matrix (e.g. Stevens Tensiomark). When the porous matrix is in contact with the soil, the water potential inside the matrix should be the same as that of the soil. By assuming a unique water retention curve for the matrix, its water content, as derived from by the heat pulse, becomes an indirect measure of the soil water potential (Reece, 1996). The time to reach the hydraulic equilibrium between soil and porous cup is in the order of minutes to tens of minutes (Antonucci et al., 2011). The time depends on the soil water content and the hydraulic conductivity of the soil. Compared to tensiometers, heat-pulse systems for measuring matrix potential work even in very dry soils (high tensions) and are not destroyed by frost.

A less common approach is to place the heat generator and the thermometer directly in the soil (Bristow, 1998). This arrangement of the sensor elements should allow estimation of the thermal properties, volumetric heat capacity, thermal conductivity and thermal diffusivity. After applying the heat pulse, the temperature probe records temperature as a function of time. Thermal properties rely on the water content, bulk density and the heat capacity of the organic and mineral fraction (Abu-Hamdeh, 2001). It is not straightforward to use these data to estimate volumetric moisture content since soil texture, mineralogy and other factors cause interference (Bristow, 1998). As a result, thermo-time domain reflectometry, a sensor combination of heat pulse and TDR, was developed for measuring water content, air-filled porosity and bulk density (Ochsner et al., 2001).

Due to the sensitivity to soil contact and their slow response, the heat-pulse sensors described so far are mainly used for stationary measurements. Currently, rather little information on heat-pulse sensor systems for continuous mobile mapping has been found in the literature. Antonucci et al. (2011) described an approach where an IR thermometer was used to capture the temperature changes of a soil due to warming up by solar irradiation immediately after ploughing.

8 Electrochemical soil sensors

The category of electrochemical sensors includes potentiometric, amperometric and electromechanical biosensors. The sensing principle is based on a specific (selective) chemical interaction between the sensor interface and a chemical constituent of a liquid.

A voltage or current output is generated in response to the activity of the constituent in the solution. The most popular electrochemical sensor is the pH electrode. The pH electrode falls into the category of ion-selective electrodes (ISE), which, together with ion-selective field-effect transistors (ISFET or ChemFET), forms the group of potentiometric sensors. Potentiometric sensors are currently the most relevant electrochemical sensors for precision agriculture and they have been comprehensively reviewed by Kim et al. (2009) and Lobsey et al. (2010). Their main advantage is that they assess the concentration of ions in a liquid more directly than most of the other usable soil sensors to date.

An ISE system consists of a selectively sensitive membrane and a reference electrode. Both are usually assembled in one probe. The difference in the potential between the membrane and the reference is measured and the concentration of the selected ions in the solution is estimated based on the Nernst equation. An ISFET combines the ion selectivity of ISE with field-effect transistor technologies. The current between two semiconductor electrodes (source and drain) is controlled by a gate electrode formed by the ion-selective membrane. As ions of interest interact with the gate, their charge controls the source-drain current, which is proportional to the ion activity.

The main differences between an ISFET and an ISE are that an ISFET does not contain an internal solution and the ion-selective membrane is affixed directly on the gate surface of the ISFET. ISFET can be produced very efficiently by using semiconductor technology. However, they tend to be more fragile than ISE. The sensitive membranes in both ISE and ISFET can be made of glass (H^+, Na^+), polyvinyl chloride (K^+, NO_3^-, Ca^{2+}, Mg^{2+}) or metal (H^+). Based on research by Adamchuk et al. (1999), the pH ISE based on an antimony membrane has been successfully integrated into commercial sensor systems for on-the-go mapping (pH manager by Veris Technologies). Due to its mechanical robustness, the antimony ISE can be put directly into the soil. The readings are linearly correlated with standardized laboratory measurements as shown by Schirrmann et al. (2011).

9 Soil radioactive radiation sensors

9.1 Gamma ray sensors

Gamma rays carry the highest energy and have the shortest wavelengths within the electromagnetic radiation spectrum (IAEA, 2003). They have frequencies of 3*1019 Hz (30 EHz) or higher and wavelengths of 10–11 m (10 pm) or lower. They are produced from the decay of excited nuclei of radioactive elements. Passive and active gamma ray sensors are available. However, due to the high energy and the ability to penetrate materials, large doses of gamma rays are harmful. As a result, active sensors, which come along with their own gamma ray sources, are avoided. Gamma ray detectors are made of a scintillation crystal (often made of caesium iodine) which transforms the incoming radioactive radiation into light photons. The light is amplified by a photomultiplier and measured by a photodetector.

Passive gamma ray sensors detect gamma photons that are released from the decay of radioactive substances in the soil. Radioactive nuclides occur naturally in soils and rocks. They emit gamma rays that are detected within a range of about 0.4–2.81 MeV. Simple sensors detect the bulk radiation within this range (total counts per second) while others can distinguish different energy levels and provide spectral data (Fig. 10).

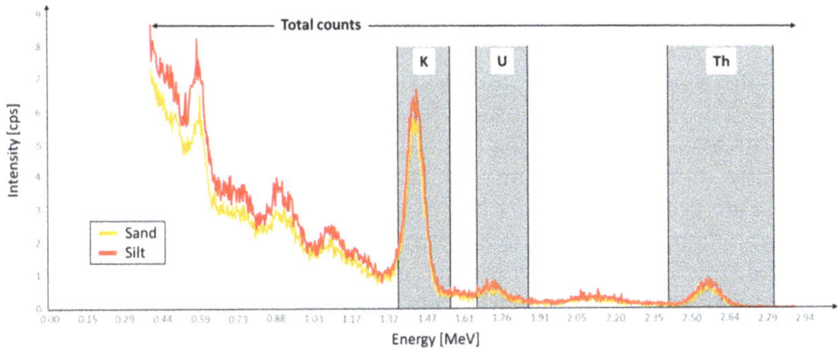

Figure 10 Gamma spectra of two soils and regions of ^{40}K, ^{238}U and ^{232}Th peaks (by S. Pätzold, University of Bonn, Germany).

There are three regions in the gamma spectra that provide particularly useful information because they are strongly related to the emissions from three decay series:

^{40}K: 1.37–1.57 MeV, peak at 1.460 MeV
^{238}U: 1.66–1.86 MeV, peak at 1.765 MeV
^{232}Th: 2.41–2.81 MeV, peak at 2.614 MeV

Radiation from other radionuclides is often not strong enough to generate distinct peaks in the spectrum and it is thus regarded as background (noise). The main sources of background noise are atmospheric radiation (^{222}Rn), cosmic radiation and sensor noise (Mahmood et al., 2013). Gamma ray intensity values for the three peak regions and the whole spectrum of the examples shown in Fig. 10 can be found in Table 5.

Gamma sensors for proximal soil sensing are employed as non-invasive, *ex situ* systems in continuous or stop-and-go mode. In this above-ground measurement, about 90% of the radiation originates from the upper 0.3 to 0.5 m of soil. Gamma rays in soils are mainly attenuated by water and mineral fractions, which means that bulk density can affect readings. The abundance of radionuclides in soils depends on the mineralogy and geochemistry of the parent material as well as on the paedogenesis (Heggemann et al., 2017). These properties can be reflected in the energy spectrum of gamma rays. Gamma ray spectroscopy provides a direct measurement of ^{40}K and thus can be correlated with total K and plant-available K (Wong and Harper, 1999). Gamma counts usually correlate indirectly with clay (soil texture) because K, U and Th are absorbed and incorporated by clay minerals, which dominate the clay fraction (Megumi and Mamuro, 1977; Heggemann et al., 2017). Other indirect relationships have been observed for pH, organic carbon, moisture and gravel (Heggemann et al., 2017). These indirect relationships depend locally

Table 5 Example of gamma ray measurements of two soil textural classes (by S. Pätzold, University of Bonn, Germany)

	Total counts (cps)	Potassium (cps)	Uranium (cps)	Thorium (cps)
Sand	1223.53	180.27	27.11	28.94
Silt	1469.26	209.61	34.32	36.66

on the strength of the correlation between total K and other soil properties (Wong and Harper, 1999). The relationships can vary from region to region or even from field to field, which necessitates site-specific calibrations. However, Heggemann et al. (2017) showed how to obtain a national prediction model for soil texture fractions (clay, silt, sand) based on non-linear, instance-based machine learning approach (SVM).

In many studies with gamma spectrometers, just the counts from the three regions and the total counts are evaluated (Viscarra Rossel et al., 2007; Heggemann et al., 2017). Some researchers recommend full spectral evaluation (Van Egmond et al., 2010). However, others have shown that there is no gain in prediction performance when spectral resolution is increased and all parts of the spectrum are considered (Viscarra Rossel et al., 2007; Mahmood et al., 2013). Like geo-electrical methods, passive gamma ray sensing is very suitable for rapid soil mapping because the technology and its theoretical background are relatively well established, the instruments are robust and can collect data quickly. As with geo-electrical methods, gamma ray data have to be calibrated carefully due to the multitude of influencing factors. Compared to geo-electrical sensors, gamma ray sensors are more expensive, which probably has hindered a wider adoption in precision agriculture.

Active gamma ray sensors are rarely used in soil investigations. Jacobs et al. (2009) describe their Medusa RhoC system which was designed to measure bulk density and texture of aquatic sediments but which could probably be used for analysis of soils as well.

9.2 X-ray fluorescence spectroscopy

X-ray fluorescence spectroscopy (XRF) is a direct method for analysing the elemental composition of a sample. X-ray fluorescence is based on the excitation of electrons by high-energy β-rays, emitted by the X-ray tube of the instrument, and the measurement of secondary β-rays emitted due to the substitution of the excited electrons by electrons from a higher energy level (orbital) (Beckhoff et al., 2006). The term fluorescence refers to the fact that the absorption of radiation of a specific energy results in the re-emission of radiation of a lower energy.

The detector measures the energy spectrum of the emitted electrons. The XRF energy spectrum contains peaks that are characteristic for each element. This is due to the fact that the electronic orbitals of each element have characteristic (unique) energy levels and so the emitted energy from the transition is also unique. The energy of the emission depends on the diameter of the atom (Moseley's law) and thus of the atomic weight. Consequently, XRF can better detect elements with higher atomic numbers. Robust portable instruments can quantify up to 42 elements. Their elemental detection range is limited to elements from atomic number 12 (Mg) onwards. More sophisticated instruments use vacuum or noble gas purge to extend the detectable range of light elements. XRF is an established laboratory method in earth sciences, in particular in geology. Due to the growing interest of industry (metal production, metal recycling, quality control in automobile industry, safety of consumer goods) and the improvement of technology, portable XRF (PXRF) instruments are available from several manufacturers at decreasing prices. The availability of these small mobile systems for handheld applications makes XRF an interesting candidate for proximal soil sensing. XRF directly quantifies several elements relevant for agriculture including P, Ca, S, Fe, Cu and Zi. Due to the correlation of elemental composition with mineralogy and with soil processes, other soil properties can be predicted by XRF after calibration. Such indirect relationships were found for soil texture (Zhu et al., 2011), plant-available phosphate and pH.

The drawbacks of XRF include the restriction to elements of higher atomic weight, the danger of operating an X-ray tube, the time for one measurement (in the range of 1–2 min), and small sample support which makes it sensitive to micro-variances within the sample. A potential alternative to XRF for elemental analysis could be LIBS.

9.3 Neutron sensors (neutron probes and cosmic ray probes)

Probes with their own neutron sources (active sensors) were the main soil moisture sensors in soil science before electrical sensors became available. Neutron probes utilize the moderator properties of water for neutrons. They emit high-energy neutrons that collide with H nuclei, which are mostly from water. These neutrons are thermalized and quantified by a detector. The more water content in the soil, the more neutrons are scattered back at the device. Neutron probes are very accurate, but they are relatively expensive and, due to the radioactive neutron source, operation is usually strictly regulated by law. Thus, neutron probes are rarely used today.

Active neutron sensors are also used in borehole geophysics (Knödel et al., 2005). For neutron generation, natural nuclides (e.g. Californium-252), Am or Pu mixes with Be and deuterium-tritium tubes are used (Knödel et al., 2005; IAEA, 2012). The detectors capture gamma rays, thermal and/or epithermal neutrons and provide energy spectra or time-dependent relaxation curves. In particular, the pulsed neutron-neutron method is of interested because several elements can be detected and its deuterium-tritium tube can be switched off, which makes handling easier (Buckup and Buckup, 2010).

Cosmic ray probes are passive sensors that measure low-energy neutrons (1 keV) that are generated within the soil by moderating incoming cosmic ray neutrons (Zreda et al., 2008). The moderation (slowing down) of the cosmic ray neutrons is mainly controlled by hydrogen atoms and, thus, the water content of the soil. The low-energy neutrons are measured non-invasively above the ground. The common stationary cosmic ray probes have usually a relatively large footprint (about 670 m in diameter). The depth of exploration depends strongly on soil moisture and varies from 0.76 m in dry soils to 0.12 m in wet soils (Zreda et al., 2008). Epithermal and fast energy neutrons (1–106 eV) are evaluated. To obtain absolute neutron intensity the geographical variations due to the geomagnetic shielding of neutron-generating cosmic rays have to be taken into account. Neutron intensities are converted into soil moisture based on universal functions that are relatively insensitive to changes in soil chemistry (Zreda et al., 2008). More recently, mobile probes have been designed that can be mounted on a vehicle and enable a more detailed spatial investigation (Chrisman and Zreda, 2013).

10 Mechanical soil sensors

10.1 Force sensors

Vertical cone penetrometers measure penetration resistance (soil strength) as they are inserted into the soil. The cone index is calculated from the penetration force per unit base area of the cone (typically in MPa). They have been used for a long time in agriculture to detect soil compaction (Stafford and Ambler, 1990; Stafford and Hendrick, 1988). Factors affecting the cone index include bulk density, soil moisture, soil type and soil structure

(Ayers and Perumpral, 1982; Domsch et al., 2006). Efforts have been made to design an automated cone penetrometer with GPS registration for soil mapping which operates in a stop-and-go mode (Domsch et al., 2006). Veris Technologies was the first to commercialize this type of equipment called 'Profiler 3000' in 2000 (Drummond et al., 2000). However, due to high variability of penetration resistance and the requirement to probe within short distances, mapping of a field is uneconomic (Domsch et al., 2006). Consequently, systems for continuous mapping are favoured. There are three main approaches: (1) horizontal penetrometers which measure the horizontal penetration resistance (Andrade-Sanchez et al., 2007), (2) draft force sensors and vertical force sensors which are attached between the tractor and a tillage implement (Hemmat and Adamchuk, 2008; Tsiropoulos et al., 2015) and (3) registration of fuel consumption of the tractor during tillage operations (Boon et al., 2005; Kichler et al., 2007). Hemmat and Adamchuk (2008) have provided a thorough review of penetrometer systems and draft force/vertical force sensors. As they emphasize, variation in penetration resistance or draft/vertical force can only be properly interpreted in terms of bulk density if soil moisture and soil texture is known. In addition, speed of operation has to be considered. This makes accurate assessment of soil compaction very complex. Shamal et al. (2015) have described an integrated system of three sensors for bulk density mapping. Measuring fuel consumption is an elegant approach which implements the 'tractor as a sensor'. However, as with the other approaches, translating fuel consumption into soil compaction and prescription maps is difficult. Up to now, none of the approaches for continuous mobile mechanical sensing has been transferred into practice.

10.2 Tensiometers, pF sensors

Water content is a measurement of the amount of water in the soil either by weight or volume and is defined as the water lost from soil upon drying to a constant mass at 105°C. It is expressed in units of either mass of water per unit mass of dry soil (kg/kg) or in units of volume of water per volume of soil (m^3/m^3) (Kirkham, 2005).

A tensiometer consists of a porous (e.g. ceramic) cup connected through a water-filled tube to a manometer, vacuum gauge, pressure transducer or other measuring devices (Kirkham, 2005). The tensiometer directly measures the soil matric water potential, which is a part of the total water potential. Soil water potential describes the relative energy status of a small parcel of soil water compared to pure, unbounded (free) water. It is defined as a negative pressure or tension or suction. The plant has to overcome this tension to absorb/pull water from soil. The energy status of soil water results mainly from four factors: the matric (capillary) potential, the gravitation potential, the solute potential (negative osmotic pressure) and the pressure potential. The potentials are usually expressed in terms of head (the unit is length in cm) or pressure (atmosphere, kPa or bar). The head can be seen as the height of a column of water required to produce the necessary pressure to overcome the energy difference between soil water and free water. Instead of using cm the head is often given in logarithms of cm (log(cm)) termed pF, hence the name pF meter. Under non-saline, unsaturated conditions, the matric and the gravitational potential are most important for water movement. As a result, tensiometer readings are also used to derive hydraulic conductivity. Advantages of tensiometers are that they provide direct measurements of water content and are easy to interpret. Unfortunately, maintenance requirements are high and the measurement range is limited. Tensiometers must be protected against frost and they must be imbedded very carefully into the soil to provide a good contact between the porous cup and the soil matrix. They cannot operate at tensions below −85 kPa because air may enter the porous cup and the connection to the pressure sensor breaks. The establishment of the equilibrium between soil matrix and tensiometer cup often takes several minutes. As a result, tensiometers should only be used as stationary sensors.

11 Other sensors

11.1 Capillary electrophoresis

Capillary electrophoresis (CA) is based on the separation of solute ions in an electric field applied to a liquid soil extract filled in a capillary. Ions are identified from their time of travel within the capillary. The ions are detected and quantified while passing the capacitor electrodes attached to the outer wall of the capillary. The electrolyte solution for separating the ions must be selected carefully. Other relevant tuning parameters of the instrument's setup include the length of the capillary and the voltage applied to the electrolyte solution, which generates the electrophoretic of the ions. Capillary electrophoresis is a common method in the lab. However, portable systems are relatively recent developments (Smolka et al., 2016). The iMETOS MobiLab (METOS® NPK) for measuring NO_3^-, K^+ and PO_4^- based on a very small and cost-efficient microfluidic chip CA was recently released by Pessl Instruments GmbH (Weiz, Austria).

11.2 Gas sensors

Gas sensors have gained some interest for detecting acetylene which is emitted by plants under unfavourable conditions, such as drought or fungal infections. While acetylene is an unspecific stress indicator, it is known that plants and infecting fungi emit a number of other molecules. To assess biological activity in soils it is relatively straightforward to analyse the CO_2 emission. Kaur et al. (2015) reported experiments with a mobile system that they have designed based on a non-dispersive infrared (NDIR) CO_2 sensor. NDIR sensors can detect simpler gaseous molecules like H_2O, CO_2, SO_2 and NO_2. However, NDIR sensors suffer from cross-sensitivity of these gases in low concentrations. To assess more complex molecules, 'electronic noses' are used (Wilson and Baietto, 2009). Unfortunately, these sensors are not very specific and tend to drift over time (Eifler et al., 2011). In the search for more specific and stable detectors of gaseous emissions from plants and soils, novel approaches based on antibodies and aptamers have been considered. Aptamers are short DNA or RNA molecules which bind specifically to a target molecule in a similar way to antibodies.

11.3 Nuclear magnetic resonance spectroscopy

Nuclear magnetic resonance (NMR) spectroscopy is based on the magnetic resonance between the nucleus of an atom and an external magnetic field. It identifies, and in many cases quantifies, the chemical forms of the target nuclei in a sample based on chemical bonding environments (Kizewski et al., 2011). Recently, a mobile NMR spectrometer was developed for analysing nitrogen, phosphorus and potassium content of animal slurry on a manure applicator (Sørensen et al., 2015). These results indicate that it may be feasible to integrate NMR instruments into a mobile soil sensing system.

12 Integration of proximal soil sensing into agricultural management

12.1 Sample preparation for soil sensors

Ideally, proximal sensing does not require any preparation of samples. As said before, sensing is regularly affected by several soil properties, which degrade the selectivity of a sensor. In some cases, unfavourable soil conditions can become a major obstacle for conducting measurements. For example galvanic coupled geo-electrical sensors do not provide meaningful results when soils are very dry; soil samplers can be blocked by plant residues; ISE need a liquid solution and protection from abrasive particles. In particular for ISE, sample extraction and sample preparation can solve this problem (Li et al., 2017).

12.2 Calibration and measurement errors in soil sensing

Sensors are quite often affected by several soil properties at the same time (see selectivity in Table 2). For example the EC, which is frequently used in soil mapping, is sensitive to water, salt and clay content, among others. Due to this ambiguity of the sensor response, data have to be interpreted with care. In many cases, it is advisable to collect samples and analyse them by conventional laboratory methods to calibrate the sensor readings. Under field conditions several soil properties and other environmental factors can vary beyond control.

12.3 Sampling design for calibration samples for soil sensing

Sensors usually require calibration (reference) samples to obtain the best results in the translation of the primary readings into soil fertility properties. Several studies have pointed out the necessity of calibration in general and the advantages of field-wise calibration in particular. This cannot be overemphasized. While determining the locations of the calibration samples for stationary sensors is relatively easy, the selection of sampling sites for calibrating mobile mapping systems is more challenging. Samples for referencing stationary sensors can be taken during the installation of these sensors and/or next to the particular plot. The main issue with stationary sensors is selecting the optimal position for their installation. If relevant ancillary spatial data are available, the task of optimizing positioning of stationary sensors converts to the problem of selecting sample positions for calibrating area-wide sensor data. Prescribing sampling locations for calibrating mobile sensor data, probably from multiple sensors, is not trivial since only a limited number of samples should be taken and the sites should be optimal with regard to the statistical calibration.

The excellent article of Adamchuk et al. (2011) describes the criteria for optimizing a sampling scheme and provides an algorithmic solution. The main criteria for optimizing the sampling positions are (1) spatial separation, (2) spread of values across the sets of measurements and (3) local spatial homogeneity within each of the sets of measurements. These criteria have the following background:

1 Accounting for spatial separation is relevant in order to avoid auto-correlation. Auto-correlation implies repeated sampling from the same phenomenon (spatial unit)

and thus the information gain from auto-correlated samples is less than that from independent samples. Some statistical procedures that are used for calibration, for example ordinary linear or non-linear regression, require independence within the observations of one variable. Increasing the spatial separation between the samples can help to avoid or reduce auto-correlation.

2 Increasing the spread of values of the observations can improve the stability of the fitted calibration model.
3 Local spatial homogeneity can tell about the representativeness of the sample. If the sample is taken at a location of a sensor's measurement where neighbouring data are similar, it is more likely that the particular sensor reading is not an outlier.

Adamchuk et al. (2011) showed how these criteria can be transferred into a mathematical function as a basis for numerical optimization.

12.4 Data alignment for fusion of data from soil sensors

It is common in precision agriculture that data are collected with different sensors at different locations using different units (Fig. 11). To allow comparison and analysis, such data need to be aligned.

12.5 Robustness, safety, ease of handling and economic efficiency

Proximal soil sensors are deployed in a rough environment. Robustness of the sensor systems, for example shielded cables and watertight plugs, is an important prerequisite for everyday use. Some sensors like XRF systems and GCR can cause serious injuries and safety measures have to be applied including training of the operators, shielding and safety shutdown. Ease of handling (ergonomic design of physical and software interfaces) not only contributes to work efficiency but to the safety of the user as well.

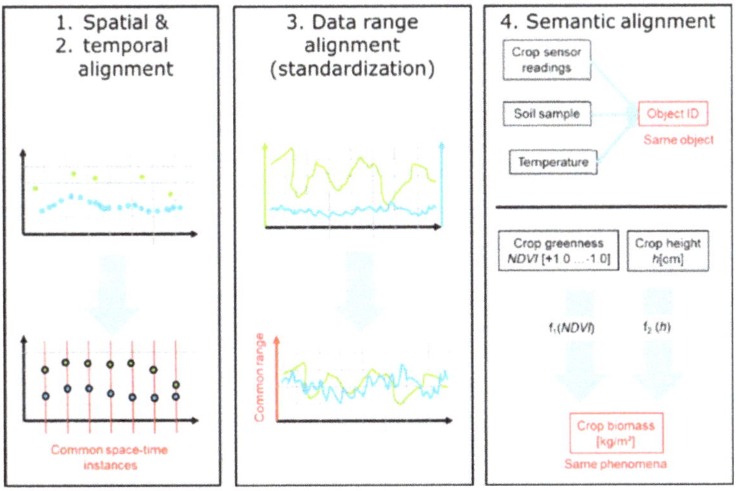

Figure 11 Data alignment.

12.6 Integration into algorithms for decision making

Sensor data only make sense and become informative if they can be integrated into a decision-making algorithm (Fig. 12). Decision support algorithms in precision agriculture rely on biophysical and socio-economic models which link certain input data (e.g. temperature, soil moisture, soil nutrient content) to outputs like crop yield or profitability. With the help of these models, the effects of agronomic measures can be approximated (e.g. increase of yield by liming) and a cost–benefit analysis can be carried out. Ultimately, after the decision, prescription maps for site-specific applications can be generated. The input requirements of the decision support system and its underlying algorithms dictate which sensor data are relevant. Relevance here implies the selection of target parameters, accuracy requirements and acceptable costs.

13 Summary

Since soil is a fundamental resource in plant production and can exhibit considerable spatial and temporal variation, detailed assessment of soils by sensors can be of great benefit for efficient and sustainable land use by precision agriculture. If soil fertility is known in greater detail, that is at a finer spatial and temporal scale and by more parameters (physical, chemical, biological), it is possible to better adjust agricultural measures to the actual requirements (Table 6).

While conventional sampling and laboratory analysis has imposed restrictions on the spatio-temporal resolution of soil monitoring, mobile proximal soil sensors can provide more timely soil data and at lower costs. In sensor-based soil assessment, three main strategies can be distinguished: a) remote sensing with space- or airborne platforms, b) proximal sensing with mobile terrestrial platforms and c) proximal sensing with stationary sensors. Mobile (kinetic) terrestrial sensor systems are currently the most common approach for

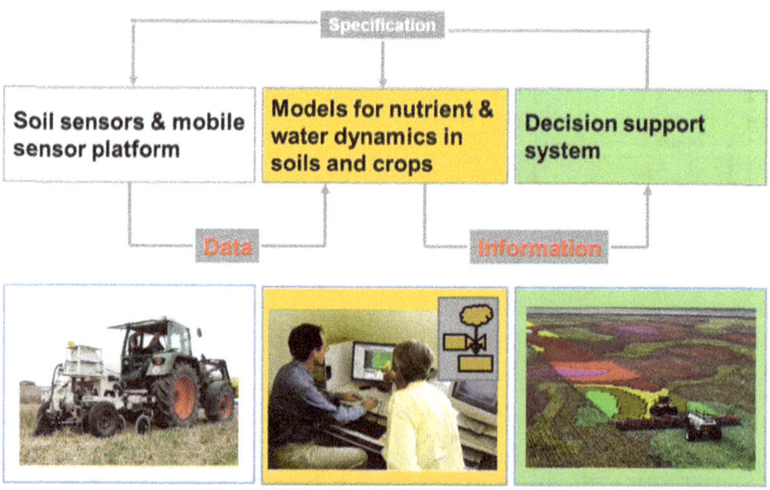

Figure 12 Integration into decision support systems.

Table 6 Soil fertility parameters and corresponding proximal soil sensors

Soil fertility parameter	Stationary and mobile sensors (D = direct; I = indirect)
Texture	D: camera I: PGRS, XRF, MIRS, NIRS, EMI, GCR
Bulk density	D: Penetrometers, draft force I: GPR, GCR, EMI, GRS
Water content	D: thermalized neutron methods I: TDR, FDR, GCR, GNSS-R, cosmic probe, GPR, EMI, GCR, PGRS, NIRS, cosmic probe
Water potential	D: Tensiometers, pF meters I: EC, see water content
Temperature	D: Thermocouples, thermal radiation sensors I:
Aggregates	D: Camera (stereo vision), TOF cameras, lasers distance, structured light I:
Permeability, hydraulic properties	D: Permeameter I: GCR, GPR
SOM	D: MIRS, Raman I: NIR
Carbonates	D: MIRS, Raman I:
Salinity	D: I: EMI, GCR
pH	D: ISE, ChemFET I: EMI, GCR, Vis, NIRS, MIRS, XRF, LIBS
CEC	D: I: MIRS, EMI, GCR, XRF, LIBS
N_{tot}	D: I: MIR, NIR, Vis, LIBS, XRF
NO_3	D: UVS, EP, ISE, ChemFET, CE I: MIRS, NIRS
NH_4	D: ISE, ChemFET, CE I: MIRS, NIRS
P_{tot}	D: XRF, LIBS I:
PO_4	D: Raman, EP, ISE I: MIRS, NIR, XRF, LIBS
K_{tot}	D: XRF, LIBS, PGRS I: MIRS, NIRS
K^+	D: ISE, ChemFET I: XRF, LIBS, PGRS, MIRS
Mg_{tot}	D: LIBS I: MIRS, NIRS

(Continued)

Table 6 (*Continued*)

Soil fertility parameter	Stationary and mobile sensors (D = direct; I = indirect)
Mg^{+}	D: ISE, ChemFET I: MIRS, NIRS
S_{tot}	D: XRF, LIBS I: MIRS, NIRS
Ca_{tot}	D: XRF, LIBS I: MIRS, NIRS, Raman
Ca^{2+}	D: ISE, ChemFET I: XRF, LIBS, MIRS, Raman, NIRS
B_{tot}	D: I:
Cu_{tot}	D: XRF, LIBS I:
Mn_{tot}	D: XRF, LIBS I:
Mo_{tot}	D: XRF, LIBS I:
Zn_{tot}	D: XRF, LIBS I:
Heavy metals	D: XRF, LIBS I:
Biological activity/respiration	D: Gas sensors I:

efficiently capturing soil spatial variation in large fields. Due to the advances in sensor and data processing technologies over the last 20 years, the assessment of the soil's physical and biochemical attributes with high spatial and temporal resolution has become feasible. The importance of understanding soil, and the complexity of soil and soil science, gave rise to the new discipline of proximal soil sensing. Viscarra Rossel and Bouma (2016) describe proximal soil sensing as a new paradigm in agriculture. It can enable us to better understand the dynamics of soils, their interactions with the environment while considering their large spatial heterogeneity. The data from sensors can help us to improve soil models and make them applicable under practical conditions. Based on the better understanding of soil processes and the high spatio-temporal information, innovative management practices and policies can be implemented that contribute to sustainable development, climate adaptation and intensified production which matches the needs of a growing population.

However, there is still a long way to go. Almost all sensors suffer from the problem that they cannot detect a soil property unambiguously. It is very important for researchers and practitioners to know the limitations of their sensors, how to calibrate them, how to secure data quality and how to transfer sensor data into useful information. Only a few measurement principles are used in practice. Electrical sensors dominate the field because they are relatively cheap and robust, and they can be related to important soil properties like soil texture, water content and salinity. However, the relationship between soil electrical conductivity/resistivity and these other soil parameters varies between fields and regions.

Three other basic types are commercially available and used in practice. These include pH measurement by ISE, gamma ray activity and optical reflectance in the Vis-NIR range.

14 Future trends

14.1 Design of new sensors

Currently, commercially available proximal soil sensor systems for precision agriculture can only address a limited number of soil fertility properties with an accuracy that is sometimes insufficient. It is necessary to develop sensor systems that capture a wider range of soil properties with higher accuracy. Among these properties are soil strength, bulk density, porosity, phosphate, sorbtion isotherms, biological activity, beneficial soil biota (earthworms), soil-borne pests and pathogens. Important parameters for new types of sensor are discussed below.

14.2 User-friendliness and cost-efficient design

Sensor systems need to be affordable and manageable in order to be adopted by farmers and service suppliers. This might be achieved by using technologies from the consumer market such as smartphones and sensors from electronic toys. In general, low-cost handheld sensors can provide a solution for smallholder farmers and those who want to start sensor-based soil sensing at a low-cost/entry level.

14.3 Sensor fusion and multi-sensor platforms

There appears to be no single sensor that can capture all relevant soil properties at the same time. The combination of sensors (sensor fusion) potentially overcomes this limitation. However, the integration of several sensors on a single platform is currently both expensive and technically challenging.

14.4 Improvement of databases and mathematical methods for sensor data evaluation

For practical applications, the sensor system needs to become smarter. Routines for self-checking, calibration and removal of erroneous data are required. In many situations, online control of an application (e.g. fertilization, tillage) by sensors would be of great benefit. This requires robust and fast evaluation of sensor data. The establishment of representative and reliable calibration databases and protocols is a key issue for evaluating sensor data, in particular for spectral sensors. The design and proliferation of reference samples (standards) can help to make sensor readings more comparable. In sensor data fusion, the weighting of the data from different sensors is an important issue.

14.5 Recommendation algorithms and decision support

Only a very few countries worldwide offer their farmers officially acknowledged, scientific sound, robust and accessible algorithms for decision making in the sense of best management practices, for example on soil tillage, fertilization and irrigation. Most of these algorithms require inputs from lab-based methods of soil analysis. For the use of sensors and sensor data in precision agriculture often only 'ad hoc' approaches exist which have not been validated systematically over a wide range of conditions by independent institutions. Authorities should support the establishment of these guidelines in order to promote the reliability and the adoption of sensor-based farming.

14.6 Online sensing for real-time control

Since crop management sometimes requires quick responses, farmers often ask for real-time control. In contrast to commercial crop sensors, most of the mobile soil sensor data are currently evaluated offline. The integration of soil sensors into tillage implements, seeders or fertilizer spreaders requires robust sensors and algorithms.

14.7 Quantifying benefits

The transfer of soil sensors into practice is difficult. Van Erp and Beens (2017) have described the challenges of introducing a business concept for fast and reliable soil sensing. Only companies with sufficient capital and staying power are able to introduce new sensors. The main problem is that, under current conditions, the immediate return of investment for farmers cannot be guaranteed and there is a lot of uncertainty about the proper use of soil sensors. However, the use of soil sensors in precision agriculture should not only be rated on its economic potential at the farm level. Precision farming can be beneficial for the whole society, for example by avoiding pollution and erosion. These positive impacts on the environment need to be quantified to support farmers in using precision agriculture technologies.

14.8 Better integration into decision support systems

Agricultural advisors and farmers need tools that help them to handle soil sensor data. They don't want to waste their time with data cleaning, calibration and integration into DSS. Instead, the data need to be transferred smoothly into meaningful information for decision making. There is a requirement to develop robust and user-friendly software which is flexible enough to cope with many different data formats.

Common decision support algorithms, for example for fertilizer recommendations (AHDB, 2017), are based on relatively simple models which require rather few input parameters. As a first step, research and development for proximal soil sensing should focus on matching the input requirements of these simple models. However, proximal soil sensing has the potential to provide information that is neglected in the standard algorithms for best management practices. This information might be integrated into advanced soil-crop models that are currently only used for research.

15 Acknowledgements

This contribution was funded by the 'Leibniz-Institute of Agricultural Engineering and Bioeconomy' and the 'German Federal Ministry of Education and Research', grant 031A564, I4S – Integrated System for Site-Specific Soil Fertility Management.

16 Where to look for further information

16.1 Introductions and reviews

General proximal soil sensing: Stafford (1988), Hummel et al. (1996), Sudduth et al. (1997), Adamchuk et al. (2004), Shibusawa (2006), Sinfield et al. (2010), Mulder et al. (2011), Viscarra Rossel et al. (2011), Kuang et al. (2012), Adamchuk et al. (2015) and Grunwald et al. (2015).

Vis-NIR spectroscopy of soils: Shepherd and Walsh (2007), Stenberg et al. (2010a,b), Stenberg and Viscarra Rossel (2010) and Horta et al. (2015).

Spectroscopy in chemistry in general/mainly optical spectroscopy: Gauglitz and Moore (2014).

Mobile electrical methods (Geo-electrical, radar, magnetometry, self-potential): Allred et al. (2008a–c), Mohamed (2008), Dearing et al. (1996), Santamarina et al. (2009), Corwin and Lesch (2005), Gebbers et al. (2009), Loke et al. (2013), Doolittle and Brevik (2014), Heil and Schmidhalter (2017) and Muñoz-Carpena (2005).

Staionary electrical sensors: Durner et al. (2017).

Electrochemical methods: Gründler (2006), Kim et al. (2009) and Lobsey et al. (2010).

Mechanical sensors: Hemmat and Adamchuk (2008).

Stationary sensors for soil water sensing: Kirkham (2005).

Statistical methods for sensor calibration and sensor data fusion: Brereton (2018), Clark et al. (2009), Hastie et al. (2009), Mitchell (2012), Reimann et al. (2008), Varmuza and Filzmoser (2009), Viscarra Rossel and Behrens (2010a,b) and Witten et al. (2016).

16.2 Key societies

Global Working Group on Proximal Soil Sensing of the International Union of Soils Sciences (IUSS, http://www.proximalsoilsensing.org/). International Society for Precision Agriculture (ISPA, www.ispag.org)

16.3 Journals

Precision Agriculture, Computers and Electronics in Agriculture, Sensors, Geoderma, Near Surface Geophysics, Soil Science Society of America Journal, European Journal of Soil Science, Transactions ASABE.

16.4 Conferences

Global Workshop on Proximal Soil Sensing, Conferences on Precision Agriculture (ECPA, ICPA, ACPA, South American Conference on Precision Agriculture), Pedometrics, EGU.

16.5 Important research centres

- China Agricultural University (Beijing, PR China), College of Information and Electrical Engineering, Key Laboratory on Modern Precision Agriculture System Integration Research & Key Laboratory of Agricultural Information Acquisition Technology, Ministry of Education

- Ghent University (Ghent, Belgium), Department of Soil Management, Professor Marc Van Meirvenne, Professor Abdul Mouazen
- Harper Adams University (Professor Simon Blackmore)
- Leibniz Institute for Agricultural Engineering and Bioeconomy ATB (Potsdam, Germany), I4S – Integrated System for Site-Specific Soil Fertility Management
- McGill University (Montreal, Canada), Bioresource Engineering, Professor Viacheslav Adamchuk
- Swedish University of Agricultural Science (Skara, Sweden), The Department of Soil and Environment, Precision Agriculture and Pedometrics Unit, Professor Bosse Stenberg
- The Commonwealth Scientific and Industrial Research Organisation (CSIRO), (Canberra, Australia), Department Division of Land and Water, Dr Raphael Viscarra Rossel
- Tokyo University of Agriculture and Technology (Tokyo, Japan), Department of Environmental and Agricultural Engineering. Professor Sakae Shibusawa
- U.S. Department of Agriculture, Columbia University (Columbia, Missouri, USA); Engineering technologies for precision agriculture, Dr Kenneth A. Sudduth
- University of Sydney (Sydney, Australia), Sydney Institute of Agriculture, Digital Agriculture & Soil Science, Professor Alex McBratney
- Zhejiang University (Hangzhou, PR China), Institute of Soil and Water Resources and Environmental Science, Professor Zhou Shi

17 References

Abu-Hamdeh, N. H. 2001. Measurement of the thermal conductivity of sandy loam and clay loam soils using single and dual probes. *Journal of Agricultural Engineering Research*, 80(2), 209–16.

Adamchuk, V. I., Morgan, M. T. and Ess, D. R. 1999. An automated sampling system for measuring soil pH. *Transactions of the ASAE*, 42, 885–91.

Adamchuk, V. I., Hummel, J. W., Morgan, M. T. and Upadhyaya, S. K. 2004. On-the-go soil sensors for precision agriculture. *Computers and Electronics in Agriculture*, 44(1), 71–91.

Adamchuk, V. I., Hempleman, C. R. and Jahraus, D. G. 2009. On-the-go capacitance sensing of soil water content. Paper No. MC09-201. St. Joseph, MI, USA: ASABE.

Adamchuk, V. I., Viscarra Rossel, R. A., Marx, D. B. and Samal, A. K. 2011. Using targeted sampling to process multivariate soil sensing data. *Geoderma*, 63(1–2), 63–73.

Adamchuk, V. I., Allred, B., Doolittle, J., Grote, K. and Viscarra Rossel, R. A. 2015. Tools for proximal soil sensing. In: Ditzler, C. and West, L. (Eds), *Soil Survey Manual*, Supplement to Chapter 4, USDA Handbook 18. Washington DC: USDA Natural Resources Conservation Service.

AHDB. 2017. Nutrient Management Guideline (RB209). Warwickshire, UK: Agriculture and Horticulture Development Board.

Allred, B. J., Ehsani, M. R. and Saraswat, D. 2006. Comparison of electromagnetic induction, capacitively-coupled resistivity, and galvanic contact resistivity methods for soil electrical conductivity measurement. *Applied Engineering in Agriculture*, 22(2), 215–30.

Allred, B. J., Rogers, M., Ehsani, M. R. and Daniels, J. J. 2008a. Resistivity methods. In: Allred, B. J., Daniels, J. J. And Ehsani, M. R. (Eds), *Handbook of Agricultural Geophysics*. Boca Raton, FL, USA: CRC Press, pp. 85–128.

Allred, B. J., Groom, D., Ehsani, M. R. and Daniels, J. J. 2008b. Magnetometry, self-potential, and seisc: Additional geophysical methods having potentially significant future use in agriculture. In: Allred, B. J., Daniels, J. J. and Ehsani, M. R. (Eds), *Handbook of Agricultural Geophysics*. Boca Raton, FL, USA: CRC Press.

Allred, B. J., Daniels, J. J. and Ehsani, M. R. (Eds). 2008c. *Handbook of Agricultural Geophysics*. Boca Raton, FL, USA: CRC Press.

Anabitarte, F., Cobo, A. and Lopez-Higuera, J. M. 2012. Laser-induced breakdown spectroscopy: Fundamentals, applications, and challenges. *ISRN Spectroscopy*, 2012(ID 285240), 1–12.

Andrade-Sanchez, P., Upadhyaya, S. K. and Jenkins, B. M. 2007. Development, construction, and field evaluation of a soil compaction profile sensor. *Transactions of the ASABE*, 50, 719–25.

Antonucci, F., Pallottino, F., Costa, C., Rimatori, V., Giorgi, S., Papetti, P. and Menesatti, P. 2001. Development of a rapid soil water content detection technique using active infrared thermal methods for in-field applications. *Sensors*, 11(11), 10114–28.

Ayers, P. D. and Perumpral, J. V. 1982. Moisture and density effect on cone index. Paper No. 81-1047. St Joseph, MI, USA: ASAE.

Bănică, F.-G. 2012. *Chemical Sensors and Biosensors: Fundamentals and Applications*. Chichester, UK: John Wiley & Sons.

Barrios, M. D. R., Marques Junior, J., Matias, S. S. R, Panosso, A. R., Siqueira, D. S. and Scala Junior, N. 2017. Magnetic susceptibility as indicator of soil quality in sugarcane field. *Revista Caatinga*, 30(2), 287–95.

Beckhoff, B., Kanngießer, B., Langhoff, N., Wedell, R. and Wolff, H. 2006. *Handbook of Practical X-ray Fluorescence Analysis*. Berlin, Germany: Springer.

Bittelli, M., Salvtorelli, F. and Pisa, P. R. 2008. Correction of TDR-based soil water content measurements in conductive soil. *Geoderma*, 143, 133–42.

Boon, N. E., Yahya, A., Kheiralla, A. F., Wee, B. S. and Gew, S. K. 2005. A tractor-mounted, automated soil penetrometer–shearometer unit for mapping soil mechanical properties. *Biosystems Engineering*, 90(4), 381–96.

Brereton, R. (2018). *Chemometrics Second edition*. Oxford, UK: Wiley-Blackwell Publishing Ltd.

Bristow, K. L. 1998. Measurement of thermal properties and water content of unsaturated sandy soil using dual-probe heat-pulse probes. *Agricultural and Forest Meteorology*, 89, 75–84.

Buckup, P. and Buckup, M. 2010. A high sensitivity impulse neutron method for real-time detection of elements and compounds in soil and subsoil water. *Near Surface 2010–16th EAGE European Meeting of Environmental and Engineering Geophysics*, 06 September 2010.

Buddenbaum, H. and Steffens, M. 2011. Laboratory imaging spectroscopy of soil profiles. *Journal of Spectral Imaging*, 2(2a), 1–11.

Camargo, L. A., Marques, J., Pereira, G. T., Alleoni, L. R. F., de S. Bahia, A. S. R. and De B. Teixeira, D. 2016. Pedotransfer functions to assess adsorbed phosphate using iron oxide content and magnetic susceptibility in an Oxisol. *Soil Use and Management*, 32(2), 172–82.

Campbell, G. S., Calissendorf, C. and Williams, J. H. 1991. Probe for measuring soil specific heat using heat-pulse method. *Soil Science Society of America Journal*, 55, 291–3.

Charlesworth, P. 2005. Soil water monitoring. An information package. Irrigation Insights No. 1. 2nd ed. Canberra, Australia: Land & Water Australia.

Chrisman, B. and Zreda, M. 2013. Quantifying mesoscale soil moisture with the cosmic-ray rover. *Hydrology and Earth Sciences*, 17, 5097–108.

Christy, C. D. 2008. Real-time measurement of soil attributes using on-the-go near infrared reflectance spectroscopy. *Computers and Electronics for Agriculture*, 61, 10–19.

Clark, B., Fokoué, E. and Zhang, H. H. 2009. *Principles and Theory for Data Mining and Machine Learning*. Dordrecht, The Netherlands: Springer.

Corwin, D. L. 2008. Past, present, and future trends of soil electrical conductivity measurement using geophysical methods. In: Allred, B. J., Daniels, J. J. and Ehsani, M. R. (Eds), *Handbook of Agricultural Geophysics*. Boca Raton, FL, USA: CRC Press, pp. 17–44.

Corwin, D. L. and Lesch, S. M. 2003. Application of soil electrical conductivity to precision agriculture: Theory, principles, and guidelines. *Agronomy Journal*, 95, 455–71.

Corwin, D. L. and Lesch, S. M. 2005. Apparent soil electrical conductivity measurements in agriculture. *Computers and Electronics in Agriculture*, 46, 11–43.

Cosh, M. H., Ochsner, T. E., McKee, L., Dong, J., Basara, J. B., Evett, S. R., Hatch, C. E., Small, E. E., Steele-Dunne, S. C., Zreda, M. and Sayde, C. 2016. The soil moisture active passive marena,

Oklahoma, *In Situ* sensor testbed (SMAP-MOISST): Testbed design and evaluation of *In Situ* sensors. *Vadose Zone Journal*, 15. doi:10.2136/vzj2015.09.0122.

Daniels, D. J., Gunton, D. J. and Scott, H. F. 1988. Introduction to subsurface radar. *IEEE Proceedings F – Radar Signal Process*, 135, 278–320.

Dearing, J. A., Hay, K., Baban, S., Huddleston, A. S., Wellington, E. M. H. and Loveland, P. J. 1996. Magnetic susceptibility of topsoils: A test of conflicting theories using a national database. *Geophysics Journal International*, 127, 728–34.

De Gruiter, J., Brus, D., Bierkens, M. and Knotters, M. 2006. *Sampling for Natural Resource Monitoring*. Berlin, Germany: Springer.

De Gruijter, J. J., McBratney, A. B. and Taylor, J. 2010. Sampling for high resolution soil mapping. In: Viscarra Rossel, R. A., McBratney, A. and Minasny, B. (Eds), *Proximal Soil Sensing*. Dordrecht, The Netherlands: Springer.

Domsch, H, Ehlert, D., Giebel, A., Witzke, K. and Boess, J. 2006. Evaluation of the soil penetration resistance along a transect to determine the loosening depth. *Precision Agriculture*, 7(5), 309–26.

Doolittle, J. A. and Brevik, E. C. 2014. The use of electromagnetic induction techniques in soils studies. *Geoderma*, 223–225, 33–45.

Drummond, P. E., Christy, C. D. and Lund, E. D. 2000. Using an automated penetrometer and soil EC probe to characterize the rooting zone. In: *Proceedings of the Fifth International Conference on Precision Agriculture*, Bloomington, Minnesota, USA, 16–19 July 2000. Madison, WI, USA: ASA, CSSA, SSSA, pp. 1–9.

Durner, W., Germer, K., Jackisch, C., Andrä, I., Schulz, K., Schiedung, M., Haller-Jans, J., Schneider, J., Jaquemotte, J., Helmer, P., Lotz, L., Gräff, T., Bauer, A., Hahn, I., Sanda, M., Kumpan, M., Dorner, J., de Rooij, G., Wessel-Bothe, S., Kottmann, L. and Schittenhelm, S. 2017. Feldstudie zur Bodenfeuchtesensorik. In: *Höhere Bundeslehr- und Forschungsanstalt für Landwirtschaft Raumberg-Gumpenstein. 17.* Gumpensteiner Lysimetertagung 2017, 227–34.

Eifler, J., Martinelli, E. and Santonico, M. 2011. Differential detection of potentially hazardous fusarium species in wheat grains by an electronic nose. *PLoS ONE*, 6(6), e21026.

Evett, S. R. and Parkin, G. W. 2005. Advances in soil water content sensing: The continuing maturation of technology and theory. *Vadose Zone Journal*, 4, 986–91.

Gauglitz, G. and Moore, D. S. 2014. *Handbook of Spectroscopy*. 2nd ed. Weinheim, Germany: Wiley-VCH, pp. 31–8.

Gebbers, R., Lück, E., Dabas, M. and Domsch, H. 2009. Comparison of instruments for geoelectrical soil mapping at the field scale. *Near Surface Geophysics*, 7, 179–90.

Grant, B. 2011. *Field Guide to Radiometry*. Bellingham, WA, USA: SPIE Press.

Gründler, P. 2006. *Chemical Sensors. An Introduction for Scientists and Engineers*. Berlin, Germany: Springer.

Grunwald, S., Vasques, G. M. and Rivero, R. G. 2015. Fusion of soil and remote sensing data to model soil properties. In: Sparks, D. L. (Ed.), *Advances in Agronomy*, 131, pp. 1–109.

Hannam, J. A., Van Dam, R. L. and Harmon, R. S. 2009. Workshop on soil magnetism: Multi-disciplinary perspectives, emerging applications and new frontiers: Report. ERDC Report.

Haskin, L. A., Wang, A., Rockow, K. M., Jolliff, B. L., Korotev, R. L. and Viskupic, K. M. 1997. Raman spectroscopy for mineral identification and quantification for *in situ* planetary surface analysis: A point count method. *Journal of Geophysical Research: Planets*, 102, 19293–306.

Hastie, T., Tibshiriani, R. and Friedman, J. 2009. *The Elements of Statistical Learning. Data Mining, Inference and Prediction*. 2nd ed. Dordrecht, The Netherlands: Springer.

Heggemann, T., Welp, G., Amelung, W., Angst, G., Frans S. O., Koszinski, S., Schmidt, K. and Pätzold, S. 2017. Proximal gamma-ray spectrometry for site-independent *in situ* prediction of soil texture on ten heterogeneous fields in Germany using support vector machines. *Soil & Tillage Research*, 168, 99–109.

Heil, K. and Schmidhalter, U. 2017. The application of EM38: Determination of soil parameters, selection of soil sampling points and use in agriculture and archaeology. *Sensors*, 17, 1–44.

Hemmat, A. and Adamchuk, V. I. 2008. Sensor systems for measuring spatial variation in soil compaction. *Computers and Electronics in Agriculture*, 63, 89–103.

Hendrickx, J. M., Borchers, B., Corwin, D. L., Lesch, S., Hilgendorf, A. C. and Schlue, J. 2002. Inversion of soil conductivity profiles from electromagnetic induction measurements: Theory and experimental verification. *Soil Science of America Journal*, 66(3), 673–85.

Hof, M. and Machán, R. 2014. Basics of spectroscopy. In: Gauglitz, G. and Moore, D. S. (Eds), *Handbook of Optical Spectroscopy*. 2nd ed. Weinheim Germany: Wiley-VCH, pp. 31–8.

Horta, A., Malone, B., Stockmann, U., Minasny, B., Bishop, T. F. A., McBratney, A. B., Pallasser, R. and Pozza, L. 2015. Potential of integrated field spectroscopy and spatial analysis for enhanced assessment of soil contamination: A prospective review. *Geoderma*, 241-242, 180–209.

Huisman, J. A., Hubbard, S. S., Redman, J. D. and Annan, A. P. 2003. Measuring soil water content with ground penetrating radar: A review. *Vadose Zone Journal*, 2, 476–91.

Hummel, J. W., L. D. Gaultney and K. A. Sudduth. 1996. Soil property sensing for site-specific crop management. *Computers and Electronics in Agriculture*, 14, 121–36.

IAEA. 2003. Guidelines for radioelement mapping using gamma ray spectrometry data. Vienna, Austria: International Atomic Energy Agency.

IAEA. 2012. Neutron generators for analytical purposes. Vienna, Austria: International Atomic Energy Agency.

Inoue, Y., Watanabe, T. and Kitamura, K. 2001. Prototype time-domain reflectometry probes for measurement of moisture content near the soil surface for applications to 'on the move' measurements. *Agricultural Water Management*, 50, 41–52.

Inoue, M., Ahmed, O., Saito, T. and Irshad, M. 2008. Comparison of twelve dielectric moisture probes for soil water measurement under saline conditions. *American Journal of Environmental Sciences*, 4(4), 367–72.

Jacobs, W., Eelkema, M., Limburg, H. and Winterwerp, J. C. 2009. A new radiometric instrument for *in situ* measurements of physical sediment properties. *Marine and Freshwater Research*, 60, 727–36.

Jenny, H. 1980. *The Soil Resource*. Berlin, Germany: Springer.

Ji, W., Adamchuk, V., Biswas, A., Dhawale, N., Sudarsan, B., Zhang, Y., Viscarra Rossel, R. A. and Shi, Z. 2016. Assessment of soil properties *in situ* using a prototype portable MIR spectrometer in two agricultural fields. *Biosystems Engineering*, 152, 14–27.

Jordanova, D., Jordanova, N. and Werban, U. 2013. Environmental significance of magnetic properties of Gley soils near Rosslau (Germany). *Environmental Earth Sciences*, 69(5), 1719–32.

Jung, A. and Vohland, M. 2014. Snapshot hyperspectral imaging for soil diagnostics – results of a case study in the spectral laboratory. *PFG – Journal of Photogrammetry, Remote Sensing and Geoinformation Science*, 6, 511–22.

Käthner, J., Ben-Gal, A., Gebbers, R., Peeters, A., Herppich, W. and Zude-Sasse, M. 2017. Evaluating spatially resolved influence of soil and tree water status on quality of European plum grown in semihumid climate. *Frontiers in Plant Science*, 8(1053), 1–10.

Kaur, J., Adamchuck, V., Whalen, J. and Ismail, A. 2015. Development of an NDIR CO_2 sensor-based system for assessing soil toxicity using substrate-induced respiration. *Sensors*, 15(3), 4734–48.

Kichler, C. M., Fulton, J. P., Raper, R. L., Zech, W. C., McDonald, T. P. and Brodbeck, C. J. 2007. Spatially monitoring tractor performance to evaluate energy requirements of variable depth tillage and implement selection. Paper 071028. St. Joseph, MI, USA: ASABE.

Kim, H.-J., Sudduth, K. A. and Hummerl, J. W. 2009. Soil macro nutrient sensing for precision agriculture. *Journal of Environmental Monitoring*, 11, 1810–24.

Kirkham, M. B. 2005. *Principles of Soil and Plant Water Relations*. Burlington, MA, USA: Elsevier.

Kizewski, F., Liu, Y. T. and Morris, A. 2011. Spectroscopic approaches for phosphorus speciation in soils and other environmental systems. *Journal of Environmental Quality*, 40, 751–66.

Knight, R. 2001. Ground penetrating radar for environmental applications. *Annual Review of Earth and Planetary Sciences*, 29, 229–55.

Knödel, K., Krummel, H. and Lange, G. 2005. Handbuch zur Erkundung des Untergrundes von Deponien. Bd. 3. Geophysik. (*Handbook of Bedrock Survey of Landfill. Vol. 3. Geophysics*) 2nd ed. Berlin, Germany: Springer.

Kuang, B., Mahmood, H. S.; Quraishi, M. Z., Hoogmoed, W. M., Mouazen, A. M. and van Henten, E. J. 2012. Sensing soil properties in the laboratory, *in situ*, and on-line: A review. *Advances in Agronomy*, 14, 155–223.

Kundler, P. (Ed.) 1989. *Erhöhung der Bodenfruchtbarkeit* (*Improving Soil Fertility*). Berlin, DDR: VEB Landwirtschaftsverlag.

Kuras, O., Beamish, D., Meldrum, P. I. and Ogilvy, R. D. 2006. Fundamentals of the capacitive resistivity technique. *Geophysics*, 71(3), G135–52.

Lambot, S., Rhebergen, J., van den Bosch, I., Slob, E. C. and Vanclooster, M. 2004. Measuring the soil water content profile of a sandy soil with an off-ground monostatic ground penetrating radar. *Vadose Zone Journal*, 3, 1063–71.

Larkin, P. J. 2011. *IR and Raman Spectroscopy: Principles and Spectral Interpretation*. Amsterdam, The Netherlands: Elsevier.

Legendre, P. and Legendre, L. 2012. *Numerical Ecology*. 3rd ed. Amsterdam, The Netherlands: Elsevier.

Lesmes, D. P. and Friedman, S. P. 2005. Relationships between the electrical and the hydrogeological properties of rocks and soils. In: Rubin, Y. and Hubbard, S. (Eds). *Hydrogeophysics*. Dordrecht, The Netherlands: Springer, pp. 87–128.

Li, Y., Zhang, M., Zheng, J., Pan, L., Kong, P. and Lei, Z. 2017. Design and experiment of prototype soil pretreatment device for ISE-based soil nitrate-nitrogen detection. *Transactions of the Chinese Society of Agricultural Engineering*, 33(1), 120–5.

Lobsey, C. R., Viscarra Rossel, R. A. and McBratney, A. B. 2010. Proximal soil nutrient sensing using electrochemical sensors. In: Viscarra Rossel, R. A., McBratney, A. B. and Minasny, B. (Eds), *Proximal Soil Sensing*. Berlin, Germany: Springer, pp. 29–47.

Loke, M. H. and Barker, R. D. 1995. Least-squares deconvolution of apparent resistivity pseudosections. *Geophysics*, 60(6), 1682–90.

Loke, M. H., Chambers, J., Rucker, D. F., Kuras, O. and Wilkinson, P. B. 2013. Recent developments in the direct-current geoelectrical imaging method. *Journal of Applied Geophysics*, 95, 135–56.

Long, D. S., Wraith, J. M. and Kegel, G. 2002. A heavy-duty TDR soil moisture probe for use in intensive field sampling. *Soil Science Society of America Journal*, 66, 396–401.

Lueck, E. and Ruehlmann, J. 2012. Resistivity mapping with GEOPHILUS ELECTRICUS – Information about lateral and vertical soil heterogeneity. *Geoderma*, 199, 2–11.

Lütticken, R. E. 2000. Automation and standardisation of site specific soil sampling. *Precision Agriculture*, 2(2), 179–88.

Mahmood, H. S., Hoogmoed, W. B. and van Henten, E. J. 2013. Proximal gamma-ray spectroscopy to predict soil properties using windows and full-spectrum analysis methods. *Sensors*, 12, 16263–80.

Maiwald, M., Müller, A. and Sumpf, B. 2016. A portable shifted excitation Raman difference spectroscopy system: Device and field demonstration. *Journal of Raman Spectroscopy*, 47, 1180–4.

Martin, M. Z., Wullschleger, S. D., Garten Jr., C. T., Anthony V. Palumbo, A. V. and Smith, J. G. 2005. Elemental analysis of environmental and biological samples using laser-induced breakdown spectroscopy and pulsed Raman spectroscopy. *Journal of Dispersion Science and Technology*, 25(5), 687–94.

McNeil, J. D. 1980. Electromagnetic terrain conductivity measurements at low induction numbers. Technical Note TN-6. Mississauga, Ontario, Canada: Geonics Ltd.

Megumi, K. and Mamuro, T. 1977. Concentration of uranium series nuclides in soil particles in relation to their size. *Journal of Geophysical Research*, 82, 353–6.

Melquiades, F. L. and Appoloni, C. R. 2004. Application of XRF and field portable XRF for environmental analysis. *Journal of Radioanalytical and Nuclear Chemistry*, 262, 533–41.

Mitchell, H. B. 2012. *Data Fusion: Concepts and Ideas*. 2nd ed. Heidelberg, Germany: Springer.

Mittelbach, H., Lehner, I. and Seneviratne, S. I. 2012. Comparison of four soil moisture sensor types under field conditions in Switzerland. *Journal of Hydrology*, 430–431, 39–49.

Mohamed, A. M. O. 2008. Impact of soil magnetic permeability on water content prediction using TDR. In: *The 12th International Conference of International Association for Computer Methods and Advances in Geomechanics (IACMAG)*, 1–6 October 2008, Goa, India, pp. 1–8.

Mouazen, A. M., Maleki, M. R. and Cockx, L. 2009. Optimum three-point link set up for optimal quality of soil spectra collected during on-line measurement. *Soil and Tillage Research*, 103, 144–52.

Mulder, V. L., DeBruin, S., Schaepman, M. E. and Mayr, T. R. 2011. The use of remote sensing in soil and terrain mapping – A review. *Geoderma*, 162, 1–19.

Muñoz-Carpena, R. 2005. Field devices for monitoring soil water content. BUL343, IFAS Extension. FL, USA: Agricultural and Biological Engineering Department, UF/IFAS ExtensionUniversity of Florida.

Noborio, K. 2001. Measurement of soil water content and electrical conductivity by time domain reflectometry: A review. *Computers and Electronics in Agriculture*, 31(3), 213–37.

Ochsner, T. E., Horton, R. and Ren, T. 2001. Simultaneous water content, air-filled porosity, and bulk density measurements with thermo-time domain reflectometry. *Soil Science Society of America Journal*, 65, 1618–22.

Pan, L., Adamchuk, V. I., Prasher, S., Gebbers, R., Taylor, R. S. and Dabas, M. 2014. Vertical soil profiling using a galvanic contact resistivity scanning approach. *Sensors*, 14(7), 13243–55.

Pellerin, L. and Wannemaker, P. E. 2005. Multi-dimensional electromagnetic modelling and inversion with application to near-surface earth investigations. *Computers and Electronics in Agriculture*, 46, 71–102.

Pouzar, M., Černohorský, T. and Průšová, M. 2009. LIBS analysis of crop plants. *Journal of Analytical Atomic Spectrometry*, 24, 953–8.

Reece, C. F. 1996. Evaluation of a line heat dissipation sensor for measuring soil matric potential. *Science Society of America Journal*, 60(4), 1022–8.

Reimann, C., Filzmoser, P., Garret, R. and Dutter, R. 2008. *Statistical Data Analysis Explained: Applied Environmental Statistics with R*. Chichester, UK: John Wiley & Sons.

Robinson, A. A., Jones, S. B., Wraith, J. M., Or, D. and Friedman, S. P. 2003. A review of advances in dielectric and electrical conductivity measurements in soils using time domain reflectometry. *Vadose Zone Journal*, 2(4), 444–75.

Robinson, D. A., Campbell, C. S., Hopmans, J. W., Hornbuckle, B. K., Jones, S. B., Knight, R., Ogden, F., Selker, J. and Wendroth, O. 2008. Soil moisture measurement for ecological and hydrological watershed-scale observatories: A review. *Vadose Zone Journal*, 7(1), 358–89.

Robles-Kelly, A. and Huynh, C. P. 2013. *Imaging Spectroscopy for Scene Analysis*. London, UK: Springer.

Rodriguez-Alvarez, N., Bosch-Lluis, X., Camps, A, Voall-Ilossera, M., Valencia, E., Marcha-Hernandez, J. F., and Ramos-Perez, I. 2009. Soil moisture retrieval using GNSS-R techniques: Experimental results over a bare soil field. *IEEE Transactions on Geoscience and Remote Sensing*, 47(11), 3616–24.

Rodriguez-Alvarez, N., Bosch-Lluis, X., Camps, A., Aguasca, A., Vall-Ilossera, M., Valencia, E., Ramos-Perez, I. and Park, H. 2011. Review of crop growth and soil moisture monitoring from a ground-based instrument implementing the interference pattern GNSS-R technique. *Radio Science*, 46(RS0C03), 1–11.

Roy, A. and Apparao, A. (1971). Depth of investigation in direct current methods. *Geophysics*, 36(5), 943–59.

Samouëlian, A., Cousin, I., Tabbagh, A., Bruand, A. and Richard, G. 2005. Electrical resistivity survey in soil science: A review. *Soil and Tillage Research*, 83, 173–93.

Santamarina, C. J., Rinaldi, V. A., Fratta, D., Klein, K. A., Wang, Y.-H., Cho, G. C. and Cascante, G. 2009. A survey of elastic and electromagnetic properties of near-surface soils. In: Buttler, D. K. (Ed.), *Near-Surface Geophysics*. Tulsa, OK, USA: Society of Exploration Geophysics Press, pp. 71–88.

Schirrmann, M., Gebbers, R., Kramer, E. and Seidel, J. 2011. Soil pH mapping with an on-the-go sensor. *Sensors*, 11(1), 573–98.
Schirrmann, M., Gebbers, R. and Kramer, E. 2013. Performance of automated near-infrared reflectance spectrometry for continuous in-situ mapping of soil fertility at field scale. *Vadose Zone Journal*, 12(3), 1–7.
Schulz, H. and Baranska, M. 2006. Identification and quantification of valuable plant substances by IR and Raman spectroscopy. *Vibrational Spectroscopy*, 43, 13–25.
Shamal, S. A. M., Alhwaimel, S. A. and Mouazen, A. M. 2015. Application of an on-line soil bulk density sensor for site specific cultivation. *Soil and Tillage Research*, 162, 78–86.
Shao, L., Han, J., Kohli, P. and Zhang, Z. (Eds). 2014. *Computer Vision and Machine Learning with RGB-D Sensors*. Cham, Switzerland: Springer.
Shepherd, K. D. and Walsh, M. G. 2007. Infrared spectroscopy – enabling an evidence based diagnostic surveillance approach to agricultural and environmental management in developing countries. *Journal of Near Infrared Spectroscopy*, 15, 1–19.
Shibusawa, S. 2006. Soil sensors for precision agriculture. In: Srinivasan, A. (Ed.), *Handbook of Precision Agriculture: Principles and Applications*. New York, NY, USA: Food Products Press, pp. 57–90.
Shibusawa, S., Sato, H., Sasao, A., Hiroko, S. and Otomo, A. 2000. A revised soil spectrophotometer. *IFAC Proceedings*, 33(29), 231–6.
Sinfield, J. V., Fagerman, D. and Colic, O. 2010. Evaluation of sensing technologies for on-the-go detection of macro-nutrients in cultivated soils. *Computer and Electronics in Agriculture*, 70(1), 1–18.
Siqueira, D. S., Marques Jr., J., Matias, S. S. R., Barrón, V., Torrent, J., Baffa, O. and Oliveira, L. C. 2010. Correlation of properties of Brazilian Haplustalfs with magnetic susceptibility measurements. *Soil Use and Management*, 26(4), 425–31.
Smolka, M., Puchberger-Enengl, D. and Bipoun, M. 2016. A mobile lab-on-a-chip device for on-site soil nutrient analysis. *Precision Agriculture*, 1, 1–17.
Sørensen, M. K., Jensen, O. and Bakharev, O. N. 2015. NPK NMR sensor: Online monitoring of nitrogen, phosphorus, and potassium in animal slurry. *Analytical Chemistry*, 87, 6446–50.
Souffaché, B., Cosenza, P., Flageul, S., Pencole, J. P., Seladji, S. and Tabbagh, A. 2010. Electrostatic multipole for electrical resistivity measurements at the decimetric scale. *Journal of Applied Geophysics*, 71(1), 6–12.
Stafford, J. V. 1988. Remote, non-contact and in-situ measurement of soil moisture content: A review. *Journal of Agricultural Engineering Research*, 41(3), 151–72.
Stafford, J. V. and Ambler, B. 1990. Computer vision as a sensing system for soil cultivator control. *Proceedings of the Institution of Mechanical Engineers*, C419, 123–9.
Stafford, J. V. and Hendrick, J. G. 1988. Dynamic sensing of soil hard pans. *Transactions of the ASAE*, 31(1), 9–13.
Stenberg, B. and Viscarra Rossel, R. A. 2010. Diffuse reflectance spectroscopy for high-resolution soil sensing. In: Viscarra Rossel, R. A., McBratney, A. B. and Minasny, B. (Eds), *Proximal Soil Sensing*. Berlin, Germany: Springer, pp. 29–47.
Stenberg, B., Viscarra Rossel, R. A., Mouazen, A. M. and Wetterlind, J. 2010a. Visible and near infra-red spectroscopy in soil science. *Advances in Agronomy*, 107, 163–215.
Stenberg, B., Viscarra Rossel, R. A., Mouazen, A. M. and Wetterlind, J. 2010b. Visible and near infrared spectroscopy in soil science. In: Sparks, D. L. (Ed.), *Advances in Agronomy*. Vol. 107. Burlington, USA: Academic Press, pp. 163–215.
Stokes, D. L., Wullschleger, S., Martin, M. and Vo-Dinh, T. 2003. Raman spectroscopy and instrumentation for monitoring soil carbon systems. Report No: R03-119247. TN, USA: Environmental Science Division, Oak Ridge National Laboratory.
Sudarsan, B., Ji, W., Adamchuk, V. and Biswas, A. 2018. Characterizing soil particle sizes using wavelet analysis of microscope images. *Computers and Electronics in Agriculture*, 148, 217–25.
Sudduth, K. A., Hummel, J. W. and Birrell, S. J. 1997. Chapter 10: Sensors for site-specific management. In: Pierce, F. T. and Sadler, E. J. (Eds), *The State of Site-Specific Management for Agriculture*. Madison, Wisconsin, USA: ASA-CSSA-SSSA, pp. 183–210.

Sudduth, K., Kitchen, N., Bollero, G., Bullock, D. and Wiebold, W. 2003. Comparison of electromagnetic induction and direct sensing of soil electrical conductivity. *Agronomy Journal*, 95, 472–82.

Tekin, Y., Kuang, B. and Mouazen, A. M. 2013. Potential of on-line visible and near infrared spectroscopy for measurement of pH for deriving variable rate lime recommendations. *Sensors*, 13, 10177–90.

Thomsen, A., Schelde, K., Drøscher, P. and Steffensen, F. 2005. Mobile TDR for geo-referenced measurement of soil water content and electrical conductivity. *Precision Agriculture*, 8(4–5), 213–23.

Tomić, Z., Makreski, P. and Gajić, B. 2010. Identification and spectra–structure determination of soil minerals: Raman study supported by IR spectroscopy and X-ray powder diffraction. *Journal of Raman Spectroscopy*, 41, 582–6.

Tsiropoulos, Z., Fountas, S., Gravalos, I., Augoustis, A., Arslan, S., Misiewicz, P. and Gemtos, T. 2015. Importance of measuring tillage implement forces for reduced fuel consumption and increased efficiency without affecting tillage depth. In Stafford, J. (Ed.), *Precision Agriculture '15*. Proceedings of 10th European Conference on Precision Agriculture. Wageningen, The Netherlands: Wageningen Academic Publishers, pp. 353–60.

Van Egmond, F. M., Loonstra, E. H. and Limburg, J. 2010. Gamma ray sensor for topsoil mapping: The mole. In: Viscarra Rossel, R. A., McBratney, A. B. and Minasny, B. (Eds), *Proximal Soil Sensing*. Berlin, Germany: Springer, pp. 323–32.

Van Erp, P. and Beens, R. 2017. Towards a successful introduction of business concepts for fast and reliable soil sensing: Scientific and business related challenges. Presentation. Pedometrics 2017.

Varmuza, K. and Filzmoser, P. 2009. *Introduction to Multivariate Statistical Analysis in Chemometrics*. Boca Raton, FL, USA: CRC Press Taylor and Francis Group.

VDLUFA. 1997. Phosphordüngung nach Bodenuntersuchung und Pflanzenbedarf. *(Phosphorus management based on soil analysis and crop requirement)* VDLUFA Standpunkt. Kerschberger, M., Hege, U. and Jungk, A. (Bearbeiter). VDLUFA-Verlag, Darmstadt, Germany.

Viscarra Rossel, R. A. and Behrens, T. 2010a. Using data mining to model and interpret soil diffuse reflectance spectra. *Geoderma*, 158, 46–54.

Viscarra-Rossel, R. and Behrens, T. 2010b. A comparison of data mining techniques to model and interpret soil diffuse reflectance spectra. *Geoderma*, 158(1–2), 46–54.

Viscarra Rossel, R. and Bouma, J. 2016. Soil sensing: A new paradigm for agriculture. *Agricultural Systems*, 148, 71–4.

Viscarra Rossel, R. A., Walvoort, D. J. J. and McBratney, A. B. 2005. Visible, near infrared, mid infrared or combined diffuse reflectance spectroscopy for simultaneous assessment of various soil properties. *Geoderma*, 131, 59–75.

Viscarra Rossel, R. A., Taylor, H. J. and McBratney, A. B. 2007. Multivariate calibration of hyperspectral γ-ray energy spectra for proximal soil sensing. *European Journal of Soil Science*, 58, 343–53.

Viscarra Rossel, R. A., McBratney, A. and Minasny, B. (Eds). 2010. *Proximal Soil Sensing*. Dordrecht, The Netherlands: Springer.

Viscarra Rossel, R. A., Adamchuk, V. I., Sudduth, K. A., McKenzie, N. J. and Lobsey, C. 2011. Proximal soil sensing: An effective approach for soil measurements in space and time. In: Sparks, D. L. (Ed.), *Advances in Agronomy*, Vol. 113. Burlington, USA: Academic Press, pp. 237–82.

Waiser, T. H., Morgan, C. L. S. and Brown, D. J. 2007. *In situ* characterization of soil clay content with visible near-infrared diffuse reflectance spectroscopy. *Soil Science Society of America Journal*, 71, 389–96.

Webster, J. G. and Eren, H. (Eds). 2014. *Measurement, Instrumentation, and Sensors Handbook*. 2nd ed. Boca Raton, FL, USA: CRC Press Taylor & Francis Group.

White, R. E. 2006. *Principles and Practice of Soil Science*. 4th ed. Malden, MA, USA: Blackwell Publishing.

Wilson, A. D. and Baietto, M. 2009. Applications and advances in electronic-nose technologies. *Sensors*, 2009(7), 5099–148.

Witten, I. H., Frank, E., Hall, M. A. and Pal, C. J. 2016. *Data Mining: Practical Tools and Machine Learning Techniques*. 4th ed. Amsterdam, The Netherlands: Morgan Kaufmann.

Wong, M. T. F. and Harper, R. J. 1999. Use of on-ground gamma-ray spectrometry to measure plant-available potassium and other topsoil attributes. *Australian Journal of Soil Research*, 37(2), 267–78.

Wraith, J. M., Robinson, D. A., Jones, S. B. and Long, D. S. 2005. Spatially characterizing apparent electrical conductivity and water content of surface soils with time domain reflectometry. *Computers and Electronics in Agriculture*, 46, 239–61.

Zheng, L., Lee, W. Z., Li, M., Katti, A., Yang, C., Li, H. and Sun, H. 2012. Analysis of soil phosphorus concentration based on Raman spectroscopy. Proc. of SPIE Vol. 8527, Multispectral, Hyperspectral, and Ultraspectral Remote Sensing Technology, Techniques and Applications IV 852718.

Zhu, Y., Weindorf, D. C. and Zhang, W. 2011. Characterizing soils using a portable X-ray fluorescence spectrometer: 1. Soil texture. *Geoderma*, 167–168, 167–77.

Zreda, M., Desilets, D., Ferré, T. P. A. and Scott, R. L. 2008. Measuring soil moisture content non-invasively at intermediate spatial scale using cosmic-ray neutrons. *Geophysical Research Letters*, 35(21).

Chapter 3

Airborne and satellite remote sensors for precision agriculture

Chenghai Yang, USDA-ARS, USA

1 Introduction

Remote sensing is the science and technology of acquiring information about the earth's surface without physically touching it. It uses sensors to measure and record the reflected and emitted electromagnetic radiation from the target area in the field of view of the sensor. The detecting and recording instruments are generally referred to as remote sensors. Remote sensors include all the instruments that detect and measure reflected and emitted electromagnetic radiation from a distance. These instruments fall into two broad categories: non-imaging (i.e. spectroradiometers) and imaging (i.e. cameras).

Remote sensors are typically carried on manned aircraft and earth-orbiting satellites, but some sensors can be handheld or mounted on ground-based vehicles. Recently, unmanned aircraft systems (UASs) have become a popular remote sensing platform that, due to their low cost, low flying altitude and ease of operation, fill the gap between manned and ground-based platforms for high-spatial resolution imagery. Manned aircraft-based

http://dx.doi.org/10.19103/AS.2017.0032.03
Published by Burleigh Dodds Science Limited, 2019.

sensors are commonly referred to as airborne sensors, while satellite- or spacecraft-based sensors are also known as satellite or spaceborne sensors. Both non-imaging and imaging sensors can be used in all four types of platforms, though non-imaging sensors, including multispectral radiometers and hyperspectral spectroradiometers, are primarily used for ground-based applications.

Imaging sensors are designed to provide views of a target area from vertical (nadir) perspectives. Depending on the source of energy, imaging sensors can be classified into passive and active sensors. Passive remote sensors, represented by electro-optical sensors and passive microwaves, detect radiation emanating naturally from the surface, including reflected sunlight or emitted thermal infrared and microwave energy. Active remote sensors, represented by imaging radar (radio detecting and ranging), provide their own energy source to a target and record the radiation reflected from the target. Advantages of active sensors include their ability to obtain measurements regardless of the time of day or season.

Electro-optical sensors are the main imaging sensors used today. These sensors use detectors to convert the reflected and/or emitted radiation from a ground scene to proportional electrical signals, which are then recorded on magnetic, optical and/or solid-state media and can be viewed as two-dimensional images on a computer or television monitor. Detectors have been designed with sensitivities for many of the spectral intervals of interest in remote sensing, including regions of the near ultraviolet (UV), visible to near-infrared (VNIR), shortwave infrared (SWIR), mid-wave infrared (MWIR) and thermal infrared or longwave infrared (LWIR). The total span of the wavelengths that can be detected by electro-optical detectors extends from 0.3 μm to 15 μm. The sensors operating in the visible to SWIR region can only be used under sunlit conditions. In contrast, thermal sensors exhibit a day or night capability.

Airborne and satellite remote sensors used for precision agriculture are predominantly electro-optical sensors. In the following sections, advances in both airborne and satellite remote sensors will be briefly reviewed and some commonly used remote sensors or imaging systems will be described. Examples will be provided to illustrate how airborne multispectral and hyperspectral imagery and high-resolution satellite imagery have been used in precision agriculture applications. Some of the advantages and limitations of using different types of remote sensing imagery for precision agriculture will also be discussed.

2 Airborne imaging systems: overview and multispectral systems based on industrial cameras

2.1 Overview

Imaging sensors can be categorized as multispectral and hyperspectral based on the number of spectral bands and band widths. Multispectral imaging sensors typically measure reflected or emitted energy in 3–12 different spectral bands. In contrast, hyperspectral imaging sensors measure radiation in tens to hundreds of narrow spectral bands across the electromagnetic spectrum. Imagery produced from hyperspectral sensors contains far more spectral details than imagery from multispectral sensors and have a great potential to detect subtle differences among similar ground objects and features.

Multispectral imaging systems typically employ multiple industrial cameras and filters to achieve multispectral bands. For example, a three-band colour-infrared (CIR) imaging system will consist of three identical cameras equipped with three different bandpass filters to obtain green, red and NIR bands. More recently, with quality improvement and price reduction, consumer-grade digital cameras have been increasingly used as a multispectral remote sensing tool. The main advantage of consumer-grade cameras is that a single camera can be directly used to capture normal colour images or the three visible bands (red, green, blue or RGB). Moreover, it can be modified to capture one NIR band or a combination of two visible bands and one NIR band. In this section, airborne multispectral imaging systems based on industrial cameras and consumer-grade cameras as well as airborne hyperspectral imaging systems will be discussed.

2.2 Multispectral systems based on industrial cameras

With advances in imaging sensor technology, cameras used in airborne multispectral imaging systems have evolved from simple, small-frame, analogue video cameras in the 1980s to sophisticated, large-frame, high-resolution digital cameras in use today. These multispectral systems have been widely used as a versatile remote sensing tool for many applications. Among the advantages of airborne imaging systems are their relatively low cost, high spatial resolution and real-time/near-real-time availability of imagery for visual assessment, and their ability to obtain data in narrow spectral bands in the visible to SWIR region of the electromagnetic spectrum (Mausel et al., 1992; King, 1995).

Over the years, numerous commercial and custom-built multispectral imaging systems have been developed and used for diverse remote sensing applications, including rangeland and cropland assessment, precision agriculture and pest management (Moran et al., 1997, 2003; Pinter et al., 2003; Yang et al., 2016). Most airborne multispectral imaging systems can provide 8–16-bit image data with submeter resolution at 3–12 narrow spectral bands in the visible to NIR region of the spectrum (Everitt et al., 1995; Escobar et al., 1998; Gorsevski and Gessler, 2009; Yang, 2012; Yang et al., 2014).

Airborne multispectral imaging systems typically employ monochrome charge coupled device (CCD) industrial cameras, though complementary metal oxide semiconductor (CMOS) sensors have been increasingly displacing CCD sensors in recent years because they tend to be less expensive and consume less power. Each camera in a multispectral system is usually equipped with a different bandpass filter. This approach has the advantage that each camera can be individually adjusted for optimum focus and aperture settings, but has the disadvantage that the band images from the individual cameras have to be properly aligned with each other. Most multispectral imaging systems are based on this approach. Another approach is to use a beam-splitting prism and multiple sensors built in one single camera to achieve multispectral imagery. One such system is the CS-MS1920 multispectral 3-CCD camera (Teledyne Optech, Inc., Vaughan, Ontario, Canada), which uses a beam-splitting prism and three CCD sensors to acquire images in 3–5 spectral bands within the 400–1000-nm spectral range. The advantage of this approach is that the band images are optically and mechanically aligned.

The USDA-ARS Aerial Application Technology Research Unit in College Station, Texas, has devoted considerable efforts to the development and evaluation of airborne imaging systems for agricultural applications. Figure 1 presents a four-camera multispectral imaging system (lower left) along with three other imaging systems developed/assembled at the U.S. Department of Agriculture, Agricultural Research Service (USDA-ARS) in

Figure 1 A four-camera imaging system (lower left) mounted along with a hyperspectral camera (upper left), a thermal camera (upper right) and a multispectral imaging system based on two consumer-grade cameras (central to lower right) owned by the USDA-ARS at College Station, Texas. The four imaging systems can be installed over the two camera ports on the USDA-ARS Cessna 206 to capture images simultaneously.

College Station, Texas. The four-camera imaging system consists of four monochrome CCD cameras and a ruggedized PC equipped with a frame grabber and image acquisition software (Yang, 2012). The cameras are sensitive in the 400–1000-nm spectral range and provide 2048 × 2048 active pixels with 12-bit data depth. The four cameras are equipped with blue (430–470 nm), green (530–570 nm), red (630–670 nm) and NIR (810–850 nm) bandpass interference filters, respectively.

Airborne Data Systems, Inc. (Redwood Falls, Minnesota, USA) offers several multispectral camera systems for airborne remote sensing. Spectra-View is an airborne multispectral camera system designed to accommodate up to eight different cameras. The cameras can vary in size, format and wavelength from UV to LWIR. The Spectra-View system is designed with a fully integrated global positioning system/inertia navigation system (GPS/INS) unit providing precise, automated geo-registration of the images. For example, the Spectra-View 5WT system can capture 12-bit images in six spectral bands: blue (480–510 nm), green (520–610 nm), red (620–730 nm), NIR (770–970 nm), MWIR (3–5 μm) and LWIR (8–14 μm). The images for the four VNIR bands have 4872 × 3248 pixels, while the images for the MWIR and LWIR bands have 640 × 512 pixels. Spectra-View 9 W captures 12-bit images with 9000 × 4872 pixels in the blue, green, red and NIR bands and in a panchromatic band (400–1000 nm). Agri-View is a low-cost airborne multispectral camera system designed to provide only the same green, red and NIR bands as in Spectra-View 5WT. All the systems integrate proprietary pre-flight, mission planning and image processing software.

Tetracam's multispectral imaging systems come in two product families (Tetracam, Inc., Chatsworth, California, USA). Each system in the agricultural digital camera (ADC) family consists of a single camera equipped with fixed filters at green, red and NIR wavelengths. The ADC, ADC Lite and ADC Micro systems offer 8- or 10-bit images with 2048 × 1536 pixels, while the ADC Snap captures images with 1280 × 1024 pixels. Unlike the ADC Micro system that is equipped with a rolling shutter sensor, the ADC Snap exposes the entire image at the same instant. Systems in the multiple camera array (MCA) family contain four, six or twelve registered and synchronized cameras to simultaneously capture user-specified band images in visible to NIR wavelengths. The MCA systems are available in the standard Micro-MCA and Micro-MCA Snap versions. Each camera in the MCA systems can capture 8- or 10-bit images with 1280 × 1024 pixels.

Based on the Micro-MCA4 architecture, Tetracam's RGB+3 system consists of four cameras: one RGB colour camera that senses three broad visible bands and three monochrome cameras that sense three narrow bands centred at 680, 700 and 800 nm, respectively. Tetracam's Micro-MCA systems can also be configured with one or two FLIR TAU thermal sensors (FLIR Systems, Inc., Nashua, New Hampshire, USA) for imaging in the range of 7.5–13 μm. The latest variant of the MCA series, the Macaw (MCAW: Multiple Camera Array Wireless), upgrades the MCA electronics to a full-featured Linux computer system and a SATA solid state drive for computation-intensive missions. It contains six 1280 × 1024 global snap shutter sensors with individual filtering. All Tetracam's cameras are lightweight and compact and permit manual, auto-timed or remotely triggered exposures. All store images on memory cards (except the Macaw) are able to continuously capture GPS co-ordinates for geotagging images. Each system is shipped with Tetracam's PixelWrench2 software for managing and editing images.

In addition to its CS-MS1920 3-CCD multispectral imaging camera mentioned earlier, Teledyne Optech offers several medium-frame RGB colour cameras and thermal cameras either as a stand-alone or integrated with a lidar (light detection and ranging) system. For example, the RGB CS-10000 camera and the thermal CS-LW640 camera integrated with the Orion M300 lidar allow an efficient and effective data acquisition method for colour and thermal imagery and 3D surface data. With 10320 by 7760 pixels, the CS-10000 delivers imagery for detailed feature visualization. The CS-LW640 is based on an uncooled microbolometer sensor with a resolution of 640 × 480 pixels. Integrated with the Orion C lidar sensor platform, it is a powerful tool for mapping thermal features in 3D.

FLIR's thermal cameras come in many series with sensor arrays up to 1280 × 1024 pixels in typical spectral ranges of 1–5, 3–5 and 7.5–13 μm. For example, FLIR T600-series cameras with uncooled microbolometer detectors feature 14-bit images with 640 × 480 pixels in the 7.5- to 13-μm range. The SC8000-series cameras feature cooled indium antimonide (InSb) detectors and high spatial resolution (1024 × 1024 or 1344 × 784 pixels) in a range of 1.5–5 or 3–5 μm.

ITRES Research Ltd. (Calgary, Alberta, Canada) offers the TABI-1800 thermal camera that is a broadband thermal imager with excellent sensitivity due to its Stirling cycle–cooled MCT (mercury cadmium telluride) detector, allowing users to distinguish temperature differences as low as one-tenth of a degree while diminishing thermal drift as compared to bolometer-based systems. It captures 14-bit broadband images with a swath of 1800 pixels, the industry's widest array, in the 3.7–4.8-μm spectral range.

3 Airborne imaging systems: multispectral systems based on consumer-grade cameras

Technological advances in imaging sensors have made consumer-grade digital cameras an attractive option for remote sensing due to their low cost, compact size, compact data storage and ease of use. Consumer-grade cameras are fitted with either a CCD or CMOS sensor. These cameras employ a Bayer colour filter mosaic to obtain true-colour RGB images using a single sensor (Bayer, 1976; Hirakawa and Wolfe, 2008).

A Bayer filter mosaic is a colour filter array for arranging RGB colour filters on the pixel array of a CCD or CMOS. Since each pixel is filtered to record only one of the three primary colours, various demosaicing algorithms can be used to interpolate a set of complete red, green and blue values for each pixel. Although pixel interpolation lowers the effective spatial resolution of the band images, the three band images are perfectly aligned. Consequently, consumer-grade digital colour cameras have been increasingly used by researchers for agricultural applications (Sakamoto et al., 2012; Akkaynak et al., 2014; Yang et al., 2014; Song et al., 2016).

The use of images with visible and NIR bands is very common in remote sensing, especially for vegetation monitoring. Many vegetation indices such as the normalized difference vegetation index (NDVI) require spectral information in the NIR and red bands, even though the three visible bands could be sufficient for some applications. Since most consumer-grade cameras only provide the three broad visible bands, NIR filtering techniques can be used to convert an RGB colour camera to an NIR camera. Most digital cameras are fitted with filters to block UV and infrared light. Therefore, it is possible to replace the blocking filter by a long-pass infrared filter on standard CCD or CMOS sensors for obtaining NIR images.

Several companies provide infrared camera conversion services, including Life Pixel (Mukilteo, Washington, USA) and LDP LLC (Carlstadt, New Jersey, USA). The NIR blocking filter is typically replaced by a 720-nm or 830-nm long-pass filter. Since the Bayer filter mosaic is fused to the sensor substrate, the transmission profiles of the Bayer colour channels remain when using RGB cameras with the long-pass NIR filter. All three channels in the converted camera only record NIR radiation. Although any of the three channels can be used as the NIR channel, the red channel usually has the best sensitivity. Studies have been conducted on the use of NIR-converted digital cameras for monitoring plant conditions, and results from these studies support their use as simple and affordable tools for plant stress detection and crop monitoring (Nijland et al., 2014; Rabatel et al., 2014; Zhang et al., 2016; Wu et al., 2017).

For both research and practical applications, low-cost and user-friendly imaging systems are needed. These systems should be able to capture geotagged images at varying altitudes and normal airplane speed. In the last few years, several single- and dual-camera imaging systems based on consumer-grade cameras have been developed for use on any traditional aircraft or agricultural aircraft at USDA-ARS, College Station, Texas (Yang et al., 2014; Yang and Hoffmann, 2015). Agricultural aircraft provide a readily available and versatile platform for airborne remote sensing. Aerial applicators are highly trained pilots that use these aircraft to apply crop production and protection materials. If these aircraft are equipped with an imaging system, they can be used to acquire aerial imagery for monitoring crop growing conditions, detecting crop pests (i.e. weeds, diseases and insect damage) and assessing the performance and efficacy of aerial application treatments. This additional imaging capability will increase the usefulness of these aircraft and help aerial applicators generate additional revenues from remote sensing services. To avoid the contamination to the cameras from spray drift, aerial imaging and aerial spray should not be conducted concurrently.

The USDA-ARS single-camera imaging system consists of a Nikon D90 digital CMOS camera with a Nikon AF Nikkor 24-mm f/2.8D lens (Nikon Inc., Melville, New York, USA) to capture the colour image with up to 4288 × 2848 pixels, a Nikon GP-1A GPS receiver (Nikon Inc., Melville, New York, USA) to geotag the image, an AUVIO 7″ portable LCD video monitor (Ignition L.P., Dallas, Texas, USA) to view the live image and a VelloFreeWave wireless remote shutter release (Gradus Group LLC, New York, New York, USA) to trigger the camera. The fixed focal length (24 mm) is selected to be about the same as the longer dimension of the camera sensor area (23.6 × 15.8 mm) so that the longer dimension of the image is about the same as the flight height and the shorter dimension of the image is about 2/3 of the flight height. For example, when the image is acquired at 300 m above ground level (AGL), the image will cover approximately a ground area of 300 × 200 m. The total cost of the single-camera imaging system is about $1500.

To assemble the dual-camera system, a second identical Nikon D90 camera was modified to an NIR camera. To trigger the two cameras simultaneously, one hahnel Giga T Pro II wireless timer remote receiver (hahnel Industries Ltd, Bandon, Co. Cork, Ireland) was attached to each camera and one single hahnel remote transmitter was used to start and stop image acquisition. The remote control allows the user to take images automatically at set time intervals between consecutive shots. Since the NIR image has the same centre GPS coordinates as the RGB image, only one GPS receiver is needed. The two cameras can be attached via a mount to an aircraft with minimal or no modification to the aircraft. Both the single- and dual-camera systems have been evaluated for crop pest detection and crop type identification (Yang and Hoffmann, 2015; Zhang et al., 2016).

Two other systems with higher-end consumer-grade cameras were also developed at USDA-ARS, College Station, Texas. One system consisted of two Canon EOS 5D Mark III digital cameras with a 5760 × 3840 pixel array (Canon USA Inc., Lake Success, New York, USA); one camera captured normal RGB colour images, while the other camera was equipped with an 830-nm long-pass filter to obtain NIR images. The other system consisted of two Nikon D810 digital cameras with a 7360 × 4912 pixel array; one of these cameras captured RGB colour images, while the other camera was equipped with an

830-nm long-pass filter to obtain NIR images. These two imaging systems have the same sensor size (36 × 24 mm) and focal length (20 mm). The dimension of the images from each system is 1.8 by 1.2 times the flight height. At a flight height of 1000 m AGL, the image covers a ground area of 1800 × 1200 m with a pixel size of 31 cm for the Canon system and 24 cm for the Nikon system. The cost for each of the two-camera systems was about $6500. The two-camera systems have also been evaluated for precision agricultural applications (Yang et al., 2014; Wu et al., 2017).

As the demand for low-cost and lightweight imaging systems is increasing, several companies offer modified consumer-grade cameras to capture blue–green–NIR or green–red–NIR (i.e. CIR) imagery with a single sensor. For example, LDP LLC provides both modified cameras and modification services to convert existing consumer-grade cameras to either blue–green–NIR 680–800-nm or green-red-NIR 800–900-nm cameras. Furthermore, the two types of modified cameras can be tied together to capture images simultaneously. The pairs of images from the two cameras can then be automatically aligned and registered to create five-band images. Unlike the modified NIR-only cameras that are not affected by the visible light, the modified blue–green–NIR or CIR cameras could have the light contamination issue, depending on the filters and algorithms being used to achieve band separation.

Rabatel et al. (2014) modified a standard RGB camera by removing the NIR-blocking filter and adding an external long-pass filter in front of the lens that blocks the blue light. The red and NIR bands were computed as specific linear combinations of the three resulting channels. However, the smoothing effect involved in the process reduced image spatial resolution and the linear combinations to create the NIR and red bands amplified image noise. This approach is also used in Tetracam's single camera-based ADC series imaging systems that can capture CIR images with NIR, red and green bands. Although modified CIR cameras have their advantages and potential for remote sensing applications, more research is needed to improve the modification methods and more field experiments are needed to evaluate the performance of this type of cameras.

4 Airborne imaging systems: hyperspectral sensors

Hyperspectral imaging sensors or imaging spectrometers are a newer generation of electro-optical sensors that can collect image data in tens to hundreds of very narrow, continuous spectral bands throughout the visible to thermal infrared region of the spectrum. These systems offer new opportunities for better differentiation and estimation of biophysical attributes for a variety of remote sensing applications.

Many commercial airborne hyperspectral sensors have been developed and used for various remote sensing applications since the late 1980s. The Airborne Visible/Infrared Imaging Spectrometer (AVIRIS) was the first hyperspectral imaging sensor that delivered calibrated images of upwelling spectral radiance in 224 contiguous spectral bands with wavelengths from 400 to 2500 nm. AVIRIS was developed by the Jet Propulsion Laboratory (JPL) (Pasadena, California, USA) and has been collecting hyperspectral images for scientific research and applications since 1987.

The AVIRIS instrument contains different detectors arranged in four separate panels that cover the nominal spectral ranges of 400–700, 700–1300, 1300–1900 and 1900–2500 nm, respectively. AVIRIS provides 12-bit spectral data. The pixel size and swath width of the

AVIRIS data depend on the altitude from which the data are collected. When collected by the NASA's ER-2 jet at 20 km above sea level at 730 km/h, imagery has a pixel size of 20 m and a ground swath of 11 km. When collected by the Twin Otter International's turboprop at 4 km above the ground at 130 km/h, the pixel size is 4 m and the swath is 2 km wide. The AVIRIS system has been upgraded and improved in a continuous effort to meet the requirements of investigators using AVIRIS spectral images for scientific research and applications. Hundreds of papers have been written for the AVIRIS workshops (http://aviris.jpl.nasa.gov/). There have also been a comparable number of articles related to AVIRIS throughout the remote sensing literature. Green et al. (1998) provided an excellent overview of the AVIRIS sensor and examples of scientific research and applications using this unique sensor.

HyMap is an airborne hyperspectral imaging sensor developed and manufactured by Integrated Spectronics Pty Ltd (Sydney, Australia). After the first sensor became operational in 1996, the HyMap series of airborne hyperspectral scanners have been deployed in many countries in North America, Europe, Africa and Australasia. The initial HyMap sensor had 96 channels in the 550- to 2500-nm range and was designed primarily for mineral exploration (Cocks et al., 1998). The current HyMap sensor, operated by HyVista Corporation Pty Ltd (Sydney, Australia), provides 128 bands in the 450- to 2500-nm spectral range. The sensor operates in a 3-axis gyro stabilized platform to minimize image distortion due to aircraft motion. The system can be rapidly installed in any aircraft with a standard aerial camera port. Geolocation and image geocoding are achieved with differential GPS and an integrated inertial measurement unit (IMU). The sensor can be configured to capture images with 512 pixels with a spatial resolution of 3–10 m for different environmental, agricultural and urban applications (Dehaan and Taylor, 2002; Galvao, et al., 2004; Schiefer et al., 2006).

The Compact Airborne Spectrographic Imager (CASI) was first introduced in 1989 by ITRES Research Ltd. (Calgary, Alberta, Canada). CASI was one of the first programmable image spectrometers, allowing the user to program the sensor to collect hyperspectral data in specific bands and bandwidths that match needs for particular applications. Currently, ITRES offers a suite of airborne hyperspectral sensors covering the visible to thermal spectral range. The CASI-1500H maintains the performance edge of the original sensor, but features a smaller and lighter design with an embedded instrument controller and integrated GPS/IMU. It captures 14-bit hyperspectral images at up to 288 bands in the 380- to 1050-nm spectral range with 1500 across-track pixels. The SASI-1000A offers images with 1000 pixels at 200 bands in the range of 950–2450 nm to allow continuous VNIR-SWIR coverage together with the CASI-1500H. The MASI-600 is the first commercially available MWIR airborne hyperspectral sensor with 600 pixels and 64 bands in a spectral range of 3–5 μm. The TASI-600 is the airborne hyperspectral thermal sensor with 600 spatial pixels and 32 bands in a spectral range of 8–11.5 μm. ITRES also offers smaller versions of the sensors for both ground use and with UAS.

As hyperspectral imagery is gaining more interest, more commercial airborne hyperspectral imaging sensors have become available in recent years with improved spatial and spectral resolutions and high-performance GPS/INS units for increased position accuracy. For example, the Specim's AISA hyperspectral systems (Spectral Imaging Ltd., Oulu, Finland) cover VNIR (380–1000 nm), SWIR (1000–2500 nm) and thermal LWIR (7.7–12.3 μm) spectral ranges. The family of the airborne hyperspectral systems includes AisaKESTREL 10 in a range of 400–1000 nm, AisaKESTREL 16 in a range of 600–1640 nm, AisaFENIX and AisaFENIX 1K in a range of 380–2500 nm and AisaOWL in a range of

7.7–12.3 μm. The AisaFENIX sensor can capture images with a swath of 384 pixels at up to 348 VNIR bands and 274 SWIR bands, while the AisaFENIX 1K sensor can capture images with a swath of 1024 pixels at up to 348 VNIR bands and 256 SWIR bands. AisaOWL can obtain images with a 384-pixel swath at 96 bands. AisaKESTREL 10 and AisaKESTREL 16 can also fit to a UAS platform with a swath of 2040 and 640 pixels, respectively. All the sensors are equipped with a GPS/INS unit for monitoring the aircraft position and attitude.

Headwall Photonics, Inc. (Fitchburg, Massachusetts, USA) is another hyperspectral instrument manufacturer that currently offers a suite of hyperspectral sensors covering the spectral range of 250–2500 nm. Separate hyperspectral imaging sensors are available for UV to visible (250–500 nm), VNIR (380–1000 nm), NIR (900–1700 nm) and SWIR (950–2500 nm) spectral ranges. It also offers a co-registered VNIR-SWIR imaging sensor with full hyperspectral coverage from 400 to 2500 nm. Headwall's sensors can be configured to capture images with a swath of up to 1600 pixels at hundreds of bands. Similarly, a high-accuracy GPS/INS unit is integrated into each hyperspectral sensor. As an example, the hyperspectral imaging system at USDA-ARS in College Station, Texas, consists of a Headwall HyperSpec VNIR E-Series imaging spectrometer (upper left of Fig. 1), an integrated GPS/INS unit and a hyperspectral data processing unit. The spectrometer can capture 16-bit images with up to 923 spectral bands and a swath of 1600 pixels in the wavelength range of 380–1000 nm. At 1000 m AGL, the hyperspectral camera covers a swath of 615 m with a pixel size of 38 cm.

5 Airborne imaging systems: application examples

Airborne multispectral imagery has been widely used in precision agriculture for assessing soil variability, mapping crop growth and yield variability, and detecting crop insect infestations and disease infections. Moran et al. (1997) reviewed some of the applications and addressed the potential of image-based remote sensing to provide spatially and temporally distributed information for precision agriculture applications. More recently, Mulla (2013) reviewed the key advances of remote sensing in precision agriculture and the remaining knowledge gaps. Image-based remote sensing plays a unique role in monitoring seasonally variable crop and soil conditions and providing crop development stage information for time-specific crop management and near-real-time information for time-critical crop management.

Barnes et al. (2003) reviewed the use of remotely sensed imagery and ground-based sensors for mapping soil properties including soil organic matter, texture, salinity, moisture and nitrogen levels. Although airborne imagery is valuable in providing a means for directed sampling and useful field-specific relationships for mapping some soil properties, the use of remotely sensed data alone has limitations in developing robust, quantitative assessments of soil properties that do not require local calibration. Therefore, a combination of multispectral imagery, ground-based sensor data and other ancillary information integrated through appropriate models could yield accurate and detailed soil maps with limited direct sampling.

Varvel et al. (1999) examined the relationship between spectral data from an aerial image and soil organic matter and phosphorus levels. Yang and Everitt (2000) evaluated the relationships between yield monitor data and airborne multi-date multispectral digital imagery for grain sorghum. Pinter et al. (2003) examined the approaches that have been developed, refined and tested for management of water, nutrients and pests in agricultural crops and assessed the role of remote sensing in yield prediction. Inman et al. (2008) examined relationships between yield and NDVI for delineating management zones in irrigated maize. Sui et al. (2011) evaluated aerial multispectral imagery for estimating high-biomass sorghum yield. Backoulou et al. (2015) used multispectral imagery to differentiate wheat crop stress caused by greenbug from other causes. Yang et al. (2016) examined the consistency and variation of cotton root rot infections within cotton fields over a 10-year interval using airborne multispectral imagery and to assess the feasibility to use historical imagery to create prescription maps for site-specific management of the disease. Cohen et al. (2016) evaluated aerial FLIR thermal imagery for mapping water status in commercial fields.

Airborne hyperspectral imagery has been evaluated for assessing soil fertility, mapping crop yield variability and detecting crop pests. Bajwa and Tian (2005) characterized soil fertility in agricultural fields using airborne hyperspectral imagery. Goel et al. (2003) estimated crop biophysical parameters using airborne and field hyperspectral remote sensing. Zarco-Tejada et al. (2005) evaluated temporal and spatial relationships between within-field yield variability in cotton and high-resolution hyperspectral imagery. Yang et al. (2007) used airborne hyperspectral imagery and linear spectral unmixing techniques for mapping variation in crop yield. Miao et al. (2009) combined chlorophyll meter readings and AISA hyperspectral images for in-season site-specific nitrogen management of corn.

Fitzgerald et al. (2004) detected infestations of spider mites in cotton using hyperspectral imagery and spectral mixture analysis. Huang et al. (2007) identified yellow rust in wheat using *in situ* spectral reflectance measurements and airborne hyperspectral imaging. Mewes et al. (2011) analysed an airborne hyperspectral HyMap image for its potential to detect plant stress symptoms in wheat. Kumar et al. (2012) evaluated aerial hyperspectral and multispectral imaging techniques for citrus greening disease detection. Li et al. (2014) evaluated airborne hyperspectral imaging and extended spectral angle mapping (ESAM) for detecting the citrus greening disease. MacDonald et al. (2016) evaluated Headwall hyperspectral imagery for grapevine leafroll virus in cabernet sauvignon vineyards.

5.1 Example 1: multispectral imagery for creating management zones and yield map

To illustrate how airborne multispectral imagery can be used for delineating within-field management zones and for mapping yield variability, the work conducted by Yang and Anderson (1999) is used as an example. An airborne multispectral imaging system described by Everitt et al. (1995) was used for image acquisition. The system consisted of three CCD cameras equipped with a green (555–565 nm) filter, a red (625–635 nm) filter and an NIR (845–857 nm) filter, respectively. The NIR, red and green signals from the cameras were digitized and combined to produce CIR composite images with 640 × 480 pixels and 256 grey levels.

Figure 2a shows a CIR image for a 6-ha grain sorghum field in South Texas. The image was acquired at an altitude of 1300 m when plants were fully expanded, corresponding to the bloom to soft dough stages of plant growth. The pixel size of the image was 1.4 m. On the CIR image, healthy plants have a reddish-magenta colour, while chlorotic plants and areas with large soil exposure have a whitish and greyish tone. The CIR image was classified into four spectral zones using an unsupervised classification procedure (Fig. 2b).

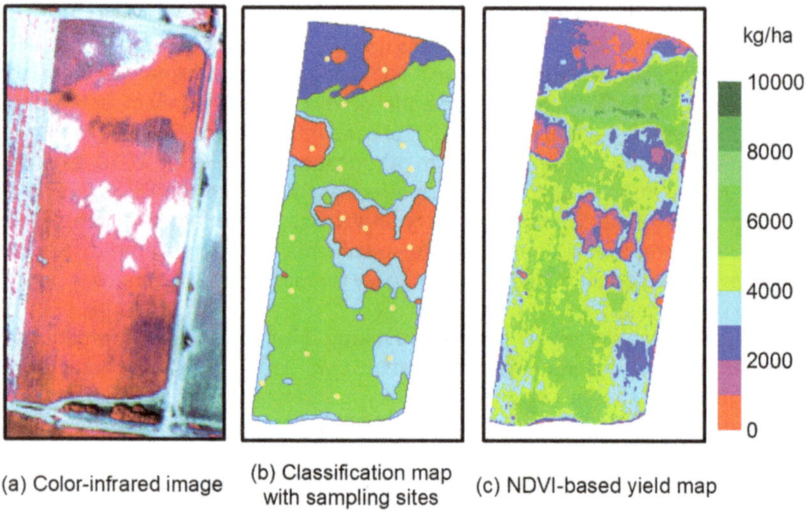

Figure 2 (a) Colour-infrared image, (b) four-zone classification map and (c) yield map generated from the CIR image based on a regression equation relating yield to NDVI for a 6-ha grain sorghum field in South Texas.

Two of the zones identified were soil related: the first zone (blue) represented areas with insufficient soil moisture and the second (red) depicted areas where plants suffered severe chlorosis due to iron deficiency. The other two zones represented areas with different production levels due to a combination of soil and environmental factors.

Unsupervised classification is an iterative process of grouping pixels into a finite number of classes or zones of similar spectral characteristics. The advantage of classifying image data into discrete zones is the reduced variance within each zone and therefore the reduced number of ground samples required to identify the cause of the variation within fields. These zones can be used as management zones for plant and soil sampling and for other farming operations. A stratified random sampling approach was used to generate the sampling sites for the field. This approach assured that every zone of the map was sampled, while simultaneously preserving an element of randomness. The sampling sites are also shown in the classification map (Fig. 2b).

Correlation analysis showed that yield was significantly related to NDVI with a correlation coefficient of 0.95. The important implication of this strong correlation is that grain yield can be estimated from NDVI for every pixel of the image. Figure 2c shows a grain yield map generated from the digital image based on the regression equation between yield and NDVI for the field. This map clearly shows the spatial variability in yield within the field. Compared with the CIR image and classification map, the yield map has a similar spatial pattern but reveals more variations within the zones.

This study demonstrates how airborne multispectral imagery can be used in conjunction with ground sampling, GPS, geographic information systems (GIS) and image processing techniques for mapping within-field yield variability. The unsupervised classification technique is effective to separate the imagery into spectral zones with different production levels. Multispectral images have also been proved to be instrumental in modelling the spatial variability of yield. Significant correlations exist between yield and image data. These results indicate that images acquired during the growing season can be used not only to identify management zones, but also to map the variation in yield within fields.

5.2 Example 2: thermal imagery for crop disease detection

As an example of thermal imagery for crop disease detection, Fig. 3 shows a thermal image captured from a cotton field infected with cotton root rot, which is caused by the soilborne fungus *Phymatotrichopsis omnivora*. Cotton root rot is a serious and destructive disease that has affected cotton production in the southwestern and south central United States for over a century. A FLIRSC660 camera owned by USDA-ARS in College Station, Texas, was used for image acquisition. The camera is sensitive in the 7.5–13-μm range to measure temperatures in the range of −40°C to 1500°C. It captures 14-bit thermal images with a 640 × 480 pixel array. At 1000 m AGL, the thermal camera covers a ground area of 426 × 318 m with a pixel size of 61 cm. Because infected plants tend to lose moisture content and have higher canopy temperatures, thermal imagery can be used to map the distribution and severity of the disease. In the thermal image, blue indicates healthy plants with lower temperature, while green, yellow and red within the irrigated circular area indicate infected plants with higher temperature and different levels of infection. Although other stressors can affect crop canopy temperature, the higher temperature in this case was mainly due to cotton root rot infection based on ground observations.

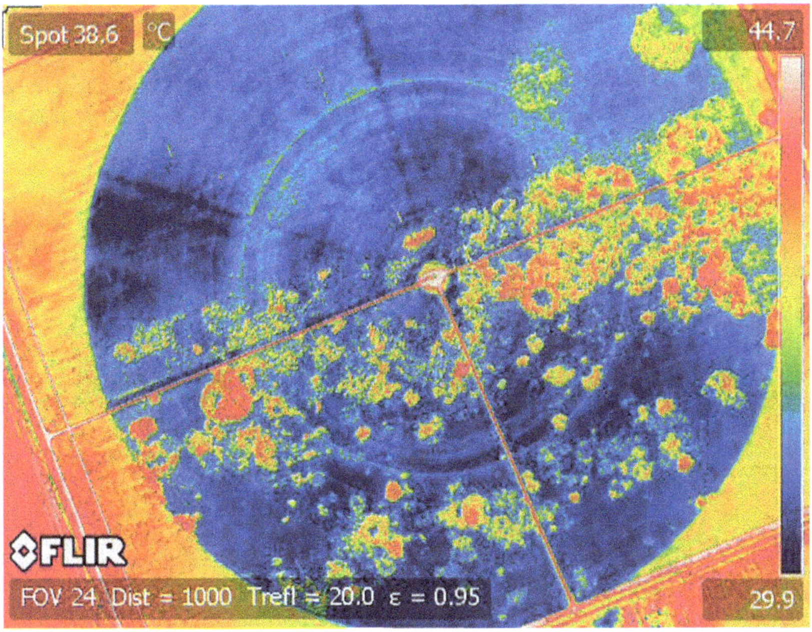

Figure 3 A thermal image of a cotton field infected with cotton root rot. Blue indicates healthy plants with lower temperature, while green, red and yellow within the irrigated circular area indicate infected plants with higher temperature and different levels of infection.

5.3 Example 3: image mosaicking for crop monitoring

If a single image cannot cover the area of interest with the required pixel size, then multiple images can be taken along one or more flight lines. For example, to map a 6.4 × 12.8-km sq cropping area near College Station, Texas, the USDA-ARS dual-camera system consisting of two Nikon D90 cameras was mounted on an Air Tractor 402B agricultural aircraft (Yang and Hoffmann, 2016). The Nikon cameras were flown at 1500 m AGL along 11 flight lines spaced 600 m apart. With a ground speed of 240 km/h and an imaging interval of 6 s, a total of 418 pairs of geotagged RGB and NIR images were acquired with a side overlap of 60% and a forward overlap of 56% (left of Fig. 4). The images were then mosaicked and processed using Pix4Dmapper software (Pix4D SA, Lausanne, Switzerland), and a 2D orthomosaic image (right of Fig. 4) and a 3D digital surface model were created. The 2D mosaicked image can be used for crop identification, within-field variability assessment, crop pest detection and other applications. The 3D surface model can be used to determine ground elevation and potential crop height.

5.4 Example 4: hyperspectral imagery for mapping crop yield variability

The work conducted by Yang et al. (2004) is used as an example to illustrate how airborne hyperspectral imagery can be used for mapping crop yield variability as compared with yield monitor data. An airborne hyperspectral imaging system described by Yang et al. (2003) was used for acquiring imagery from a 14-ha grain sorghum field in South Texas. The system consisted of a digital CCD camera, a prism-grating-prism imaging spectrograph and a PC equipped with a frame grabbing board and camera utility software. The system was configured to capture 12-bit images with a swath of 640 pixels and 128 bands from 457.2 to 921.7 nm at 3.63-nm intervals. The first five bands and the last 21 bands were

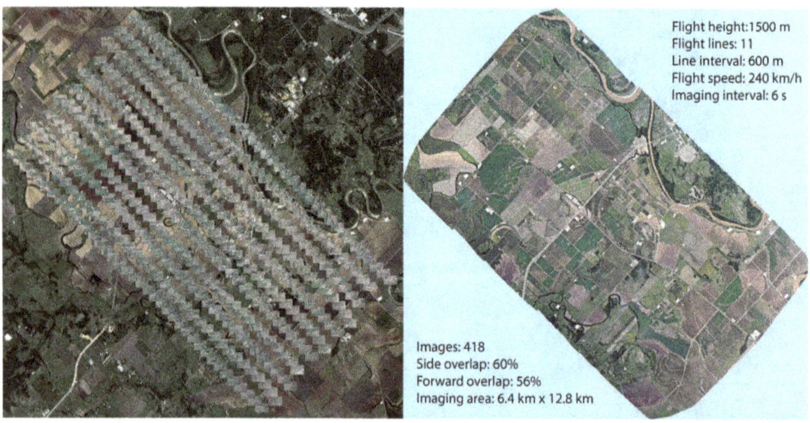

Figure 4 Thumbnails (not actual size) of 418 geotagged RGB images taken using a consumer-grade camera (left) and a mosaicked image for a 6.4 km by 12.8 km cropping area near College Station, Texas.

removed from the image and the remaining 102 bands with wavelengths from 475 to 845 nm were used for analysis. Yield monitor data were collected from the field using a Yield Monitor 2000 system (Ag Leader Technology, Ames, Iowa, USA).

Figure 5 shows a normal colour image and a CIR composite extracted from the 102-band hyperspectral image for the field. The image was taken around the peak plant development for the crop on 27 April 2000. Poor plant canopies in the problem areas were mainly due to the very sandy soil. Plants in those areas had poor stand and low canopy cover because of the low water and nutrient-holding capability of the sandy soil.

Grain yield was significantly negatively related to the visible bands and positively related to the NIR bands. Correlation coefficients among the 102 bands varied from −0.80 to 0.84. To eliminate the redundancy in the image data, the original 102-band image was transformed into a set of unrelated, independent principal components. Stepwise regression results for relating grain yield to the first ten principal components show that five of the ten principal components were found to be significant and explained 80% of the variability in yield. To identify the bands or combinations of bands that were particularly

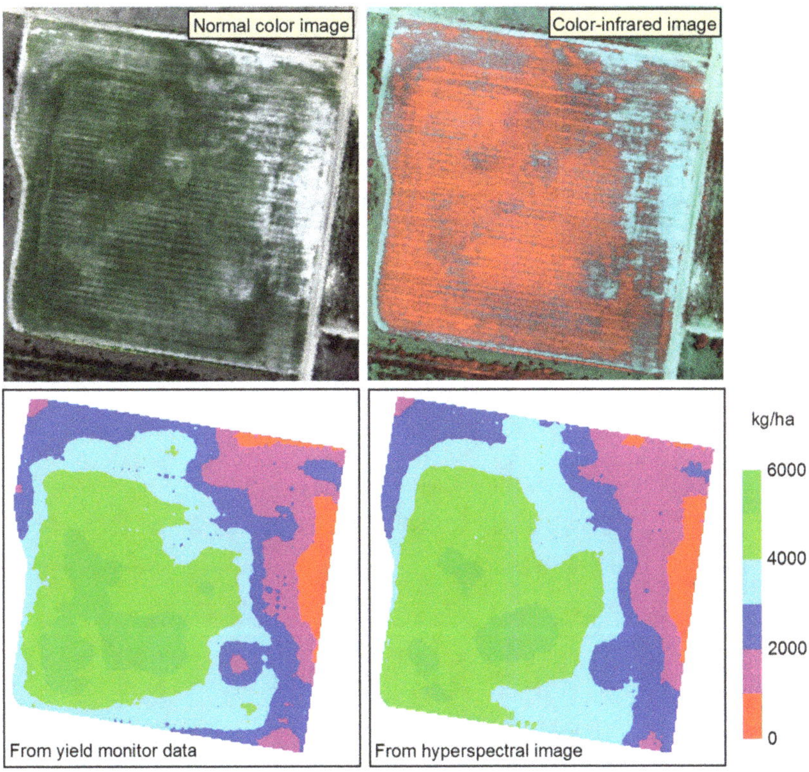

Figure 5 Normal colour image, colour-infrared image and yield maps generated from yield monitor data and based on seven significant bands in a 102-band hyperspectral image for a 14-ha grain sorghum field.

Published by Burleigh Dodds Science Limited, 2019.

Table 1 High-resolution satellite sensors with spatial resolutions of 10 m or finer

Sensor name	Year launched	Number of multispectral bands[2]	Multispectral pixel size (m)[9]	Panchromatic pixel size (m)[9]	Radiometric resolution (bit)	Revisit time (day)
IKONOS	1999[1]	4	3.2	0.82	11	3
QuickBird	2001[1]	4	2.62	0.65	11	1–3.5
SPOT 5	2002[1]	4[3]	10	2.5	8	2–3
KOMPSAT-2	2006	4	4	1	10	1
RapidEye	2008	5[4]	6.5	n/a	12	1–5.5
GeoEye-1	2008	4	1.65 (1.84)[10]	0.41 (0.46)[10]	12	3
WorldView-2	2009	8[5]	1.84	0.46	11	1.1–3.7
Pléiades-1A	2011	4	2	0.5	12	1
KOMPSAT-3	2012	4	2.8	0.7	14	1
SPOT 6	2012	4	6	1.5	12	1
Pléiades-1B	2012	4	2	0.5	12	1
SkySat-1	2013	4	2	0.9	10	1
SPOT 7	2014	4	6	1.5	12	1
SkySat-2	2014	4	2	0.9	10	1
WorldView-3	2014	28[6]	1.24	0.31	11	1–4.5
Gaofen-2	2014	4	3.2	0.8	14	5

Published by Burleigh Dodds Science Limited, 2019.

KOMPSAT-3A	2015	5[7]	2.2	0.55	14	1
Sentinel-2A	2015	13[8]	10	n/a	12	5
TripleSat	2015	4	3.2	0.8	10	1
WorldView-4	2016	4	1.24	0.31	11	1–4.5
Sentinel-2B	2017	13[8]	10	n/a	12	5

[1]Retired in 2015.
[2]All sensors except the ones with a superscript have four standard spectral bands (blue, green, red and near-infrared).
[3]Three standard spectral bands (green, red and near-infrared) at 10 m and one short-wave infrared band at 20 m.
[4]Four standard spectral bands and one red edge band.
[5]Eight spectral bands (coastal, blue, green, yellow, red, red edge, near-infrared 1 and near-infrared 2).
[6]The same eight spectral bands as WorldView-2 at 1.24 m, eight shortwave bands at 3.7 m and 12 CAVIS bands at 30 m for improved atmospheric corrections for clouds, aerosols, vapours, ice and snow.
[7]Four standard spectral bands at 2.2 m and one mid-wave infrared at 5.5 m.
[8]Four standard spectral bands at 10 m, six bands at 20 m and three bands at 60 m.
[9]At nadir.
[10]Orbit altitude was raised in 2013.

responsive to yield variability, stepwise regression was performed directly on the yield data and the 102-band hyperspectral image data. Seven bands (481, 543, 713, 731, 735, 771 and 818 nm) were identified to be significant in the final regression equation and explained about 82% of the variability in yield. Figure 5 shows the yield maps generated from yield monitor data and from the seven significant bands in the image. The spatial patterns displayed on both yield maps are similar, indicating that yield patterns can be estimated from airborne hyperspectral imagery taken during the growing season.

6 High-resolution satellite sensors: overview and varieties

6.1 Overview

Traditional satellite systems such as LandSat and SPOT have been widely used for agricultural applications over large geographic areas, but they have limited use for precision agriculture because of their relatively coarse spatial resolution. Recent developments in high-resolution satellite sensors have significantly narrowed the gap in spatial resolution between traditional satellite imagery and airborne imagery. Moreover, the high revisit frequency and fast data turnaround time combined with their relatively large aerial coverage make high-resolution satellite sensors attractive for many applications, including precision agriculture.

Table 1 summarizes most of the high-resolution satellite sensors that have been launched since 1999 with spatial resolutions of 10 m or finer for multispectral imagery. All the satellite sensors except RapidEye and Sentinel-2A and -2B listed in the table contain a panchromatic band, which has much higher spatial resolution than the corresponding multispectral bands.

6.2 IKONOS

IKONOS was the world's first high-resolution satellite sensor launched into space on 24 September 1999. It provided 82-cm panchromatic images in the 450–900-nm spectral range and 4-m multispectral imagery in the blue (450–520 nm), green (510- to 600 nm), red (630–700 nm) and NIR (760–850 nm) bands. The radiometric resolution was 11 bits and the image swath was 11.3 km at nadir. IKONOS was originally owned by Space Imaging and then acquired by GeoEye, Inc. (Herndon, Virginia, USA) to become part of its constellation of high-resolution satellites. GeoEye was merged into DigitalGlobe, Inc. (Longmont, Colorado, USA) on 29 January 2013. IKONOS was deactivated on 31 March 2015.

6.3 QuickBird

QuickBird, the world's second commercial high-resolution satellite, was launched by DigitalGlobe on 18 October 2001. QuickBird provided panchromatic and multispectral data in essentially the same spectral ranges as those of IKONOS, but at finer spatial resolutions: 0.6-m for panchromatic data and 2.4-m for multispectral data. Pixel depth was 11 bits, but the image swath at nadir was wider at 16.4 km. QuickBird was deactivated in early 2015.

6.4 SPOT 5, 6, 7 and Pléiades-1A and -1B

The satellites operated by Airbus Defence and Space in Toulouse, France, include SPOT 5, SPOT 6, SPOT 7, Pléiades-1A and Pléiades-1B. SPOT 5 was placed into orbit on 4 May 2002 and decommissioned on 31 March 2015. It offered panchromatic data at 2.5–5 m and multispectral data at 10 m for the green, red and NIR bands and 20 m for the SWIR band (1.58–1.75 μm). SPOT 5 covered 60 × 60 km or 60 × 120 km in twin-instrument mode and provided an ideal balance between high-resolution and wide-area coverage.

Pléiades-1A was launched on 16 December 2011. Joined by its twin, Pléiades-1B, in space on 2 December 2012, they provide 50-cm panchromatic and 2-m four-band multispectral (i.e. blue, green, red and NIR) products with daily revisit time and 20-km footprints. It should be noted that the native resolution of Pléiades-1A and -1B is 70-cm panchromatic and 2.8-m multispectral before it is processed for delivery.

SPOT 6 was launched on 9 September 2012, while SPOT 7 was launched on 30 June 2014. SPOT 6 and 7 are capable of imaging the earth with a resolution of 1.5 m in a panchromatic band and a resolution of 6 m in four standard multispectral bands. The imaging swath of the SPOT twins is 60 km at nadir. SPOT 6 and 7 are phased in the same orbit, 90° apart from one another. They form a constellation of earth-imaging satellites designed to provide continuity of high-resolution, wide-swath data up to 2024.

6.5 KOMPSAT-2, -3 and -3A

KOMPSAT-2, also referred to as Arirang-2, was developed and launched by the Korea Aerospace Research Institute in Daejeon, South Korea, on 28 July 2006. KOMPSAT-2 acquires 1-m panchromatic imagery and 4-m multispectral imagery in blue, green, red and NIR bands. KOMPSAT-3, launched on 17 May 2012, is capable of providing 70-cm panchromatic and 2.8-m multispectral resolution. KOMPSAT-3A was launched on 25 March 2015 to provide 55-cm resolution in the panchromatic band, 2.2-m resolution in the four spectral bands and 5.5-m resolution in an additional MWIR band (3.3–5.2 μm).

6.6 RapidEye

RapidEye, launched on 29 August 2008, is the constellation of five satellites currently owned and operated by Planet Labs, Inc. (San Francisco, California, USA). Each of RapidEye's five satellites contains identical sensors and travel on the same orbital plane. The five satellites together are capable of collecting multispectral imagery in five spectral bands: blue (440–510 nm), green (520–590 nm), red (630–690 nm), red edge (690–730 nm) and NIR (760–880 nm). The nominal resolution on the ground is 6.5 m, but the imagery is resampled to 5 m. The swath width is 77 km and revisit time is from daily off-nadir to 5.5 days at nadir.

6.7 GeoEye-1

GeoEye-1, launched by GeoEye on 6 September 2008, offers unprecedented spatial resolution by simultaneously acquiring 0.41-m panchromatic and 1.65-m four-band multispectral imagery with a revisit time of less than three days. The spectral ranges are similar to those of IKONOS. The pixel dynamic range is also 11 bits. The image swath is increased to 15.2 km. During late summer of 2013, the orbit altitude of the GeoEye-1 satellite sensor was raised from 680 to 770 km. As a result, the nadir spatial resolutions of GeoEye-1 were changed to 46 cm for panchromatic data and 1.84 m for multispectral data. GeoEye became part of DigitalGlobe in January 2013.

6.8 WorldView-2, -3 and -4

WorldView-2, launched into space on 8 October 2009, was the most technologically advanced high-resolution satellite ever launched before its twin, WorldView-3, reached orbit. WorldView-2 collects 46-cm panchromatic data in 450–800 nm and 1.85-m multispectral imagery in eight spectral bands: coastal (400–450 nm), blue (450–510 nm), green (510–580 nm), yellow (585–625 nm), red (630–690 nm), red edge (705–745 nm), NIR 1 (770–895 nm) and NIR 2 (860–1040 nm). The image swath at nadir is 16.4 km and the average revisit time is 1.1 days.

WorldView-3, launched on 13 August 2014, collects 31-cm panchromatic, 1.24-m eight-band VNIR, 3.7-m eight-band SWIR and 30-m 12-band CAVIS imagery with an average revisit time of less than one day. The high-resolution panchromatic and multispectral bands, which have the same spectral ranges as those of WorldView-2, are ideal for visual analysis and plant health assessments. The eight SWIR bands (1195–2365 nm) are ideal for mineral and hydrocarbon exploration and 12 CAVIS bands (405–2245 nm) are ideal for improved atmospheric corrections. WorldView-4, previously known as GeoEye-2, was launched on 11 November 2016. It provides 31-cm panchromatic and 1.24-m four-band multispectral imagery with a revisit time of less than one day. WorldView-4 has the same spectral bands as GeoEye-1 and similar spatial resolutions as WorldView-3.

6.9 SkySat-1 and -2

SkySat-1 and SkySat-2 are the first two micro-satellites built by Google's Terra Bella (formerly called Skybox Imaging) in Mountain View, California, USA, to collect high-resolution panchromatic and multispectral imagery of Earth. The satellites operate in a polar inclined, circular orbit at approximately 450 km above the earth. SkySat-1 was launched on 21 November 2013, while SkySat-2 was launched on 8 July 2014. The optical imager on each satellite covers a panchromatic band (450–900 nm) at 90 cm and four multispectral channels (blue 450–515, green 515–595, red 605–695 and NIR 740–900 nm) at a resolution of 2 m at nadir. A ground swath of 8 km is covered at nadir. Stereo imaging is supported by the satellites. The sensor can also capture high-definition video in its panchromatic channel.

Terra Bella ultimately plans a 24-satellite constellation occupying four different polar-orbit planes that will provide high-resolution imagery and full-motion video for commercial sale. SkySat-3, -4, -5, -6 and -7 were launched in 2016 and more will be launched in the coming years. The company's goal is to be able to provide high-resolution satellite imagery of any place on Earth multiple times a day. On 3 February 2017, Google announced that it would sell Terra Bella and its SkySat satellite constellation to Planet Labs.

6.10 Gaofen-2

The Gaofen-2 satellite, designed and developed by the China Academy of Space Technology in Beijing, China, was launched on 19 August 2014. It is capable of collecting imagery with a spatial resolution of 80 cm in a panchromatic band and 3.2 m in four standard multispectral bands on a swath of 45 km. It has a revisit time of five days. Gaofen-2 is one of the seven satellites in the China High-resolution Earth Observation System (CHEOS) constellation to be completed by 2020. So far, four Gaofen satellites have been launched with Gaofen-2 providing the highest resolution panchromatic and multispectral imagery.

6.11 Sentinel-2A and -2B

The Sentinel-2 mission comprises two polar-orbiting satellites in the same orbit, phased at 180° to each other. The twin satellites, Sentinel-2A and Sentinel-2B, were launched on 23 June 2015 and 7 March 2017, respectively. They are the first two optical Earth observation satellites in the European Copernicus programme that were developed and built under the industrial leadership of Airbus Defence and Space for the European Space Agency (ESA). Sentinel-2A and 2B each offer 13 spectral bands in the 443–2190-nm range with four standard bands at a spatial resolution of 10 m, four red edge bands and two SWIR bands at 20 m, and three atmospheric correction bands at 60 m. The orbital swath width is 290 km. Revisit time is five days at the equator and 2–3 days at mid-latitudes with the twin satellites.

6.12 TripleSat

The TripleSat constellation, built by Surrey Satellite Technology Ltd. (SSTL) in Guildford, UK, was launched on 10 July 2015. TripleSat consists of three identical satellites in the same orbital plane exactly 120° apart, making it possible to target any location on earth once per day. The TripleSat constellation delivers 80-cm visible panchromatic (450–650 nm) and 3.2-m four-band multispectral imagery with a 24-km swath.

Imagery from the satellite sensors listed in Table 1 can be ordered from multiple websites such as https://apollomapping.com and www.satimagingcorp.com/satellite-sensors. A tasking order can be made by defining a custom polygon on the planet to be imaged by the next available high-resolution satellite. Your area of interest (AOI) can be specified by WGS-84 latitude/longitude or UTM grid co-ordinates with zone number. You can upload your AOI as one of the file formats (txt, pdf, kmz/kml and ESRI shapefile). Some websites allow you to draw your AOI online. The time to collect the AOI is controlled by local weather conditions and competition from surrounding tasking orders. An archive order can also be made from a historic database of high-resolution satellite imagery with specific dates.

7 High-resolution satellite sensors: application examples

IKONOS and QuickBird were the first high-resolution satellite sensors that have been evaluated for diverse precision agricultural applications. Johnson et al. (2003) evaluated IKONOS multispectral imagery for mapping leaf area index for vineyard canopy management. Dobermann and Ping (2004) integrated IKONOS imagery with yield monitor data for improving the accuracy of yield maps. Sullivan et al. (2005) used IKONOS imagery to estimate surface soil property variability in two Alabama physiographies.

Yang et al. (2006) compared QuickBird imagery and airborne imagery for mapping grain sorghum yield patterns. Franke and Menz (2007) evaluated QuickBird imagery for multi-temporal wheat disease detection. Bausch et al. (2008) determined correlations between QuickBird satellite data and agronomic parameters of irrigated maize grown in small plots. Song et al. (2009) delineated agricultural management zones with high-resolution QuickBird imagery. Santoso et al. (2011) used QuickBird imagery to map and identify basal stem rot disease in oil palms. De Castro et al. (2013) evaluated QuickBird imagery for broad-scale cruciferous weed patch classification in winter wheat for in-season site-specific control.

SPOT-5 imagery with a coarser spatial resolution was available shortly after QuickBird was launched. Yang et al. (2009) evaluated SPOT 5 imagery for estimating crop yield. Soderstrom et al. (2010) predicted protein content in malting barley using proximal sensing and SPOT 5 imagery. The 10-m SPOT 5 sensor was decommissioned in 2015, but the 6-m SPOT 6 and 7 sensors continue to provide imagery. Yuan et al. (2016) used SPOT 6 high-resolution satellite imagery for mapping powdery mildew at a regional scale.

Other satellite sensors have also been used for various precision agriculture applications. Wagner and Hank (2013) evaluated the suitability of aerial and RapidEye satellite data for calculation of site-specific nitrogen fertilization compared to ground-based sensor data. Gomez-Candon et al. (2012) assessed the errors in input prescription maps based on high-spatial resolution GeoEye-1 images. Caturegli et al. (2015) compared GeoEye-1 satellite and ground-based multispectral data for estimating nitrogen status of turfgrasses. Li et al. (2015) studied the feasibility of Huanglongbing (citrus greening) detection based on WorldView-2 satellite imagery.

To illustrate how high-resolution satellite imagery can be used for site-specific crop management, cotton root rot is used as an example. Not until 2015 was the commercial Topguard Terra Fungicide registered to control the disease. Historical airborne imagery taken from infected cotton fields has demonstrated that cotton root rot tends to occur in the same general areas within fields over recurring years (Yang et al., 2016). The infection patterns within fields can generally be predicted based on historical remote sensing imagery. Based on aerial surveys, root rot-infected areas vary from 0% to 75% within fields,

Figure 6 (a) A colour-infrared composite image, (b) a corresponding two-zone classification map and (c) a prescription map with a 5-m buffer (right) for a 41-ha cotton field near Edroy, Texas.

but most fields contain about 20–40% infected areas. Therefore, site-specific treatment makes more economic and environmental sense, even though uniform treatment can be used for severely infected fields.

Figure 6 shows a GeoEye-1 CIR image acquired on 27 July 2009 for a cotton field infected with cotton root rot, a corresponding two-zone image classification map, and a two-zone prescription map. On the CIR image, infected areas had a greenish colour, while the non-infected areas had a reddish tone. Approximately 11% of the field was infected based on the classification map (Fig. 6b). To accommodate the potential expansion and temporal variation of the disease, buffer zones around the infected areas should be added as part of the treatment areas. With a 5-m buffer added to the infected areas, about 25% of the field (green areas) needs to be treated. With a 10-m buffer, treatment areas increase to 37%. The prescription map with 10-m buffer was loaded into a variable rate control system mounted on a tractor for site-specific application of Topguard Terra Fungicide at planting in 2016. Aerial imagery and ground observations showed that the disease was effectively controlled in the field.

8 Challenges and future trends

Although tremendous progress has been made in the development of airborne imaging systems and high-resolution satellite sensors, there are many challenges concerning the selection, acquisition and processing of various types of airborne and satellite imagery for precision agriculture applications. The first challenge is related to the selection of appropriate types of imagery for particular applications. Although growers are generally aware of the availability of airborne and satellite imagery, most of them do not keep up with the various types of imagery in this fast-changing market. They are not clear what type of imagery to select and how to order new or archived imagery for their particular applications. Therefore, image providers and vendors need to develop better instructions for growers and customers to select and order image products.

The second challenge is timely acquisition and delivery of task-ordered imagery. Although many satellite sensors have a revisit time of 1–5 days, the actual image acquisition time can vary greatly, depending on the local weather conditions and the competition with other customers in the similar geographic area. As more satellite sensors are becoming available, customers are able to order from more than one satellite sensor owned by a company. This will increase the chances of obtaining timely imagery. In this regard, airborne sensors have the advantage of taking imagery anytime that weather permits. However, unlike satellite imagery that is typically available anywhere on the earth, airborne imagery can only be acquired in areas where airborne imaging services are provided.

Other challenges are related to the processing and conversion of the imagery into useful maps or information. Although numerous publications are available in the literature, there are currently no standard procedures available for converting imagery to classification maps, vegetation index maps, and prescription maps for precision agriculture. Software used for image processing and analysis comes in different capabilities, complexities and prices. Growers may have difficulty selecting appropriate software for image processing. Delivered imagery is usually georeferenced, but its positional accuracy varies. For most applications, the delivered imagery can be directly used for processing and analysis. However, accurate ground control points are needed to improve its positional accuracy.

For applications that require time series image data, radiometric calibration is important, but there is no standard, accepted procedure for operational reflectance and temperature retrieval.

Researchers and extension specialists need to develop practical guides for growers and other end-users so that they will be able to use available techniques and software to convert imagery to proper maps for precision agriculture applications. For users who have some experience with GIS and image processing, they should be able to use appropriate software to perform image processing and analysis with the help of software documentations and tutorials. For those who are new to GIS and image analysis, but have good computer skills and experience, they may be able to learn to use the software to perform basic image analysis. After all, image processing is a specialized field and requires more advanced computer skills and a basic understanding of the techniques involved. If this is not practical for some growers, they can always use a commercial image processing service to process their images and create relevant maps. Some agricultural dealers are providing precision agriculture services for image acquisition, prescription map creation and variable rate application.

9 Summary

Remote sensing applications in precision agriculture have been steadily increasing in the last two decades due to improvements in spatial, spectral and temporal resolution of airborne and satellite remote sensors. This chapter provides an overview of manned aircraft-based airborne imaging systems and high-resolution satellite remote sensors that have been used for precision agriculture. Numerous airborne multispectral and hyperspectral imaging systems and high-resolution satellite sensors are discussed, and examples are provided to illustrate how airborne and satellite imagery has been used in precision agricultural applications.

Although all types of remote sensing platforms and imaging systems are available, various factors need to be considered as to which is more appropriate for a particular application. Some of the factors include the size of the area to be imaged, complexity of associated crop types and time and cost constraints. Airborne imaging systems offer some advantages, including the immediate availability for real-time assessment and the flexibility to change filters for desired wavelengths and bandwidths. High-resolution satellite systems can cover large areas with relatively fine spatial resolution. UASs are becoming a useful remote sensing platform to fill the gap between manned-aircraft and ground-based platforms, but there are still some restrictions on their use for commercial applications. Different platforms and remote sensors should be evaluated for their suitability and effectiveness for different precision agriculture applications.

10 Where to look for further information

A good introduction to the subject of remote sensing can be found at: http://www.nrcan.gc.ca/earth-sciences/geomatics/satellite-imagery-air-photos/satellite-imagery-products/educational-resources/9309. The online tutorial on fundamentals of remote sensing and its applications is offered by the Canada Centre for Mapping and Earth Observation (formerly Canada Centre for Remote Sensing). The tutorial can also be downloaded as a PDF file.

The following three websites provide detailed information about the high-resolution satellite sensors. They also provide information on aerial multispectral imagery. Both archived and newly tasked imagery can be ordered from these websites.

https://apollomapping.com
www.satimagingcorp.com/satellite-sensors and
http://landinfo.com/satellite-imagery-pricing.html

The two articles by Moran et al. (1997) and Mulla cited in this chapter provide useful information on the use of remote sensing for precision agriculture and will help readers explore the subject further.

11 Disclaimer

Mention of trade names or commercial products in this chapter is solely for the purpose of providing specific information and does not imply recommendation or endorsement by the U.S. Department of Agriculture. USDA is an equal opportunity provider and employer.

12 References

Akkaynak, D., Treibitz, T., Xiao, B., Gürkan, U. A., Allen, J. J., Demirci, U., et al. (2014), 'Use of commercial off-the-shelf digital cameras for scientific data acquisition and scene-specific color calibration', *Journal of the Optical Society of America*, 31, 312–21.

Backoulou, G. F., Elliott, N. C., Giles, K. L. and Mirik, M. (2015), 'Processed multispectral imagery differentiates wheat crop stress caused by greenbug from other causes', *Computers and Electronics in Agriculture*, 115, 34–39.

Bajwa, S. G. and Tian, L. F. (2005), 'Soil fertility characterization in agricultural fields using hyperspectral remote sensing', *Transactions of the ASAE*, 48(6), 2399–406.

Barnes, E. M., Sudduth, K. A., Hummel, J. W., Lesch, S. M., Corwin, D. L., Yang, C., et al. (2003), 'Remote- and ground-based sensor techniques to map soil properties', *Photogrammetric Engineering & Remote Sensing*, 69(6), 619–30.

Bausch, W. C., Halvorson, A. D. and Cipra, J. (2008), 'QuickBird satellite and ground-based multispectral data correlations with agronomic parameters of irrigated maize grown in small plots', *Biosystems Engineering*, 101, 306–15.

Bayer, B. E. (1976), *Color Imaging Array*, US Patent 3971065, Eastman Kodak Company, Rochester, New York, USA.

Caturegli, L., Casucci, M., Lulli, F., Grossi, N., Gaetani, M., Magni, S., et al. (2015), 'GeoEye-1 satellite versus ground-based multispectral data for estimating nitrogen status of turfgrasses', *International Journal of Remote Sensing*, 36(8), 2238–51.

Cocks, T., Jenssen, R., Stewart, A., Wilson, I. and Shields, T. (1998), 'The HYMAP airborne hyperspectral sensor: The system, calibration and performance', In: Schaepman, M., Schläpfer, D. and Itten, K. (Eds), *Proceedings of the 1st EARSeL Workshop on Imaging Spectroscopy, European Association of Remote Sensing Laboratories (EARSeL)*, European Association of Remote Sensing Laboratories, Strasbourg, France, pp. 37–42.

Cohen, Y., Alchanatis, V., Saranga, Y., Rosenberg, O., Sela, E. and Bosak, A. (2016), 'Mapping water status based on aerial thermal imagery: Comparison of methodologies for upscaling from a single leaf to commercial fields', *Precision Agriculture*, 18(5), 801–22.

de Castro, A. I., Lopez-Granados, F. and Jurado-Exposito, M. (2013), 'Broad-scale cruciferous weed patch classification in winter wheat using QuickBird imagery for in-season site-specific control', *Precision Agriculture*, 14, 392–413.

Dehaan, R. L. and Taylor, G. R. (2002), 'Field-derived spectra of salinized soils and vegetation as indicators of irrigation-induced soil salinization', *Remote Sensing of Environment*, 80, 406–17.

Dobermann, A. and Ping, J. L. (2004), 'Geostatistical integration of yield monitor data and remote sensing improves yield maps', *Agronomy Journal*, 96, 285–97.

Escobar, D. E., Everitt, J. H., Noriega, J. R., Cavazos, I. and Davis, M. R. (1998), 'A twelve-band airborne digital video imaging system (ADVIS)', *Remote Sensing of Environment*, 66, 122–8.

Everitt, J. H., Escobar, D. E., Cavazos, I., Noriega, J. R. and Davis, M. R. (1995), 'A three-camera multispectral digital video imaging system', *Remote Sensing of Environment*, 54, 333–7.

Fitzgerald, G. J., Maas, S. J. and Detar, W. R. (2004), 'Spidermite detection in cotton using hyperspectral imagery and spectral mixture analysis', *Precision Agriculture*, 5, 275–89.

Franke, J. and Menz, G. (2007), 'Multi-temporal wheat disease detection by multi-spectral remote sensing', *Precision Agriculture*, 8, 161–72.

Galvao, L. S., Ponzoni, F. J., Epiphanio, J. C. N., Rudorff, B. F. T. and Formaggio, A. R. (2004). 'Sun and view angle effects on NDVI determination of land cover types in the Brazilian Amazon region with hyperspectral data', *International Journal of Remote Sensing*, 25, 1861–79.

Goel, P. K., Prasher, S. O., Landry, J. A., Patel, R. M., Viau, A. A. and Miller, J. R. (2003), 'Estimation of crop biophysical parameters through airborne and field hyperspectral remote sensing', *Transactions of the ASAE*, 46(4), 1235–46.

Gomez-Candon, D., Lopez-Granados, F., Caballero-Novella, J. J., Pena-Barragan, J. M. and Garcia-Torres, L. (2012), 'Understanding the errors in input prescription maps based on high spatial resolution remote sensing images', *Precision Agriculture*, 13, 581–93.

Gorsevski, P. V. and Gessler, P.E. (2009), 'The design and the development of a hyperspectral and multispectral airborne mapping system', *ISPRS Journal of Photogrammetry and Remote Sensing*64, 184–92.

Green, R. O., Eastwood, M. L., Sarture, C. M., Chrien, T. G., Aronsson, M., Chippendale, B. J., et al. (1998), 'Imaging Spectroscopy and the Airborne Visible/Infrared Imaging Spectrometer (AVIRIS)', *Remote Sensing of Environment*, 65, 227–48.

Hirakawa, K. and Wolfe, P. J. (2008), 'Spatio-spectral sampling and color filter array design', In: *Single-Sensor Imaging: Methods and Applications for Digital Cameras* (Lukac, R. (Ed.)),CRC Press, Boca Raton, Florida, USA,pp. 137–51.

Huang, W., David, W. L., Niu, Z., Zhang, Y., Liu, L. and Wang, J. (2007), 'Identification of yellow rust in wheat using *in situ* spectral reflectance measurements and airborne hyperspectral imaging', *Precision Agriculture*, 8, 187–97.

Inman, D., Khosla, R., Reich, R. and Westfall, D. G. (2008), 'Normalized difference vegetation index and soil color-based management zones in irrigated maize', *Agronomy Journal*, 100, 60–6.

Johnson, L. F., Roczen, D. E., Youkhana, S. K., Nemani, R. R. and Bosch, D. F. (2003), 'Mapping vineyard leaf area with multispectral satellite imagery', *Computers and Electronics in Agriculture*, 38, 33–44.

King, D. J. (1995), 'Airborne multispectral digital camera and video sensors: A critical review of systems designs and applications', *Canadian Journal of Remote Sensing*, 21, 245–73.

Kumar, A., Lee, W. S., Ehsani, M. R., Albrigo, L. G., Yang, C. and Mangan, R. L. (2012), 'Citrus greening disease detection using aerial hyperspectral and multispectral imaging techniques', *Journal of Applied Remote Sensing*, 6, 063542.

Li, H., Lee, W. S., Wang, K., Ehsani, R. and Yang, C. (2014), 'Extended spectral angle mapping (ESAM) for citrus greening disease detection using airborne hyperspectral imaging', *Precision Agriculture*, 15, 162–83.

Li, X., Lee, W. S., Li, M., Ehsani, R., Mishra, A. R., Yang, C., et al. (2015), 'Feasibility study on Huanglongbing (citrus greening) detection based on WorldView-2 satellite imagery', *Biosystems Engineering*, 132, 28–38.

MacDonald, S. L., Staid, M., Staid, M. and Cooper, M. L. (2016), 'Remote hyperspectral imaging of grapevine leafroll-associated virus 3 in cabernet sauvignon vineyards', *Computers and Electronics in Agriculture*,130,109–17.

Mausel, P. W., Everitt, J. H., Escobar, D. E. and King, D. J. (1992), 'Airborne videography: Current status and future perspectives', *Photogrammetric Engineering & Remote Sensing*, 58, 1189–95.

Mewes, T., Franke, J. and Menz, G. (2011), 'Spectral requirements on airborne hyperspectral remote sensing data for wheat disease detection', *Precision Agriculture*, 12, 795–812.

Miao, Y., Mulla, D. J., Randall, G. W., Vetsch, J. A. and Vintila, R. (2009), 'Combining chlorophyll meter readings and high spatial resolution remote sensing images for in-season site-specific nitrogen management of corn', *Precision Agriculture*, 10, 45–62.

Moran, M. S., Fitzgerald, G., Rango, A., Walthall, C., Barnes, E. M., Bausch, W., et al. (2003), 'Sensor development and radiometric correction for agricultural applications', *Photogrammetric Engineering & Remote Sensing*, 69(6), 705–18.

Moran, M. S., Inoue, Y. and Barnes, E. M. (1997), 'Opportunities and limitations for image-based remote sensing in precision crop management', *Remote Sensing of Environment*, 61, 319–46.

Mulla, D. J., (2013), 'Twenty five years of remote sensing in precision agriculture – Key advances and remaining knowledge gaps', *Biosystems Engineering*, 114 (4), 358–71.

Nijland, W., de Jong, R., de Jong, S. M., Wulder, M. A., Bater, C. W. and Coops, N. C. (2014), 'Monitoring plant condition and phenology using infrared sensitive consumer grade digital cameras', *Agricultural and Forest Meteorology*, 184, 98–106.

PinterJr., P.J., Hatfield, J.L., Schepers, J.S., Barnes, E.M., Moran, M.S., Daughtry, C. S. T., et al. (2003), 'Remote sensing for crop management', *Photogrammetric Engineering & Remote Sensing*, 69(6), 647–64.

Rabatel, G., Gorretta, N. and Labbé, N. (2014), 'Getting simultaneous red and near-infrared band data from a single digital camera for plant monitoring applications: Theoretical and practical study', *Biosystems Engineering*, 117, 2–14.

Sakamoto, T., Gitelson, A. A., Nguy-Robertson, A. L., Arkebauer, T. J., Wardlow, B. D., Suyker, A. E., et al. (2012), 'An alternative method using digital cameras for continuous monitoring of crop status', *Agricultural and Forest Meteorology*, 154–155, 113–26.

Santoso, H., Gunawan, T., Jatmiko, R. H., Darmosarkoro, W. and Minasny, B. (2011), 'Mapping and identifying basal stem rot disease in oil palms in North Sumatra with QuickBird imagery', *Precision Agriculture*, 12, 233–48.

Schiefer, S., Hostert, P. and Damm, A. (2006), 'Correcting brightness gradients in hyperspectral data from urban areas', *Remote Sensing of Environment*, 101, 25–37.

Soderstrom, M., Borjesson, T., Pettersson, C.-G., Nissen, K. and Hagner, O. (2010), 'Prediction of protein content in malting barley using proximal and remote sensing', *Precision Agriculture*, 11, 587–99.

Song, H., Yang, C., Zhang, J., Hoffmann, C. W., He, D. and Thomasson, J. A. (2016), 'Comparison of mosaicking techniques for airborne images from consumer-grade cameras', *Journal of Applied Remote Sensing*, 10(1), 016030.

Song, X., Wang, J., Huang, H., Liu, L., Yan, G. and Pu, R. (2009), 'The delineation of agricultural management zones with high resolution remotely sensed data', *Precision Agriculture*, 10, 471–87.

Sui, R., Hartley, B. E., Gibson, J. M., Yang, C., Thomasson, J. A. and Searcy, S. W. (2011), 'Yield estimate of biomass sorghum with aerial imagery', *Journal of Applied Remote Sensing*,5, 053523.

Sullivan, D. G., Shaw, J. N. and Rickman, D. (2005), 'IKONOS imagery to estimate surface soil property variability in two Alabama physiographies', *Soil Science Society of America Journal*, 69, 1789–98.

Varvel, G. E., Schlemmer, M. R. and Schepers, J. S. (1999), 'Relationship between spectral data from an aerial image and soil organic matter and phosphorus levels', *Precision Agriculture*, 1, 291–300.
Wagner, P and Hank, K. (2013), 'Suitability of aerial and satellite data for calculation of site-specific nitrogen fertilisation compared to ground based sensor data', *Precision Agriculture*,14, 135–50.
Wu, M., Yang, C., Song, X., Hoffmann, W. C., Huang, W., Niu, Z., et al. (2017), 'Evaluation of orthomosaics and digital surface models derived from aerial imagery for crop type mapping', *Remote Sensing*, 9, 239.
Yang, C. (2012), 'A high resolution airborne four-camera imaging system for agricultural applications', *Computers and Electronics in Agriculture*, 88, 13–24.
Yang, C. and Anderson, G. L. (1999), 'Airborne videography to identify spatial plant growth variability for grain sorghum', *Precision Agriculture*, 1(1), 67–79.
Yang, C. and Everitt, J. H. (2000), 'Relationships between yield monitor data and airborne multidate multispectral digital imagery for grain sorghum', *Precision Agriculture*, 3(4), 373–88.
Yang, C., Everitt, J. H. and Bradford, J. M. (2004), 'Airborne hyperspectral imagery and yield monitor data for estimating grain sorghum yield variability', *Transactions of the ASAE*, 47(3), 915–24.
Yang, C., Everitt, J. H. and Bradford, J. M. (2006), 'Evaluating of high resolution QuickBird satellite imagery for estimating cotton yield', *Transactions of the ASAE*, 49(5), 1599–606.
Yang, C., Everitt, J. H. and Bradford, J. M. (2007), 'Airborne hyperspectral imagery and linear spectral unmixing for mapping variation in crop yield', *Precision Agriculture*, 8(6), 279–96.
Yang, C., Everitt, J. H. and Bradford, J. M. (2009), 'Evaluating high resolution SPOT 5 satellite imagery to estimate crop yield', *Precision Agriculture*, 10, 292–303.
Yang, C., Everitt, J. H., Davis, M. R. and Mao, C. (2003), 'A CCD camera-based hyperspectral imaging system for stationary and airborne applications', *Geocarto International*, 18(2), 71–80.
Yang, C. and Hoffmann, W. C. (2015), 'Low-cost single-camera imaging system for aerial applicators', *Journal of Applied Remote Sensing*, 9, 096064.
Yang, C. and Hoffmann, W. C. (2016), 'Converting aerial imagery to application maps', *Agricultural Aviation*, 43(4), 72–4.
Yang, C., Odvody, G. N., Thomasson, J. A., Isakeit, T. and Nichols, R. L. (2016), 'Change detection of cotton root rot infection over 10-year intervals using airborne multispectral imagery', *Computers and Electronics in Agriculture*, 123, 154–62.
Yang, C., Westbrook, J. K., Suh, C. P., Martin, D. E., Hoffmann, W. C., Lan, Y., et al. (2014), 'An airborne multispectral imaging system based on two consumer-grade cameras for agricultural remote sensing', *Remote Sensing*, 6, 5257–78.
Yuan, L., Pu, R., Zhang, J., Wang, J. and Yang, H. (2016), 'Using high spatial resolution satellite imagery for mapping powdery mildew at a regional scale', *Precision Agriculture*, 17, 332–48.
Zarco-Tejada, P. J., Ustin, S. L. and Whiting, M. L. (2005), 'Temporal and spatial relationships between within-field yield variability in cotton and high-spatial hyperspectral remote sensing imagery', *Agronomy Journal*, 97, 641–53.
Zhang, J., Yang, C., Song, H., Hoffmann, W. C., Zhang, D. and Zhang, G. (2016), 'Evaluation of an airborne remote sensing platform consisting of two consumer-grade cameras for crop identification', *Remote Sensing*, 8, 257.

Chapter 4

The use of unmanned aerial systems (UASs) in precision agriculture

Chunhua Zhang, Algoma University, Canada; and John M. Kovacs and Dan Walters, Nipissing University, Canada

1 Introduction

Agricultural production is particularly sensitive to both environmental conditions and management practices. Furthermore, farmers are often confronted with large variations in soil moisture, plant nutrition, crop growth and crop yields due to terrain, weather, tillage, drainage and other factors. Fast crop canopy changes make continuous monitoring of crop canopy necessary (Schirrmann et al., 2016). It is thus critical that farmers know, in a timely fashion, where these variations exist in their fields in order to adjust their practices accordingly.

Remotely collected images have been applied to identify soil and crop conditions (Zhang and Kovacs, 2012) and numerous types of satellite imagery (e.g. IKONOS, QuickBird, GeoEye-1 and WorldView- 2 and -3) have been used successfully in crop yield prediction (e.g. Shanahan et al., 2001; Chang et al., 2003; Seelan et al., 2003; Yang et al., 2006a,b, 2013; Inman et al., 2008). However, the shortcomings of satellite images which include availability and coarse spatial and temporal resolutions make their application in precision agriculture quite limited. With finer spatial resolution and real-time monitoring capability,

http://dx.doi.org/10.19103/AS.2017.0032.04

airborne multispectral (e.g. Yang et al., 2000, 2004, 2013) and hyperspectral (e.g. De Tar et al., 2008) sensors have been applied to monitor vineyard and crop condition and yield. Aerial imagery has been shown to be as effective as high-resolution satellite imagery in monitoring spatial variation of crop condition and crop yield.

The recent technical developments in small unmanned aerial systems (UASs) (<25 kg), also known as unmanned aerial vehicles (UAVs), unmanned aircraft systems, drones and remotely piloted vehicles (Nex and Remondino, 2014; Whitehead and Hugenholtz, 2014), make their application for precision agriculture more practical than ever. Decreasing costs, ultra-high spatial imagery resolution (in cm) and flexibility of imagery acquisition make the UAS an important alternative and perhaps the ideal tool for monitoring and mapping crop conditions and for field scouting. Although there remains concern (e.g. privacy, security) regarding the increased use of UAS for civilian purposes, the public has already accepted that UASs could greatly benefit society, particularly in their potential role for precision agriculture (Freeman and Freeland, 2016).

Agriculture is the largest commercial market for UAS with a recent telephone survey indicating that there would be an increasing market share (80–90%) for UAS applications in agriculture (Association for Unmanned Vehicle Systems International, 2013). Presently, UASs have been applied primarily to monitor crop physiological parameters, leaf area index (LAI), yield, diseases and crop water stress with more applications to be discussed later. UASs have been shown to be particularly effective in monitoring crop growth during the growing season (e.g. Bendig et al., 2013; Geipel et al., 2014; Tilly et al., 2014; Kyratzis et al., 2015; Vega et al., 2015). Additionally UASs have also been used for air broadcasting seeds (Li et al., 2016), removing rainwater (Zhou et al., 2016), and spraying pesticide, herbicide and fungicide (Faiçal et al., 2014; Pan et al., 2016; Xue et al., 2016). UASs have been designed to also be able to measure and map environmental variables (temperature, humidity, luminosity and CO_2 concentration) within a greenhouse environment (Roldán et al., 2015) and to monitor the evolution of landslides in an olive grove (Fernández et al., 2016).

2 Platforms and sensors

The main components of a UAS are the UAV(s) and the sensor(s). Essentially, the UAV acts as the platform for the sensor(s) as well as other equipment such as an altimeter and a global navigation and satellite system (GNSS). Generally, UASs can be categorized as either fixed wing or rotary wing (Sankaran et al., 2015). Fixed-wing UASs generally tend to have a longer range and a faster speed than rotary-wing UASs. However, it is often a challenge to find an ideal take-off and landing spot for fixed-wing UAS. In contrast, rotary-wing UASs normally have a shorter flight duration and a limited range but better manoeuvrability. Most UASs have a GNSS and an inertial measurement unit (IMU). While the GNSS provides the positional information for the platform, the IMU provides the attitude data of the platform (yaw, pitch and roll). Information of the position, altitude and attitude of the UAV is often integrated with the autopilot system, which adjusts the course of the UAS according to the flight planning as well as controlling the acquisition of images at preset locations. However, if necessary, the operator can, at any time, override the autopilot to take back control of the UAS via the ground control station.

Current UAS regulations can considerably restrict the operation of UASs. In Canada, for example, it is generally required that UASs do not fly higher than 90 m and that the UASs remain at all times in unaided (i.e. no binoculars) visual range. Low flight height results in an image of ultra-high spatial resolution but of small footprint. Consequently, it is typically a challenge to map large crop fields, particularly since the average Canadian crop farm was 315 ha in 2010 (Statistics Canada, 2016). Therefore several flights may be necessary to collect a large enough array of images with adequate overlap to create an image mosaic for any given farm area. This can also pose some challenges for photogrammetric processing including changing light conditions (i.e. clouds) during lengthy flight times.

There are a variety of sensors/cameras available for UAS. Optical (either metric or commercial scale) or infrared (metric or commercial with modified filter to record near-infrared radiation) sensors are the most commonly used ones for crop monitoring (Table 1). Thermal infrared sensors have been shown to be useful for monitoring soil moisture and crop water stress (Table 1). Most recently, hyperspectral sensors on board a UAV have also been applied to examine leaf physiological parameters (Table 1). Chlorophyll fluorescence could also potentially be a useful tool for monitoring plant photosynthesis and water use efficiency under water stress (Table 1). However, such elaborate sensors are very expensive in comparison to optical sensors and hence will likely remain limited to scientific investigations on crop phenotyping.

When comparing the performance of a commercial scale camera to a multispectral metric camera for measuring ground spectral responses, overall good performance has been recorded for both types of camera (Haghighattalab et al., 2016). Although there are some concerns about the data quality from commercial scale cameras (e.g. poor calibration limiting conversion from digital numbers to radiance and reflectance) (Whitehead and Hugenholtz, 2014), the low cost of data acquisition still makes it an appealing choice for many agricultural applications. Specifically, results from various studies indicate that simple RGB cameras are not only cost-effective but also powerful tools for monitoring plant condition and plant phenology (e.g. Geipel et al., 2014; Gómez-Candón et al., 2014; Jannoura et al., 2015; Schirrmann et al., 2016). Moreover, qualities of point cloud and digital terrain model (DTM) from low-cost commercial scale cameras were found to be comparable to those derived

Table 1 Sensor types and crop monitoring variables

Sensor	Variable monitored	Example references
Optical (RGB and infrared)	Crop conditions	Hunt et al. (2005), Bendig et al. (2014), Diaz-Varela et al. (2014), Torres-Sánchez et al. (2014), Verger et al. (2014), Pérez-Ortiz et al. (2015) and Rasmussen et al. (2016)
Thermal	Soil moisture and crop stress	Sujiura et al. (2007), Berni et al. (2009), Baluja et al. (2012), Gonzalez-Dugo et al. (2013), Bellvert et al. (2014), Hoffmann et al. (2016), Sepúlveda-Reyes et al. (2016) and Martínez et al. (2017)
Hyperspectral	Leaf physiological parameters	Zarco-Tejada et al. (2012, 2013), Uto et al. (2013), Lucieer et al. (2014), Duan et al. (2014), Aasen et al. (2015), Torres-Sánchez et al. (2015b) and Willkomm et al. (2016)
Chlorophyll fluorescence	Plant photosynthesis and water use efficiency	Zarco-Tejada et al. (2012) and Gago et al. (2015)

from DSLRs (Micheletti et al., 2015; Smith et al., 2016). However, consumer-grade cameras are sensitive to changes in illumination, and thus they should be used either in stable light conditions or when incoming light is measured concurrently in order to make adjustments in accordance with the variable illumination (Rasmussen et al., 2016).

3 Flight planning and imagery acquisition

The majority of UASs that adopted autopilot flight planning methods for acquiring images/photos are based on predetermined flight plans. Although it is possible to fly the UAS through manual control, it may cause issues for image post-processing. For example, an image gap for the field may be caused by limited overlap between images.

For autopilot, flight planning software is generally utilized to create a predetermined flight plan. A flight plan includes, for example, the flying area, the flight height, the forward lap, the side lap, the flight course, the flight speed and the camera settings. Normally high-resolution imagery or a large-scale map is shown as the background for the flight planning. High-resolution images for the study area may be downloaded/cached for use in the field prior to the field trip especially when Internet access is not feasible, which is likely the case for many applications. Flight planning software is either provided with the UAS or is acquired as a third-party application. In recent years, a few free downloadable apps have become available for field mapping for both Android and Apple cell phones (e.g. DroneDeploy and Pix4DCapture). In addition to the imagery, GNSS and IMU data are also typically recorded throughout the flight. These data can be used to assist in the image stitching stage for determining image centre position and camera orientation estimations (Whitehead and Hugenholtz, 2014; Smith et al., 2016). For most UASs, the acquired images and log data are stored in a peripheral storage device located on the UAV and, consequently, need to be downloaded to a computer for further processing.

The flight altitude of the UAS helps in determining the spatial resolution of the imagery; images collected with lower flight altitudes providing higher spatial resolution. It is not uncommon to set the UAS flight altitude somewhere between 100 and 120 m. A much lower flight altitude, such as 30 m, would result in much higher spatial resolution that may be required for particular tasks such as weed mapping. However, for many applications, aviation regulations determine the maximum flight altitude (e.g. 90 m in Canada and 120 m in Spain). Although the flight altitude generally does not impact the calculation of vegetation indices (VIs), it can greatly influence image segmentation with more mixed pixels present in those images collected at higher altitudes (Rasmussen et al., 2016). It is necessary to have a minimum of four pixels to identify one ground object.

Another crucial factor to consider in the flight planning stage is the image overlap setting. Normally it is suggested that the forward lap and side lap be set at a minimum of 80% and 60%, respectively (e.g. Colomina and Molina, 2014). This is partly because the low flight height and low accuracy of on-board navigational sensors can cause mismatching between the estimated image footprint and the actual ground coverage of the image (Haala et al., 2011; Whitehead and Hugenholtz, 2014). Additionally the greater overlap assists the photogrammetric software in identifying common points between image pairs. Finally, high overlap between images also helps to minimize bidirectional reflectance impacts by allowing the image processing software to extract only the central portion of each image for the image mosaic (Hunt et al., 2010).

Prior to the UAS image data acquisition, many researchers position artificial targets in the field. These targets can be used as ground control points (GCPs) or for spectral calibration. The GCP quality is critical for overall final product quality (position and elevation) (Smith et al., 2016). The position of these GCPs should be measured using a total station or a differential GNSS (d'Oleire-Oltmanns et al., 2012). This would guarantee the positional accuracy of subsequent image mosaics since the positional accuracies of the GPS on board the UAS could be as coarse as several metres. Normally it is suggested that a minimum of three GCPs be used in the photogrammetric adjustment. Moreover, it is recommended that these GCPs cover the entire study area and be well distributed (Smith et al., 2016). If used for spectral calibration, the surface of the targets could be painted in a different albedo whose reflectance would be measured using a ground spectroradiometer. An empirical line calibration could then be later applied to the image mosaic to convert the digital numbers to reflectance (e.g. Shi et al., 2016). The targets may also be applied to adjust digital numbers from image mosaics of different dates.

4 Image processing: stitching and ortho-rectification

Applying photogrammetry algorithms to rectify and mosaic images is a common task in the use of UAS. After downloading the images and log files, photogrammetric software is often used to generate ortho-rectified mosaics. Generally a semi-automated workflow needs to be followed in order to finish block bundle adjustment and ortho-rectification. The log file, previously mentioned, is used to provide initial estimates for the position and orientation of each image. There is a variety of software available for the photogrammetric processing including, for example, Trimble Inpho (e.g. Haala et al., 2011; Whitehead et al., 2013), LPS (Laliberte and Rango, 2011; d'Oleire-Oltmanns et al., 2012; Gómez-Candón et al., 2016), ENVI (Gómez-Candón et al., 2011), GeoLink (Selsam et al., 2017) and MicMac (Ballesteros et al., 2014a; Candiago et al., 2015).

In recent years, Structure from Motion (SfM) photogrammetry has also been applied in many UAS applications. SfM uses bundle adjustment algorithms to establish the structure of a scene, the internal orientation and the external orientation (Aasen et al., 2015; Smith et al., 2016). The main advantage of the SfM method is the ability to calculate the camera position, the orientation and the scene geometry purely from the overlapping images provided; this offers a simple processing workflow as compared to the alternative photogrammetry techniques. Specifically, SfM does not require camera calibration parameters as these parameters are estimated from the image tags and the large array of images available (Smith et al., 2016). This approach may be most convenient since many of these parameters are not readily available for many commercial scale cameras placed on UAVs. Furthermore, SfM photogrammetry can extract height information from two-dimensional images and there have been many examples of applications of SfM photogrammetry for geomorphology and physical geography in the production of high-resolution DTM (Smith et al., 2016). Such models are now being used to monitor crop growth during the growing season (e.g. Bendig et al., 2013, 2014). The most commonly used commercially available SfM software include Pix4dmapper (e.g. Zhang et al., 2014; Mesas-Carrascosa et al., 2015; Rasmussen et al., 2016; Shi et al., 2016), Agisoft PhotoScan Pro (e.g. Mathews and Jensen, 2013; Bendig et al., 2014; Geipel et al., 2014; Candiago et al., 2015; Fernandez et al., 2016; López-Granados et al., 2016) and Automatic Photogrammetric Processing Station (Caturegli et al., 2016).

There are also freely available SfM Web services including Autodesk 123D Catch (e.g. Micheletti et al., 2015) and Microsoft Photosynth. However, post-processing in other software for cleaning and editing is first needed (Smith et al., 2016). Unfortunately, both services were discontinued in late 2016 or early 2017. Open-source SfM packages are currently available. Some examples include Bundler and Patch-based Multi-View Stereo (PMVS2), VisualSFM and Ecosynth (Smith et al., 2016). Although there have been some criticisms regarding the computational time, the complicated applications and the reliability of these freeware, their performance has sometimes been found to be comparable with that of commercial ones (Smith et al., 2016). Moreover, Microsoft ICE has also been reported to be useful for panoramic image stitching (Wulfsohn and Lagos, 2014; Rasmussen et al., 2016).

5 UAS imagery applications

Within the past few years researchers have been able to extract a plethora of agricultural data from UAS imagery including: plant height, crop biophysical parameters, plant stress, and thematic information extraction for many crop species (Table 2) such as barley, cassava, chickpea, corn, cotton, onion, opium poppy, potato, rice, rye, sorghum, soybean, sugar beets, sugarcane, sunflower, tomato and wheat. In particular, there has been considerable work with UAS in vineyard monitoring most likely due to the high cash value of these crops.

5.1 Plant height

Plant height is a key indicator of crop biomass and yield potential prediction, plant cultivars separation, growth stages monitoring, treatments and stress monitoring (Bendig et al., 2013). It can also provide information about underlying ecological, hydrological and biophysical processes (Zarco-Tejada et al., 2014). Extracted crop height data have been used for the construction of crop surface models for biomass/yield monitoring (Geipel et al., 2014) and, as a time series, crop surface models can be used to accurately monitor crop growth (e.g. Bendig et al., 2013, 2014; Holman et al., 2016; Willkomm et al., 2016). In combination with VIs, plant height data have also been used to improve the accuracy of biomass estimation for barley crops (Bendig et al., 2014). Currently, plant height can be derived either from LiDAR data or from SfM-based photogrammetry. Terrestrial laser scanning has been applied to accurately measure plant height for crop surface modelling and growth monitoring (e.g. Tilly et al., 2014; Hoffmeister et al., 2016; Shi et al., 2016). Although there are numerous examples of UAS-based LiDAR applications to map forests and geomorphological properties (e.g. Hugenholtz et al., 2013; Wallace et al., 2014), no similar applications can be found for crop mapping.

While the direct DEM product of SfM algorithms is a digital surface model (with canopy information), it is possible to filter out vegetation canopy points from the point cloud and thus obtain the ground surface elevation (DTM). The separation of ground and non-ground points facilitates the estimation of biomass and other relevant measurements (Geipel et al., 2014; Aasen et al., 2015; Zahawi et al., 2015; Holman et al., 2016; Smith et al., 2016). However, efforts to filter out vegetation information may only be partially successful (Rokhmana, 2015; Fernández et al., 2016). Alternatively, a DTM could be extracted from UAS images taken before or after the growing season (e.g. Bendig et al.,

Table 2 Crop species being monitored using UAS images

Crop species	Example references
Barley	Bendig et al. (2014), Hoffmann et al. (2016) and Rasmussen et al. (2016)
Cassava	Selsam et al. (2017)
Chickpea	Sankaran et al. (2015)
Corn	Hunt et al. (2005), Link et al. (2013), Ballesteros et al. (2014b), Maresma et al. (2016) and Shi et al. (2016)
Cotton	Shi et al. (2016)
Onion	Córcoles et al. (2013) and Ballesteros et al. (2014b)
Opium poppy	Calderón et al. (2014)
Potato	Sankaran et al. (2015), Zhou et al. (2016), Hunt and Rondon (2017) and Hunt et al. (2017)
Rice	Inoue et al. (2000), Swain et al. (2010) and Uto et al. (2013)
Rye	Hunt et al. (2014)
Sorghum	Shi et al. (2016)
Soybean	Inoue et al. (2000) and Zhang et al. (2014)
Sugar beets	Martínez et al. (2017)
Sugarcane	Luna and Lobo (2016) and Duan et al. (2017)
Sunflower	Torres-Sánchez et al. (2013), Pérez-Ortiz et al. (2015), Vega et al. (2015), López-Granados et al. (2016) and Pérez-Ortiz et al. (2016)
Tomato	Candiago et al. (2015)
Wheat	Hunt et al. (2010), Honkavaara (2013), Geipel et al. (2014), Gómez-Candón et al. (2014), Lucieer et al. (2014), Torres-Sánchez et al. (2014) and Shi et al. (2016)
Vineyard	Baluja et al. (2012), Primicerio et al. (2012), Guillen-Climent et al. (2012), Garcia-Ruiz et al. (2013), Gonzalez-Dugo et al. (2013), Mathews and Jensen (2013), Zarco-Tejada et al. (2013), Bellvert et al. (2014), Zarco-Tejada et al. (2014), Burgos et al. (2015), Candiago et al. (2015), Comba et al. (2015), Matese et al. (2015) and Gómez-Candón et al. (2016)

2013, 2014). When compared with height information based on TLS, SfM-based height measurements have been found to provide high levels of accuracies (Holman et al., 2016). It has also been found that a lower flight (40 m) altitude creates more accurate crop height measurements (Holman et al., 2016).

5.2 Crop biological parameters

Variations in leaf pigment contents, leaf size, leaf angles and other canopy properties can also be detected using UAS-based imagery (Table 3). For example, UAS images have been applied to estimate crop emergence, crop lodging, planting gap, crop vigour, biomass, LAI, yield, pigment contents, protein and nitrogen and vegetation fraction. Moreover, UAS images have been applied to identify soil parameters and soil carbon emissions, as well as to identify the location of tile drains.

Yet another important UAS application is for crop phenotyping. Information extracted from UAS images has been used for evaluating plant growth, estimating biomass and determining physiological changes caused by biotic/abiotic stresses where high-yielding, stress-tolerant varieties could be selected (Li et al., 2014; Sankaran et al., 2015; Holman et al., 2016). Expensive LiDAR, hyperspectral and thermal sensors could be applied in this aspect of research (Tables 3 and 4).

5.3 Plant stress

Crop biological parameters should show abnormal measurements if the crop is stressed due to environmental/abiotic (e.g. drought and heat waves) or biotic (e.g. pests, fungus, weeds) factors. Not surprisingly, there are now numerous examples of UAS imagery for monitoring plant stress including crop water stress, crop diseases, crop pests and pest management, weed detection and weed management.

Irrigated cropland (16% of world cropland area) yields 36% of global harvest (Food and Agriculture Organization, 1996). Furthermore, nearly 20% of irrigated land in the world has higher than normal salt concentration (Metternicht and Zinck, 2003). Crop water, salinity and heat stress can cause stomatal closure, decreased stomatal conductance, increased leaf temperature and decreased photosynthesis (Zarco-Tejada et al., 2012; Sankaran et al., 2015). Periodical thermal and visible to near-infrared UAS imaging throughout the growing

Table 3 Crop biological parameters being monitored using UAS images

Crop biological parameters	Example references
Biomass	Hunt et al. (2005, 2010), Honkavaara et al. (2012, 2013), Bendig et al. (2014) and Jannoura et al. (2015)
Crop emergence	Sankaran et al. (2015)
Crop lodging	Liu et al. (2014) and Zhang et al. (2014)
Crop vigour	Primicerio et al. (2012), Zhang et al. (2014) and Sankaran et al. (2015)
Leaf area index	Hunt et al. (2010), Córcoles et al. (2013), Mathews and Jensen (2013), Jannoura et al. (2015) and Ballesteros et al. (2014b)
Yield	Hunt et al. (2010), Swain et al. (2010), Geipel et al. (2014), Wulfsohn and Lagos (2014), Kyratzis et al. (2015), Maresma et al. (2016) and Khot et al. (2016)
Pigment contents	Hunt et al. (2005), Zarco-Tejada et al. (2013) and Hunt et al. (2014)
Planting gap	Luna and Lobo (2016)
Protein and nitrogen	Jensen et al. (2007), Caturegli et al. (2016) and Hunt et al. (2017)
Vegetation fraction	Torres-Sánchez et al. (2014) and Ballesteros et al. (2014b)
Soil parameters	d'Oleire-Oltmanns et al. (2012), Hassan-Esfahani et al. (2014) and Quiquerez et al. (2014)
Soil carbon emissions	Wehrhan et al. (2016)
Drainage pipes identification	Zhang et al. (2014)

Table 4 Plant stress being monitored using UAS images

Plant stress	Example references
Crop water stress	Sullivan et al. (2007), Berni et al. (2009), Suarez et al. (2010), Zarco-Tejada et al. (2012) and Gago et al. (2015)
Crop diseases	Garcia-Ruiz et al. (2013), Sankaran et al. (2010) and Calderón et al. (2014)
Crop pests and pest management	Huang et al. (2009), Yue et al. (2012), Severtson et al. (2016) and Hunt et al. (2017)
Weed detection	Herwitz et a. (2004), Peña et al. (2013), Rasmussen et al. (2013), Torres-Sánchez et al. (2013), Pérez-Ortiz et al. (2015, 2016), Shi et al. (2016) and Castaldi et al. (2017)
Weed management	López-Granados et al. (2016) and Castaldi et al. (2017)

season can provide qualitative and quantitative data for crop stress monitoring (Sankaran et al., 2015).

Leaf temperature can be an indicator of stomatal conductance and response of plants to water stress. Temperature measurements from thermal imagery taken from UAS (Berni et al., 2009; Zarco-Tejada et al., 2012; Gago et al., 2015) have been used to calculate the Crop Water Stress Index or Water Deficit Index for leaf/canopy anomalies (Gonzalez-Dugo et al., 2013; Bellvert et al., 2014; Hoffmann et al., 2016; Sepúlveda-Reyes et al., 2016). Other crop biophysical parameters that may be monitored by thermal remote sensing include soil moisture, soil texture, evapotranspiration, crop residue cover, tillage, field drainage tiles and yield (Khanal et al., 2017). Although water stress is primarily monitored using thermal remote sensing techniques, optical remote sensing can also be applied to identify plant/canopy condition (e.g. chlorophyll content and biomass) (Sankaran et al., 2015).

Plant nutrient deficiencies can result in visual symptoms such as lower chlorophyll content, lower growth speed, tissue necrosis and higher disease susceptibility (Sankaran et al., 2015). Most importantly, nutrient deficiencies can make crops more susceptible to herbivorous pests (Severtson et al., 2016) and optical UAS images have been applied to identify such symptoms (Cilia et al., 2014; Li et al., 2015; Severtson et al., 2016).

Weeds compete with crops for solar radiation, water and nutrition and consequently impact the production potential. The mapping of weeds could provide data for site-specific weed management, which would ideally lower the cost of production and minimize environmental impacts. UAS images have been shown to provide ultra-high resolution images, which are necessary to identify weed species, weed density and weed patches (e.g. Peña et al., 2013; Borra-Serrano et al., 2015; Pérez-Ortiz et al., 2015; López-Granados et al., 2016). Such applications are available for early-season weed detection and late-season weed detection (e.g. Torres-Sánchez et al., 2013; López-Granados et al., 2016). In particular, hyperspectral sensors are believed to be more advantageous than multispectral sensors in detecting weeds due to the spectral similarity of weeds and crops (López-Granados, 2011). The analytical results could be further integrated into a herbicide spraying plan for the current or upcoming year. It has been shown that patch spraying can be a considerably cheaper alternative to blanket applications (e.g. López-Granados et al., 2016; Castaldi et al., 2017).

UASs have been reported to be used in assessing pest/disease development and symptoms and in assessing and monitoring pathogen in the atmosphere (e.g. Schmale et al., 2008; Aylor et al., 2011) and for precision spraying. There are a few reports of

UAS-based sensing for disease monitoring in fruit tree production (e.g. Garcia-Ruiz et al., 2013; Calderón et al., 2014). However, the application of UAS-based sensing for evaluating disease severity and susceptibility of different varieties to diseases is still relatively undeveloped (Sankaran et al., 2015).

6 Image analysis

6.1 Qualitative (thematic information extraction) analysis of UAS images

In many situations, it is necessary to extract categorical information (e.g. crop type, stressed crop, lodged crop) from UAS images prior to further analysis. Although visual interpretation proves to be straightforward for many image analysts (e.g. Selsam et al., 2017), more researchers adopt a variety of classification methods for assigning class labels. The most commonly adopted method is object-based image segmentation, which is quite useful given the ultra-high spatial resolution of UAS images (e.g. Gonzalez-Dugo et al., 2013; Peña et al., 2013; Diaz-Varela et al., 2014; Torres-Sánchez et al., 2015a,b; Borra-Serrano et al., 2015; Comba et al., 2015; López-Granados et al., 2016; Pérez-Ortiz et al., 2016; Selsam et al., 2017). Other methods, for example, include supervised classification – support vector machine (Garcia-Ruiz et al., 2013; Castaldi et al., 2017; Pérez-Ortiz et al., 2015), K-nearest neighbour (Pérez-Ortiz et al., 2015), linear discriminant analysis (Luna and Lobo, 2016), maximum likelihood (Vega et al., 2015; Haghighattalab et al., 2016) and random forest (Yu et al., 2016). However, it is not uncommon for researchers to use unsupervised classification approaches, such as K-means (Pérez-Ortiz et al., 2015; Shi et al., 2016) and Iterative Self-Organizing Data Analysis Technique (Quiquerez et al., 2014) to determine pixel classes. Other machine learning algorithms have also been developed to segment vegetation and bare soil (e.g. Duan et al., 2017).

6.2 Quantitative analysis of UAS images

VIs are commonly computed to identify the relationships between spectral data and field biological data (e.g. Hunt et al., 2005, 2010; Torres-Sánchez et al., 2014; Candiago et al., 2015). Digital numbers, reflectance and VIs can all be linked to crop biological and environmental variables through a variety of methods. The most commonly used methods include Pearson's correlation, linear regression (Hunt et al., 2005, 2010; Swain et al., 2010), partial least squares regression (Jensen et al., 2007) and k-nearest neighbour estimator (Honkavaara et al., 2013). In several cases, biomass/yield has been modelled by further incorporating plant height (Bendig et al., 2014; Huang et al., 2016; Maresma et al., 2016) or fractional vegetation cover (Geipel et al., 2014; Duan et al., 2017).

The most commonly used VI is the Normalized Difference Vegetation Index (NDVI), which has been applied to measure biomass, chlorophyll content, nitrogen content and many other biological variables. Another useful VI is the Soil Adjusted Vegetation Index which is often employed to enhance spectral differences between vegetation and soil (Shi et al., 2016) as well as sensitivity to crop damage (Zhou et al., 2016). VIs based on RGB cameras [e.g. Excess Green Index (ExG), Green NDVI and Green Ratio Vegetation Index] have also been found to be good indicators of pigment contents, leaf area and canopy structure and, indirectly, of biomass (Erdle et al., 2011; Rasmussen et al., 2016). Various VIs (e.g. NDVI, TCARI/OSAVI) have also been used to identify vineyard water stress (e.g. Baluja et al., 2012). Most importantly, studies (Primicerio et al., 2012; Matese et al., 2015) have even shown that UAS-derived VIs were found to be comparable to those calculated from satellite, manned-aircraft images and from ground measurements and that they also provided more spatial details.

7 Case study

7.1 Study area

The case study took place in a small farming community of Verner, Ontario, Canada (46° 24′ N, 80° 07′ W). The region has an annual mean temperature of 4.7°C and the annual mean length of the growing season is 180 days with roughly 110 frost-free days. The annual precipitation is 1008 mm of which 273 mm is in the form of snow. The short growing season makes timely monitoring of the field crop conditions of critical importance. Moreover, the large area and scattered distribution of the fields, typical of this region, make personal visits and ground scouting a difficult task.

During the 2013 growing season, the authors were contacted by a local producer (Ferme Roberge) to monitor fertilizer field trials (spring wheat, *Triticum* spp.). Permission to fly the UAS over this region of Ontario was provided to Nipissing University, Canada, by Transport Canada (Special Flight Operations Certificate, SFOC, # 5812-15-33-2012-1).

7.2 Data collection and methods

A quadrocopter UAV, Aeryon Scout (Aeryon Labs Inc., Waterloo, Canada), was utilized to acquire images for the study area. The Aeryon Scout is equipped with an infrared camera, ADC lite camera (Tetracam, Chatsworth, USA). It is electrically powered by rechargeable lithium polymer batteries and has a maximum flight time of 25 minutes per battery. The control station allows the user to create flight plans that can be reused at a later date. The Aeryon Scout collects GPS/INS data for each photograph. The area of study was photographed on 21 July and on 30 July 2013.

The ADC lite has three bands: near infrared, red and green. The images captured with the ADC lite are stored directly on a flash card located on the camera. The flight altitude was set at 120 m as per the approved SFOC. Consequently, the spatial resolution for the NIR images was set at 5 cm. The front overlap and side overlap were set at 85% and 65%, respectively.

GCP were set up to help in the ortho-rectification of the final mosaic images. Each GCP consisted of a 30 cm × 30 cm colour foam pad placed on the top of a wood stake at a height of 1.5 m. Six GCPs were dispersed throughout the field prior to each UAS flight. Locations of the GCPs were recorded using a Trimble GeoXH GPS (Trimble, Sunnyvale, United States). Positional accuracy of this unit is sub-decimetre after real-time differential correction.

A variable nitrogen application experiment was conducted on the field to examine the impacts of N application on wheat grain yield. Specifically, fertilizer was applied at rates of 0, 50, 100 and 150 kg/ha to various sections of the field (Fig. 1). The treatments were arranged in a randomized complete block design with three blocks of replicate treatments. Regularly spaced field sampling sites were predetermined prior to the field season using ArcGIS (ESRI, Redlands, USA). Coordinates of these points were input into a Garmin GPSMAP 62s (Garmin, Olathe, United States) receiver so that these points could be easily located during the field season. The study area was visited once a week during the growing season (early May to mid-August). For each visit, the LAI was acquired using an AccuPAR LP-80 Ceptometer (Decagon Devices, Pullman, USA). Five LAI measurements were collected in a circle around the central point and then averaged to provide one LAI value per sample point. Plant phenology was measured using a BBCH scale (Zadoks et al., 1974). A one-metre long strip of the crop was harvested and the wet and dry biomass weighed. A CCM-200 Chlorophyll Content Meter (Opti-Sciences, Hudson, USA) was also used to measure chlorophyll content from still intact wheat leaves. At every sample location, five readings were acquired around the centre of each sample point. Images captured were exported and vignetting effect correction conducted using the PixelWrench2 software (Tetracam, Chatsworth, USA). Images were then mosaicked and ortho-rectified using Pix4dmapper (Pix4D, Lausanne, Switzerland). A log file with geotagged information and GCP locations were integrated during the image processing stage. An NDVI band [(near infrared-red)/(near infrared+red)] was then created using the Pix4D software.

A polygon shapefile layer with doughnut-shaped polygons (circles of 1 and 2 m radius, which corresponded to the field plot size and shape) was created using ArcGIS, which was used to extract digital numbers for each sampling point. No pixels within the central part of the donut were extracted as these corresponding areas were trampled by the field staff during the *in situ* measurements. Statistical data (mean, min, max, range and standard deviation) for each plot were then calculated using the extracting pixel values. Analysis of variance was then conducted using IBM SPSS (IBM, North Castle, USA) to identify whether significant differences between different treatments existed. Finally, correlation and regression analyses were used to identify any relationships between the spectral responses and the field data using IBM SPSS.

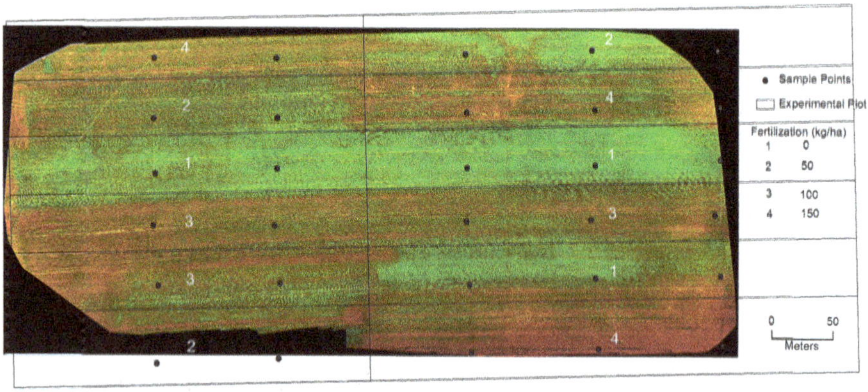

Figure 1 Nitrogen experiment design and spatial distribution of sampling points. Background is the false colour-composite (RGB: near infrared, red and NDVI) image mosaic for 21 July 2013 flights.

7.3 Results and discussion

An adequate amount of nitrogen is critical for crop growth and thus to support a profitable crop yield. However, nitrogen use efficiency is generally low (Ma and Biswas, 2016) and unused nitrogen can cause water pollution and eutrophication. Consequently, it is critical that nitrogen application rates match crop demand. As expected, in this investigation there was a contrasting pattern between plots of various fertilization levels in the image (red band, $P = 0.048$; near-infrared band, $P = 0.011$) (Fig. 1). The largest differences were found between fertilization 1 (0 kg/ha) and 4 (150 kg/ha) (red, $P = 0.046$; near infrared, $P = 0.002$). However, there were not many significant differences in the spectral responses between treatments 1, 2 and 3.

The crop was flowering and in the early ripening stages when both UAS flights were taken. The average height of the wheat crop on 24 July and 30 July 2013 was 93 cm and 92 cm, respectively. The corresponding BBCH stages for these two sampling dates were 65–69 and 83, respectively. Various relationships between the spectral indicators and the crop biological parameters were recorded. Although the correlation between crop biological parameters and spectral indicators could be moderate (Fig. 2a and b), there were strong correlations between red reflectance, biomass and chlorophyll content (Fig. 2c and d). The results are consistent with those reported in previous studies. Normally correlations between VIs and ground measurements are strong to very strong (e.g. Jannoura et al., 2015; Shi et al., 2016; Zhou et al., 2016). However, it is not uncommon to record correlations that are weak or medium in strength (e.g. Jannoura et al., 2015; Shi et al., 2016). Zarco-Tejada et al. (2013) even noted that coefficients of determination (r^2) between some crop biological variables and structural indices (e.g. NDVI) can be generally low. As a matter of fact, the strength of the correlation may be impacted by phenology, canopy coverage (i.e. the fraction of soil and shadows), illumination condition variations and bidirectional reflectance.

8 Future trends and conclusion

From this review, it is clear that in the past few years, there has been a quick adoption of UAS for agricultural sciences with many research papers with promising results having been published in various academic journals. It is thus safe to say that UAS imagery can and will continue to be applied for monitoring a variety of crop biological and environmental variables. However, for many scientists and farmers without considerable expertise of UAS and remote sensing, it may appear that this field of work is somewhat overwhelming to consider. It is suggested that the main barriers to UAS adoption in precision agriculture are:

1. UASs are still relatively expensive and require trained technicians to use them effectively. This could certainly be overcome in the near future with advances in UAS technology and with decreasing costs.
2. There is no agreed autonomous procedure/tool to finish all steps in UAS remote sensing. Specifically, the image processing, interpretation and analysis are not straightforward and require highly qualified technicians (Ballesteros et al., 2014a). The analysis and interpretation of UAV images lie behind the interpretation of satellite and ground-remote sensing and thus need to be fairly straightforward in order to be adopted by the entire agricultural sector (Rasmussen et al., 2013, 2016). An automated procedure/tool would allow for a much broader adoption (Shi et al., 2016).

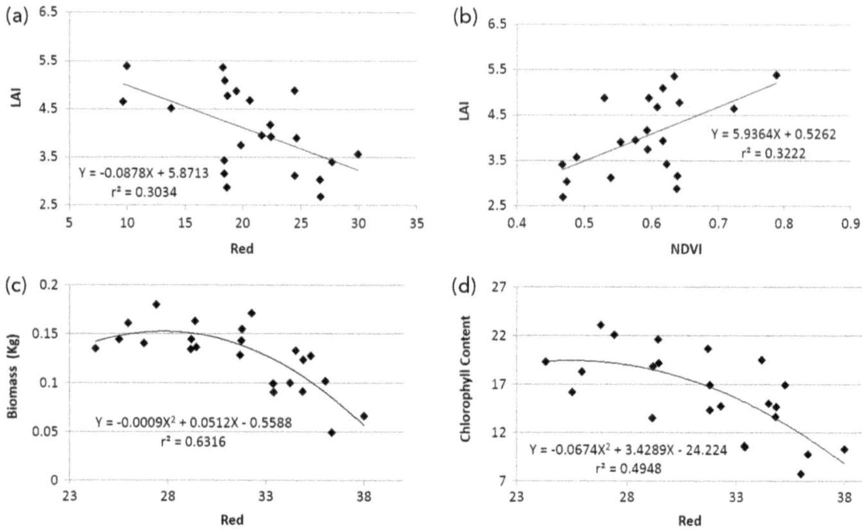

Figure 2 Statistical relationships between crop biological parameters and spectral indicators. Images for (a) and (b) were acquired on 12 July 2013, and images for (c) and (d) were acquired on 30 July 2013. Ground measurements were collected on 16 July 2013 and 25 July 2013. Digital numbers for red band were applied for the regression analyses.

3 The area covered by most UASs may not be large enough which can result in many technical difficulties as mentioned in Section 3. A loosening of the aviation regulation and further development of UAS technology (allowing longer flight time) may help in this direction.

It is anticipated that there will be a continued interest in the application of UAS remote sensing, testing of new sensors and expansion of study topics in the agricultural sciences. However, it is suggested that the next step is to inform the local producers on what UAS technology is available for them and to convince them as to what versions/options would be most cost-effective for them. Certainly, the cost of acquiring relevant equipment is expected to continue to decrease and it will be much easier to use the equipment in the near future. However, it is unlikely that farmers will understand the whole process of UAS remote sensing. Because of the involvement of a variety of equipment and the need for research expertise from different academic disciplines, it is recommended that a group of researchers work together in attaining these tasks (Zhang et al., 2014; Shi et al., 2016). Further, it is also quite feasible that farmers simply contact a UAS consulting company either individually or as a group in order to receive proper service (Huang et al., 2013; Zhang et al., 2014).

9 Acknowledgements

This research was supported by a grant (project #920161) provided to John M. Kovacs and Dan Walters from the Northern Ontario Heritage Fund Corporation of Canada.

10 Where to look for further information

This chapter provides a comprehensive review of the application of UASs in precision agriculture. Readers can find more detailed information on the topics covered in this chapter by reading papers listed in the References. For a general review of UAS remote sensing and SfM photogrammetry, we suggest reading Colomina and Molina (2014), Whitehead and Hugenholtz (2014) and Smith et al. (2016).

11 References

Aasen, H., Burkart, A., Bolten, A. and Bareth, G. (2015), 'Generating 3D hyperspectral information with lightweight UAV snapshot cameras for vegetation monitoring: From camera calibration to quality assurance', *Journal of Photogrammetry & Remote Sensing*, 108, 245–59.

Association for Unmanned Vehicle Systems International (AUVSI). (2013), 'The economic impact of unmanned aircraft systems integration in the United States'. https://higherlogicdownload.s3.amazonaws.com/AUVSI/958c920a-7f9b-4ad2-9807-f9a4e95d1ef1/UploadedImages/New_Economic%20Report%202013%20Full.pdf (Accessed 4 May 2017).

Aylor, D. E., Schmale, D. G., Shields, E. J., Newcomb, M. and Nappo, C. J. (2011), 'Tracking the potato late blight pathogen in the atmosphere using unmanned aerial vehicles and Lagrangian modeling', *Agricultural and Forest Meteorology*, 151(2), 251–60.

Ballesteros, R., Ortega, J. F., Hernández, D. and Moreno, M. A. (2014a), 'Applications of georeferenced high-resolution images obtained with unmanned aerial vehicles, Part I: Description of image acquisition and processing', *Precision Agriculture*, 15(6), 579–92.

Ballesteros, R., Ortega, J., Hernández, D. and Moreno, M. (2014b), 'Applications of georeferenced high-resolution images obtained with unmanned aerial vehicles. Part II: Application to maize and onion crops of a semi-arid region in Spain', *Precision Agriculture*, 15(6), 593–614.

Baluja, J., Diago, M. P., Balda, P., Zorer, R., Meggio, F., Morales, F. and Tardaguila, J. (2012), 'Assessment of vineyard water status variability by thermal and multispectral imagery using an unmanned aerial vehicle (UAV)', *Irrigation Science*, 30(6), 511–22.

Bellvert, J., Zarco-Tejada, P. J., Girona, J. and Fereres, E. (2014), 'Mapping crop water stress index in a 'Pinot-noir' vineyard: Comparing ground measurements with thermal remote sensing imagery from an unmanned aerial vehicle', *Precision Agriculture*, 15(4), 361–76.

Bendig, J., Willkomm, M., Tilly, N., Gnyp, M. L., Bennertz, S., Qiang, C., Miao, Y., Lenz-Wiedemann, V. I. S. and Bareth, G.(2013), 'Very high resolution crop surface models (CSMs) from UAV-based stereo images for rice growth monitoring in Northeast China', *The International Archives of the Photogrammetry, Remote Sensing and Spatial Information Sciences*, 40, 45–50.

Bendig, J., Bolten, A., Bennertz, S., Broscheit, J., Eichfuss, S. and Bareth, G. (2014), 'Estimating biomass of barley using crop surface models (CSMs) derived from UAV-based RGB imaging', *Remote Sensing*, 6(11), 10395–412.

Berni, J., Zarco-Tejada, P. J., Suarez, L. and Fereres, E. (2009), 'Thermal and narrowband multispectral remote sensing for vegetation monitoring from an unmanned aerial vehicle', *Institute of Electrical and Electronics Engineers Transactions on Geoscience and Remote Sensing*, 47, 722–38.

Borra-Serrano, I., Peña, J. M., Torres-Sánchez, J., Mesas-Carrascosa, F. J. and López-Granados, F. (2015), 'Spatial quality evaluation of resampled unmanned aerial vehicle-imagery for weed mapping', *Sensors (Basel, Switzerland)*, 15(8), 19688–708.

Burgos, S., Mota, M., Noll, D. and Cannelle, B. (2015), 'Use of very high-resolution airborne images to analyse 3D canopy architecture of a vineyard', *The International Archives of the Photogrammetry, Remote Sensing and Spatial Information Sciences*, 40(3/W3), 399.

Calderón, R., Montes-Borrego, M., Landa, B., Navas-Cortes, J. and Zarco-Tejada, P., (2014), 'Detection of downy mildew of opium poppy using high-resolution multi-spectral and thermal imagery acquired with an unmanned aerial vehicle', *Precision Agriculture*, 15, 639–61.

Candiago, S., Remondino, F., De Giglio, M., Dubbini, M. and Gattelli, M. (2015), 'Evaluating multispectral images and vegetation indices for precision farming applications from UAV images', *Remote Sensing*, 7(4), 4026–47.

Castaldi, F., Pelosi, F., Pascucci, S. and Casa, R. (2017), 'Assessing the potential of images from unmanned aerial vehicles (UAV) to support herbicide patch spraying in maize', *Precision Agriculture*, 18(1), 76–94.

Caturegli, L., Corniglia, M., Gaetani, M., Grossi, N., Magni, S., Migliazzi, M. I. Angelini, L., Mazzoncini, M., Silvestri, N., Fontanelli, M., Raffaelli, M., Peruzzi, A. and Volterrani, M. (2016), 'Unmanned aerial vehicle to estimate nitrogen status of turfgrasses', *PLoS ONE*, 11(6), 1–13.

Chang, J., Clay, D. E., Dalsted, K., Clay, S. and O'Neill, M. (2003), 'Corn (Zea mays L.) yield prediction using multispectral and multidate reflectance'. *Agronomy*, 95, 1447–53.

Cilia, C., Panigada, C., Rossini, M., Meroni, M., Busetto, L., Amaducci, S., Boschetti, M., Picchi, V. and Colombo, R. (2014), 'Nitrogen status assessment for variable rate fertilization in maize through hyperspectral imagery', *Remote Sensing*, 6, 6549–65.

Colomina, I. and Molina, P. (2014), 'Unmanned aerial systems for photogrammetry and remote sensing: A review', *ISPRS Journal of Photogrammetry and Remote Sensing*, 92, 79–97.

Comba, L., Gay, P., Primicerio, J. and Ricauda Aimonino, D. (2015), 'Vineyard detection from unmanned aerial systems images', *Computers and Electronics in Agriculture*, 114, 78–87.

Córcoles, J. I., Ortega, J. F., Hernández, D. and Moreno, M. A. (2013), 'Use of digital photography from unmanned aerial vehicles for estimation of leaf area index in onion (Allium cepa L.)', *European Journal of Agronomy*, 45, 96–104.

De Tar, W. R., Chesson, J. H., Penner, J. V. and Ojala, J. C. (2008), 'Detection of soil properties with airborne hyperspectral measurements of bare fields', *Transactions of the American Society of Agricultural and Biological Engineers*, 51, 463–70.

Diaz-Varela, R. A., Zarco-Tejada, P. J., Angileri, V. and Loudjani, P. (2014), 'Automatic identification of agricultural terraces through object-oriented analysis of very high resolution DSMs and multispectral imagery obtained from an unmanned aerial vehicle', *Journal of Environmental Management*, 134, 117–26.

d'Oleire-Oltmanns, S., Marzolff, I., Peter, K. D. and Ries, J. B. (2012), 'Unmanned aerial vehicle (UAV) for monitoring soil erosion in Morocco', *Remote Sensing*, 4(11), 3390–416.

Duan, S., Li, Z., Wu, H., Tang, B., Ma, L., Zhao, E. and Li, C. (2014), 'Inversion of the PROSAIL model to estimate leaf area index of maize, potato, and sunflower fields from unmanned aerial vehicle hyperspectral data', *International Journal of Applied Earth Observation and Geoinformation*, 26, 12–20.

Duan, T., Zheng, B., Guo, W., Ninomiya, S., Guo, Y. and Chapman, S. C. (2017), 'Comparison of ground cover estimates from experiment plots in cotton, sorghum and sugarcane based on images and ortho-mosaics captured by UAV', *Functional Plant Biology*, 44(1), 169.

Erdle, K., Mistele, B. and Schmidhalter, U., 2011, 'Comparison of active and passive spectral sensors in discriminating biomass parameters and nitrogen status in wheat cultivars', *Field Crops Research*, 124, 74–84.

Faiçal, B. S., Costa, F. G., Pessin, G., Ueyama, J., Freitas, H., Colombo, A., Fini, P. H., Villas, L., Osório, F. S., Vargas, P. A. and Braun, T. (2014), 'The use of unmanned aerial vehicles and wireless sensor networks for spraying pesticides', *Journal of Systems Architecture*, 60(4), 393–404.

Fernández, T., Pérez, J. L., Cardenal, J., Gómez, J. M., Colomo, C. and Delgado, J. (2016), 'Analysis of landslide evolution affecting olive groves using UAV and photogrammetric techniques', *Remote Sensing*, 8(10), 1–29.

Food and Agriculture Organization. (1996), 'Water and food security'. http://www.fao.org/docrep/x0262e/x0262e01.htm (Accessed: 7 June 7 2017).

Freeman, P. K. and Freeland, R. S. (2016), 'Media framing the reception of unmanned aerial vehicles in the United States of America', *Technology in Society*, 44, 23.

Gago, J., Douthe, C., Coopman, R. E., Gallego, P. P., Ribas-Carbo, M., Flexas, J., Escalona, J. and Medrano, H. (2015), 'UAVs challenge to assess water stress for sustainable agriculture', *Agricultural Water Management*, 153, 9–19.

Garcia-Ruiz, F., Sankaran, S., Maja, J. M., Lee, W. S., Rasmussen, J. and Ehsani, R., (2013), 'Comparison of two aerial imaging platforms for identification of Huanglongbing-infected citrus trees', *Computers and Electronics in Agriculture*, 91, 106–15.

Geipel, J., Link, J. and Claupein, W. (2014), 'Combined spectral and spatial modeling of corn yield based on aerial images and crop surface models acquired with an unmanned aircraft system', *Remote Sensing*, 6(11), 10335–55.

Gómez-Candón, D., López-Granados, F., Caballero-Novella, J. J., Gómez-Casero, M., Jurado-Expósito, M. and García-Torres, L. (2011), 'Geo-referencing remote images for precision agriculture using artificial terrestrial targets', *Precision Agriculture*, 12(6), 876–91.

Gómez-Candón, D., De Castro, A. I. and López-Granados, F. (2014), 'Assessing the accuracy of mosaics from unmanned aerial vehicle (UAV) imagery for precision agriculture purposes in wheat', *Precision Agriculture*, 15(1), 44–56.

Gómez-Candón, D., Virlet, N., Labbé, S., Jolivot, A. and Regnard, J.-L. (2016), 'Field phenotyping of water stress at tree scale by UAV-sensed imagery: New insights for thermal acquisition and calibration', *Precision Agriculture*, 17(6), 786–800.

Gonzalez-Dugo, V., Zarco-Tejada, P., Nicolás, E., Nortes, P., Alarcón, J., Intrigliolo, D. S. and Fereres, E. (2013), 'Using high resolution UAV thermal imagery to assess the variability in the water status of five fruit tree species within a commercial orchard', *Precision Agriculture*, 14(6), 660–78.

Guillen-Climent, M., Zarco-Tejada, P., Berni, J., North, P. and Villalobos, F. (2012), 'Mapping radiation interception in row-structured orchards using 3D simulation and high-resolution airborne imagery acquired from a UAV', *Precision Agriculture*, 13(4), 473–500.

Haala, N., Cramer, M., Weimer, F. and Trittler, M. (2011), 'Performance test on UAV-based photogrammetric data collection'. *Proceedings of the International Archives of the Photogrammetry, Remote Sensing and Spatial Information Sciences*, 38(1/C22), 7–12.

Haghighattalab, A., Pérez, L. G., Mondal, S., Singh, D., Schinstock, D., Rutkoski, J., Ortiz-Monasterio, I., Singh, R. P., Goodin, D. and Poland, J. (2016), 'Application of unmanned aerial systems for high throughput phenotyping of large wheat breeding nurseries', *Plant Methods*, 12, 1–15.

Hassan-Esfahani, L., Torres-Rua, A., Ticlavilca, A. M., Jensen, A. and McKee, M. (2014), 'Topsoil moisture estimation for precision agriculture using unmmaned aerial vehicle multispectral imagery', in *Geoscience and Remote Sensing Symposium (IGARSS)*. Institute of Electrical and Electronics Engineers International, Quebec, QC, Canada, pp. 3263–6.

Herwitz, S., Johnson, L., Dunagan, S., Higgins, R., Sullivan, D., Zheng, J., Lobitz, B. M., GLeung, J., Gallmeyer, B. A., Aoyagi, M., Slye, R. E. and Brass, J. A. (2004), 'Imaging from an unmanned aerial vehicle: Agricultural surveillance and decision support', *Computers and Electronics in Agriculture*, 44, 49–61.

Hoffmann, H., Jensen, R., Thomsen, A., Nieto, H., Rasmussen, J. and Friborg, T. (2016), 'Crop water stress maps for entire growing seasons from visible and thermal UAV imagery', *Biogeosciences Discussions*, 13, 6545–63.

Hoffmeister, D., Waldhoff, G., Korres, W., Curdt, C. and Bareth, G. (2016), 'Crop height variability detection in a single field by multi-temporal terrestrial laser scanning', *Precision Agriculture*, 17(3), 296–312.

Holman, F. H., Riche, A. B. Michalski, A., Castle, M., Wooster, M. J. and Hawkesford, M. J. (2016), 'High throughput field phenotyping of wheat plant height and growth rate in field plot trials using UAV based remote sensing', *Remote Sensing*, 8(12), 1–23.

Honkavaara, E., Kaivosoja, J., Mäkynen, J., Pellikka, I., Pesonen, L., Saari, H., Salo, H., Hakala, T., Marklelin, L. and Rosnell, T. (2012), 'Hyperspectral reflectance signatures and point clouds for precision agriculture by light weight UAV imaging system', *ISPRS Annals of Photogrammetry, Remote Sensing and Spatial Information Sciences*, I-7, 353–8.

Honkavaara, E., Saari, H., Kaivosoja, J., Pölönen, I., Hakala, T., Litkey, P., Mäkynen, J. and Pesonen, L. (2013), 'Processing and assessment of spectrometric, stereoscopic imagery collected using a lightweight UAV spectral camera for precision agriculture.', *Remote Sensing*, 5(10), 5006–39.

Huang, Y., Hoffmann, W., Lan, Y., Wu, W. and Fritz, B. (2009), 'Development of a spray system for an unmanned aerial vehicle platform', *Applied Engineering in Agriculture*, 25, 803–9.

Huang, Y., Thomson, S. J., Hoffmann, W. C., Lan, Y. and Fritz, B. K. (2013), 'Development and prospect of unmanned aerial vehicle technologies for agricultural production management', *International Journal of Agricultural and Biological Engineering*, 6(3), 1–10.

Huang, Y., Thomson, S. J., Brand, H. J. and Reddy, K. N. (2016), 'Development and evaluation of low-altitude remote sensing systems for crop production management', *International Journal of Agricultural and Biological Engineering*, 9(4), 1–11.

Hugenholtz, C. H., Whitehead, K., Brown, O. W., Barchyn, T. E., Moorman, B. J., LeClair, A., Riddell, K. and Hamilton, T. (2013), 'Geomorphological mapping with a small unmanned aircraft system (sUAS): Feature detection and accuracy assessment of a photogrammetrically-derived digital terrain model', *Geomorphology*, 194, 16–24.

Hunt, E. R. and Rondon, S. I. (2017), 'Detection of potato beetle damage using remote sensing from small unmanned aircraft systems', *Journal of Applied Remote Sensing*, 11(2), 026013–026013.

Hunt, E. R., Cavigelli, M., Daughtry, C. S. T., McMurtrey, J. E. and Walthall, C. L. (2005), 'Evaluation of digital photography from model aircraft for remote sensing of crop biomass and nitrogen status', *Precision Agriculture*, 6, 359–78.

Hunt, E. R., Hively, W. D., Fujikawa, S. J., Linden, D. S., Daughtry, C. S. T. and McCarty, G. W. (2010), 'Acquisition of NIR-green-blue digital photographs from unmanned aircraft for crop monitoring', *Remote Sensing*, 2, 290–305.

Hunt, E. R., Daughtry, C. S., Mirsky, S. B. and Hively, W. D. (2014), 'Remote sensing with simulated unmanned aircraft imagery for precision agriculture applications', *Institute of Electrical and Electronics Engineers Select Topic Applications*, 7(11), 4566–71.

Hunt, E. R., Horneck, D. A., Spinelli, C. B., Turner, R. W., Bruce, A. E., Gadler, D. J., Brungardt, J. J. and Hamm, P. B. (2017), 'Monitoring nitrogen status of potatoes using small unmanned aerial vehicles', *Precision Agriculture*, 1–20.

Inman, D., Khosla, R., Reich, R. and Westfall, D. G. (2008), 'Normalized difference vegetation index and soil color-based management zones in irrigated maize', *Journal of Agronomy*, 100, 60–6.

Inoue, Y., Morinaga, S. and Tomita, A. (2000), 'A blimp-based remote sensing system for low-altitude monitoring of plant variables: A preliminary experiment for agricultural and ecological applications', *International Journal Remote Sensing*, 21, 379–85.

Jannoura, R., Brinkmann, K., Uteau, D., Bruns, C. and Joergensen, R. G. (2015), 'Monitoring of crop biomass using true colour aerial photographs taken from a remote controlled hexacopter', *Biosystems Engineering*, 129, 341–51.

Jensen, T., Apan, A., Young, F. and Zeller, L. (2007), 'Detecting the attributes of a wheat crop using digital imagery acquired from a low-altitude platform', *Computers and Electronics in Agriculture*, 59(1), 66–77.

Khanal, S., Fulton, J. and Shearer, S. (2017), 'An overview of current and potential applications of thermal remote sensing in precision agriculture', *Computers and Electronics in Agriculture*, 139, 22–32.

Khot, L. R., Sankaran, S., Carter, A. H., Johnson, D. A. and Cummings, T. F. (2016), 'UAS imaging-based decision tools for arid winter wheat and irrigated potato production management', *International Journal of Remote Sensing*, 37(1), 125–37.

Kyratzis, A., Skarlatos, D., Fotopoulos, V., Vamvakousis, V. and Katsiotis, A. (2015), Investigating correlation among NDVI index derived by unmanned aerial vehicle photography and grain yield under late drought stress conditions', *Procedia Environmental Sciences*, 29, 225.

Laliberte, A. S., Goforth, M. A., Steele, C. M. and Rango, A. (2011), 'Multispectral remote sensing from unmanned aircraft: Image processing workflows and applications for rangeland environments', *Remote Sensing*, 3(11), 2529–51.

Li, L., Zhang, Q. and Huang, D. (2014), 'A review of imaging techniques for plant phenotyping', *Sensors*, 14(11), 20078–111.

Li, J., Zhang, F., Qian, X., Zhu, Y. and Shen, G. (2015), 'Quantification of rice canopy nitrogen balance index with digital imagery from unmanned aerial vehicle', *Remote Sensing Letters*, 6, 183–9.

Li, J., Lan, Y., Zhou, Z., Zeng, S., Huang, C., Yao, W., Yang, Z. and Qiuyang, Z. (2016), 'Design and test of operation parameters for rice air broadcasting by unmanned aerial vehicle', *International Journal of Agriculture & Biological Engineering*, 9(5), 24–32.

Link, J., Senner, D. and Claupein, W. (2013), 'Developing and evaluating an aerial sensor platform (ASP) to collect multispectral data for deriving management decisions in precision farming', *Computers and Electronics in Agriculture*, 94, 20–8.

Liu, H., Yang, G and Zhu, H. (2014), 'The extraction of wheat lodging area in UAV's image used spectral and texture features', *Applied Mechanics and Materials*, 651–653, 2390–3.

López-Granados, F. (2011), 'Weed detection for site-specific weed management: Mapping and real-time approaches', *Weed Research*, 51(1), 1–11.

López-Granados, F., Torres-Sánchez, J., Serrano-Perez, A., Castro, A. I., Mesas-Carrascosa, F.-J. and Pena, J.-M. (2016), 'Early season weed mapping in sunflower using UAV technology: Variability of herbicide treatment maps against weed thresholds', *Precision Agriculture*, 17(2), 183–99.

Lucieer, A., Malenovský, Z., Veness, T. and Wallace, L. (2014), 'HyperUAS—Imaging spectroscopy from a multirotor unmanned aircraft system', *Journal of Field Robotics*, 31(4), 571–90.

Luna, I. and Lobo, A. (2016), 'Mapping crop planting quality in sugarcane from UAV imagery: A pilot study in Nicaragua', *Remote Sensing*, 8(6), 1–18.

Ma, B. L. and Biswas, D. K. (2016), 'Field-level comparison of nitrogen rates and application methods on maize yield, grain quality and nitrogen use efficiency in a humid environment', *Journal of Plant Nutrition*, 39(5), 727–41.

Maresma, Á., Ariza, M., Martínez, E., Lloveras, J. and Martínez-Casasnovas, J. A. (2016), 'Analysis of vegetation indices to determine nitrogen application and yield prediction in maize (zea mays l.) from a standard UAV service', *Remote Sensing*, 8(12), 973–87.

Martínez, J., Egea, G., Agüera, J. and Pérez-Ruiz, M. (2017), 'A cost-effective canopy temperature measurement system for precision agriculture: A case study on sugar beet', *Precision Agriculture*, 18(1), 95–110.

Matese, A., Toscano, P., Di Gennaro, S., Genesio, L., Vaccari, F., Primicerio, J., Belli, C., Zaldei, A., Bianconi, R. and Gioli, B. (2015), 'Intercomparison of UAV, aircraft and satellite remote sensing platforms for precision viticulture', *Remote Sensing*, 7(3), 2971–90.

Mathews, A. J. and Jensen, J. L. R. (2013), 'Visualizing and quantifying vineyard canopy LAI using an unmanned aerial vehicle (UAV) collected high density structure from motion point cloud', *Remote Sensing*, 5(5), 2164–83.

Mesas-Carrascosa, F. J., Torres-Sánchez, J., Clavero-Rumbao, I., García-Ferrer, A., Peña, J. M., Borra-Serrano, I. and López-Granados, F. (2015), 'Assessing optimal flight parameters for generating accurate multispectral orthomosaicks by UAV to support site-specific crop management', *Remote Sensing*, 7(10), 12793–814.

Metternicht, G. I. and Zinck, J. A. (2003), 'Remote sensing of soil salinity: Potentials and constraints', *Remote Sensing of Environment*, 85(1), 1–20.

Micheletti, N., Chandler, J. H. and Lane, S. N. (2015), 'Investigating the geomorphological potential of freely available and accessible Structure-from-Motion photogrammetry using a smartphone', *Earth Surface Processes and Landforms Journal*, 40(4), 473–86.

Nex, F. and Remondino, F. (2014), 'UAV for 3D mapping applications: A review', *Applied Geomatics*, 6(1), 1–15.

Pan, Z., Deng, L., Lyu, Q., He, S., Yi, S., Liu, Y., Yongxu, Y. and Haiyang, P. (2016), 'Effects of citrus tree-shape and spraying height of small unmanned aerial vehicle on droplet distribution', *International Journal of Agriculture & Biological Engineering*, 9(4), 45–52.

Peña, J. M., Torres-Sánchez, J., de Castro, A. I., Kelly, M. and López-Granados, F. (2013), 'Weed mapping in early-season maize fields using object-based analysis of unmanned aerial vehicle (UAV) images', *PLoS ONE*, 8(10), 77151.

Pérez-Ortiz, M., Peña, J. M., Gutiérrez, P. A., Torres-Sánchez, J., Hervás-Martínez, C. and López-Granados, F. (2015), 'A semi-supervised system for weed mapping in sunflower crops using unmanned aerial vehicles and a crop row detection method', *Applied Soft Computing*, 37, 533–44.

Pérez-Ortiz, M., Peña, J. M., Gutiérrez, P. A., Torres-Sánchez, J., Hervás-Martínez, C. and López-Granados, F. (2016), 'Selecting patterns and features for between- and within-crop-row weed mapping using UAV-imagery', *Expert Systems with Applications*, 47, 85–94.

Primicerio, J., Di Gennaro, S. F., Fiorillo, E., Genesio, L., Lugato, E., Matese, A. and Vaccari, F. P. (2012), 'A flexible unmanned aerial vehicle for precision agriculture', *Precision Agriculture*, 13(4), 517–23.

Quiquerez, A., Chevigny, E., Allemand, P., Curmi, P., Petit, C. and Grandjean, P. (2014), 'Assessing the impact of soil surface characteristics on vineyard erosion from very high spatial resolution aerial images (Côte de Beaune, Burgundy, France)', *Catena*, 116, 163–72.

Rasmussen, J., Nielsen, J., Garcia-Ruiz, F., Christensen, S. and Streibig, J. C. (2013), 'Potential uses of small unmanned aircraft systems (UAS) in weed research', *Weed Research*, 53(4), 242–8.

Rasmussen, J., Ntakos, G., Nielsen, J., Svensgaard, J., Poulsen, R. N. and Christensen, S. (2016), 'Are vegetation indices derived from consumer-grade cameras mounted on UAVs sufficiently reliable for assessing experimental plots?' *European Journal of Agronomy*, 74, 75–92.

Rokhmana, C. A. (2015), 'The potential of UAV-based remote sensing for supporting precision agriculture in Indonesia', in *The 1st International Symposium on LAPAN-IPB Satellite (LISAT) for Food Security and Environment Monitoring*. Elsevier, Singapore, Volume 24, pp. 245–53.

Roldán, J. J., Joossen, G., Sanz, D., del Cerro, J. and Barrientos, A. (2015), 'Mini-UAV based sensory system for measuring environmental variables in greenhouses', *Sensors*, 15(2), 3334–50.

Sankaran, S., Mishra, A., Ehsani, R. and Davis, C. 2010, 'A review of advanced techniques for detecting plant diseases', *Computers and Electronics in Agriculture*, 72, 1–13.

Sankaran, S., Khot, L. R., Espinoza, C. Z., Jarolmasjed, S., Sathuvalli, V. R., Vandemark, G. J., Miklas, P. N., Carter, A. H., Pumphrey, M. O., Knowles, N. R. and Pavek, M. J. (2015), 'Low-altitude, high-resolution aerial imaging systems for row and field crop phenotyping: A review', *European Journal of Agronomy*, 70, 112–23.

Schirrmann, M., Giebel, A., Gleiniger, F., Pflanz, M., Lentschke, J. and Dammer, K.-H. (2016), Monitoring agronomic parameters of winter wheat crops with low-cost UAV imagery', *Remote Sensing*, 8(9), 1–19.

Schmale, D. G., Dingus, B. R. and Reinholtz, C. (2008), 'Development and application of an autonomous unmanned aerial vehicle for precise aerobiological sampling above agricultural fields', *Journal of Field Robotics*, 25(3), 133–47.

Seelan, S. K., Laguette, S., Casady, G. M. and Seielstad, G. A. (2003), 'Remote sensing applications for precision agriculture: A learning community approach', *Remote Sensing Environment*, 88, 157–69.

Selsam, P., Schaeper, W., Brinkmann, K. and Buerkert, A. (2017), 'Acquisition and automated rectification of high-resolution RGB and near-IR aerial photographs to estimate plant biomass and surface topography in arid agro-ecosystems', *Experimental Agriculture*, 53(1), 144–57.

Sepúlveda-Reyes, D., Zúñiga, M., Ortega-Farías, S., Ingram, B., Bardeen, M. and Poblete-Echeverría, C. (2016), 'Selecting canopy zones and thresholding approaches to assess grapevine water status by using aerial and ground-based thermal imaging', *Remote Sensing*, 8(10), 1–16.

Severtson, D., Callow, N., Flower, K., Neuhaus, A., Olejnik, M. and Nansen, C. (2016), 'Unmanned aerial vehicle canopy reflectance data detects potassium deficiency and green peach aphid susceptibility in canola', *Precision Agriculture*, 17(6), 659–77.

Shanahan, J. F., Schepers, J. S., Francis, D. D., Varvel, G. E., Wilhelm, W. W., Tringe, J. M., Schlemmer, M. R. and Major, D. J. (2001), 'Use of remote sensing imagery to estimate corn grain yield', *Journal of Agronomy*, 93, 583–9.

Shi, Y., Thomasson, J. A., Murray, S. C., Pugh, N. A., Rooney, W. L., Shafian, S., Rajan, N., Rouze, G., Morgan, C. L. S., Neely, H. L., Rana, A., Bagavathiannan, M. V., Henrickson, J., Bowden, E., Valasek, J., Olsenholler, J., Bishop, M. P., Sheridan, R., Putman, E. B., Popescu, S., Burks, T., Cope, D., Ibrahim, A., McCutchen, B. F., Baltensperger, D. D., Avant Jr., R. V., Vidrine, M. and Yang, C. (2016), 'Unmanned aerial vehicles for high-throughput phenotyping and agronomic research', *PLoS ONE*, 11(7), 0159781.

Smith, M. W., Carrivick, J. L. and Quincey, D. J. (2016), 'Structure from motion photogrammetry in physical geography', *Progress in Physical Geography*, 40(2), 247–75.

Statistics Canada. (2016), 'Agriculture is growing and evolving'. http://www.statcan.gc.ca/pub/95-640-x/2011001/p1/p1-01-eng.htm (Accessed: 17 May 2017).

Suarez, L., Zarco-Tejada, P. J., Gonzalez-Dugo, V., Berni, J., Sagardoy, R., Morales, F. and Fereres, E. (2010), 'Detecting water stress effects on fruit quality in orchards with time-series PRI airborne imagery', *Remote Sensing Environment*, 114, 286–98.

Sujiura, R., Naguchi, N. and Ishii, K. (2007), 'Correction of low-altitude thermal images applied to estimating of soil water status', *Biosystems Engineering*, 96, 301–13.

Sullivan, D., Fulton, J., Shaw, J. and Bland, G., (2007), 'Evaluating the sensitivity of an unmanned thermal infrared aerial system to detect water stress in a cotton canopy', *Transactions of the American Society of Agricultural and Biological Engineers*, 50, 1955–62.

Swain, K. C., Thomson, S. J. and Jayasuriya, H. P. W. (2010), 'Adoption of an unmanned helicopter for low altitude remote sensing to estimate yield and total biomass of a rice crop', *Transactions of the American Society of Agricultural and Biological Engineers*, 53, 21–7.

Tilly, N., Hoffmeister, D., Cao, Q., Huang, S., Lenz-Wiedemann, V., Miao Y. and Bareth, G. (2014), 'Multitemporal crop surface models: Accurate plant height measurement and biomass estimation with terrestrial laser scanning in paddy rice', *Journal of Applied Remote Sensing*, 8(1), 083671.

Torres-Sánchez, J., López-Granados, F., De Castro, A. I. and Peña-Barragán, J. M. (2013), 'Configuration and specifications of an unmanned aerial vehicle (UAV) for early site specific weed management', *PLoS ONE*, 8(3), 58210.

Torres-Sánchez, J., Peña, J. M., De Castro, A. I. and López-Granados, F. (2014), 'Multi-temporal mapping of the vegetation fraction in early-season wheat fields using images from UAV', *Computers and Electronics in Agriculture*, 103, 104–13.

Torres-Sánchez, J., López-Granados, F. and Peña, J. M. (2015a), 'An automatic object-based method for optimal thresholding in UAV images: Application for vegetation detection in herbaceous crops', *Computers and Electronics in Agriculture*, 114, 43–52.

Torres-Sánchez, J., López-Granados, F., Serrano, N., Arquero, O. and Peña, J. M. (2015b), 'High-throughput 3-D monitoring of agricultural-tree plantations with unmanned aerial vehicle (UAV) technology', *PLoS ONE*, 10(6), 0130479.

Uto, K., Seki, H., Saito, G. and Kosugi, Y. (2013), 'Characterization of rice paddies by a UAV-mounted miniature hyperspectral sensor system', *Institute of Electrical and Electronics Engineers Journal of Select Topics Application*, 6(2), 851–60.

Vega, F. A., Ramírez, F. C., Saiz, M. P. and Rosúa, F. O. (2015), 'Multi-temporal imaging using an unmanned aerial vehicle for monitoring a sunflower crop', *Biosystems Engineering*, 132, 19–27.

Verger, A., Vigneau, N., Chéron, C., Gilliot, J. M., Comar, A. and Baret, F. (2014), 'Green area index from an unmanned aerial system over wheat and rapeseed crops', *Remote Sensing Environment*, 152, 654–64.

Wallace, L., Lucieer, A. and Watson, C. S. (2014), 'Evaluating tree detection and segmentation routines on very high resolution UAV LiDAR data', *Institute of Electrical and Electronics Engineers Transactions Geoscience Remote Sensing*, 52(12), 7619–28.

Wehrhan, M., Rauneker, P. and Sommer, M. (2016), 'UAV-based estimation of carbon exports from heterogeneous soil landscapes–a case study from the carboZALF experimental area', *Sensors*, 16, 255.

Whitehead, K. and Hugenholtz, C. H. (2014), 'Remote sensing of the environment with small unmanned aircraft systems (UASs), part 1: A review of progress and challenges', *Journal of Unmanned Vehicle Systems*, 2(3), 69–85.

Whitehead, K., Moorman, B. J. and Hugenholtz, C. H. (2013), 'Brief Communication: Low-cost, on-demand aerial photogrammetry for glaciological measurement', *The Cryosphere*, 7(6), 1879.

Willkomm, M., Bolten, A. and Bareth, G. (2016), 'Non-destructive monitoring of rice by hyperspectral in-field spectrometry and UAV-based remote sensing: Case study of field-grown rice in North Rhine-Westphalia, Germany', *International Archives of the Photogrammetry, Remote Sensing*, 41(B1), 1071–7.

Wulfsohn, D. and Lagos, I. Z. (2014), 'The use of a multirotor and high-resolution imaging for precision horticulture in Chile: An industry perspective', in *Proceedings of the 12th International conference on precision agriculture*. https://www.ispag.org/proceedings/?action=abstract&id=1688&search=years (Accessed: 11 August 2017).

Xue, X., Lan, Y., Sun, Z., Chang, C. and Hoffmann, W. C. (2016), 'Develop an unmanned aerial vehicle based automatic aerial spraying system', *Computers and Electronics in Agriculture*, 128, 58–66.

Yang, C., Everitt, J. H., Bradford, J. M. and Escobar, D. E. (2000), 'Mapping grain sorghum growth and yield variations using airborne multispectral digital imagery', *Transactions of the American Society of Agricultural and Biological Engineers*, 43, 1927–38.

Yang, C., Everitt, J. H., Bradford, J. M. and Murden, D. (2004), 'Airborne hyperspectral imagery and yield monitor data for mapping cotton yield variability', *Precision Agriculture*, 5, 445–61.

Yang, C., Everitt, J. H. and Bradford, J. M. (2006a), 'Comparison of QuickBird satellite imagery and airborne imagery for mapping grain sorghum yield patterns', *Precision Agriculture*, 7, 33–44.

Yang, C., Everitt, J. H. and Bradford, J. M. (2006b), 'Evaluating high resolution QuickBird satellite imagery for estimating cotton yield', *Transactions of the American Society of Agricultural and Biological Engineers*, 49, 1599–606.

Yang, C., Everitt, J., Qian, D., Luo, B. and Chanussor, J. (2013), 'Using high-resolution airborne and satellite imagery to assess crop growth and yield variability for precision agriculture', *Proceedings of the Institute of Electrical and Electronics Engineers*, 101, 582–92.

Yu, N., Li, L., Schmitz, N., Tian, L. F., Greenberg, J. A. and Diers, B. W. (2016), 'Development of methods to improve soybean yield estimation and predict plant maturity with an unmanned aerial vehicle based platform', *Remote Sensing of Environment*, 187, 91–101.

Yue, J., Lei, T., Li, C. and Zhu, J. (2012), 'The application of unmanned aerial vehicle remote sensing in quickly monitoring crop pests', *Intelligent Automation and Soft Computing Journal*, 18, 1043–52.

Zadoks, J. C., Chang, T. T. and Konzak, C. F. (1974), 'A decimal code for the growth stages of cereals.', *Weed Research*, 14(6), 415–21.

Zahawi, R. A., Dandois, J. P., Holl, K. D., Nadwodny, D., Reid, J. L. and Ellis, E. C. (2015), 'Using lightweight unmanned aerial vehicles to monitor tropical forest recovery', *Biological Conservation*, 186, 287–95.

Zarco-Tejada, P. J., Gonzalez-Dugo, V. and Berni, J. A. J. (2012), 'Fluorescence, temperature and narrowband indices acquired from a UAV platform for water stress detection using a micro-hyperspectral imager and a thermal camera', *Remote Sensing Environment*, 117, 322–37.

Zarco-Tejada, P. J., Guillén-Climent, M. L., Hernández-Clemente, R., Catalina, A., González, M. R. and Martín, P. (2013), 'Estimating leaf carotenoid content in vineyards using high resolution hyperspectral imagery acquired from an unmanned aerial vehicle (UAV)', *Agricultural and Forest Meteorology*, 171, 281–94.

Zarco-Tejada, P. J., Diaz-Varela, R., Angileri, V., and Loudjani, P. (2014), 'Tree height quantification using very high resolution imagery acquired from an unmanned aerial vehicle (UAV) and automatic 3D photo-reconstruction methods', *European Journal of Agronomy*, 55, 89–99.

Zhang, C. and Kovacs, J. M. (2012), 'The application of small unmanned aerial systems for precision agriculture: A review', *Precision Agriculture*, 13, 693–712

Zhang, C., Walters, D. and Kovacs, J. M. (2014), 'Applications of low altitude remote sensing in agriculture upon farmers' requests– a case study in Northeastern Ontario, Canada', *PLoS ONE*, 9(11), e112894.

Zhou, J., Khot, L. R., Peters, T., Whiting, M. D., Zhang, Q. and Granatstein, D. (2016). 'Efficacy of unmanned helicopter in rainwater removal from cherry canopies', *Computers & Electronics in Agriculture*, 124, 161–7.

Chapter 5

Key challenges and methods in identifying management zones

Spyros Fountas, Evangelos Anastasiou and Zisis Tsiropoulos, Agricultural University of Athens, Greece; Aristotelis Tagarakis, BioSense Institute - Research Institute for Information Technologies in Biosystems, Serbia; and Athanasios Balafoutis, Centre for Research and Technology Hellas, Institute of Bioeconomy & Agro-technology, Greece

1 Introduction

The ultimate goal of precision agriculture (PA) is to identify homogeneous regions within a field that display similar behaviour. Therefore, a vast amount of spatial and temporal data within a field are collected and analysed to define homogeneous areas with common characteristics that can be managed separately. These areas are called management zones (MZs) and are zones that show differences in major factors such as soil type, topography and availability of water and nutrients (Bullock et al., 2009). The MZ should be large enough to permit variable rate application (VRA) of inputs but small enough to achieve precise application. MZ formation can affect rural development positively by increasing the economic output of farms, especially small-scale farms, due to the increased efficiency of inputs and higher yield and quality of production (Koch et al., 2004).

The delineation of MZs is based on a large variety of attributes that are mainly soil-based through soil sampling (i.e. texture, fertility, depth) or sensing [i.e. soil electrical conductivity (EC_a), resistivity, reflectance], or plant-based (i.e. yield, quality, plant tissue sampling, vegetation indices, chlorophyll content) as well as a combination of soil- and plant-based attributes. Any type of information, which can be obtained for different properties and for successive years, opens new and interesting possibilities in crop analysis and management (Arno et al., 2009). The most popular attributes used in MZ delineation

http://dx.doi.org/10.19103/AS.2017.0032.05

are automatically recorded using sensors and data loggers that provide the ability to attain high-spatial resolution data.

More temporally stable zones are formed using soil-based measurements. Among those, EC_a mapping has been extensively used to delineate MZs (Kitchen et al., 2005; Molin and Castro, 2008; Moral et al., 2010) assisted by elevation mapping using real-time kinematic-global positioning system (Fridgen et al., 2004; Schepers et al., 2004; Peralta et al., 2015). Soil EC_a measurements are suitable to analyse spatial variability for static soil properties, such as salinity, texture (Corwin and Plant, 2005) and organic matter (Shaner et al., 2008). However, soil-based zones do not always explain the spatial variability of yield or quality, indicating the importance of including in the procedure for MZ delineation a combination of parameters to achieve a more meaningful result. Schepers et al. (2004) concluded that the static soil-based MZ concept alone is inadequate for variable application of inputs across temporal variability, suggesting combining the MZ concept with crop-based in-season remote sensing as a better strategy.

Vegetation indices are used as a means to assess plant growth and development and have been extensively used for MZ delineation. Among those, normalized difference vegetation index (NDVI) is related to canopy properties such as chlorophyll content, total biomass, leaf area (Hansen and Schoerring, 2003), quality (Johnson et al., 2003) and crop yield (Tagarakis et al., 2017; Tagarakis and Ketterings, 2017), allowing it to be a useful tool in MZ delineation (El Nahry et al., 2011). Most of the research on vegetation indices using remote sensing has involved analysing multispectral satellite or aerial images (Lamb et al., 2004; Acevedo-Opazo et al., 2008; Hall et al, 2011) or using ground-based sensors measuring in close proximity to the plant canopy to assess canopy reflectance (Raun et al., 2002; Bramley et al., 2011; Bourgeon et al., 2017; Tagarakis et al., 2017; Tagarakis and Ketterings, 2017).

Several types of MZs can be delineated depending on the application of inputs: irrigation MZs that are designated for variable rate irrigation (Vellidis et al., 2008) or nutrient MZs that are used to perform variable rate fertilization (Basso et al., 2012) or MZs for pesticides and herbicide applications (Timmermann et al., 2003; Vasileiadis et al., 2013).

Zaman and Schumann (2006) produced nutrient MZs in a citrus orchard, based on the variation of soil properties and tree performance. Nutrient MZs were also delineated in wheat crop based on soil EC_a, elevation and soil depth mapping and were used for variable rate nitrogen application, leading to increased nitrogen use efficiency (Peralta et al., 2015). Mann et al. (2010) created productivity zones using fruit yield, ultrasonically measured tree canopy volume, NDVI, elevation and EC_a. Citrus fruit yield was positively correlated with canopy volume, NDVI and EC_a, and negatively correlated with elevation. Although all the properties were strongly correlated with yield and were able to explain the productivity of the orchard, citrus tree canopy volume was most strongly correlated ($r = 0.85$) with yield, explaining 72% of its variation. Tree canopy volume was used to classify the citrus grove into five productivity zones termed as 'very poor', 'poor', 'medium', 'good' and 'very good'.

Lawrence et al. (2015) used a spatio-temporal Bayesian framework for optimizing nitrogen management in spring wheat taking into account various data from annual on-site experiments including VRA prescriptions, crop prices, input cost and climatic conditions. On the other hand, Haghverdi et al. (2015) delineated irrigation zones using soil available water, EC_a satellite images and yield data. They concluded that soil EC_a exhibited a considerable potential in forming irrigation MZs designed for variable rate irrigation applications in pivot systems. Verdugo-Vasquez et al. (2016) delineated MZs for taking decisions of production activities based on the within-field spatial variability of phenology in two irrigated vineyards. An Opportunity Index, along the lines proposed by McBratney et al. (2000) and Pringle et al. (2003), was used to assess the suitability of site-specific management. The calculation of a Technical Opportunity Index was also suggested to assess the benefit from applying VRA to a field (Tisseyre et al., 2007, 2008).

2 Methods to delineate management zones

To delineate meaningful MZs, overlaying multiple layers of information is a common practice, while in most cases the attributes measured in the field have different spatial resolution. One of the main constraints in the application of PA is the lack of a widely acceptable method to analyse the multiple layers of information derived from the field and delineate MZs (Guastaferro et al., 2010). Therefore, a key challenge facing PA is the evaluation of different algorithms and methods for the delineation of MZs. A variety of methods have been used to delineate MZs in the framework of PA. Soft computing techniques have been employed to define correlation between the measured properties and permit a forecast of the results. Neural networks, fuzzy logic and fuzzy cognitive maps have been used recently to analyse data and explain yield variation in field crops and orchards with encouraging results (Papageorgiou et al., 2009, 2013). Some of the most commonly used methods are described in the following sections.

2.1 Empirical methods

The easiest and simplest method that many farmers use to delineate MZs employs visual inspection of the attribute maps or the field aerial images, seeking areas with common characteristics and combining the findings with their personal experience concerning the field properties. This method lacks accuracy since it is based on field observations that are not scientifically validated and the result has a high risk of being biased since personal judgement plays an important role in the procedure.

2.2 Geostatistical methods

Given the spatial dependence of the values, interpolation between the sampling points can be made using geostatistical methods. Maps covering the whole field can be produced and indicate the variability of the properties. Using interpolation all the layers of information can be referred to the same spatial resolution and overlaid on top of each other for further geostatistical analysis. Geostatistical methods have been used to analyse and combine different parameters and produce smoothed surface MZ maps. Factor co-kriging determines multivariate indices of spatial variation as a linear combination of the attributes of neighbouring observations, and therefore it can be used to subdivide fields into homogeneous MZs using georeferenced observations of different variables (Castrignanò et al., 2000). Shaddad et al. (2016) indicated the need for delineating MZs using a combination of data fusion and multivariate geostatistical methods. Aggelopoulou et al. (2010, 2013) used multivariate analysis for MZ delineation. They used yield, quality and soil parameters to delineate MZs, but the results were not very clear. Landrum et al. (2015) used multivariate geostatistics to delineate soil moisture MZs for water management under stressed conditions. They proposed the use of multi-collocated co-kriging, a multivariate geostatistical technique, for integrating a variety of sensor data sets combined with ground-based observations at different spatial scales.

Casa and Castrignano (2008) recommended factorial kriging analysis as a multivariate geostatistical method that allows processing of several layers of information on spatially and temporally variable crop and soil properties simultaneously.

2.3 Clustering methods (fuzzy k-means)

Even though there is no widely accepted method, clustering techniques may be a basis for delineating MZs. Cluster analysis is an explorative technique which classifies data in to different combinations of many variables in discrete classes or clusters. The non-hierarchical cluster analysis has been extensively used for MZ delineation. The most important non-hierarchical clustering is k-means (also known as c-means) where multidimensional data are classified into k classes (clusters) (Hartigan, 1975). The centroid in each class has the minimum Euclidean distance from each data point. Fuzzy k-means is an extension of k-means clustering that accounts for uncertainties associated with class boundaries and membership (Dobermann et al., 2003).

Fuzzy c-means and non-parametric density algorithms are some of the most commonly used clustering techniques (Guastaferro et al., 2010). Ping et al. (2005) used k-means cluster analysis, multivariate analysis of variance and discriminant analysis in an effort to evaluate a framework of procedures for delineating potential MZs in cotton. They proposed cluster analysis of yield and soil properties as the basis for delineating MZs.

In Brazil, fuzzy clustering techniques were applied to establish MZs in a field cultivated with arable crops (Molin and Castro 2008). The authors concluded that principal component analysis (PCA) and fuzzy logic application on EC_a and soil data may lead to the delineation of reliable soil MZs. Yan et al. (2007) delineated MZs using fuzzy c-means clustering algorithm on NDVI, salinity and yield data in cotton, providing an effective decision-making support tool for VRA.

During the last decade, software that performs fuzzy k-means clustering has been developed by different institutes. Management zone analyst (MZA) software developed by the University of Missouri (Fridgen et al., 2004) and FUZME developed by the University of Sydney, Australia, have been extensively used by researchers worldwide for data analysis and MZ delineation. Recently, there has been an effort to develop a web-based decision support tool that performs zone mapping for PA. The software called ZoneMAP has been developed to automatically delineate MZs using satellite imagery and field data provided by users and processed with the fuzzy c-means algorithm (Zhang et al., 2010).

Recent research used a common grouping technique based on minimizing the sum of squares between groups to create an open source tool called EZZone for MZ delineation. The tool is accessible at ezzone.pythonanywhere.com and allows users to upload their data for delineation (Lowrance et al., 2016). This tool has been successfully tested in five fields in Greece and in Georgia, USA, with very promising results. The tool was developed in Python programming language (Python.org), which supports numerous GIS modules for the language.

2.4 Degree of agreement

Kitchen et al. (2005) used unsupervised fuzzy c-means clustering to delineate soil productivity zones (SPZs) by analysing EC_a and elevation data of research fields cultivated with corn and soybean. They also delineated yield productivity zones (YPZs) by analysing yield mapping data. Outcomes of YPZs and SPZs were matched and the degree of agreement was calculated with an overall accuracy statistic (the number of correctly matched cells divided by the total number of cells in the data set).

A similar concept was applied in a study on grapevines with the aim of developing a simplified procedure for the delineation of MZs in vineyards (Tagarakis et al., 2013). To achieve this objective, field data were collected and analysed using sensors of both soil and canopy properties, and mono-parametric zone maps were produced using fuzzy clustering. In addition, yield and quality zone maps were produced and set as reference maps and compared with each mono-parametric map at pixel level. The degree of agreement of soil, topography and canopy parameters with yield and grape composition (reference) was used to select the best-suited parameters for final MZ delineation in terms of both grape yield and winemaking quality. This data analysis methodology can be further developed to include comparisons between a greater number of parameters, providing a way to manage information during PA experiments and assist with MZ delineation.

2.5 Other methods

Other studies have used a combination of statistical methods such as correlation analysis and PCA and geostatistical methods with fuzzy clustering methods to subdivide fields into homogeneous zones. Yield, EC_a, pH and NDVI data were examined in an effort to establish chlorosis MZs in maize and soybean (Kyaw et al., 2008). Data were interpolated to a 10 × 10-m grid and regression analysis was used to determine the properties with the strongest relationship to yield. They selected NDVI and EC_a to delineate MZs using MZA (Fridgen et al., 2004) which performs fuzzy c-means clustering. Urretavizcaya et al. (2014) used two different zoning algorithms for delineating MZs based on the different winery requirements. For this reason, they used a combination of fuzzy k-means analysis and a zoning procedure that takes into account membership degree and neighbourhood criteria.

3 Case study on delineation of management zones in wine grapes

Delineation of MZs in wine grapes to apply precision viticulture (PV) techniques can be quite a complicated task since the goal is to maximize quality and, concurrently, achieve reasonable yields. Measuring EC_a as an indicator of soil variability is quite common in many crops; however, EC_a measurements do not always correlate well with vine parameters such as yield, and additional measurements are necessary to explain their spatial variation (Corwin and Plant, 2005). On the other hand, vine canopy properties can be a good indicator of grape yield and quality (Hall et al., 2011). Grape composition improvement has been the focus of recent studies on PV applications. In contrast to yield, no commercial on-the-go sensor or any index of grape composition is currently available for PV use (Bramley, 2010); therefore, grape quality zones can be identified only indirectly. Generally, grape composition spatial patterns follow those of yield, although the ranked order of the zones with respect to quality may not be the same every year (Bramley, 2005). The NDVI has also been reported to correlate well with both yield and composition of grapes (Best et al., 2005). Lamb et al. (2004) found a negative correlation between grape phenolics and colour and canopy NDVI, showing that low-vigour zones presented the best quality for winemaking. However, NDVI zoning is not always related to the spatial variation of grape composition, especially in non-irrigated vineyards (Acevedo-Opazo et al., 2008).

A four-year study (2009–13) was conducted in a commercial vineyard (1 ha area) at Mikrothives, Central Greece (Latitude 39.26°, Longitude 22.73°). The vineyard was planted with *Vitis vinifera* cv. Agiorgitiko, the most widely cultivated indigenous red cultivar in Greece, grafted onto 1103P rootstock. Vines were spaced 1.0 m apart within rows and 2.6 m apart between rows and trained to a bilateral cordon system. The study included measurements of soil and landscape parameters (EC_a, elevation, slope and soil depth), canopy reflectance using proximal sensing (NDVI), grape quality (Brix, pH and must acidity) and yield. The focus of the study was the development of a method to analyse the variability of yield, canopy and soil properties for MZ delineation in vineyards. The proposed method utilizes fuzzy c-means clustering to produce non-parametric zone maps for each of the measured parameters. Then, pixel-by-pixel comparison between maps of EC_a, elevation, slope, soil depth and NDVI with yield and grape composition maps, set as reference parameters, allowed for the calculation of the degree of agreement, that is, the percentage of pixels belonging to the same zone. The procedure for the delineation of MZs is described in Fig. 1.

Following the procedure, the degree of agreement was used to select the best-suited parameters for final MZ delineation. For the year 2009, soil depth, early and mid-season NDVIs were used for yield-based MZs, while for quality-based MZs, EC_a, early and mid-season NDVIs were utilized. For the year 2010, EC_a, elevation and NDVI acquired during flowering and veraison were used for the delineation of yield-based MZs, while for quality-based MZs, EC_a and NDVI acquired during flowering and harvest were utilized.

Following the same procedure, an irrigation zone map was produced and the irrigation network was separated according to the two irrigation zones (Fig. 2).

Soil moisture sensors were installed in each zone and variable rate irrigation was performed in 2012 based on the soil moisture content and the stem water potential of the vines in each zone. According to the results, yield and sugar content did not seem to show

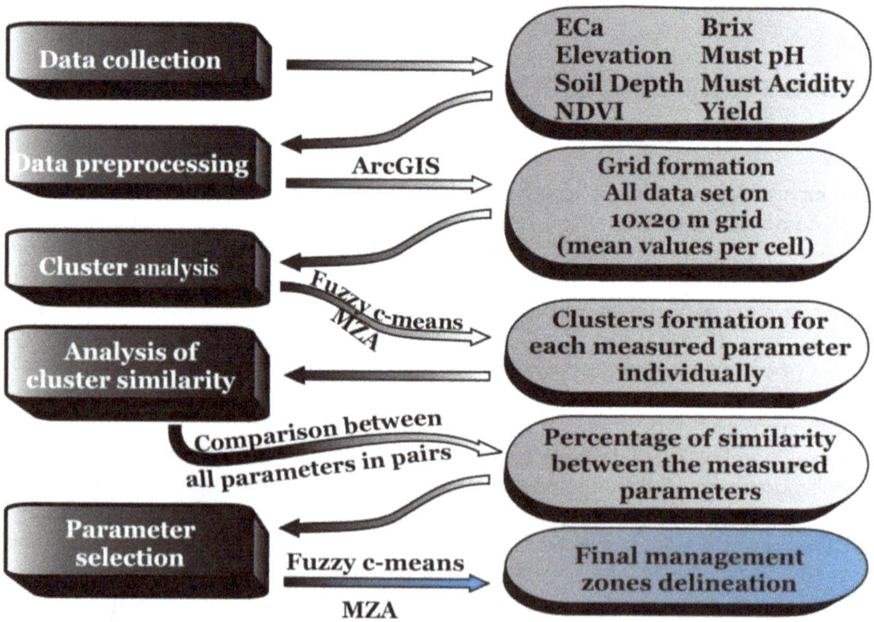

Figure 1 General flowchart of the procedure for the delineation of MZs (from data collection to data analysis).

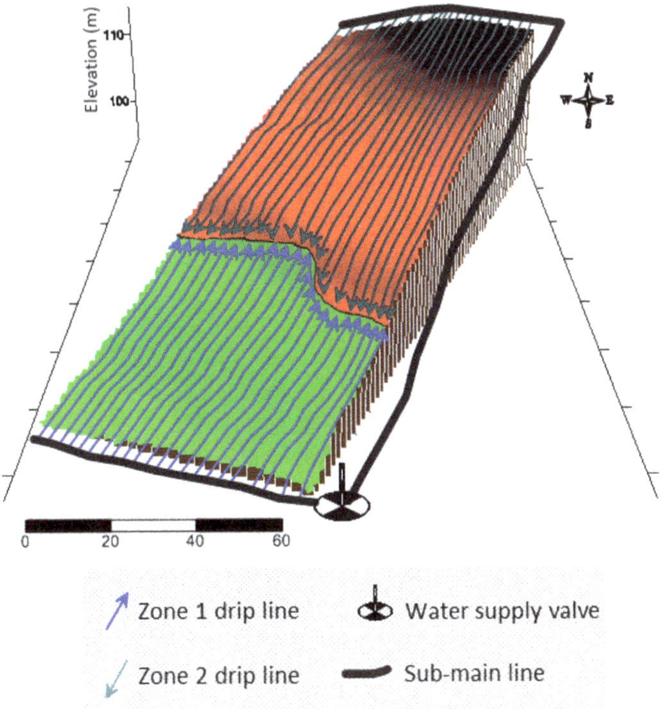

Figure 2 Irrigation zone map used to separate the irrigation network into two separate zones to perform variable rate irrigation.

significant spatial differences between the two zones compared to the previous years. However, the must total acidity (g/L) showed considerably lower spatial variability in 2012 than the previous years, indicating that water management was a driving factor influencing this quality characteristic (Table 1).

Moreover, in 2010 and 2011, approximately half of the vineyard had must acidity outside of the optimal range (5–7 g/L), while in 2012, the total area had values within the optimal range (Fig. 3).

4 Case study on delineation of management zones in table grapes

Table grapes are considered one of the most important fruit crops globally. There exist a vast number of table grapes cultivars with differences in berry size, shape, colour, chemical composition and seedlessness and consequently different acceptability by the consumers and thus different market price (Strik, 2011; Maante et al., 2015; Rolle et al., 2015). However, these quality characteristics differ in each cultivar and are highly correlated with the farming practices that growers follow such as pruning, irrigation,

Table 1 Descriptive statistics of yield, sugar content and must total acidity for the irrigation zones and the total field in 2010–12

		2010			2011			2012		
	N	Mean	Std. dev.	C.V.%	Mean	Std. dev.	C.V.%	Mean	Std. dev.	C.V.%
Yield (t/ha)										
Zone 1	18	18.78	3.55	19	9.90	2.18	22	12.56	2.68	21
Zone 2	30	14.09	3.68	26	9.13	2.59	28	10.18	2.82	28
Total	48	15.85	4.27	27	9.42	2.45	26	11.07	2.98	27
Brix										
Zone 1	18	21.48	1.19	6	23.67	1.29	05	21.00	1.15	5
Zone 2	30	21.63	1.55	7	22.36	1.19	05	21.71	1.18	5
Total	48	21.58	1.42	7	22.85	1.37	06	21.45	1.21	6
Must total acidity (g/L)										
Zone 1	18	7.03	1.45	21	6.53	0.41	06	5.96	0.29	5
Zone 2	30	5.80	1.44	25	5.96	0.50	08	5.70	0.29	5
Total	48	6.26	1.55	25	6.17	0.54	09	5.80	0.31	5

spraying and fertilizing, which have significants effect not only on the final products but also on their shelf life (Champa, 2015). Table grapes can differentiate their market position in terms of demand and price according to quality parameters that are affected by in-field variability. Different parts of the field produce different products in terms of quantity and quality, and the application of PA practices according to delineated MZs could make a difference in the final product. Many studies have indicated the benefits of using remote sensing technologies for delineating MZs on table grapes. Alsina et al. (2013) used an airborne platform with a multispectral sensor for detecting plant canopy water status on

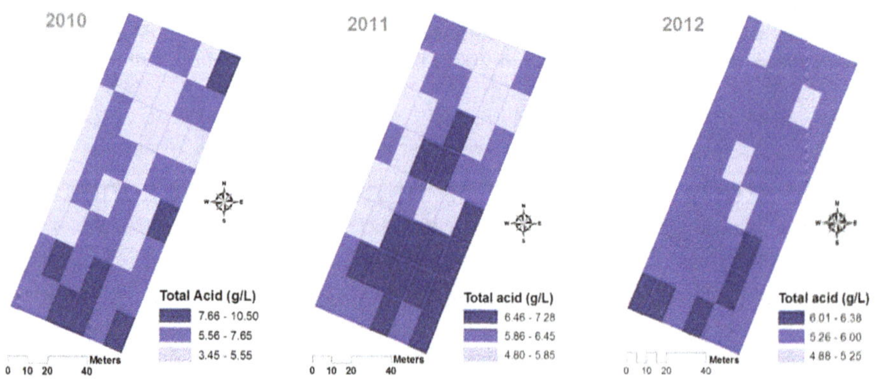

Figure 3 Must total acidity maps for 2010 and 2011 (uniform irrigation) and 2012 (variable rate irrigation).

table grapes and thus delineating MZs for irrigation. Oldoni and Bassoi (2016) delineated irrigation MZs on table grapes by using different combinations of physical and hydraulic soil properties and a fuzzy c-means clustering algorithm.

Anastasiou et al. (2016) executed a field experiment in a commercial table grape farm in Corinth, southern Greece, in 2015. The field was 1.4 ha and was planted in 2006 with *V. vinifera* L. cv. Thompson seedless (37° 54.532′ N, 22° 44.798′ E). The field presents variation in soil texture, having two different soil types (sandy clay loam and clay loam). Table grapes were trained on a double cross-arm trellis system and spaced 1.8 × 2.6 m apart. Table grape management requires numerous operations. It was irrigated ten times during the season, receiving a total of 240 mm of water. It received numerous vineyard operations (trimming, leaf plucking) to stimulate more vigorous vegetative growth. In addition, there was excessive use of plant growth regulators and fertilizers for achieving the commercial standards in respect of berry diameter and sugar content along with high yield. Anastasiou et al. (2016) delineated two MZs for irrigation purposes as shown in Fig. 4. These MZs were based on soil EC_a and accordingly irrigation scheduling was differentiated using soil moisture sensors.

In the same study, it was found that there is high variability in crop yield (CV = 21%), titratable acidity (CV = 28%) and total soluble solids-to-titratable acidity ratio (CV = 31%) which was induced by the in-field variability. They also used a proximal canopy sensor for assessing different crop yields and quality zones in a table grapes field established with the 'Thompson seedless' cultivar (Fig. 5).

Therefore, the use of multiple parameters for MZ delineation in table grapes vineyards including quality (total soluble solids, titratable acidity, berry diameter, berry deformation and detachment from peduncle), yield and canopy indices (NDVI, GNDVI, and NDRE) could provide more informative MZs for other agricultural practices. Such MZs are of profound significance in terms of yield manageability and thus farm profitability. It should be noted that the delineation of such MZs and their application could be affected by the farming practices that farmers follow and thus have low reliability in terms of providing trustworthy estimations.

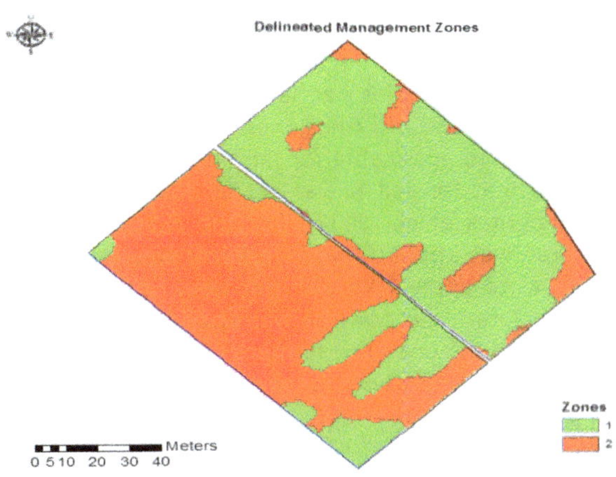

Figure 4 Management zone map to perform variable rate irrigation.

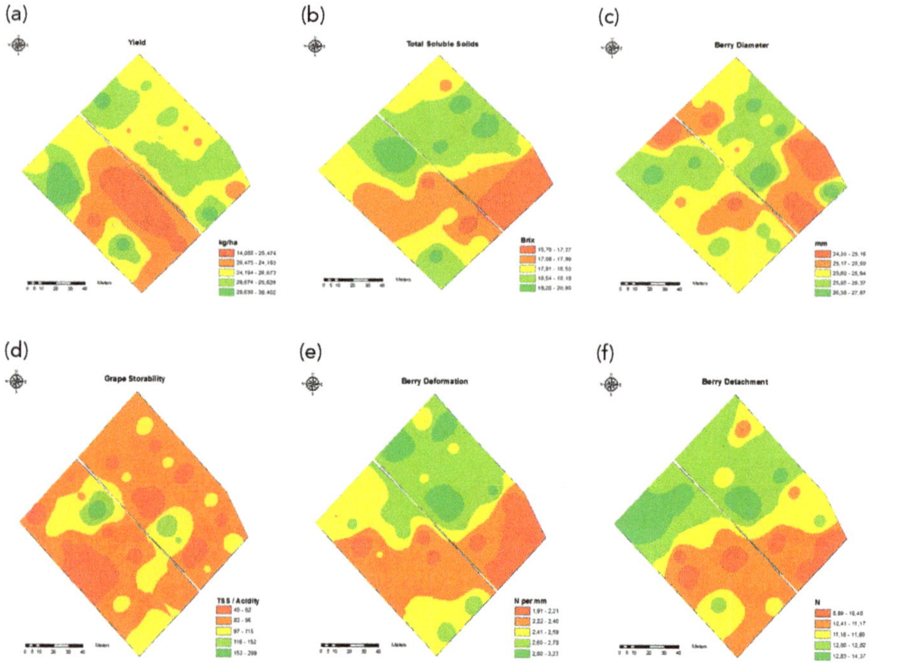

Figure 5 Maps of (a) yield, (b) total soluble solids, (c) berry diameter, (d) grape storability, (e) berry deformation and (f) berry detachment of a table grapes vineyard (Anastasiou et al., 2016).

5 Case study on delineation of management zones in olive oil tree plantation

The delineation of MZs for VRA of P, K and lime based on soil sampling was carried out in a 9.1-ha olive oil tree plantation in southern Greece in 2007–10, where a dense regular grid sampling with ten samples per ha was performed. The field was sectioned according to the delineated MZs and the generated contours. For P and K, only two MZs were created to be more applicable for farming practices. For soil P, a threshold was set at 45 ppm and that for soil K was set at 350 ppm, and the MZs for 2018 are shown in Fig. 6 (Fountas et al., 2011). For the P application, in the low-P MZ, 1 kg of fertilizer per tree was applied. No P fertilizer was applied for the high-P MZ. The farmer's practice was to apply 1 kg of P per tree. For K application, in the low-K management zone, 2 kg K fertilizer per tree was applied, whereas 1 kg K fertilizer per tree was applied in the high-K zone. The farmer's practice was to apply 2 kg of K per tree. The differences in the applications between the VRA and the farmer's practice were 59% for P and 31% for K.

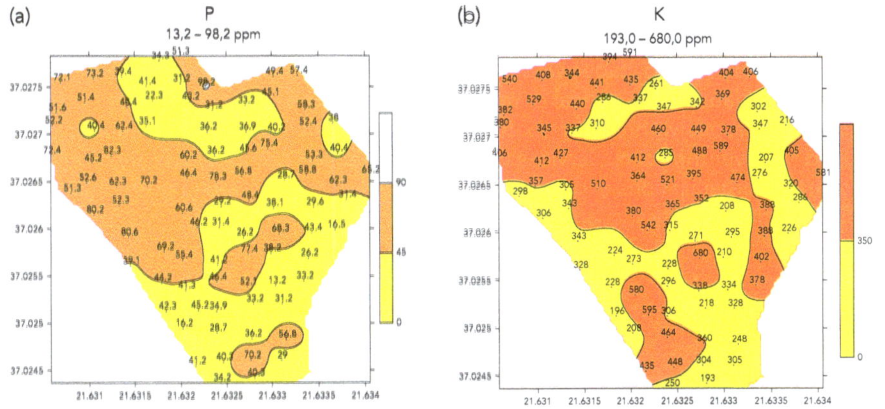

Figure 6 Delineated management zones for (a) P using a 45-ppm threshold; (b) K using a 350-ppm threshold (Fountas et al., 2011).

6 Summary

Delineation of MZs is crucial for maximizing the benefits of site-specific applications in terms of yield, quality and productivity. The spatial variability of important crop parameters is vast, and therefore if the farmer has the right tools and capabilities to divide a field into zones with the same or similar characteristics, a set of MZs could be created for each treatment, leading to spatial application of inputs according to the crop needs. This may improve crop production and certainly enhance land sustainability. MZs offer the application of targeted inputs avoiding over application of inputs, such as excess of N fertilization which causes leaching to aquifers or excess of irrigation water in sandy and sloping terrains which can cause soil erosion. There are several methods to delineate MZs, the selection of which depends on the nature of data sources, the inter-dependencies of the different data variables and agronomic knowledge of the field. The most commonly used methods consist of empirical methods using farmers' tacit knowledge, geostatistical methods, clustering techniques based on fuzzy logic and the degree of agreement method. There are no fixed methods to delineate MZs, but it depends on the application, and different MZs can be generated using data variables that best represent the specific application of inputs. However, there are some soil characteristics that seem to best characterize the fields to provide an overall classification of MZs, mainly EC_a and soil topography.

7 Where to look for further information

MZs are essential in site-specific application of fertilizers in crops. With the plethora of new sensors, both remote and proximal, the available data are becoming huge and there is an urgent need to analyse them. New tools and algorithms coming from the big data analytics domain should be employed to engage all new sensor data for producing more accurate MZs over a series of crop stages within a growing season or within years. In particular, the use of satellite data, either optical or radar, will provide very valuable insights both during the growing seasons with higher spatial resolution, as well as historical data to test the produced MZ over the years. Services that provide fast and accurate MZs are required and both research organizations and private companies will be involved in this domain for the benefit of both crop advisors and farmers.

8 References

Acevedo-Opazo, C., Tisseyre, B., Guillaume, S. and Ojeda, H. (2008), 'The potential of high spatial resolution information to define within-vineyard zones related to vine water status', *Precision Agriculture*, 9, 285–302.

Aggelopoulou, K., Castrignanò, A., Gemtos, T. A. and De Benedetto, D. (2013), 'Delineation of management zones in an apple orchard in Greece using a multivariate approach', *Computers and Electronics in Agriculture*, 90, 119–30.

Aggelopoulou, K. D., Pateras, D., Fountas, S., Gemtos, T. A. and Nanos, G. D. (2010), 'Soil spatial variability and site-specific fertilization maps in an apple orchard', *Precision Agriculture*, 12, 118–29.

Alsina, M. M., Cheng, T., Riaño, D., Whiting, M., Ustin, S. and Smart, D. R. (2013), 'Water status detection in California table grapes: From leaf to airborne', In: Stafford J. V. (Eds) *Precision Agriculture '13*, Wageningen Academic Publishers. https://doi.org/10.3920/978-90-8686-778-3_26.

Anastasiou, E., Balafoutis, A., Lentzou, D., Tsiropoulos, Z., Xanthopoulos, G. and Fountas, S. (2016), 'Validation of 5 spectral vegetation indices in estimating yield and quality parameters on table grapes using a side canopy sensor', In: *Proceedings of CIGR-AgEng Conference*, Aarhus, Denmark, June 2016, CIGR International Commission of Agricultural and Biosystems Engineering, Kyoto, Japan.

Arno, J., Martinez-Casasnovas, J. A., Ribes-Dasi, M. and Rosell, J. R. (2009), 'Review: Precision viticulture. Research topics, challenges and opportunities in site-specific vineyard management', *Spanish Journal of Agricultural Research*, 7, 779–90.

Basso B., Sartori, L., Cammarano, D., Grace, P. R., Fountas, S. and Sorensen, C. (2012), 'Environmental and economic evaluation of N fertilizer rates in a maize crop in Italy: A spatial and temporal analysis', *Biosystems Engineering*, 113, 103–11.

Best, S., León L. and Claret, M. (2005), 'Use of precision viticulture tools to optimize the harvest of high quality grapes', In: *Proceedings of the Fruits and Nuts and Vegetable Production Engineering TIC (Frutic05)*, (pp. 249–58), Cemagref, Montpellier, France.

Bourgeon, M. A., Gee, C., Debuison, S., Villette, S., Jones, G. and Paoli, J. N. (2017), 'On-the-go multispectral imaging system to characterize the development of vineyard foliage with quantitative and qualitative vegetation indices', *Precision Agriculture*, 18(3), 293–308.

Bramley, R. G. V. (2005), 'Understanding variability in winegrape production systems 2. Within vineyard variation in quality over several vintages', *Australian Journal of Grape and Wine Research*, 11, 33–45.

Bramley, R. G. V. (2010), 'Precision Viticulture: Mapping vineyard variability for improved quality outcomes', In: A. G. Reynolds (Ed.), *Managing Wine Quality, Vol 1: Viticulture and Wine Quality* (pp. 445–80), Cambridge, UK: Woodhead Publishing Limited.

Bramley, R. G. V., Trought, M. C. T. and Praat, J. P. (2011), 'Vineyard variability in Marlborough, New Zeland: Characterizing variation', *Australian Journal of Grape and Wine Research*, 17, 72–8.

Bullock, D. S., Ruffo, M. L., Bullock, D. G. and Bollero, G. A., 2009, 'The value of variable rate technology: An information-theoretical approach', *American Journal of Agricultural Economics*, 21, 209–23.

Casa R. and Castrignanò, A. (2008), 'Analysis of spatial relationships between soil and crop variables in a durum wheat field using a multivariate geostatistical approach', *European Journal of Agronomy*, 28, 331–42.

Castrignanò, A., Giugliarini, L., Risaliti, R. and Martinelli, N. (2000), 'Study of spatial rela-tionships among some soil physio-chemical properties of a field in Central Italyusing multivariate geostatistics', *Geoderma*, 97(1–2), 39–60.

Champa, W. A. H. (2015), 'Pre and postharvest practices for quality improvement of table grapes (Vitis vinifera L.)', *Journal of the National Science Foundation of Sri Lanka*, 43(1), 3–9.

Corwin, D. L. and Plant, R. E. (2005), 'Applications of apparent soil electrical conductivity in precision agriculture', *Computers and Electronics in Agriculture*, 46, 1–10.

Dobermann, A., Ping, J. L., Adamchuk, V. I., Simbahan, G. C. and Ferguson, R. B. (2003), 'Classification of crop yield variability in irrigated production fields', *Agronomy Journal*, 95, 1105–20.

El Nahry, A. H., Ali, R. R. and El Baroudy, A. A. (2011), 'An approach for precision farming under pivot irrigation system using remote sensing and GIS techniques', *Agricultural Water Management*, 98, 517–31.

Fountas, S., Aggelopoulou, K., Bouloulis, C., Nanos, G., Wulfsohn, D., Gemtos, T., et al. (2011), 'Site-specific management in olive tree plantation', *Precision Agriculture*, 12(2), 179–95.

Fridgen, J. J., Kitchen, N. R., Sudduth, K. A., Drummond, S. T., Wiebold, W. J. and Fraisse, C. W. (2004), 'Management Zone Analyst (MZA): Software for subfield management zone delineation', *Agronomy Journal*, 96, 100–8.

Guastaferro, F., Castrignano, A., De Benedetto, D., Sollitto, D., Troccoli, A. and Cafarelli, B. (2010), 'A comparison of different algorithms for the delineation of management zones', *Precision Agriculture*, 11, 600–20.

Haghverdi, A., Leib, B. G., Washington-Allen, R. A., Ayers, P. D. and Buschermohle, M. J. (2015), 'Perspectives on delineating management zones for variable rate irrigation', *Computers and Electronics in Agriculture*, 117, 154–67.

Hall, A., Lamb, D. W., Holzapfel, B. P. and Louis, J. P. (2011), 'Within-season temporal variation in correlations between vineyard canopy and winegrape composition and yield', *Precision Agriculture*, 12, 103–17.

Hansen, P. M. and Schjoerring, J. K. (2003), 'Reflectance measurement of canopy biomass and nitrogen status in wheat crops using normalized difference vegetation indices and partial least squares regression', *Remote Sensing of the Environment*, 86, 542–53.

Hartigan, J. A. (1975), '*Clustering Algorithms*', New York: Wiley

Johnson, L. F., Roczen, D. E., Youkhana, S. K., Nemani, R. R. and Bosch, D. F. (2003), 'Mapping vineyard leaf area with multispectral satellite imagery', *Computers and Electronics in Agriculture*, 38, 33–44.

Kitchen, N. R., Sudduth, K. A., Myers, D. B., Drummond, S. T. and Hong, S. Y. (2005), 'Delineating productivity zones on claypan soil fields apparent soil electrical conductivity', *Computers and Electronics in Agriculture*, 46, 285–308.

Koch, B., Khosla, R., Frasierb, W. M., Westfalla, D. G. and Inmana, D. (2004), 'Economic feasibility of variable-rate nitrogen application utilizing site-specific management ones', *Agronomy Journal*, 96, 1572–80.

Kyaw, T., Ferguson, R. B., Adamchuk, V. I., Marx, D. B., Tarkalson, D. D. and McCallister, D. L. (2008), 'Delineating site-specific management zones for pH-induced iron chlorosis', *Precision Agriculture*, 9, 71–84.

Lamb, D. W., Weedon, M. M. and Bramley, R. G. V. (2004), 'Using remote sensing to predict phenolics and colour at harvest in a Cabernet Sauvignon vineyard: Timing observations against vine phenology and optimising image resolution', *Australian Journal of Grape and Wine Research*, 10, 46–54.

Landrum, C., Castrignanò, A., Mueller, T., Zourarakis, D., Zhu, J. and De Benedetto, D. (2015), 'An approach for delineating homogeneous within-field zones using proximal sensing and multivariate geostatistics', *Agricultural Water Management*, 147, 144–53.

Lawrence, P. G., Rew, L. J. and Maxwell, B. D. (2015), 'A probabilistic Bayesian framework for progressively updating site-specific recommendations', *Precision Agriculture*, 16 (3), 275–96.

Lowrance, C., Fountas, S., Liakos, V. and Vellidis, G. (2016), 'EZZone – An online tool for delineating management zones', In: *Proceedings of the 13th International Conference on Precision Agriculture*. (unpaginated, online), Monticello, IL: International Society of Precision Agriculture.

Maante, M., Vool, E., Ratsep, R. and Karp, K. (2015), 'The effect of genotype on table grapes soluble solids content', *Agronomy Research*, 13, 141–7.

Mann, K. K., Schumann, A. W. and Obreza, T. A. (2010), 'Delineating productivity zones in a citrus grove using citrus production, tree growth and temporally stable soil data', *Precision Agriculture*, 12, 457–72.

McBratney, A. B., Odeh, I. O. A., Bishop, T. F. A., Dunbar, M. S. and Shatar, T. M. (2000), 'An overview of pedometric techniques for use in soil survey', *Geoderma*, 97, 293–327.

Molin, J. P. and Castro, C. N. (2008), 'Establishing management zones using soil electrical conductivity and other soil properties by the fuzzy clustering technique', *Scientia Agricola*, 65(6), 567–73.

Moral, F. J., Terrón, J. M. and Marques da Silva, J. R. (2010), 'Delineation of management zones using mobile measurements of soil apparent electrical conductivity and multivariate geostatistical techniques', *Soil and Tillage Research*, 106(2), 335–43.

Oldoni, H. and Bassoi, L. H. (2016), 'Delineation of irrigation management zones in a Quartzipsamment of the Brazilian semiarid region', *Pesquisa Agropecuária Brasileira*, 51, 1283–94.

Papageorgiou, E., Markinos, A. and Gemtos, T. (2009), 'Application of fuzzy cognitive maps for cotton yield management in precision farming', *Expert Systems with Applications*, 36, 12399–413.

Papageorgiou, E. I., Aggelopoulou, K., Gemtos, T. A. and Nanos, G. D. (2013), 'Yield prediction in apples using Fuzzy Cognitive Map learning approach', *Computers and Electronics in Agriculture*, 91, 19–29.

Peralta, N. R., Costa, J. L., Balzarini, M., Franco, M. C., Cordoba, M. and Bullock, D. (2015), 'Delineation of management zones to improve nitrogen management of wheat', *Computers and Electronics in Agriculture*, 110, 103–13.

Ping, J. L., Green, C. J., Bronson, K. F., Zartman, R. E. and Dobermann, A. (2005), 'Delineating potential management zones for cotton based on yields and soil properties', *Soil Science*, 170(5), 371–85.

Pringle, M. J., McBratney, A. B., Whelan, B. M. and Taylor, J. A. (2003), 'A preliminary approach to assessing the opportunity for site-specific crop management in a field, using yield monitor data', *Agricultural Systems*, 76, 273–92.

Raun, W. R., Solie, J. B., Johnson, G. V., Stone, M. L., Mullen, R. W., Freeman, K. W., et al. (2002), 'Improving nitrogen use efficiency in cereal grain production with optical sensing and variable rate application', *Agronomy Journal*, 94, 815–20.

Rolle, L., Torchio, F., Giacosa, S. and Segade, S. R. (2015), 'Berry density and size as factors related to the physicochemical characteristics of Muscat Hamburg table grapes (Vitis vinifera L.)', *Food Chemistry*, 173, 105–13.

Schepers, A. R., Shanahan, J. F., Liebig, M. A., Schepers, J. S., Johnson, S. H. and Luchiari, A. (2004), 'Appropriateness of management zones for characterizing spatial variability of soil properties and irrigated corn yields across years', *Agronomy Journal*, 96, 195–203.

Shaddad, S. M., Madrau, S., Castrignano, A. and Mouazen, A. M. (2016), 'Data fusion techniques for delineation of site-specific management zones in a field in UK', *Precision Agriculture*, 17(2), 200–17.

Shaner, D. L., Khosla, R., Brodahl, M. K., Buchleiter, G. W. and Farahani, H. J. (2008), 'How well does zone sampling based on soil electrical conductivity maps represent soil variability', *Agronomy Journal*, 100(5), 1472–80.

Strik, B. C. (2011), '*Growing Table Grapes. State University*', Corvallis, May, USA: Extension Service. http://smallfarms.oregonstate.edu/sites/default/files/publications/growing_table_grapes_ec1639_may_2011.pdf (Accessed 23 March 2017).

Tagarakis, A., Liakos, V., Fountas S., Koundouras, S. and Gemtos, T. (2013), 'Management zones delineation using fuzzy clustering techniques in grapevines', *Precision Agriculture*, 14(1), 18–39.

Tagarakis, A. C. and Ketterings, Q. M. (2017), 'In-season estimation of corn yield potential using proximal sensing', *Agronomy Journal*, 107, 1–8.

Tagarakis, A. C., Ketterings, Q. M., Lyons, S. and Godwin, G. (2017), 'Proximal sensing to estimate yield of brown midrib forage sorghum', *Agronomy Journal*, 109(1), 107–14.

Timmermann, C., Gerhards, R. and Kühbauch, W. (2003), 'The economic impact of site-specific weed control', *Precision Agriculture*, 4, 249–60.

Tisseyre B., Mazzoni, C. and Fonta, H. (2008), 'Within-field temporal stability of some parameters in viticulture: potential toward a site specific management', *International Journal of Wine and Vine Research*, 42(1), 27–39.

Tisseyre B., Ojeda, H. and Taylor, J. (2007), 'New technologies and methodologies for site-specific viticulture', *International Journal of Wine and Vine Research*, 41(2), 63–76.

Urretavizcaya I., Santesteban, L. G., Tisseyre, B., Guillaume, S., Miranda, C. and Royo, J. B. (2014), 'Oenological significance of vineyard management zones delineated using early grape sampling', *Precision Agriculture*, 15(1), 111–29.

Vasileiadis, V. P., Moonen A. C., Sattin M., Otto S., Pons, X., Kudsk, P., et al. (2013), 'Sustainability of European maize-based cropping systems: Economic, environmental and social assessment of current and proposed innovative IPM-based systems', *European Journal of Agronomy*, 48, 1–11.

Vellidis, G., Tucker, M., Perry, C., Kvien, C. and Bednarz, C. (2008), 'A real-time wireless smart sensor array for scheduling irrigation', *Computers and Electronics in Agriculture*, 61(1), 44–50.

Verdugo-Vasquez, N., Acevedo-Opazo, C., Valdes-Gomez, H., Araya-Alman, M., Ingram, B., Garcia de Cortazar-Atauri, I., et al. (2016), 'Spatial variability of phenology in two irrigated grapevine cultivar growing under semi-arid conditions', *Precision Agriculture*, 17(2), 218–45.

Yan, L., Zhou, S., Cifang, W., Hongyi, L. and Feng, L. (2007), 'Classification of management zones for precision farming in saline soil based on multi-data sources to characterize spatial variability of soil properties', *Transactions of the Chinese Society of Agricultural Engineering*, 23(8), 84–9.

Zaman, Q. and Schuman, A. W. (2006), 'Nutrient management zones for citrus based on variation in soil properties and tree performance', *Precision Agriculture*, 7, 45–63.

Zhang, X., Shi, L., Jia, X., Seielstad, G. and Helgason, C. (2010), 'Zone mapping application for precision-farming: A decision support tool for variable rate application', *Precision Agriculture*, 11, 103–14.

Chapter 6

Modelling and decision support systems in precision agriculture

Nicolas Tremblay, Agriculture and Agri-Food Canada, Canada

1 Introduction

Decision support systems (DSSs) are computerized systems, including models and databases, for decision-making purposes. DSSs help choosing the best alternative solution from the economic, social or environmental point of view. They are structured around the database, the model base, the user support base and their corresponding management systems (Fig. 1). Manos et al. (2004) gave an overview of DSSs and their contribution to decision-making in agriculture, including in farm planning and management. DSSs in agriculture were the subject of a large number of reports in the 1990s decade (Manos et al., 2004), before the precision agriculture (PA) concept really took off. DSS adoption in PA is low, mainly because of the differences in scientists' and farmers' contexts and ways of operating.

Agricultural production is a complex endeavour, particularly with the need to insure sustainability. For a time, not much attention was given to the reduction of the environmental pressure created by agricultural production. Short- and long-term profitability was the only objective to secure. This is no longer the case. The expectations for agricultural production from a diversity of stakeholders are high. The farmer is subjected to a lot of pressure from the environmental protection agencies, the consumer, the public and the bank (Fig. 2). Protection of the environment requires farming practices to be adjusted to field

http://dx.doi.org/10.19103/AS.2017.0032.07
Published by Burleigh Dodds Science Limited, 2019.

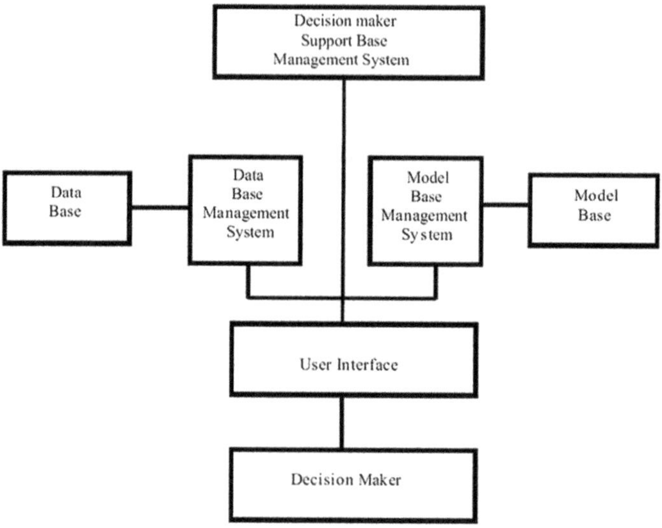

Figure 1 A typical decision support system. Source: Manos et al. (2004).

characteristics such as soil type, climate, cropping system or crop N status. The imposition on all farms of regulations or specifications concerning production methods cannot, in itself, guarantee clean agriculture (Meynard et al., 2002). According to Manos et al. (2004), 'the planning and the development processes in the agricultural sector constitute a multi-complex problem that is difficult to be solved, if it is not faced thoroughly. The interdependence between the various production activities on the one hand and between the services on the other hand is very high, making their planning difficult. The diversity of agricultural conditions may result in the same practices having different effects in different places (Meynard et al., 2002). To draw conclusions and take decisions in agriculture is a process that is not based on stable, simple, and well-defined rules, but on knowledge, information, experience and skills of the producers/decision makers'. It is therefore no wonder that one of the key challenges in the research-development-transfer-adoption continuum is to actually achieve the last steps: transfer and adoption.

In the typical 'linear' model, research results are derived from a large number of controlled experiments. In each experiment, the variation in response to a small number

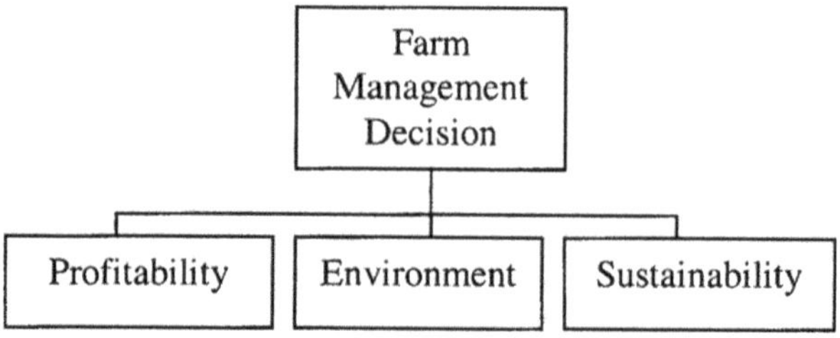

Figure 2 Decision-making structure for farm management decisions. Source: Jones and Barnes (2000).

of factors is observed, with all others being controlled. The outcome is homogeneous recommendation domains with standard recommendations suitable for large, relatively uniform areas (Cock et al., 2011). When the linear model is applied to agriculture, research data are analysed and interpreted and the extension agencies are charged with prescribing actions for farmers. In reality, however, the linear model rarely exists and few growers apply the recommendations without adapting them to their own particular circumstances.

In a short extension paper, Cook et al. (2013) summarized why scientific results from agricultural experimentations are not well translated into DSS, or transferred and really adopted by farmers. As compared to the typical industrial sector, farming is 'disorganized' because of uncontrolled variation due to many site- and time-specific effects. This uncontrolled variation was at the origin of the development, in the early twentieth century of statistical methods by Fisher and co-workers at Rothamsted Experimental Station, Harpenden, UK, to clarify experimental treatment effects in agricultural fields. Their work was so successful that the method started to define the problem: scientists began to place more value on information about the quantifiable effects of factors and discounted information about the farming environment in which they occurred. According to Cook et al. (2013), scientists and farmers deal differently with reduction of uncertainties (Fig. 3), the sources of which were described by Rowe (1994). For example, scientists are more concerned with metric uncertainty (the precision with which the outcome is stated) and farmers more with translational uncertainty (the degree to which issues around the decision are understood).

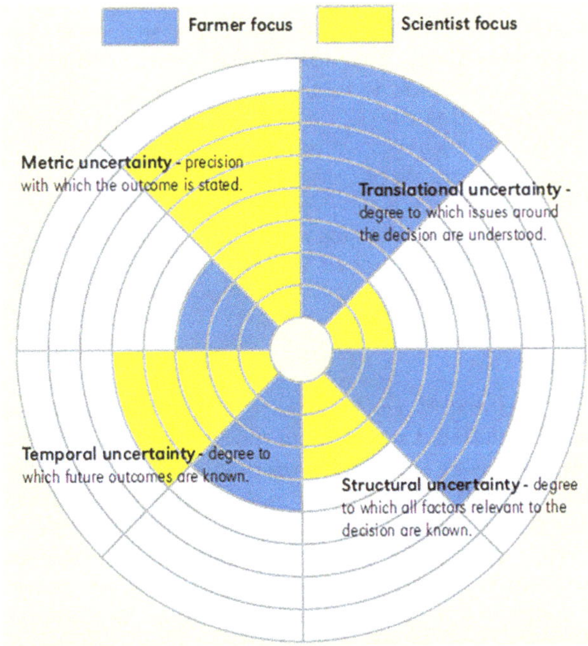

Figure 3 Farmers and scientists focus on different kinds of uncertainty. Source: Cook et al. (2013).

Most scientific publications obey strict rules about experimental design, statistical procedures and other technical requirements for paper acceptance. This difference between the world of 'scientific' factors and the more complex world of farm management means that many farmers find insights from formal experiments difficult to apply. In short, the restricted application of Fisher's statistical methodology has allowed for a simplified evaluation of papers to publish in scientific journals at the expense of the recognition of, and successful adaptation to, the particular conditions of the end users. Much research ends in peer-reviewed publications with no requirement for the scientist to demonstrate the actual use of the findings by the end user, even when the expressed goal is to find solutions to their problems. This is where modelling and validated DSSs can make the difference in bridging the gap between research and actual adoption by the final user, the farmer. By virtue of recent communication developments, most farmers in the world can now have access to key information, either through a smart phone or through sophisticated and expensive agricultural machinery. Mandel et al. (2012) described the concept and opportunities for smallholders of a cropping management system with marketing, production and crop protection features using geoinformatics and land, soil, water, meteorology and forecasting inputs.

2 Key issues

McBratney et al. (2005) estimated that the development of proper DSSs remains a major stumbling block to the adoption of PA decisions. Already, twenty years ago, Brisco et al. (1998) observed that PA systems were not well adapted to the needs of the users and were not easy to integrate into effective crop production DSSs. Although more user-friendly, the systems of today do not provide total solutions for the users. In essence, PA is 'a management strategy that uses a wide range of technology to gather and process data for the purpose of guiding targeted actions that improve the precision, efficiency and productivity of agricultural operations'. Historically, those targeted actions were focused on handling spatial variability but, increasingly, the consideration of variability in time has been included in the PA portfolio, both for crops and animal production. On top of spatial and temporal considerations, issues for DSSs in PA include how to deal with uncertainties as well as human/social behaviour. These issues are discussed below.

2.1 Soil (spatial)

Cock et al. (2011) stated that PA can be applied to whole fields, as there is a need for technology adapted to the field scale before moving to particular spots within a field. Historically, PA has been seen as a way to adapt interventions according to the spatial characteristics of fields (soils or crops). The vast majority of PA research has therefore focused exclusively on the management of spatial variability. In the past, the physico-chemical properties of soils within or among fields have normally been assessed by sampling. Soil sampling is still much compatible with PA goals, provided it is planned and executed by means of adequate spatial statistics and interpreted to derive spatially manageable information. Soil features can be assessed by sampling at known locations to provide a map of pH or nutrient status which, through calibration, can be matched to input requirements.

For the soil (spatial) information to be properly handled, it has to be brought down to useful parameters that are correctly quantified at a scale appropriate for decision-making purposes. In non-uniform fields, however, it is in many cases not possible to use interpolation between the mapped points because of limitations of the sampling resolution (Heege, 2013). It follows that the interpolation used to create the application map is misleading and therefore detrimental. Producers attempting to integrate soil information into the decision process of PA face a challenge since public soil survey maps provide only general descriptions of soil–landscape features. Remote sensing (RS) provides the data to convert point measurements into spatial information over time. Bobryk et al. (2016) used the USDA Soil Survey Geographic (SSURGO) soil maps to develop Environmental Response Units (ERUs) by integrating high-resolution elevation data and demonstrated that ERUs reduced corn yield variance. Still, it is generally acknowledged that traditional soil maps are not of a sufficient spatial resolution to accommodate PA applications. With the development of proximal or remote methods, the possibility exists to substitute sampling by sensing at a much greater spatial density. So dense, in fact, that interpolation and management zone production are no longer needed and each manageable area can be assessed and treated independently. There are two good examples of sensing for PA applications: soil electrical conductivity (SEC) and vegetation status assessed from proximal sensors or remotely sensed imagery. The rapid introduction and high turnover of the technology, however, has resulted in a gap between the technological capability to measure and apply variable rates of crop inputs, and the scientific understanding of the causal relationships between the inputs and the crop output (Brisco et al., 1998). Patzold et al. (2008) have examined approaches to extract pertinent soil-based information for site-specific crop protection purposes. Indeed, variable soil properties affect site-specific occurrence of weeds and pests as well as pesticide behaviour (particularly surface texture and soil organic carbon). According to them, the use of information on soil heterogeneity within the concept of precision crop protection (PCP) is beneficial, both economically and ecologically. Christensen et al. (2003) reported on an efficacy-based system to assist decision-makers in choosing herbicides or herbicide doses. A first generation of computerized weed management models comprised large databases with herbicide performance in different crops for a range of weed species at different growth stages. This information enabled ranking of herbicide efficacy and recommendations of the most efficient herbicides or herbicide doses to use against weed mixtures. A more recent group of models (population based) incorporate weed biology and ecology through simple deterministic models. Weed population distributions are known to be spatially aggregated in fields with a spatially varying mixture of weed species so that there is an opportunity for a decision algorithm including the spatial variation of the weed composition and density. Results of trials conducted for over five years in Denmark did not show any significant grain yield difference between decision algorithm for patch spraying (DAPS) and uniform application. However, the recommended herbicide doses were significantly lower when using DAPS in all years. At this stage in the research, the protocol did not rely on PA technologies to detect and characterize weed patches.

Roger-Estrade et al. (2009) reported that soil structure varies considerably within a field due to localized mechanical stresses such as compaction during traffic and soil fragmentation which depends on tillage depth. They state that there is a change in the topology of the soil structure with time because of several biological, chemical and physical soil property changes and that a better consideration of changes in soil structure with time is needed in crop models. In addition, nutrients such as NO_3-N are related to soil water so that they are subjected to transient changes and their spatial distribution can change considerably over time (Heege, 2013). Given the work needed to sample or assess nutrients like NO_3-N, this reduces the value for decision-making purposes of maps involving transient parameters for PA applications. Soil is only one component with the E term of the G × E × M model (Hatfield and Walthall, 2015) which describes the interactions between genetic potential (G), biophysical environment (E) and management (M) activities.

2.2 Seasonal weather (temporal)

Magri et al. (2005) concluded for precision management of crop inputs in corn that lime, P, K and organic matter showed, on the one hand, a strong or moderate spatial structure that was highly consistent over time. On the other hand, responses to hybrids and N fertilization rates were highly influenced by weather patterns so that there was little justification for their site-specific management. Farmers and advisers agree with the fact that their decision-making is impaired by seasonal variability (mostly rainfall) (McCown et al., 2009). Leaf wetness duration (LWD) is a key parameter for the prediction of fungal and bacterial crop diseases which is already included in a number of online weather products, despite the fact that it is highly spatially and temporally variable within a crop canopy. Pechenick and Rizzo (2008) have used two artificial neural networks (ANNs) to address this problem, one to accomplish spatial mapping and another to include the temporal property of the data. Prior sensitivity analyses revealed that two of the six inputs (net canopy radiation and leaf area index, LAI) did not significantly improve the predictions. The authors recommended improvements in local weather estimates of temperature, relative humidity and wind speed using more sophisticated downscaling interpolation methods and shorter prediction intervals to improve ANN forecasts and reduce error estimates. These variables are relatively easy to measure locally by growers. Tripathy et al. (2013) have mined wireless sensor data (temperature, humidity, leaf wetness, soil moisture) to extract relationships with pest and disease dynamics and develop early warning models (population buildup, initiation and severity). Sampurno et al. (2014) have developed a weed control DSS which relies heavily on the characterization of weather conditions for scheduling spraying and selecting the proper nozzle size. Those are examples of PA DSSs that do not include primarily spatial, but, rather, temporal information. Climate and weather variations become of paramount importance when management units are dispersed over larger distances, and over time differences, as opposed to within-field variation. In such cases, many more variables enter into the equation (Cock et al., 2011).

Among critical research issues to develop the PA concept to its full potential, McBratney et al. (2005) cited the insufficient consideration of temporal variation. For Whelan and McBratney (2000), 'much data remains to be gathered to test the philosophy of PA. What does appear definitive is that temporal variability in crop production indicators (i.e. yield) at the within-field scale is often larger in magnitude than spatial variability. This will increase the risk of economically and environmentally inappropriate actions if differential treatments are solely based on spatial information. Scientific analyses of treatments that consider both the spatial and temporal variability are required'. McBratney et al. (2005) later wrote: 'We have become very familiar with yield maps and analysing the spatial variation in them. We seem to have forgotten about temporal variation. A rule of thumb might say that, if we look at the variation of yield across a field and across years, half of the variation comes from year-to-year variation. Knowledge of this temporal aspect needs to be greatly increased. (...) We need to think of precision management as appropriate spatial AND temporal intervention'. Failure to properly address temporal variation in PA DSSs will undoubtedly reduce their usefulness and potential for adoption as a great part of the variation would be unaccounted for, reducing the value of the recommendations for the users. And, with climate change, the amplitude of weather variations from year to year is expected to increase in the future. Cock et al. (2011) suggested that 'operational research' is an opportunity to take into account the variation in both space and time of the production conditions as well as the variation in management practices. This opportunity will be treated in greater detail later in this chapter.

2.3 Uncertainties

Uncertainty is essentially the absence of information which may or may not be obtainable (Rowe, 1994). It is a ubiquitous component of the risk management business that agriculture constitutes and yet it is seldom incorporated in decision support tools. Rowe (1994) proposed four classes of uncertainty: temporal, structural, metrical and translational. The approach of a scientist is to provide the best estimate of risk and the range of uncertainty above and below the best estimate. It follows a tendency to measure that which is measurable (metric uncertainty) rather than that which is critical (structural uncertainty) but difficult to measure. Farmers have a completely different way of dealing with uncertainties than scientists. They are more concerned with structural and translational uncertainties than with metric uncertainty. This fundamental difference comes from the framework established by Fisher (1959, 1970) whose perspective was to apply mathematical probability to ascertain the likelihood that experimental observations are consistent with a hypothesis. Whelan and McBratney (2000) have commented on the consequences of Fisherian hypothesis testing in a PA context in these words: 'The levels of significance (against which the p-value statistic is compared) are usually chosen with little consideration given to the impact on the decisions and recommendations arising from the experiment if the null hypothesis (that the treatments have no effect that is given the large temporal variation evident in crop yield relative to the scale of a single field, then the optimal risk aversion strategy is uniform management) is rejected. Of particular importance to PA is the influence of large numbers of observations in a data set. Large numbers of observations provide high degrees of freedom. In such cases (e.g. yield sensor data), small differences may be regarded as significant at a certain probability level but could have little bearing on real-world management decisions'. To paraphrase them, framing the sources of uncertainties most critical to farmers and the adequate testing of hypotheses should be considered vital to PA because the adoption of its practices without reasonable testing may well lead to lower profitability and poorer environmental outcomes. If this should occur with a number of 'advocated' products or practices then it is easy to imagine that long-term harm may be inflicted on the concept of PA. Improvements to PA research can be made from data sets containing a rich diversity of well-documented soil and seasonal conditions *a contrario* not from demonstrating treatment effects for a limited set of fields and seasons.

In agriculture, uncertainty is due in major part to seasonal weather. Planning farming operations based on local climate conditions has been done for several centuries. Long-range weather forecasts in the *Farmer's Almanac* have been dated back to 1793 (Klemm and McPherson, 2017) and used to some extent by farmers since then. Klemm and McPherson (2017) provided examples of three-month outlook temperature probabilities and North American Multimodel Ensemble (NMME) forecasts in one- and three-month aggregates for total precipitation anomalies and farm management decisions related to different forecasts of lead times. Seasonal climate forecasts are issued for 13 lead times at one-month intervals, beginning with the season starting 0.5 months (two weeks) after the forecast issue date, through the same season one year ahead. Schneider and Garbrecht (2006) reported on the dependability and effectiveness of seasonal forecasts for agricultural applications in the continental United States for the (January-February-March) two-week forecasts over the period (1997–2005). Results depended on forecast variable (temperature or rainfall), direction of the forecast (wetter/drier, warmer/cooler), season and forecast lead time. Globally, forecasts for warmer conditions were better than those for drier conditions. The lack of utility for summer precipitation forecasts for the major agricultural areas of the Midwest and most of the Great Plains, and especially for the forecasts for drier than average conditions, was particularly disappointing to the authors. Also, they state that due to the probabilistic nature of the forecasts, their implementation in best management practices (BMPs) or use in agricultural DSSs is neither straightforward nor obvious. However, the influence of weather on the actual impact of a management decision is neither necessarily relying on forecasts nor spanning across an entire season. For instance, Tremblay et al. (2012) have shown that the maximal correlation between seasonal rainfall and the N rates effects on corn yield covers a period from 15 days prior to N application, to 30 days after. The existence of such a window for the consideration of weather information allows for the incorporation of cumulative data as well as forecast estimations. Also, the science of weather forecasts is constantly improving, particularly with the recent access to super computing power. Given the obvious importance of weather on the outcome of crop management decisions, this opportunity for the improvements of DSS should be given major consideration.

3 Human and social aspects

At the very core of all changes towards a more sustainable agriculture are the individual decision-makers as the decisions of each farmer will have positive or negative effects on sustainable intensification (Lindblom et al., 2016). Technologies that fit with the technology push strategy are characterized by non-complex issues, observable impacts, high compatibility and ease of implementation, while more complex technologies without immediately observable benefits and difficulties in testing and implementation do instead need co-development strategies in order to be adopted (Lindblom et al., 2016). DSSs have been seen as an opportunity to fill the hole left by the disappearance of rural extension services, as a bridge between the scientific theory of the researcher and the real-world practice of the farmer. Farmers confess that they know little about the resources that they know are the key to their enterprise (McCown et al., 2009). Yet, agricultural DSSs suffer from a credibility gap and have been poorly taken up worldwide, so researchers have to reconsider how they approach their work (Meynard et al., 2002; McCown et al., 2009; Mackrell et al., 2009). They often produce tools valuable in research which seemed to be of potential value to farmers in some yet-to-be-determined form (McCown et al., 2009). DSSs are sometimes seen by farmers as 'toys for scientists' (McCown et al., 2009) and rejected for reasons not necessarily anticipated by builders (Mackrell et al., 2009). Current agricultural DSSs available to farmers are not used to their full potential and are not adapted to the trade-offs and high complexity characterizing farmers' decision-making (Lindblom et al., 2016). There is a need to better account for the social and human issues that are determinant in the adoption or non-adoption of DSSs (the *problem of implementation*; Lindblom et al., 2016). Issues like the poor identification of clients and users, the lack of consideration for social and personal views, a perceived problem of complexity, lack of observability, level of knowledge of the users, lack of confidence, poor user interface design, tedious data input requirements, low adaptation to the farm situation, no frequent information update, lack of incentive to learn and adopt new practices and the fear of replacing advisors are important to consider ahead of time (Lindblom et al., 2016; Mackrell et al., 2009). Agricultural DSSs should be approached not as a technological innovation, but as a socio-technical innovation (Mackrell et al., 2009), as agricultural innovations and complex interactive processes of co-learning and negotiation, in which social learning practices are fostered (Lindblom et al., 2016).

4 Case studies with an emphasis on nitrogen management

The challenge of providing optimal recommendations for N rates in the corn crop is the object of considerable efforts with simulation models, particularly in the Corn Belt area of the United States. Many producers do not trust the general N recommendations currently made and apply more N than what would be needed on average, in order to capitalize on potential seasonal features that would be exceptionally favourable to N effect. The fact that soil properties are known to condition optimal N rate has triggered PA strategies in order to incorporate spatial adjustments. This whole domain constitutes the best example of attempts to achieve performant DSS in a PA context.

Indeed, the minimum rates of N that preserve crop productivity are difficult to predict, because they vary in space and with seasonal weather conditions. The current methods for determining the optimal N rates are not reliable, because they do not include all the important determinants in the proper priority order. When they are used, they do not repeatedly produce satisfactory results. Some are more laborious or require an amount of effort that discourages their adoption. When N is applied only at sowing, the weather conditions that prevail in the next few weeks can lead to efficiency losses. Some farmers do not use split application and, at sowing, apply all the N they anticipate will be needed, before the early season weather conditions have had their effect on N availability to the crop. To ensure a large crop, the producers prefer not to run the risk of a shortage of N. Research has shown that most farmers have the same kind of routine procedure, based on a mean rate of fertilizer application and several simple rules used for adjusting this rate as a function of cropping history and soil type and, in some cases, year (Meynard et al., 2002). This leads them to apply the N at higher than optimal rates. Nitrogen fertilizer use efficiency is therefore generally very low in agriculture (<50%). This low efficiency reduces the profitability of the businesses and leads to N loss to the environment (water contamination and greenhouse gas production). According to Ehm (2016), empirical studies with similar research questions and experimental protocols often yield substantially differing or even contradictory results, leading to resolute scepticism. It is even more of a problem in experimental crop research as it is acknowledged that if a field experiment is repeated, the results are likely to differ because important soil characteristics or, even more likely, seasonal conditions will differ from the initial experiment. That is why it is difficult to build confidence in the transferability of results obtained from geographically limited agronomic research, particularly when it comes to assessing optimal N rate for crops.

4.1 Benchmarking

Thorp et al. (2006) produced a family of curves that explain the chance of corn yield response to N rates in Iowa, given past weather conditions, which is a good example of benchmarking for helping with N management. Cock et al. (2011) suggested that a basic premise behind the use of commercial data to support site-specific agriculture is that the farmers themselves are the ones best placed to decide on the best practices under their particular circumstances. Thus, a prerequisite of the flow of information is that the growers themselves have access to the information, not in a prescriptive format but in one that allows them to interpret the information and make decisions on the most appropriate management approach for their particular circumstances. Rather than providing prescriptive solutions and recommendations derived from research results developed in the linear model, the focus is on providing growers and producers with information that they can use to make better decisions under their particular circumstances. A good example of this approach is the On-Farm Network maintained by the Iowa Soybean Association (ISA). According to the ISA website (http://www.iasoybeans.com/programs/isa-research/), this network collaborates with the other ISA research teams, university researchers, industry and government organizations on a wide variety of topics aligned with issues farmers care about, including pest and nutrient management, planting populations and cover crops among many others. These relationships allow the On-Farm Network to provide data as a trusted third-party source to enhance farmers' management decisions in an effort to improve profitability and environmental stewardship. The Analytics team works closely with the other two ISA research teams to analyse the collected data and turn it into information that can improve farmer decision-making. They also work with external partners to help improve the data farmers receive in the areas of image calibration, nitrogen efficiency (Kyveryga et al., 2013) and others. Cook et al. (2013) suggested that observations from farms are obviously most valid for where they were produced, but at the same time can be used by others under similar conditions, and may also prompt more conventional experiments. Observations of on-farm experimentation apart from their immediate and direct effect on farm management may encourage scientists to bridge the boundaries between 'formal' science and farming practice. The bridging of this gap and the direct use of the results of on-farm experimentation can reduce management uncertainties and allow farmers to make informed decisions (Cock et al., 2011).

4.2 Sensing

Proximal or remote sensing of canopies can be used to estimate crop removal of nutrients such as P at high spatial density and to replenish them in a site-specific manner (Heege, 2013). When dealing with underlying transient variables such as NO_3-N in the soil, which in turn affect plant growth and chlorophyll levels, proximal sensing offers the opportunity to sense and apply the corrective action on-the-go which is of potential interest. Samborski et al. (2009) have reviewed the strategies to make use of plant-sensor-based diagnostic information for nitrogen recommendations. Instruments like the handheld chlorophyll meter have been studied for a long time to make use of the known relationship between chlorophyll and N status, but they can hardly be used in a PA context because of their heavy sampling requirements. Samborski et al. (2009) have also summarized tractor-mounted systems such as the Yara N-Sensor/Fieldscan, the GreenSeeker or the Crop Circle which are basically radiometers relying on the characteristics of the light reflected from plant canopies. According to Heege (2013), crop sensing cannot be used directly for the control of the nitrogen application as it only provides information about the chlorophyll content of the leaves and the biomass or the LAI which somehow must be converted into fertilizer requirements. This conversion calls for detailed agronomic knowledge taking into account many factors influencing final crop yield, which, in itself, cannot be predicted precise enough at the time of fertilizing because it is affected by the weather in the growth stages that still lie ahead. Yet, to cover for this yield estimation at time of fertilizer applications, Lukina et al. (2001) have proposed an NDVI-based in-season estimate of yield (INSEY) to compute the potential N that could be removed in the grain. In-season N fertilization needs are then considered to be equal to the amount of predicted grain N uptake (potential yield times grain N) minus predicted early season plant N uptake (at the time of sensing), divided by an efficiency factor of 0.70. The relationship between INSEY and N application rate, however, was found by Roberts et al. (2013) as a poor predictor of yield response to N. They conclude that INSEY response to N cannot be used to accurately predict the economically optimal in-season top-dress N application rate. Obenauf et al. (2014) reported, over three years, mixed economic benefits using the Yara N-Sensor for variable N rate applications. In Sweden (Lindblom et al., 2016), farmers should not have any problems to grasp the usefulness of the Yara N-sensor. Nevertheless, many farmers do not use variable nitrogen application. The reasons could be higher costs, extra work, doubts of the credibility or cost benefit. In fact, not even in special crops, for example in potatoes, where varied fertilization has been shown to provide a direct added value due to better quality, the technology has been more broadly adapted so the uptake is disappointing. Samborski et al. (2016) reported inconsistent benefits of variable N application with active optical sensors in winter wheat. Boyer et al. (2011) reported that winter wheat yields from optical reflectance measurement-based treatments using a GreenSeeker were not statistically different from pre-determined uniform rates of N. Walsh et al. (2013) cited results from long-term research and on-farm trials, suggesting that farmer profits can be increased by more than \$4/ha in wheat production and \$8/ha in corn production systems by implementation of sensor-based nitrogen-rate-calculator-derived N fertilizer top-dress rates. Roberts et al. (2013) reported poor performances of optical reflectance measurements with a GreenSeeker handheld sensor for predicting N rate requirements and attributed them directly to the imprecise relationship between yield and optical reflectance responses to nitrogen.

4.3 N-balance-based DSS assisted with remote sensing

Farmstar is a commercial RS service using space-based RS and airborne images which includes nitrogen (N) fertilization management of wheat (Blondlot et al., 2005) and other crops. FarmStar proceeds from imagery normally coming from a diversity of satellite platforms (Mallavan, 2014). After the acquisition, a physical (orthorectification, positioning adjustments) and a biophysical (radiometric correction, atmospheric correction, reflectance extraction) treatment of the image is made. Then, GLCV (Green Leaf Cover Vegetation) is calculated which is claimed to be more robust to observation and solar angle than NDVI. LAI and chlorophyll estimations are made by model inversion such that the absolute data used for the agronomic calculation step are largely platform independent. This step derives biomass and population density from the LAI and N status from the chlorophyll estimation, which are then used to characterize N content in the canopy, lodging risk, yield potential and N fertilization or growth regulators needed. Validation trials conducted with FarmStar from 2008 to 2010 inclusive (Mallavan, 2014) have shown median yield increases for winter wheat (7%, with 5% more N), winter barley (4%, with same N) and colza (4%, with 2% less N). Indeed, according to Lagarde and Champolivier (2006), colza farmers in France who use FarmStar to determine their N fertilization tend to reduce the rates. Fresh biomass estimated by satellite at the exit of winter becomes an input in a DSS considering the region in France, soil type, organic manure history, average yield, biomass weight before and after (which the imagery provides) winter, and N applied at first dressing and previous crop.

4.4 Simulation models

A model is a representation of a real system and crop simulation is the implementation of a mathematical model to study the relationship between various interlinked components of the soil–plant–atmosphere continuum (Anbumozhi et al., 2003). Empirical models are used to make predictions without necessarily improving the understanding of why the results were obtained. Mechanistic models proceed from relationships explaining phenomenon. Most crop simulation models contain a mixture of empirical and mechanistic models. According to Manos et al. (2004), simulation models hold a major position in the DSS for agricultural purposes. They potentially provide a means of transferring research applications to farmers and policy makers. Anbumozhi et al. (2003) separated the history of crop simulation models into three periods of increasing complexity. The last one started in the 1990s and is still used today as simulation models aim at prediction and universality in their applications, for which mechanistic models are more conducive. According to these authors, it is partly true that the existing complex models are not well applied or sometimes fail to make good predictions because there are still gaps in basic science knowledge so that the first universal crop simulation model is yet to be developed. According to Bellocchi et al. (2010), systems with small timescale, spatial extension and complexity can be isolated; thus, models of these systems are more readily accessible to validation. For systems with increasing time and spatial scales with increasing complexity, validation of system models requires increasingly more efforts so that, at some point, the effort required to gather validation data may no longer be feasible (Fig. 4).

According to Meynard et al. (2002), parameter estimation remains a difficult task for dynamic models, due to the large number of parameters involved. In all cases, the quality of parameter estimates depends heavily on the quality of the database. The cost of acquiring the data required for parameter estimation is an important criterion that should be taken into account in the conception of models for use in decision support. Model-related uncertainty can potentially be reduced by assimilating RS data during the growing season (Fig. 5, Jones and Barnes, 2000; Ines et al., 2013; Fang et al., 2013). Models have demonstrated usefulness in evaluating the changes that will occur in the implementation of new agricultural policies and to balance their economic and environmental benefits (Fassio et al., 2005; Thorp et al., 2006; Dal Ferro et al., 2016). Carberry et al. (2009) reported four applications for Australia with the APSIM model (McCown et al., 1996): 1) benchmarking; 2) tactical planning; 3) yield forecasting and 4) scenario exploration. For them, scientifically rigorous demonstration of models' capabilities for accurate simulation is not the key to changing the attitude of those farmers who consider models to be merely 'toys for scientists'. In practice, 'goodness' of a model is judged according to experience of its performance in aiding 'thinking about' a particular management situation and what to do.

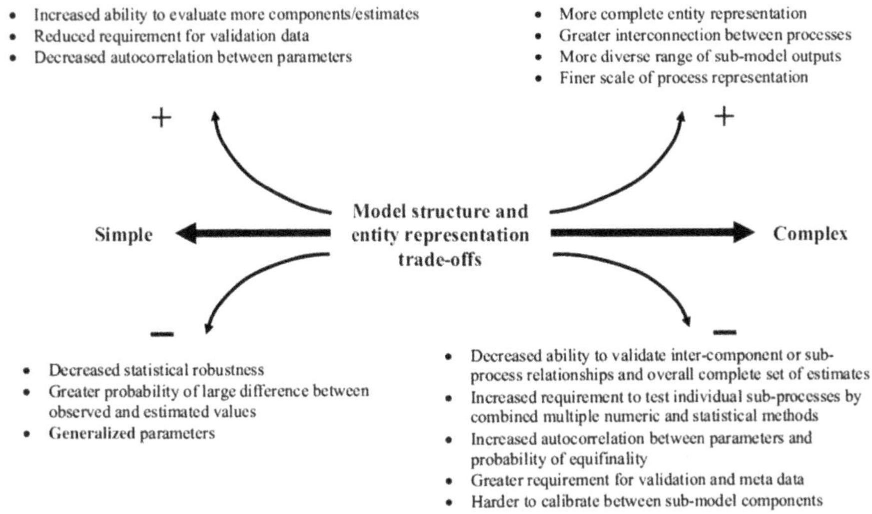

Figure 4 Trade-off between model complexity and validation. Source: Bellocchi et al. (2010).

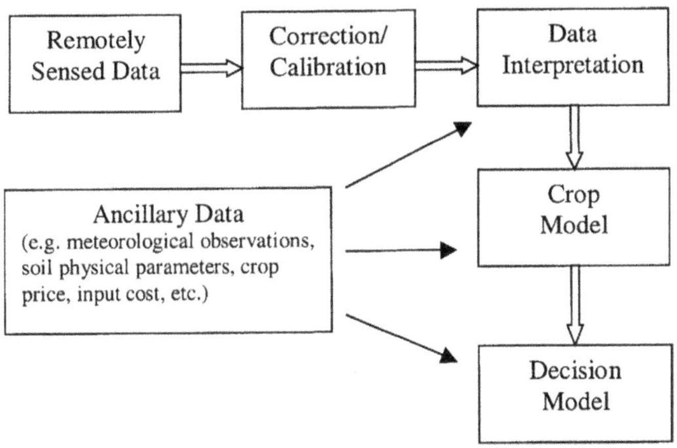

Figure 5 Decision-making framework including remotely sensed data. Source: Jones and Barnes (2000).

According to Setiyono et al. (2011), some existing crop simulation models can be used for post-season evaluation of nutrient limitations in a maize crop but they were not designed to support preplant or in-season decisions about fertilizer N management. It is for that purpose, that Maize-N (Setiyono et al., 2011) has been developed from the Hybrid-Maize model and a database with yield vs N uptake relationships from the United States, Indonesia, Vietnam, Brazil, Ecuador and China. The validation was executed from published literature originating from central and western Nebraska and eastern South Dakota. Setiyono et al. (2011) reported economically optimum N rate (EONR) estimations with greater accuracy than the other N recommendation approaches. In their study, however, it appears that it is the exact seasonal characteristics of the whole seasons that were used for the simulations, which will not be possible if the model is to be used for real-time N management decisions. Despite this lack of validation in a real farmer's context, the value of Maize-N for improving EONR estimations appears promising.

Adapt-N is based on the Precision Nitrogen Management model, which is an integrated combination of the LEACHN biogeochemistry model and a maize-N uptake, growth and yield model (Sela et al., 2016). Adapt-N is one of the few DSSs with performance openly and fully documented in peer-reviewed publications. It aims at recommending in-season N dressing for the corn crop in the United States. On average, by accounting for variable site and weather conditions, both increased grower profits and decreased environmental losses were reported from the validation of Adapt-N (Sela et al., 2016). However, it is apparent from the partial profit analysis of the 113 trials used to validate Adapt-N that there are still cases where it generated losses instead of profits and that the proportion of these failures varies among years and/or locations. Indeed, trials conducted with Adapt-N in 2014, the latest year of trials reported, were successful in New York state (where it has been developed), but not in Iowa. Yet, overall and for Iowa included, this DSS has so far been documented as highly profitable and capable of decreasing reactive N losses in water and the air. It is likely that a determinant cause of this success is the dynamic access to gridded high-resolution (4 by 4 km) weather data (precipitation, maximum and minimum temperature and solar radiation), which allows for field-specific and timely adjustments. Reports outside the peer-reviewed literature have described the ability of Adapt-N to produce within-field 'variable rate' recommendations (Fig. 6). Even if it were not for this spatial resolution feature, Adapt-N would qualify as a PA DSS as it is one of the most convincing examples of a 'management strategy that uses information technology to bring data from multiple sources to bear on decisions associated with crop production' (NRC, 1997). The achievement is even more commendable as Adapt-N has been licensed and commercialized in 2014 and is actually available as an online tool.

4.5 Artificial intelligence

Artificial intelligence tools, such as expert systems, are used as much as simulation models in driving DSS for agricultural purposes (Manos et al., 2004). In an expert system, the knowledge base concerns the space of a concrete problem and represents the knowledge and the logic that are used in order to reach a conclusion. The inference engine is the part undertaking the process of searching the knowledge base for decision-making (Manos et al., 2004).

The Soil, Crop, Atmosphere for N (SCAN) DSS (Tremblay et al., 2016) makes in-season N rate recommendations by means of a fuzzy inference system (FIS) that takes primarily into account soil surface texture, the 'abundant and well-distributed rainfall' (AWDR) parameter, and an optional indicator of the N sufficiency status (Nitrogen Sufficiency Index, NSI) of the crop. SCAN is transferred to users in the form of a WebMapper application. SCAN recommendations in commercial field trials over three years in Quebec (2013–15) resulted in average benefits ranging from \$25 to \$49/ha depending on the year. SCAN reduced nitrogen rates in grain corn for half of the commercial field trials by an average of 25%, without affecting yield. These reductions translated into equivalent fertilizer cost savings. Nitrogen losses to the environment are also reduced, leading to lower risks of aquifer contamination. Moreover, the use of SCAN resulted in increased yields for the other half of the commercial field trials when a higher nitrogen rate was justified by climatic and soil

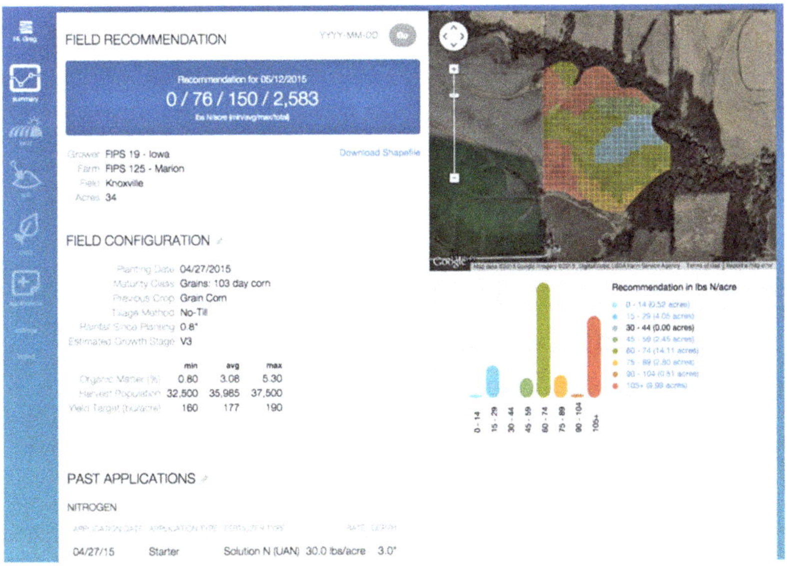

Figure 6 Example of Adapt-N VRT recommendation. Source: http://www.adapt-n.com/#section-field-modes (Date last accessed: 10 November 2017).

conditions. With a residual mean square error (RMSE) of 38 kg N/ha, SCAN has been shown to be close to Cornell's Adapt-N-DSS (RMSE = 32 kg N/ha) in predicting the economically optimal N rate (EONR), while other DSSs had poorer performance (Tremblay et al., 2016).

One of the original characteristics of SCAN is that its decision rules are the results of meta-analyses. A meta-analysis is based on the results of multiple independent studies, in order to synthesize the evidence from many possible sources in a formal quantitative manner. It is an opportunity to draw more applicable conclusions on a topic that has generated research at several locations and over a certain time. Meta-analyses are considered to be a way to avoid premature acceptance of facts leading to the application of ineffective or harmful treatments. The key is to group studies to examine the context in which a product, a rate or a management practice produces the desired effect. Indeed, in larger surveys, some variation of study-specific features is acceptable and even desirable for probing a reasonably broad range of conditions (Ehm, 2016). Meta-analyses developed originally in the social sciences in the 1970s and are now being used within many scientific disciplines, particularly in medical research. Tens of thousands of journal articles have been published on meta-analysis. Such an approach fits the complexity encountered in the social and life sciences better than the requirement of strict replicability (Ehm, 2016). After a slow start, agricultural scientists are now applying meta-analytic methods to a range of problems in order to benefit from its strengths: good predictive ability and high statistical power. As far as DSS development in agriculture is concerned, meta-analysis results are basically quantitative relationships established for different contexts of applications. In agriculture, these contexts are often tied up with soil or seasonal conditions which are known to impact the effect of management decisions. Meta-analyses can also come from the quantitative outcome of systematic literature reviews. In such a case, their main limitation is the 'file-drawer problem', or selective reporting practices which contribute to a severe overrepresentation of 'positive' results that do not survive the test of time (Ehm, 2016). Meta-analyses can also come from the analysis of original data sets collected from contributors, a situation which greatly reduces the risk of bias. This is the case for Tremblay et al. (2012) who have published the results of a meta-analysis, later enriched from hundreds of data sets assembled from North American locations that finally led to the SCAN DSS for nitrogen fertilization in corn. SCAN is the first known example in agriculture of artificial intelligence modelling based on the results of the meta-analysis.

5 Research options for decision support systems to improve productivity in a precision agriculture framework

5.1 Open science, open data and metadata

Tyrychtr et al. (2015) suggested that a knowledge base (in the form of rules) may not be applicable in full at present (due to changes of technology, growing practices, etc.) and can be more useful when coupled to knowledge gained from available open data documented in terms of time, place and weather changes, for example.

McNamara et al. (2016) suggested for livestock research that, in the past, limitations of knowledge, funding, animals, statistical methodology, and more importantly, scientific approach resulted in many smaller studies with narrow foci, the global applicability of which was often undefinable. Field experiments to evaluate management practices are costly and time and labour demanding. They are generally useful for a very limited set of management alternatives available to farmers (Anbumozhi et al., 2003). McNamara et al. (2016) stated: 'These experiments can often be aggregated to yield a more comprehensive data set, provided the needed array of inputs required by more complicated models has been reported for all of the experiments, which is often not the case. (...) The concurrent improved understanding of statistical strategies and analytical programs has facilitated design, implementation, and interpretation of more complex, multivariate experiments. The advantage of these larger experiments is that the data are typically distributed throughout the observed range, yielding complete response surfaces within an experiment as opposed to inferring a response surface from multiple independent experiments.' For Cock et al. (2011), in order to rigorously analyse the response to multiple factors and their interactions, large data sets are needed in which each individual cropping event is characterized in terms of the environment and management so that obtaining data sets from a large number of well-characterized events is the key.

The interpretation of most results in crop management trials would conclude with the significance of experimental treatments at a given probability level. Quite often, the treatment effects would be found to be dependent on site year. That is, the treatments would generate differences in yield, for example, but only in some locations or for some seasons. Such findings are good enough for scientific publication purposes, pending on novelty and the respect of methodological criteria. How much this new information is directly applicable to the intended final users is doubtful because of determinant differences in production contexts that are affecting the expression of treatment effects. In agriculture, these contexts can be summarized as G × E × M, that is the Genetics × Environment × Management interaction (Hatfield and Walthall, 2015). A parallel can be drawn with the research conducted in the animal science sector: for McNamara et al. (2016), an area of opportunity with most models involves the addition of parameters to reflect health status or environmental stress, provided that the related data be consistently reported in the literature. To quote them: 'When assessing studies individually, it is difficult to ascertain how these experimental differences can bias results; however, when assessed from a meta-analytic standpoint, these exogenous sources of variation can be identified and their relationship with the response variables can be quantified'. The consequence here is that, for PA to be successful, the effect of treatments has to be known for many different contexts of uses. In other words, is treatment effect dependent on cultivars (or hybrids), environment (soil features, weather characteristics) or management options (previous crop, tillage, ...)? This prior step to successful PA adoption is dependent on a proper handling of contextual production information and their expected effects on treatment effects outcomes. In a PA framework, the handling of the environment component of the G × E × M contextual interaction is typically reduced to soil (spatial) and seasonal weather (temporal) characteristics.

Pechenick and Rizzo (2008) stressed the importance of validating model predictions with observations over a range of crop types and climates using standard protocols. Otherwise barriers to entry into disease management strategies will be faced. Indeed, as the power of assembling data sets from different groups becomes more and more obvious, the lack of uniformity in protocols, methodologies, control treatments and reporting procedures appear particularly frustrating. McNamara et al. (2016) found that inconsistent and incomplete reporting of data, in the current body of animal science literature, frequently prohibits the efficient advance of modelling research efforts. They reviewed the status of experimentation and data reporting, developed a comprehensive checklist of data that should be included in publications and proposed four key opportunities to advance animal nutrition knowledge. Contrary to crop science, they reported that models are perceived to be 'only for applied science' as they have primarily focused on the whole-animal or farm level for practical application, instead of the molecular, cellular and organismal biology (basic biological science). They advocated the need to break the silos where science is being conducted (nutrition vs genetics vs reproduction, for example) and to encourage scientists to conduct research across disciplines and among various organizations, and summarized by saying that 'the cow does it all at the same time'! Indeed, several competencies, approaches and scientific disciplines need to act in concert to help develop a sustainable agriculture, since the overall trend in agriculture is towards a more complex, technology based crop production with increasing regulation and supervision regarding the use of fertilizers, pesticides and other chemicals (Lindblom et al., 2016). Prospective improvements in efficiency of animal agriculture from the use of more advanced models are promising but to fully realize these potential benefits, a larger quantity and more complete data are needed. For instance, a detailed description of environmental conditions useful for meta-analysis or modelling purposes is often lacking in the published papers. This failure often occurs because this level of data reporting is deemed extraneous to the objectives of the publication and may be removed by authors, reviewers or editors (McNamara et al., 2016). Thus, the consistency and completeness of data reporting and sharing must be improved and descriptions of the experimental conditions as complete as possible should be included in all publications or made available upon request. A small investment in measurement and reporting will enhance the value of the work by extending the impact of the work beyond the stated experimental objectives to inclusion in a database that could be used for future model development (McNamara et al., 2016). These authors provide examples of better ways to report experimental conditions in the animal science sector than the ones typically found in the literature.

5.2 Operational research

Cock et al. (2011) observed that every time farmers produce a crop, they face a plethora of changing conditions to which they react by adjusting their practices. Observation of the crop response to this variation could potentially provide valuable insights into the crop response to variations in management and growing conditions. Currently farmers observe the response of their crops to these conditions, but have little possibility of interpreting their observations as they only have a narrow range of experiences to analyse and rarely have sufficiently complete data sets to accurately define or characterize the growing conditions and management associated with a particular cropping event. Hence, coupled with random effects, each individual farmer (or researcher) observes a limited range of variation which makes it difficult to make sense of their observations. They further suggested that when variation is observed for management units dispersed over larger distances, and over time differences, as opposed to within-field variation, many more variables enter into the equation, with climate and weather variation becoming of paramount importance. Cock et al. (2011) described operational research as a system by which: (a) to obtain data from a series of cropping events that characterize the conditions under which each crop is grown, how it is managed and how it performs under commercial conditions (data capture), (b) to manage and analyse the data in centralized databases (data management and analysis) and (c) to make the information derived from the data analysis available to growers so that they can use it to make better informed decisions (interpretation) (Fig. 7).

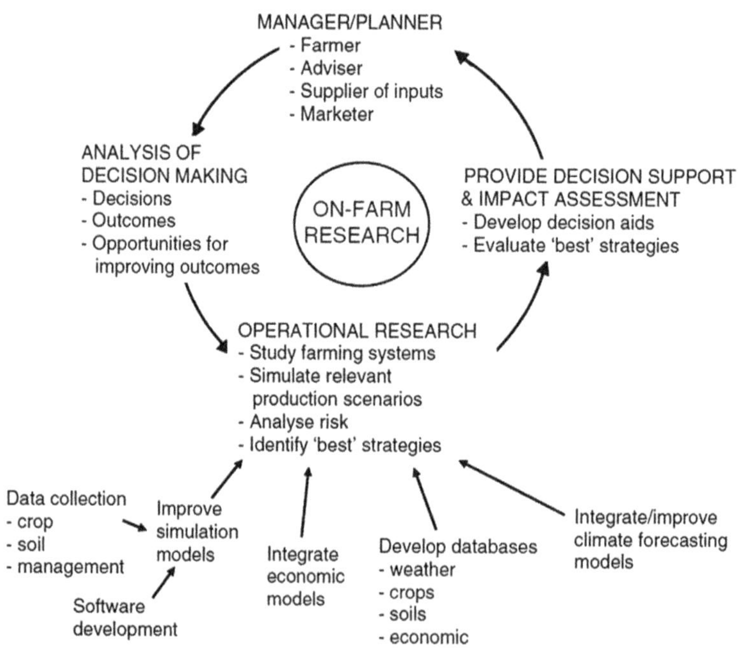

Figure 7 A framework for using models in farming systems research involving operational research.

An example of operational research in action is the Agricultural Production Systems Research Unit (APSRU), formed in 1991 by what was then the Queensland Department of Primary Industries and the CSIRO Division of Tropical Crops and Pastures (from McCown et al., 2009).

6 Conclusion

All aspects of the methodology depend on the social organization of the growers and the supply chain of which they form a part, and hence social organization is an integral part of the methodology. They also suggest that the operational research approach is particularly effective in combination with PA technologies and in heterogeneous landscapes with crops that have not been the subject of intense research. Indeed, PA is an opportunity to record and characterize a sufficiently large number of these unique events, or experiments, and to use that information to describe and predict the crop response to variations in the environment and management. Operational research is effective in determining the crop response to variables that are not readily studied in small plots and in determining optimal combinations of multiple variables. According to Cock et al. (2011), 'the methodologies and analysis required for the approach of using commercial information differ from those used in traditional agricultural research. Four major features stand out. First methodologies are required to characterize the spatial and temporal variation in factors which will affect productivity and to assess the spatial and temporal variation of productivity (including quality) down to at least the field or management unit level. This is not just a problem of physical collection of data, but also one of establishing social structures with producers and others contributing information and knowledge to the overall system in a standardized format. Secondly the data must be collected and stored in databases that are maintained over time. Thirdly novel analytical techniques and means of presentation may be needed to make sense of the mass of information collected. Finally mechanisms are needed to ensure that producers can access, in an interactive manner, the information that has been compiled in such a manner that they can ask pertinent questions about their particular lots or sites.' Producers believe in the results obtained as there are none of the problems of scaling up from experimental plots to commercial conditions. There are numerous different ways of analysing the information of operational research coupled to PA technologies in terms of crop response to the multiple variables associated with each management unit.

7 Future trends

7.1 Artificial intelligence

Data mining aims at extracting useful information from large data sets. It was considered by Mucherino et al. (2009) as an emerging research field in agriculture. It is common for commercial and proprietary data mining applications to use ANNs as more training data increase the value of classifications and predictions (Pechenick and Rizzo, 2008). But Mucherino et al. (2009) suggested that *k*-means, *k* nearest neighbour and support vector machines, on top of ANNs, were of particular interest for agricultural applications. They differ mainly in their use of a training set. *k*-means, for instance, classify new observations in one of *k* groups and *k* has to be known *a priori*. Their paper presents a good introduction to the different principles behind each method and their limitations. In the case of ANNs, for example, the authors warn about the risk of overfitting, which tends to have the network performing badly when trying to generalize using unknown entries. They give examples of data mining for the study of weather conditions and forecasts, soils and crop classifications, the detection of weed and nitrogen stress in corn and the approximation, based on per cent clay, silt, sand and organic carbon, of soil parameters needed for simulation models of soil dynamics. They conclude that mathematicians and computer scientists can help agronomists in finding solutions to their complex problems. Behmann et al. (2015) have reviewed advanced machine learning methods for the detection of biotic stress in PCP from sensor data. They suggested that traditional methods of data analysis (linear regression, linear discriminant analysis) perform better with predefined distributions and model assumptions. Advanced methods of machine learning require less prior information, are tolerant to complex data characteristics (non-linearity, non-Gaussian noise and outliers) and are applicable to a wider range of tasks normally found in PA data with high spectral, spatial and temporal resolutions. Figure 8 illustrates how support vector regression is superior to more traditional regression functions. Machine learning benefits from new information automatically. It allows the detection of patterns and the extraction of important information from raw data even if the underlying data model is unknown. The model reveals task-relevant patterns without relying on unjustifiable assumptions. While Behman et al. (2015) focused on *k*-means clustering, neural networks and support vector machines, they gave a diversity of examples of applications to PCP (detection of weeds, diseases or insects). They suggested that with weak information on data distribution or model structure, discriminative models like support vector machine (SVM) and Random Forest should be applied.

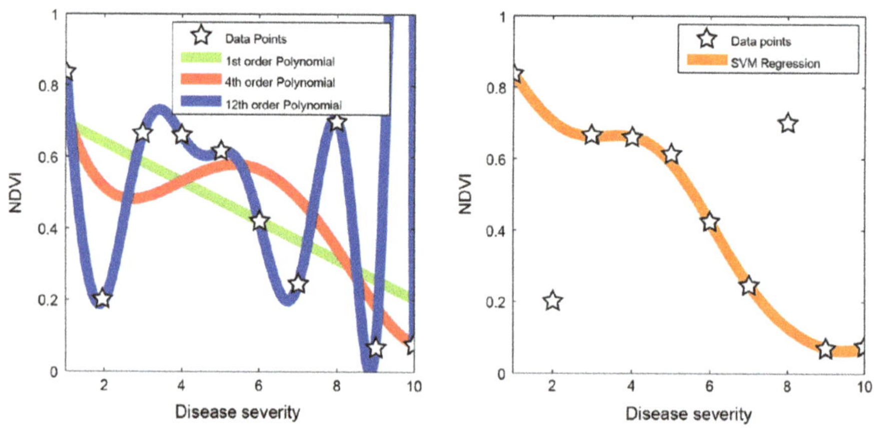

Figure 8 Support vector regression in a regression scenario involving the NDVI Vegetation Index and the disease severity, compared to polynomial models of fixed degree and least square optimization. Source: Behmann et al. (2015).

8 Where to look for further information

Several papers from the group led by Prof. McBratney (e.g. McBratney et al., 1997, 2005; Whelan and McBratney, 2000) in Australia have provided, at the time of their publication, an assessment of the status and a vision of critical issues facing successful application precision solutions to the agricultural sector. This Australian group managed to share co-authorship with scientists in Colombia (Pringle et al., 2004a,b) who later published an interesting extension paper (Cook et al., 2013) based on the work of Rowe (1994). In Australia, the human side of DSS adoption has also been addressed (McCown et al., 2009; Mackrell et al., 2009). For information on innovative statistical approaches and on-farm research applications, Dr. Peter Kyveryga's publications are well worth a read (Kyveryga et al., 2007a,b, 2011; Kyveryga and Blackmer, 2012). The book of Oliver (2010) is an essential read on the topic of geostatistical applications for PA.

The International Society of Precision Agriculture (ISPA; www.ispag.org) is a scientific society with a long history of involvement in PA. Its International Conferences on Precision Agriculture (ICPA) are held every second year and scientists dealing with PA assemble from all over the world. Complementary to the ICPA are the European Conferences on Precision Agriculture (ECPA) and the Asian Conferences on Precision Agriculture (ACPA), held on alternate years with ICPA. Regarding research data access and sharing, the Research Data Alliance (https://www.rd-alliance.org/) is the most prominent group of players with their plenaries, working groups and interest groups.

9 References

Anbumozhi, V., Reddy, V. R., Lu, Y. C. and Yamaji, E. 2003. The role of crop simulation models in agricultural research and development: A review. *International Agricultural Engineering Journal*, 12, 1–18.

Behmann, J., Mahlein, A.-K., Rumpf, T., Römer, C. and Plümer, L. 2015. A review of advanced machine learning methods for the detection of biotic stress in precision crop protection. *Precision Agriculture*, 16, 239–60.

Bellocchi, G., Rivington, M., Donatelli, M. and Matthews, K. 2010. Validation of biophysical models: Issues and methodologies: A review. *Agronomy for Sustainable Development*, 30, 109–30.

Blondlot, A., Gate, P. and Poilvé, H. 2005. Providing operational nitrogen recommendations to farmers using satellite imagery. In: Stafford, J. V. (Ed.), *Proceedings of the 4th European Conference on Precision Agriculture, Precision Agriculture '05*. the Netherlands: Wageningen Academic Publishers, pp. 345–52.

Bobryk, C. W., Myers, D. B., Kitchen, N. R., Shanahan, J. F., Sudduth, K. A., Drummond, S. T., Gunzenhauser, B. and Gomez Raboteaux, N. N. 2016. Validating a digital soil map with corn yield data for precision agriculture decision support. *Agronomy Journal*, 108, 957–65.

Boyer, C. N., Wade Brorsen, B., Solie, J. B. and Raun, W. R. 2011. Profitability of variable rate nitrogen application in wheat production. *Precision Agriculture*, 12, 473–87.

Brisco, B., Brown, R. J., Hirose, T., McNairn, H. and Staenz, K. 1998. Precision agriculture and the role of remote sensing: A review. *Canadian Journal of Remote Sensing*, 24, 315–27.

Carberry, P. S., Hochman, Z., Hunt, J. R., Dalgliesh, N. P., McCown, R. L., Whish, J. P. M., Robertson, M. J., Foale, M. A., Poulton, P. L. and van Rees, H. 2009. Re-inventing model-based decision support with Australian dryland farmers. 3. Relevance of APSIM to commercial crops. *Crop and Pasture Science*, 60, 1044–56. doi:10.1071/CP09052.

Christensen, S., Heisel, T., Walter, A. M. and Graglia, E. 2003. A decision algorithm for patch spraying. *Weed Research*, 43, 276–84.

Cock, J., Oberthür, T., Isaacs, C., Läderach, P. R., Palma, A., Carbonell, J., Victoria, J., Watts, G., Amaya, A., Collet, L., Lema, G. and Anderson, E.2011. Crop management based on field observations: Case studies in sugarcane and coffee. *Agricultural Systems*, 104, 755–69.

Cook, S. E., Cock, J., Oberthür, T. and Fisher, M. 2013. On-farm experimentation. *Better Crops With Plant Food*, 97, 17–20.

Dal Ferro, N., Cocco, E., Lazzaro, B., Berti, A. and Morari, F. 2016. Assessing the role of agri-environmental measures to enhance the environment in the Veneto Region, Italy, with a model-based approach. *Agriculture, Ecosystems and Environment*, 232, 312–25.

Ehm, W. 2016. Reproducibility from the perspective of meta-analysis. In: Atmanspacher, H. and Maasen, S. (Eds), *Reproducibility: Principles, Problems, Practices, and Prospects*. Hoboken, NJ: John Wiley & Sons, Inc., pp. 141–67. ISBN: 9781118865064.

Fang, H., Liang, S. and Hoogenboom, G. 2013. Assimilation of remote sensing data and crop simulation models for agricultural study: Recent advances and future directions. In: Liang, S., Li, X. and Xie, X. (Eds), *Land Surface Observation, Modeling and Data Assimilation*. 5 Toh Tuck Link, Singapore 596224: World Scientific, pp. 405–40.

Fassio, A., Giupponi, C., Hiederer, R. and Simota, C. 2005. A decision support tool for simulating the effects of alternative policies affecting water resources: An application at the European scale. *Journal of Hydrology*, 304, 462–76.

Fisher, R. A. 1959. *Statistical Methods and Scientific Inference*. Edinburgh, UK: Oliver and Boyd, p. 175.

Fisher, R. A. 1970. *Statistical Methods for Research Workers*, 14th ed. Edinburgh, UK: Oliver and Boyd, p. 362.

Hatfield, J. L. and Walthall, C. L. 2015. Meeting global food needs: Realizing the potential via genetics × environment × management interactions. *Agronomy Journal*, 107, 1215–26.

Heege, H. J. 2013. Site-specific fertilizing. In: Heege, H. J. (Ed.), *Precision in Crop Farming: Site Specific Concepts and Sensing Methods: Applications and Results*. Dordrecht, the Netherlands: Springer Science + Business Media, p. 356. doi: 10.1007/978-94-007-6760-7.

Ines, A. V. M., Das, N. N., Hansen, J. W. and Njoku, E. G. 2013. Assimilation of remotely sensed soil moisture and vegetation with a crop simulation model for maize yield prediction. *Remote Sensing of Environment*, 138, 149–64.

Jones, D. and Barnes, E. M. 2000. Fuzzy composite programming to combine remote sensing and crop models for decision support in precision crop management. *Agricultural Systems*, 65, 137–58.

Klemm, T. and McPherson, R. A. 2017. The development of seasonal climate forecasting for agricultural producers. *Agricultural and Forest Meteorology*, 232, 384–99.

Kyveryga, P. M. and Blackmer, T. M. 2012. On-farm evaluations to calibrate tools for estimating late-season nitrogen status of corn. *Agronomy Journal*, 104, 1284–94.

Kyveryga, P. M., Blackmer, A. M. and Morris, T. F. 2007a. Alternative benchmarks for economically optimal rates of nitrogen fertilization for corn. *Agronomy Journal*, 99, 1057–65.

Kyveryga, P. M., Blackmer, A. M. and Morris, T. F. 2007b. Disaggregating model bias and variability when calculating economic optimum rates of nitrogen fertilization for corn. *Agronomy Journal*, 99, 1048–56.

Kyveryga, P. M., Blackmer, T. M. and Caragea, P. C. 2011. Categorical analysis of spatial variability in economic yield response of corn to nitrogen fertilization. *Agronomy Journal*, 103, 796–804.

Kyveryga, P. M., Caragea, P. C., Kaiser, M. S. and Blackmer, T. M. 2013. Predicting risk from reducing nitrogen fertilization using hierarchical models and on-farm data. *Agronomy Journal*, 105, 85–94.

Lagarde, F. and Champolivier, L. 2006. Le raisonnement et les avancées techniques permettent de réduire la fertilisation azotée: Le cas de Farmstar-colza® (Reasoning and technical advances allow for a reduction of nitrogen fertilisation: The case of Farmstar-colza®). *OCL - Oleagineux Corps Gras Lipides*, 13, 384–7.

Lindblom, J., Lundström, C., Ljung, M. and Jonsson, A. 2016. Promoting sustainable intensification in precision agriculture: Review of decision support systems development and strategies. *Precision Agriculture*, 18(3), 309–31. doi: 10.1007/s11119-016-9491-4.

Lukina, E. V., Freeman, K. W., Wynn, K. J., Thomason, W. E., Mullen, R. W. and Stone, M. L. 2001. Nitrogen fertilization optimization algorithm based on in-season estimates of yield and plant nitrogen uptake. *Journal of Plant Nutrition*, 24, 885–98.

Mackrell, D., Kerr, D. and von Hellens, L. 2009. A qualitative case study of the adoption and use of an agricultural decision support system in the Australian cotton industry: The socio-technical view. *Decision Support Systems*, 47, 143–53. doi: http://dx.doi.org/10.1016/j.dss.2009.02.004.

Magri, A., Van Es, H. M., Glos, M. A. and Cox, W. J. 2005. Soil test, aerial image and yield data as inputs for site-specific fertility and hybrid management under maize. *Precision Agriculture*, 6, 87–110.

Mallavan, B. 2014. De l'observation aux opérations – Solution FarmStar. In: *Conference Agriculture v3.0 & Geolocalisation*. Toulouse, France: Automatismes et innovations pour l'agriculture de précision. https://www.guide-gnss.net/contenuguide/uploads/2014/01/140408-15H20-AGRIGNSS-B.-Mallavan-AIRBUS-DS-Observt-aux-Op%C3%A9rations-Farmstar-.pdf. Accessed: 10 November 2017.

Mandel, D., Ghosh, P. P. and Dasgupta, M. K. 2012. Appropriate precision agriculture with site-specific cropping system management for marginal and small farmers. *CAB Reviews: Perspectives in Agriculture, Veterinary Science, Nutrition and Natural Resources*, 7, 1–16.

Manos, B., Ciani, A., Bournaris, T., Vassiliadou, I. and Papathanasiou, J. 2004. A taxonomy survey of decision support systems in agriculture. *Agricultural Economics Review*, 5, 80–94.

McBratney, A. B., Whelan, B. M. and Shatar, T. M. 1997. Variability and uncertainty in spatial, temporal and spatiotemporal crop-yield and related data. *CIBA Foundation Symposia*, 210, 141–60.

McBratney, A., Whelan, B., Ancev, T. and Bouma, J. 2005. Future directions of precision agriculture. *Precision Agriculture*, 6, 7–23.

McCown, R. L., Hammer, G. L., Hargreaves, J. H. G., Holzworth, D. P. and Freebairn, D. M. 1996. APSIM: A novel software system for model development, model testing and simulation in agricultural systems research. *Agricultural Systems*, 50, 255–71.

McCown, R. L., Carberry, P. S., Hochman, Z., Dalgliesh, N. P. and Foale, M. A. 2009. Re-inventing model-based decision support with Australian dryland farmers. 1. Changing intervention concepts during 17 years of action research. *Crop and Pasture Science*, 60, 1017–30.

McNamara, J. P., Hanigan, M. D. and White, R. R. 2016. Invited review: Experimental design, data reporting, and sharing in support of animal systems modeling research. *Journal of Dairy Science*, 99, 9355–71.

Meynard, J. M., Cerf, M., Guichard, L., Jeuffroy, M. H. and Makowski, D. 2002. Which decision support tools for the environmental management of nitrogen? *Agronomie*, 22, 817–29.

Mucherino, A., Papajorgji, P. and Pardalos, P. M. 2009. A survey of data mining techniques applied to agriculture. *Operational Research*, 9, 121–40.

National-Research-Council (NRC). 1997. *Precision Agriculture in the 21st Century: Geospatial Information Technologies in Crop Management*. Washington DC: National Academy Press. http://www.nap.edu/openbook.php?record_id=5491&page=17.

Obenauf, U., Borchardt, I., Lubkowitz, C. and Kock, C. 2014. N-sensor im praktischen einsatz: Erfahrungen und Ergebnisse aus dem Projekt On Farm Research (N sensor in practical use: Experiences and results from the On Farm Research project). *Journal für Kulturpflanzen*, 66, 48–56.

Oliver, M. A. (Ed.). 2010. *Geostatistical Applications for Precision Agriculture*. the Netherlands: Springer. ISBN: 978-90-481-9132-1 (Print), 978-90-481-9133-8 (Online).

Patzold, S., Mertens, F. M., Bornemann, L., Koleczek, B., Franke, J., Feilhauer, H. and Welp, G. 2008. Soil heterogeneity at the field scale: A challenge for precision crop protection. *Precision Agriculture*, 9, 367–90.

Pechenick, A. and Rizzo, D. M. 2008. Using artificial neural networks to predict local disease risk indicators with multi-scale weather, land and crop data. In: *Proceedings iEMSs 4th Biennial Meeting - Int. Congress on Environmental Modelling and Software: Integrating Sciences and Information Technology for Environmental Assessment and Decision Making*. Manno, Switzerland: International Environmental Modelling & Software Society, pp. 336–45.

Pringle, M. J., Cook, S. E. and Mcbratney, A. B. 2004a. Field-scale experiments for site-specific crop management. Part I: Design considerations. *Precision Agriculture*, 5, 617–24.

Pringle, M. J., Mcbratney, A. B. and Cook, S. E. 2004b. Field-scale experiments for site-specific crop management. Part II: A geostatistical analysis. *Precision Agriculture*, 5, 625–45.

Roberts, D. C., Brorsen, B. W., Solie, J. B. and Raun, W. R. 2013. Is data needed from every field to determine in-season precision nitrogen recommendations in winter wheat? *Precision Agriculture*, 14, 245–69.

Roger-Estrade, J., Richard, G., Dexter, A. R., Boizard, H., De Tourdonnet, S., Bertrand, M. and Caneill, J. 2009. Integration of soil structure variations with time and space into models for crop management: A review. *Sustainable Agriculture*, 29, 135–42.

Rowe, W. D. 1994. Understanding uncertainty. Risk analysis, 14, 743–50.

Samborski, S. M., Tremblay, N. and Fallon, E. 2009. Strategies to make use of plant sensors-based diagnostic information for nitrogen recommendations. *Agronomy Journal*, 101, 800–16.

Samborski, S. M., Gozdowski, D., Stepień, M., Walsh, O. S. and Leszczyńska, E. 2016. On-farm evaluation of an active optical sensor performance for variable nitrogen application in winter wheat. *European Journal of Agronomy*, 74, 56–67.

Sampurno, R. M., Seminar, K. B. and Suharnoto, Y. 2014. Weed control decision support system based on precision agriculture approach. *Telkomnika (Telecommunication Computing Electronics and Control)*, 12, 475–84.

Schneider, J. M. and Garbrecht, J. D. 2006. Dependability and effectiveness of seasonal forecasts for agricultural applications. *Transactions of the ASABE*, 49, 1737–53.

Sela, S., Van Es, H. M., Moebius-Clune, B. N., Marjerison, R., Melkonian, J., Moebius-Clune, D., Schindelbeck, R. and Gomes, S. 2016. Adapt-N outperforms grower-selected nitrogen rates in Northeast and Midwestern United States strip trials. *Agronomy Journal*, 108, 1726–34.

Setiyono, T. D., Yang, H., Walters, D. T., Dobermann, A., Ferguson, R. B., Roberts, D. F., Lyon, D. J., Clay, D. E. and Cassman, K. G. 2011. Maize-N: A decision tool for nitrogen management in maize. *Agronomy Journal*, 103, 1276–83.

Thorp, K. R., Batchelor, W. D., Paz, J. O., Steward, B. L. and Caragea, P. C. 2006. Methodology to link production and environmental risks of precision nitrogen management strategies in corn. *Agricultural Systems*, 89, 272–98.

Tremblay, N., Bouroubi, Y. M., Bélec, C., Mullen, R. W., Kitchen, N. R., Thomason, W. E., Ebelhar, S., Mengel, D. B., Raun, W. R., Francis, D. D., Vories, E. D. and Ortiz-Monasterio, I.2012. Corn response to nitrogen is influenced by soil texture and weather. *Agronomy Journal*, 104, 1658–71. https://www.agronomy.org/publications/aj/pdfs/104/6/1658.

Tremblay, N., Bélec, C. and Bouroubi, Y. M. 2016. *SCAN: A New Decision-support System for Sidedressed N Rate Recommendation in Corn*. Madison, WI: ASA-SSSA-CSSA.

Tripathy, A. K., Adinarayana, J., Merchant, S. N., Desai, U. B., Ninomiya, S., Hirafuji, M. and Kiura, T. 2013. Data mining and wireless sensor network for groundnut pest/disease precision protection. In: *National Conference on Parallel Computing Technologies*. Bangalore, India: PARCOMPTECH, IEEE, pp. 1–8. doi:10.1109/ParCompTech.2013.6621399. URL: http://ieeexplore.ieee.org/stamp/stamp.jsp?tp=&arnumber=6621399&isnumber=6621385.

Tyrychtr, J., BroŽek, J. and Vostrovský, V. 2015. Multidimensional modelling from open data for precision agriculture. *Lecture Notes in Business Information Processing*, 231, 141–52.

Walsh, O. S., Klatt, A. R., Solie, J. B., Godsey, C. B. and Raun, W. R. 2013. Use of soil moisture data for refined GreenSeeker sensor based nitrogen recommendations in winter wheat (*Triticum aestivum* L.). *Precision Agriculture*, 14, 343–56.

Whelan, B. M. and Mcbratney, A. B. 2000. The 'null hypothesis' of precision agriculture management. *Precision Agriculture*, 2, 265–79.

Part 2

Delivery systems

Chapter 7

Variable-rate application technologies in precision agriculture

Kenneth A. Sudduth, USDA-ARS, USA; Aaron J. Franzen, South Dakota State University, USA; and Heping Zhu and Scott T. Drummond, USDA-ARS, USA

1 Introduction

Precision agriculture (PA) can be thought of as being composed of a 'menu' of available management operations, with the overall PA system being different depending on what operations are selected for a particular farm or field. For example, one farmer may only practice variable-rate application (VRA) of nitrogen (N) fertilizer, while another farmer's PA system may include VRA of all fertilizer applied, plus precision herbicide and pesticide application. Regardless, the PA approach can be viewed as cyclic in nature, involving data collection, development of management plans, implementation of those plans in the field and evaluation of the effort (Fig. 1).

Precision field operations may be of several types, including precision tillage, variable seeding or site-specific irrigation (discussed in other chapters in this book). However, the most widely practised precision field operation is VRA of fertilizer or crop protection chemicals. This chapter provides information on the basic concepts and implementation of VRA technology, including the alternative approaches available for both liquid and dry product application. It also provides two case study examples of operating VRA systems, summarizes the current status of VRA technology and explores possible future trends in the field.

Early variable flow-rate systems, such as the one described by Gebhardt et al. (1974), adjusted flow rate to achieve a constant areal application rate when ground speed of the

http://dx.doi.org/10.19103/AS.2017.0032.08
Published by Burleigh Dodds Science Limited, 2019.

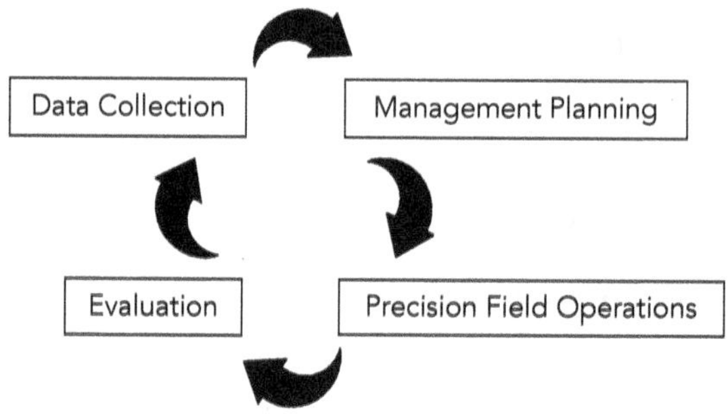

Figure 1 The precision agriculture cycle.

applicator varied. The availability of radar ground speed sensors in the 1980s enabled development of control systems that measured travel speed and adjusted flow to maintain a constant, desired application rate. These early systems provided a starting point for the VRA systems needed in PA, where the concept of rate control based on spatial variability was overlaid on the original concept of speed control.

2 Characteristics of VRA control systems

Figure 2 shows that physical connectivity and data flow in a VRA system can be quite complex. The general system consists of both office tasks and vehicle tasks. Office tasks include interpreting input data, developing management plans and determining application rate maps. Vehicle tasks include using these application rate maps in conjunction with on-board sensors and actuators to apply fertilizer, chemicals or inputs in the field. Note that in any given VRA system, some elements shown in this general schematic (Fig. 2) may not be present. For example, a system may or may not include on-line sensors and may or may not generate an actual application rate map. Also, the function of multiple blocks may be combined in a single physical unit. For example, the VRA rate processor and application controller are often combined in modern VRA systems.

2.1 Map-based vs. sensor-based control

Control decisions for VRA can be implemented either on-line or off-line. In the on-line, or sensor-based, approach, the controlled equipment incorporates on-board sensors and the sensor data are used immediately for automatic control. In the off-line, or map-based, approach, data are collected and stored in one operation, and the controlled equipment uses the information in a separate field operation. The map-based approach allows more flexibility in data manipulation and preprocessing but requires that the location of equipment in the field be precisely defined, usually with global positioning system (GPS) or more generally, global navigation satellite system (GNSS) technology. Most systems currently available are map based, but more on-line systems will likely become available as real-time sensing technologies become more mature. Hybrid systems which rely on a combination of both mapped and real-time data may also come into more widespread use. Map-based and sensor-based systems each have their own benefits (Ess and Morgan, 2010).

Published by Burleigh Dodds Science Limited, 2019.

Benefits of map-based systems:

- Allows use of data which cannot be obtained in real time.
- Amount of product needed can be calculated before application.
- Time is available between data collection and application, to permit more intensive data analysis and processing, potentially improving results.
- System is able to 'look ahead' at the application rates needed at future locations in the field to minimize errors due to lag times.

Benefits of sensor-based systems:

- Automated sensor-based systems can eliminate the need for human intervention in the data collection and processing stream (e.g. sample collection and lab analysis) required for many map-based systems.
- Dense sensor-derived data minimize spatial interpolation errors.
- Real-time sensor systems can control applications based on characteristics having high temporal variability (e.g. soil nitrate level).

Data obtained from analysis of manually collected soil samples, past years' yields, soil type maps and many other sources may be considered in development of map-based VRA plans. Previous agronomic research and knowledge provide a basis for beginning to understand the spatial relationships between yields and other measured factors. However, the use of traditional, general response relationships based on aggregation of data from numerous sites across a state or region to guide VRA is often questionable. Rather,

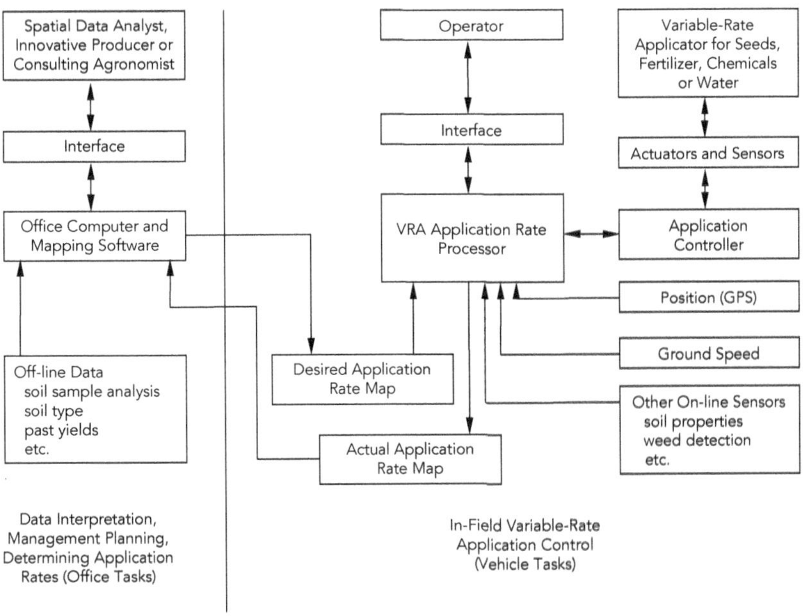

Figure 2 Generalized schematic of data flow in a VRA system.

research indicates that crop response is a site-specific parameter, and may vary from region to region and even from place to place within a single field (Kitchen et al., 2010). Integrated, computer-based decision support systems (DSS) provide a way to incorporate expert knowledge with information based on many different sources. Crop growth models have been coupled with GIS systems for spatial estimates of potential yields, yield-limiting factors and fertilizer requirements (Hartkamp et al., 1999). As an example, although on-line VRA control of fertilizer N based on in-season crop canopy sensing has been adopted in corn production (Samborski et al., 2009), Tremblay et al. (2012) found that better results could be obtained by incorporating off-line information about soils and weather.

2.1.1 On-line sensors

Data obtained from sensors on the applicator vehicle may be used to assess field conditions and to determine (in whole or in part) desired application rates. The on-line sensing technology currently most used in practice is crop canopy reflectance sensing for N status assessment and subsequent control of N application (discussed in Part 1 of this book). Initial systems, both in the United States (Stone et al., 1996) and Europe (Heege et al., 2008), used passive radiometers that were dependent on ambient sunlight for illumination. Difficulties in compensating for spatio-temporal variation in ambient illumination (Souza et al., 2010) led to the development of active sensors with their own illumination source (Holland et al., 2012), so that the sensor would respond only to reflectance based on the active illumination and not on sunlight. Active canopy reflectance sensors designed for N application control are commercially available and have been compared (Sudduth et al., 2015). Considerable research has been directed toward developing algorithms to translate commercial sensor output into N rate control decisions for crops such as wheat (Raun et al., 2002), maize/corn (Kitchen et al., 2010; Solari et al., 2008) and cotton (Oliveira et al., 2013). Many of these algorithms use sensor data in the form of relative normalized difference vegetation index (NDVI), where each NDVI value is divided by the mean NDVI obtained from an area in the field where N is not limiting. Potential for improved grower profits and reduced environmental impacts has been demonstrated (Roberts et al., 2010; Scharf et al., 2011).

Various sensing approaches have been used to detect, identify or quantify weeds and pests as a control input for real-time VRA herbicide. Such approaches have generally been based on spectral response, plant structure or imaging of the plant, which can provide information on both spectral response and structure (Peteinatos et al., 2014; Liu and Zhu, 2016). A case study describing an imaging-based VRA system can be found in an accompanying chapter on spray technologies in this volume. Other real-time sensors, such as those for soil apparent electrical conductivity (EC_a; Corwin and Lesch, 2005), have been used in an off-line mode to develop control maps for VRA of nematicide (Ortiz et al., 2012) or soil-applied herbicide (Gundy et al., 2017). Soil sensors such as those for EC_a are covered in more detail elsewhere (discussed in Part 1 of this book).

2.2 Control hardware

2.2.1 Application rate processor

The application rate processor and associated software are at the centre of the VRA system. Here, position, velocity and on-board sensor data are combined with the desired application rate map and rate commands are sent to the application rate controller. Many VRA systems combine the application rate processor, application controller and operator interface into one box located in the cab. Wiring harnesses connect this multifunction box with actuators and sensors. In other systems, the interface and application rate processor are combined and communicate with a separate controller. In this way, the application controller may be optimized for the specific control task, which may provide a higher degree of confidence in the inner-loop portion of the control system. Widespread adoption of the ISO 11783 vehicle communication standard (Stone et al., 2008) has made separate, dedicated application controllers located near the point of actuation an attractive configuration.

2.2.2 Application controller

The application controller receives set point information from the application rate processor and then controls the actuators on the application device. In general, the controllers used in VRA equipment implement closed-loop control. In closed-loop control, a feedback signal proportional to the actual application rate is compared with the desired application rate, and the resulting rate error is used to adjust the control signal, such that the error is minimized. The means by which this adjustment is carried out will vary from case to case and is beyond the scope of this article. The reader is referred to a basic text in automatic control theory, such as Kuo and Golnaraghi (2002).

Commercial VRA equipment may use either centralized or distributed control system architecture. In the centralized system, a single controller resides in the cab of the vehicle, with dedicated wiring running to each of the sensors and actuators in the system. Examples of this approach would include dedicated spray rate controllers. With the distributed architecture, a single interface console in the applicator cab communicates with a network of digital controllers, each of which monitors and controls one or a few processes. This approach has gained favour as more and more processes are being controlled electronically on agricultural vehicles. The distributed control system is more flexible, simpler to maintain, and often less expensive than providing several dedicated controller/display units in the cab.

Standardization in distributed control systems promotes component interchangeability and allows implements from one manufacturer to function with vehicles made by another company. The agricultural equipment industry has supported the development and adoption of the ISO 11783 vehicle network standard (Stone et al., 2008). This standard provides support for PA–related message information to be transferred along the data bus, in addition to the more conventional equipment operating messages. This standard has been widely adopted by manufacturers of agricultural equipment, and many current VRA systems follow the ISO 11783 standard. A more detailed functional description of the standard is given by Anderson and Humburg (1997) and Stone et al. (1999).

2.2.3 Operator interface

The operator interface is an important part of the overall VRA system. It must effectively communicate complex and rapidly changing information to the driver, particularly in the case of simultaneous multiple-product application. It must also provide an efficient means for the operator to interact with the VRA system to change operating parameters, override default settings or fix fault conditions that may be detected. Often, a combination of audible alarms, numeric readouts and graphical displays is used to provide information to the operator. Data entry is generally accomplished through a touchscreen interface.

The ISO 11783 standard provides support for a single operator interface in the cab of the vehicle, termed a universal terminal. All devices on the vehicle network are able to use this physical interface for communicating with the operator. This includes vehicle-specific devices such as vehicle sensors (hitch position, PTO shaft status, etc.) as well as application controllers. Multiple devices can present information to the operator at once through multiple panels in the universal terminal (Stone et al., 1999).

2.2.4 Equipment actuators and sensors

Most site-specific application devices are driven by hydraulic or electric motors. Liquid fertilizers and chemicals are pumped through a hose and dry fertilizers and chemicals are metered from a holding tank for broadcast or banded applications. A feedback signal proportional to actuator motor speed, and therefore indirectly to product delivery rate, is input to the application controller to allow for closed-loop control. In other cases, product metering may be accomplished by changing the size of a flow passage; for example height of a gate in a dry system or size of an orifice in a liquid system. Further information on actuators for accomplishing flow control is given separately below for liquid and dry systems.

2.2.5 Documentation of application: as-applied maps

Most VRA systems generate maps of actual application rates. At specified time or distance intervals, the current rates reported by the application controller are logged, along with a GPS position and time tag. These maps are useful for evaluating and documenting performance of the VRA system. They also provide a lasting record of what was applied, along with when and where that application occurred. As such, they may be useful to document compliance with environmental regulations that limit the application of fertilizers, chemicals or animal wastes. It should be noted that the 'actual' rates reported by the application controller are often not based on flow measurements of the product applied, but rather on the rotational speed of some shaft in the product metering system. A constant calibration between shaft speed and product delivery is then assumed. This assumption is sufficient in most cases, but may not be accurate if the equipment is not functioning as designed. Some VRA systems do incorporate flowmeters to quantify the actual product flow rate, such as the liquid manure application system described by Morris et al. (1999).

3 Liquid VRA systems

3.1 Flow control methods

To obtain the desired application rate control, either the concentration of the active ingredient (a.i.) or the total output (a.i. + carrier) must be varied. Hughes and Frost (1985) presented an early review and comparison of these two methods. The variable concentration method allows spraying of a constant total volume of solution at a constant nozzle pressure, thus maintaining uniform coverage with standard nozzle designs.

Problems which have been noted with this method include increased deposition variability and transient errors in application rate resulting from time lags in the system (Vidrine et al., 1975; Tompkins et al., 1990; Felizardo et al., 2016). Additionally, when the applicator is travelling at very low or very high speeds, this method may result in concentrations that are outside of the labelled range for the product being applied. Variations in total nozzle output are traditionally accomplished by changing the inlet pressure at the nozzle, which may affect droplet size and distribution. Other methods of varying total output have been developed and are discussed briefly below.

3.1.1 Total output control method

Historically, the output of liquid application systems was controlled by adjusting the pressure at the nozzles. This pressure, coupled with nozzles sized to deliver the required flow rate at a planned travel speed, resulted in the desired application rate in volume of liquid per unit area. Early electronic spray control systems incorporated a flow sensor, and sometimes a pressure sensor, to provide closed-loop control of the flow rate. Often, a ground speed sensor was incorporated to allow the system to change flow rate to compensate for changes in travel speed. Later, these speed-compensating systems formed the basis for VRA application by modifying the control inputs to include desired application rate changes along with travel speed.

One issue with implementing VRA in flow control systems is the large turn-down ratio (i.e. ratio of maximum to minimum flow rate) required. Han et al. (1986) estimated that a turn-down ratio of 5:1 would be required for VRA of soil-applied herbicides, and similar ratios would be needed for VRA of side-dress N fertilizer (Kitchen et al., 2010). However, because orifice (nozzle) flow is a square root function of pressure, a 5× change in flow would require an impractical 25× change in pressure. Even much smaller turn-down ratios that would require a 2× change in pressure would be problematical for VRA of crop protection chemicals, where nozzle patterns and spray droplet sizes might then vary outside of the optimum ranges for effective control.

Several alternative methods of changing total nozzle output without requiring large pressure changes have been developed. These include the use of bypass nozzles which return a portion of their inlet flow back to the supply tank (Han et al., 1986), multiple nozzle systems (Miller, 2003; Kang et al., 2011), pulse width modulation control of a fast-acting solenoid valve (Giles and Comino, 1990; Han et al., 2001; Liu et al., 2014) and variable-orifice nozzles (Womac and Bui, 2002; Luck et al., 2015; Sharda et al., 2016). Additional details can be found in an accompanying chapter on spray technologies in this volume.

3.1.2 Variable concentration method

Reports of systems to meter, or inject, concentrated pesticides into the diluent stream, thus changing the concentration of the a.i., began to appear in the research literature in the mid-1970s (Hughes and Frost, 1985). Vidrine et al. (1975) developed and tested an early laboratory system which demonstrated the feasibility of injecting concentrated pesticides. Problems noted with this system included non-uniform pesticide application and transient application errors resulting from changes in operating speed. Field sprayers that included injection of pesticides at the proper rate for variations in travel speed and/or application rate have been developed and tested (Reichard and Ladd, 1983; Krebs et al., 2015). Additional details on injection systems can be found in an accompanying chapter on spray technologies in this volume.

A major impediment to VRA with single-point injection systems is the delay time from the point of injection to the individual nozzles. Tompkins et al. (1990) injected a tracer chemical into the diluent stream at three points on a boom sprayer immediately before and after the diluent pump, and at the individual nozzles. They reported that, as the injection point was moved downstream, transient time was reduced, but variations in chemical concentration between nozzles were greater. Miller and Smith (1992) developed a direct nozzle injection system that overcame the concentration variation problems reported by previous researchers. However, incomplete mixing of concentrated pesticide and carrier has been reported for nozzle injection (Zhu et al., 1998), leading to the design (Vondricka and Schulze Lammers, 2009a) and evaluation (Vondricka and Schulze Lammers, 2009b) of mixing chambers for improved uniformity in nozzle injection systems. Another approach has been to simultaneously control both carrier flow rate and chemical injection rate for improved system performance (Steward and Humburg, 2000; Vondricka and Schulze Lammers, 2009c).

3.2 Section control

Section control refers to dividing the width of the applicator into a number of sections and then controlling each individual section independently to better match application needs within a field. A primary goal of such a system is to reduce overlaps in application, particularly for irregularly shaped fields. Luck et al. (2010) reported a 15–17% reduction in area sprayed when evaluating automatic map-based section control compared to conventional application for three such irregular fields. Another primary goal is to prevent spraying on non-target areas such as grassed waterways or adjacent water bodies, or areas outside the field boundary.

Control of individual boom sections is through the application rate processor which monitors the location of each boom section in relation to both static and dynamic field maps. Static maps would delineate zones of no application, typically determined off-line from georeferenced images or prior field surveys. Dynamic maps would continuously update those areas where product has already been applied in the field. If a boom section is in an area where application is allowed and has not previously occurred, that section is turned on; otherwise it will remain off. Dividing an applicator into more sections offers better spatial resolution of control, and therefore more potential reduction in product over-applied (Luck et al., 2010).

However, when boom sections are turned on and off, large dynamic variations in nozzle pressure and flow rate can result (Sharda et al., 2010). In an evaluation of two commercial controllers incorporating section control, significant departures from desired flow rate were noted as sections were turned on and off, with settling times to return to the desired rate as long as 20 s, suggesting a need for development of improved control algorithms (Sharda et al., 2013).

4 Dry VRA systems

Application of fertilizer, soil amendments and other crop production products in dry or granular form has become common in recent decades. These dry products range from very fine powders to granules up to multiple millimetres in diameter. Application equipment for dry VRA systems mostly falls into two categories: pneumatic systems and spinner spreaders. Both categories are capable of broadcasting product across wide swaths of a field, while pneumatic systems are also well suited to precision placement of material with respect to plant location. The potential accuracy of either type is dependent on multiple variables, including particle size, sizing uniformity, product density, product strength and machine configuration. For either category, the application accuracy is initially dependent on the accuracy of the metering system used to convey the bulk product from the storage tank or bin. Even if the material were applied in a perfectly uniform manner across the width of the application pattern, poor metering calibration would result in poor VRA system accuracy. ASAE Standard S341.4 (ASABE, 2009), 'Procedure for Measuring Distribution Uniformity and Calibrating Granular Broadcast Spreaders', provides established methods for both calibration of the metering system and for documenting and evaluating distribution across the width of the application pattern.

Most current dry material applicators store bulk material in one or more large central bins or hoppers. In current systems, the material is most often metered toward the final distribution mechanism by means of a variable-speed conveyor belt or chain, although some systems use augers. Conveyor style metering moves the product from the central storage bin through a gate to control the depth of the product on the conveyor. This results in a uniform-depth layer of product on the conveyor outside of the storage bin. Gates can either be fixed or adjustable. In the case of adjustable gates, the metering system requires separate calibration curves for different gate heights. Automatic control of gate height is an alternative metering strategy that has been used in some systems (Kweon and Grift, 2006).

Variable-rate metering is achieved using feedback control on the drive actuator of the metering system, such that the desired speed of the conveyor or auger is measured and adjusted by the control system if the desired rate is different from the current calculated application rate. The metering system determines the applied rate in mass or volume per unit time, which is adjusted by the application controller as the machine moves through the field. The resulting mass or volume per unit area is then dependent on the dispersion system, either pneumatic or mechanical. The overall accuracy of a dry system is dependent on the metering accuracy as well as the distribution or dispersion system subsequent to metering. Specifically for VRA systems, the metering system response when changing from one flow rate to another needs to be fast and consistent. Fulton et al. (2005a) found that only one in four commercial applicators exhibited such a response. In a field study considering both metering and dispersion of granular fertilizer, a spinner spreader was within ± 10% of target rate only 25–35% of the time, while a pneumatic applicator achieved that accuracy 40–45% of the time (Fulton et al., 2013). Although results would depend on the field geometries and test conditions encountered, this finding points out the need for improvements to dry applicator systems for increased accuracy.

4.1 Pneumatic systems

Pneumatic systems convey dry material from the central metering system to the final placement area by accelerating particles through hoses or ducts, referred to as conveyance lines, using a fast-moving stream of air. Typically, the dry material is metered into a distribution chamber with a set of baffles or a manifold designed to divide the product into many smaller conveyance lines that will distribute the product to its final placement. At the end of each conveyance line, the product can either be directed to a narrow area for banded placement, or it can be redirected by baffles to be broadcast in a uniform manner across the field.

For either broadcast or banded placement, pneumatic systems can use the same conveyance lines for one or multiple products concurrently. If each product is metered independently, the mixing of products can be either passive or active. In a passive mixing system, the two or more streams of product are simply added to the same mass flow stream. For instance, with two products metered by conveyor, one conveyor's material is simply dropped onto the second conveyor prior to entering the air stream. Conversely, multiple products can be actively mixed after metering. As an example, two products could be dropped from separate conveyors into an auger system that combines the two material streams. In cases where the two products are similar in density and particle size distribution, it is likely that active and passive metering would attain similar results. In the case that there were large disparities in either density or particle size between products, active mixing is likely to result in better distribution uniformity between the two products.

Banded application is often done concurrently with planting or other applications, and typically places product in close proximity to plants, either near currently growing plants or in areas close to plants that will later be placed or emerge from the soil. Newer tillage technologies that rely heavily upon GNSS employ pneumatic application of dry material in locations that will be planted to crop in the future. Broadcast pneumatic applicators are very similar to liquid VRA systems in that they employ wide booms with multiple conveyance lines for each smaller swath of the field. They should theoretically achieve more uniform application rates across a range of operating conditions than the spinner spreader alternative, but with greater cost and complexity. Fulton et al. (2005b) tested two pneumatic spreaders and two spinner spreaders, finding that one spinner spreader exhibited the best overall uniformity in an overlapping-pass analysis, while one of the pneumatic spreaders exhibited the least uniform results. However, the distributions of the two pneumatic spreaders were less influenced by application rate than were those of the two spinner spreaders.

4.2 Spinner spreaders

Spinner spreader applicators distribute product across the field by accelerating granules using spinning discs with fins via centrifugal force. Such applicators are only appropriate for broadcast application since they are not capable of directing product to a banded area. For spinner spreader applicators, product is dropped onto one or more spinning disc(s) with multiple spreading vanes. The product is spread across a wide swath to cover large areas, typically at relatively large rates of mass per unit area when compared to pneumatic dry systems or liquid VRA systems. Spinning disc spreaders are best suited to apply products that require a large volume of application per unit of field area. A typical application for a spinner spreader is applying fertilizer or pH balancing products between harvest and planting seasons.

After metering, the application of dry product with a spinner spreader is dependent on the material(s) being applied, the spinning disc configuration, the rotational velocity of the disc(s), the configuration of the disc fins and the placement of the disc with respect to the material falling from the metering device. Given the ideal use case of a) high levels of material placement per unit area, b) high levels of field coverage per unit time and c) variable but controllable distribution of product on each unit of field area, an ideal spinner spreader would be able to apply all materials uniformly across the swath, but also vary the application rate along the swath. Because distribution uniformity of spinner spreaders can be affected by application rate (Fulton et al., 2005b), maintaining uniformity under VRA may be difficult.

Spinning discs accelerate the product with fins mounted to the disc. Faster rotation of the discs and fins provides more centrifugal force on each product particle, resulting in a larger distance of spread potential for faster spinning discs. Heavier particles are spread further than lighter particles given the same centrifugal force. Thus, if a blended fertilizer is used with a spinner spreader, there may be segregation of the blend components across the spread pattern (Crozier and Robertson, 2014). This means that larger and denser particles are spread furthest by any spinning disc system at a given rotation speed, assuming that the particles remain intact. If the force imparted on the particle is great enough to break it into more than one particle, it will have less mass and will not travel as far.

The best use case for spinner spreader applicators is one in which achieving high material application rates and large land coverage per hour is important. Although research (Fulton et al., 2005b) indicates distribution patterns of some spinner spreaders are more variable with changes in application rate than those of pneumatic spreaders, other tests (Crozier and Robertson, 2014) indicate that spinner spreaders may not, as a group, exhibit poorer distribution uniformity compared to pneumatic applicators. As each of these studies examined only a few (≤5) commercial applicators, additional testing of commercial and prototype applicators is warranted. To facilitate such testing in a repeatable and scientifically sound manner, a standard, 'Procedures for Evaluating Variable-Rate Granular Material Application Accuracy' is currently under development within the American Society of Agricultural and Biological Engineers.

5 Case studies

5.1 Laser-guided VRA sprayer for orchards, vineyards and nurseries

Conventional spray application technology requires use of large amounts of pesticide to achieve effective pest control in orchard, vineyard, nursery and other horticultural crop production systems. An automated variable-rate, air-assisted sprayer employing sensor technology to match spray outputs to crop structures during pesticide applications was developed to solve this problem (Fig. 3). The sprayer characterized the presence, size, shape and foliage density of target trees and applied the optimum amount of pesticide with minimum human involvement (Chen et al., 2012; Shen et al., 2017). It integrated a high-speed, 270° radial and 30-m range laser scanning sensor in conjunction with a non-contact travel speed sensor, a sophisticated automatic nozzle flow rate controller (Liu et al., 2014), an embedded computer, a touch screen, a control switch box and 40 variable-rate nozzles on two sides of the sprayer.

The laser sensor mounted between the tractor and sprayer detected the return distance signals of the bilateral tree structure. A dedicated C++ language algorithm was developed to process the signals and translate them into tree canopy structures. The algorithm also determined the requisite flow rate for each tree sectional structure and sprayer travel speed, which was then used to control individual nozzles coupled with pulse-width-modulation-controlled solenoid valves. The automatic controller consisted of two microprocessors that generated flow rate commands to modulate the 10-Hz pulse width of each solenoid valve. All 40 nozzles could independently discharge variable flow rates to their designated canopy sections. The functional touch screen displayed the sprayer travel speed, total

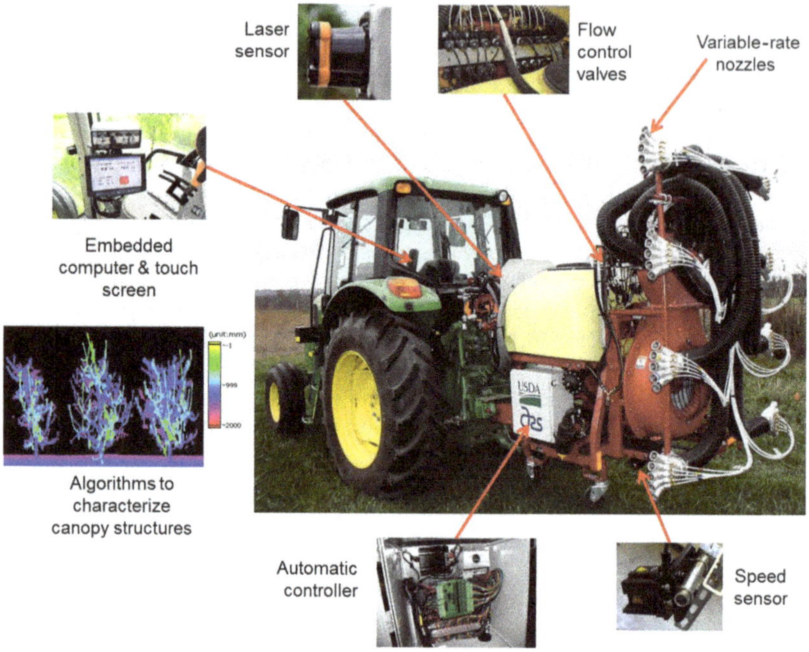

Figure 3 Laser-guided variable-rate sprayer.

discharged spray volume, spray width and active nozzles. Operators were able to modify spray parameters through the touch screen as needed.

Sprayer performance was evaluated in an apple orchard by quantifying spray deposition inside canopies at leafing, half-foliage and full-foliage growth stages, and was compared with conventional spray applications (Chen et al., 2013a,b). Spray deposition uniformity inside canopies and off-target losses to the ground, around trees, beyond trees and in the air were measured at the three growth stages. Biological control efficiency was also evaluated in ornamental nurseries and fruit farms when controlling insects and diseases with the laser-guided sprayer and conventional air-assisted sprayers (Zhu et al., 2017).

There were no significant differences in the amount of spray deposits and spray coverage inside tree canopies at different travel speeds between the laser-guided variable-rate sprayer and the conventional constant-rate sprayer; however, the laser-guided sprayer greatly reduced spray losses to the ground (Fig. 4a) and in the air (Fig. 4b). For example, reductions in the amount of spray losses to the ground and through the space between trees with the laser-guided sprayer were 85%, 80% and 71% when compared with the conventional air-assisted sprayer at the leafing, half-foliage and full-foliage growth stages, respectively. Similarly, the laser-guided sprayer reduced downwind airborne drift by 70%, 90% and 99% at 5, 15 and 35 m downwind at the half-foliage stage, and by 87% and 100% at 15 and 35 m downwind at the full-foliage stage.

Field biological control tests demonstrated that the laser-guided sprayer provided comparable insect control efficiencies (Fig. 5a) and lower disease infection rates (Fig. 5b) than the conventional air-blast sprayer. With the comparable insect control efficiency and better disease control or prevention, the average application rates during the growing

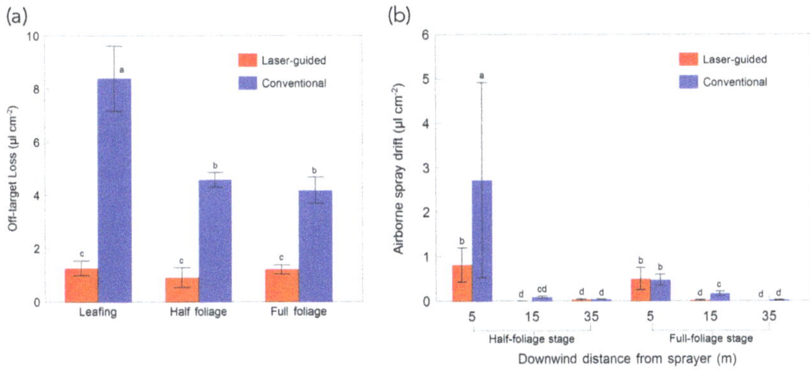

Figure 4 Comparison of (a) spray off-target losses to the ground and around trees and (b) airborne spray drift in three growth stages between the laser-guided variable-rate sprayer and a conventional constant-rate sprayer in an apple orchard.

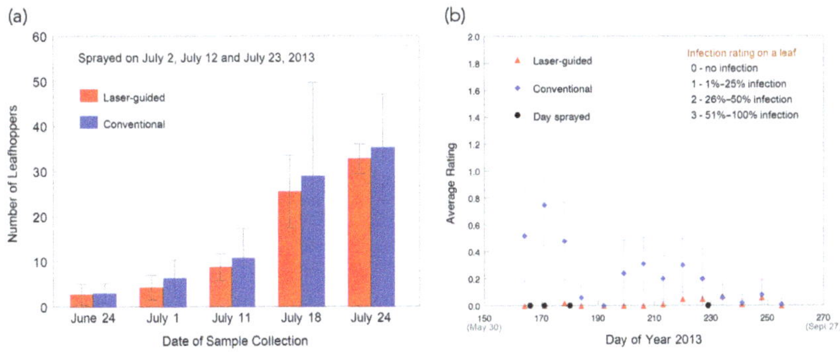

Figure 5 Comparisons of (a) leafhopper infestations on red maple trees and (b) powdery mildew infections on Norway maple trees between the laser-guided variable-rate and conventional air-blast sprayer treatments.

season with the laser-guided sprayer were 412 L ha^{-1} in farm #1, 206 L ha^{-1} in farm #2 and 309 L ha^{-1} in farm #3, while the average application rates from the conventional constant-rate sprayers were 767, 514 and 973 L ha^{-1}, respectively. The laser-guided sprayer reduced pesticide use by 46–68%.

The new variable-rate spraying system greatly reduced the excessive application of pesticides when spraying trees or similar plants with conventional constant-rate sprayers. This waste reduction is beneficial to ecosystems and saves growers money. It also saves time by requiring fewer spray tank refills to cover the same area.

5.2 VRA system for nitrogen application in agronomic research

A commercial self-propelled sprayer (Spra-Coupe 3640, Melroe Co., Bismarck, ND, USA) was modified to implement VRA top-dressing or side-dressing of N fertilizer in wheat and corn research (Fig. 6). Design requirements of the system included:

- Rapid, accurate, rate changes within 1 m of desired location
- Turn-down ratio of at least 5:1
- Integrated crop canopy sensing and associated algorithms for on-line determination and application of prescribed N rate
- Ability to use off-line maps to directly control N rate or to modify on-line sensor-based algorithms

For pressure-based flow control systems, which operate under the constraint that flow through an orifice (i.e. spray nozzle) is a function of the square root of pressure, achieving a high turn-down ratio while maintaining reasonable pressures is problematic. Therefore, this system combined a commercial pressure-regulated, flow feedback system (AgLeader Integra, Ames, Iowa, USA) with an array of three-nozzle clusters. The Integra system provided travel speed compensation and determined the maximum flow rate of the system. The three-nozzle clusters, where each nozzle in the cluster was supplied by a separate solenoid-controlled boom, provided the ability to obtain a 7:1 turn-down ratio. This was accomplished by sizing the three nozzles to provide a flow rate ratio of 4:2:1. When a maximum rate was required at a location, all three solenoids were energized and fertilizer flowed from all nozzles. When a lower rate was prescribed, a subset of the solenoids were turned off, and the flow for those booms routed back to tank through a matched set of orifices to maintain the pressure balance in the system. With this nozzle arrangement, the system could apply seven discrete rates (plus zero) evenly arranged up to the maximum rate set by the Integra system. For small-plot research, it was important to minimize any misapplication due to flow leaking past the nozzle check valve after the primary solenoid was closed. Therefore, the pressure in those lines was eliminated by three secondary solenoids that were energized when the respective primary solenoid was de-energized, allowing the pressure to escape back to tank and allowing the check valves to re-seat.

A Microsoft Visual Basic program was developed to control this system of solenoid valves (Fig. 7). The software read data from a differential GPS (DGPS) receiver, from a variety of off-line map types, and could also interface with a variety of commercially available active crop canopy sensors (Sudduth et al., 2015) as well as other sensors (e.g. ultrasonic sensors for crop height estimation, infrared thermometers for crop canopy temperature). Multiple algorithms for VRA application on corn (Solari et al., 2008; Kitchen et al., 2010) and wheat

Figure 6 Liquid VRA system for in-season N application in corn and wheat.

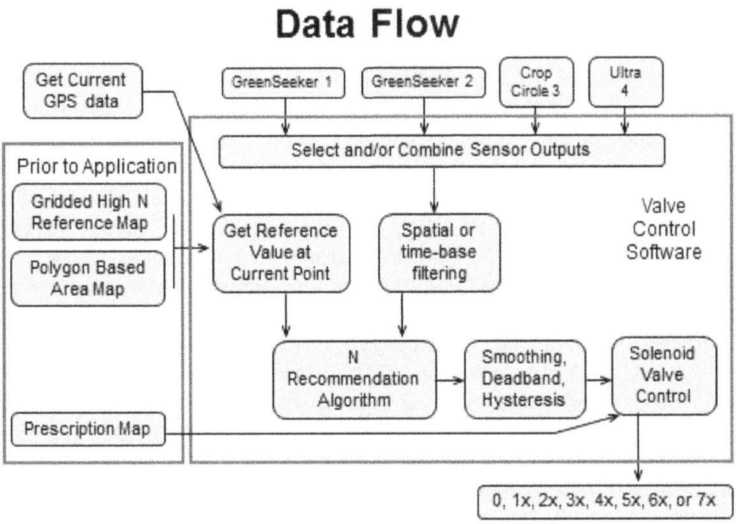

Figure 7 Data flow in the VRA valve control software.

(Raun et al., 2005) were integrated into the valve control software. The software included a number of parameters that could be modified to more precisely control flow and system behaviour, including spatial layout between the sensors and the boom tips, system latency and hysteresis. This level of accounting for system details was necessary to achieve the sub-meter spatial placement accuracy of application rate often desired in plot research applications.

The system provided the ability to implement many research designs. For example, a predefined response plot layout could be applied within a research field by loading a prescription map, setting the target 'maximum' rate, and engaging the application software. As the operator drove across the mapped area, the correct rate was placed in each cell of the grid as the applicator moved through it. For active canopy sensor-based VRA the user could select the canopy sensor(s) used, the crop and desired algorithm, and various parameters pertinent to that algorithm. In most cases, an N non-limiting area would be required, and the reading that the canopy sensors provided within that area, the 'high N reference value', would be a required parameter in most sensor-based VRA N algorithms. The system allowed flexibility in utilizing either a single reference value or a gridded map of variable high N references within each field. It also allowed definition of polygons where the N rate was non-limiting, such that the high N reference value could be updated dynamically whenever the applicator passed through one of these pre-defined polygons.

One early use of this system was VRA N application on a 20 ha field sowed to wheat in late October 2004. An N rate of 33 kg ha^{-1} was applied at planting. In the following February, an additional 100 kg N ha^{-1} was applied to a small area near the southwest corner of the field to create a high N reference area. On 6 April 2005, sensor-based VRA N was carried out using three GreenSeeker 505 sensors (Trimble, Sunnyvale, CA, USA) 3 m apart on a 9.1 m boom mounted to the front of the Spra-Coupe 3640 (Fig. 6). Data were collected at 10 Hz from the GreenSeeker sensors, while DGPS data were collected at 1 Hz. GreenSeeker data were collected over the reference area, resulting in a high N reference (NDVI) value of 0.80. The version of the Oklahoma State University wheat algorithm (Raun et al., 2005) used for this application incorporated both NDVI and a measure of the coefficient of variation (CV) of the NDVI, with a high CV indicating a sparse crop canopy. Using appropriate algorithm parameters, the maximum rate indicated by the algorithm was 90 kg N ha^{-1}, meaning the step size between applied N rates was 12.8 kg N ha^{-1}.

An aerial photo taken near the date of application (Fig. 8a) shows the condition of the field, with areas of little or no crop along the field edges where ponding was prevalent in headlands and areas along the central drainage channel which accumulated runoff from adjacent field areas but had very little slope to allow water to drain. Note that the VRA algorithm did an excellent job of applying less (and even zero) N in areas where the planted crop failed to emerge or develop (Fig. 8b). These included the areas described above and the east-west stripes in the southeastern quadrant of the field where poor stands were not explained by obvious drainage issues. Despite being spaced 3 m apart on the boom, all three GreenSeeker sensors provided very similar NDVI results (Fig. 9). While there appeared to be a slight bias between sensors, the general patterns and range of all three were consistent, and matched the spatial pattern evident in the aerial image quite well. As expected, they also showed a strong relationship to the N applied map (Fig. 8).

While field-average N use was reduced by 17% compared to conventional uniform application, the mean wheat yield of 4.6 Mg ha^{-1} was 31% higher than the county average for that year (3.5 Mg ha^{-1}). In contrast, the mean wheat yield on plots adjoining this field where N was applied at a uniform rate was 13% lower than the county average for the period 1994–2010 (Yost et al., 2016), suggesting a yield advantage with VRA N. Yield was quite variable over the field (Fig. 10) with patterns to some extent resembling spatial variations in crop stand (Fig. 8). Approximately 5% of the field where there was no viable crop stand and no potential yield received no N. In addition to the economic benefit, the lack of N application in those areas eliminated the likelihood of that N fertilizer being lost to the environment.

This research-oriented VRA system has evolved over the last 15 years with the incorporation of new sensors and algorithms supporting new approaches to in-season

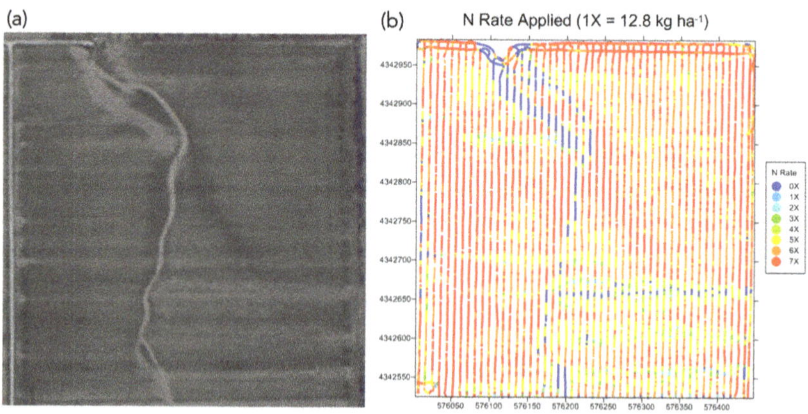

Figure 8 (a) Aerial photo and (b) applied N rate for 20 ha wheat field, 2005.

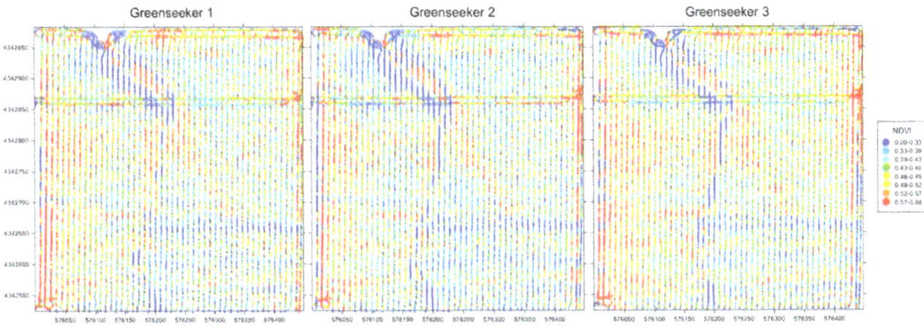

Figure 9 GreenSeeker NDVI maps of the field from three sensors spaced 3 m apart.

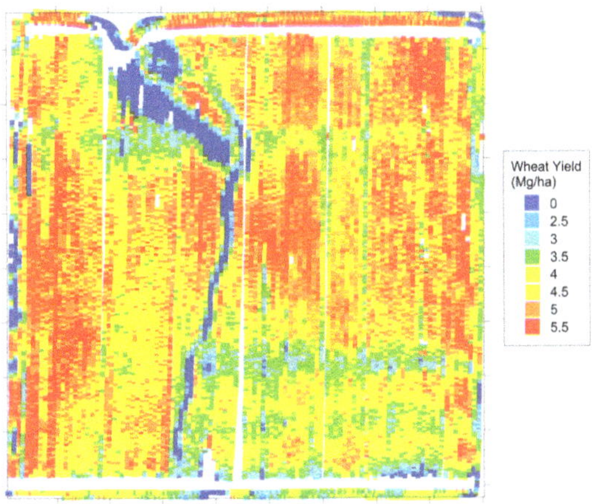

Figure 10 Yield map for wheat in 2005.

VRA N. During that period, the system has facilitated a variety of research projects. A large number of on-farm studies have documented the agronomic and economic (Kitchen et al., 2010; Scharf et al., 2011) and environmental (Roberts et al., 2010) benefits of canopy-sensor-based N application in corn. In-season, VRA N application on corn and wheat with this system was also a key part of a decade-long PA system field experiment (Yost et al., 2017).

6 Current status

Today, VRA technologies are widely used by producers practising PA. A recent survey of retail crop input dealers in the United States (Erickson and Widmar, 2015) found that, in 2015, at least 70% of the respondents offered their customers some type of VRA fertilizer application, while 27% offered VRA pesticide application. In addition, many producers implement VRA without the assistance of a dealer. The average proportions of dealer market area receiving VRA fertilizer and pesticide were 60% and 14%, respectively. Interestingly, the use of GPS-enabled sprayer section control was higher than VRA, being reported by 74% of the dealers, with use on an average of 33% of their market area. Producers can easily see the value of section control in not spraying an area more than once, and in not spraying non-cropped areas. In contrast, they may be more hesitant to adopt VRA technologies where there can be some level of uncertainty in the rate control algorithms used.

A current impediment to adoption, noted by 42% of dealers in the same survey, is incompatibility across PA equipment and technology. This incompatibility may be mechanical or electrical/electronic, but more often it is related to data formats. There are many different companies offering PA hardware, software and services, with many of these adopting their own data formats. For example, a VRA map generated within office software from one company may not be in the same data format required by a VRA spreader from another company. Generally, software will allow import and/or export of data in some other formats, but this may not be sufficient, and in any case imposes another step in the process. Various organizations are involved in developing standards for PA data interchange, including the AgGateway industry consortium in North America (Applegate et al., 2016) and the Farm-Oriented Open Data in Europe project (Řeznik et al., 2016). However, standards development and implementation is often a lengthy process, and the benefits of improved interoperability may not find their way into commercial systems for some time.

Research has shown VRA to provide environmental and economic benefits. Studies on the environmental aspects of VRA fertilizer and pesticide were reviewed by Bongiovanni and Lowenberg-DeBoer (2004). Much of the research up to that time attributed environmental advantages to VRA, including improved N use efficiency, reduction in nitrate leaching and reduction in overall fertilizer and pesticide application rates. However, it was noted that only a few of these studies actually measured environmental impacts, relying in most cases on inferred environmental benefits based on reduced chemical loadings. This caveat remains true at this time, as it is expensive and difficult to measure environmental variables at the field scale.

Articles reporting economic benefits of VRA N based on in-season canopy sensing in field crops include Scharf et al. (2011) and Godwin et al. (2003). Nawar et al. (2017), in a review of management zones for VRA fertilization, presented a case study where combining sensed soil and crop information improved profitability of VRA N in oil seed rape production. Bongiovanni and Lowenberg-DeBoer (2004) provide information on VRA N profitability from several studies. Generally, VRA N systems where control was based on real-time, in-season information (e.g. proximal canopy sensors, remote sensing) were more profitable than those based on soil variability alone. This is likely due to the ability of in-season sensing to more closely estimate economically optimal N rate, which varies greatly both temporally and spatially.

In a review of VRA herbicide studies, Swinton (2005) noted that if all costs were accounted for, VRA was rarely profitable. However, he speculated that advances in real-time, sensor-driven VRA, coupled with emerging DSS, could reduce costs and improve profitability. Tona et al. (2018) simulated the profitability of VRA fungicide in vineyards and orchards, finding mixed results among the different scenarios analysed. It is worthwhile to note that environmental regulations limiting pesticide application rates may provide an added economic advantage to VRA pesticides (Karkee et al., 2014). The same may be true in some cases for VRA fertilizer application.

Many of the technical innovations in delivery systems, control systems and data sources (i.e. sensors) described in this chapter have moved from research to commercialization.

Commercialization has happened more quickly in field crop applications than in specialty crops such as vegetables and tree crops. This is detailed by Karkee et al. (2014), who describe and contrast the current (as of 2014) state of the art in pesticide application systems for specialty crops and field crops. Currently, producers and researchers alike are challenged with applying VRA systems in ways that are profitable and increase sustainability. The key in this regard is the development and use of robust and accurate application algorithms that accurately represent fertilizer and pesticide requirements in the face of both spatial and temporal variability.

7 Future trends and conclusion

Future VRA systems will exhibit enhanced precision compared to today's systems, in terms of both the spatial placement of the pesticide or fertilizer and the appropriate rate. In VRA pesticide systems, enhanced spatial precision may result from a transition from section control to individual nozzle control. This would allow spraying to be more precisely restricted to locations where, for example, weeds were sensed or conditions were such that fungicide application was warranted. Enhanced rate precision may be facilitated by improved control systems and equipment designs that reduce transient errors due to rate changes. Improved DSS for weed and pest management are also expected to contribute to more precise rate recommendations.

New and improved sensors will provide important information to optimize future VRA systems. Sensors that detect and identify weeds will allow precise herbicide placement. Soil and crop sensors that estimate levels of plant-available nutrients will provide spatial and rate precision for fertilizer decisions. Because application rates in most current systems are indirectly inferred, new sensors that can directly measure application rate (Back et al., 2014; Cool et al., 2017) may offer enhanced accuracy in closed-loop control systems.

In future VRA fertilizer systems, particularly for N, precision will be enhanced through better forecasting of the future needs of the crop. This will likely be accomplished through a combination of real-time sensor data, mapped soil information and weather and crop growth forecast models. However, it remains to be seen how quickly such integrated systems can transition from research to commercial use. For this to occur, difficult data issues such as interoperability, connectivity, security and privacy need to be resolved.

8 Where to look for further information

Introductory material on VRA can be found in a recently published textbook chapter by Sharda et al. (2018). Another source of basic information is the online document by Grisso et al. (2011). Although somewhat dated, the overview by Anderson and Humburg (1997) is still relevant, as are the pertinent chapters in Ess and Morgan (2010). Those interested in VRA for crop protection may find the book by Oerke et al. (2010) to be a good reference.

Current advances in VRA technology are often presented at the International Conference on Precision Agriculture, European Conference on Precision Agriculture and Asian Conference on Precision Agriculture. Information about and links to proceedings papers from these three biennial conferences can be found at the website of the International Society of Precision Agriculture (www.ispag.org). International journals that often publish VRA research include

Applied Engineering in Agriculture, Biosystems Engineering, Computers and Electronics in Agriculture, Precision Agriculture and *Transactions of the ASABE.*

Specifics of available VRA technology change rapidly as new devices, implements and machines are commercialized. Several e-magazines and websites, including www.precisionag.com and www.newaginternational.com, provide reviews of advances in VRA and other technologies that may facilitate staying abreast of the latest trends.

9 References

Anderson, N. W. and Humburg, D. S. (1997), Chapter 12: Application equipment for site-specific management, *in* F. J. Pierce and E. J. Sadler (Eds), *The State of Site-Specific Management for Agriculture* (pp. 245–81). Madison, WI: ASA, CSSA and SSSA.

Applegate, D. B., Berger, A. W., Berne, D. T., Bullock, R., Craker, B. E., et al. (2016), Toward geopolitical-context-enabled interoperability in precision agriculture: AgGateway's SPADE, PAIL, WAVE, CART and ADAPT, *Proceedings of the 13th International Conference on Precision Agriculture*, https://ispag.org/proceedings/?action=abstract&id=2177 (accessed 1 May 2018).

ASABE (2009), Procedure for measuring distribution uniformity and calibrating granular broadcast spreaders, Standard S341.4, St Joseph, MI: American Society of Agricultural and Biological Engineers.

Back, S.-W., Yu, S.-H., Kim, Y.-J., Chung, S.-O. and Lee, K.-H. (2014), An image-based application rate measurement system for a granular fertilizer applicator, *Transactions of the ASABE*, 57(2), 679–87.

Bongiovanni, R. and Lowenberg-DeBoer, J. (2004), Precision agriculture and sustainability, *Precision Agriculture*, 5, 359–87.

Chen, Y., Zhu, H. and Ozkan, H. E. (2012), Development of variable-rate sprayer with laser scanning sensor to synchronize spray outputs to tree structures, *Transactions of the ASABE*, 55(3), 773–81.

Chen, Y., Ozkan, H. E., Zhu, H., Derksen, R. C. and Krause, C. R. (2013a), Spray deposition inside tree canopies from a newly developed variable-rate air-assisted sprayer, *Transactions of the ASABE*, 56(6), 1263–72.

Chen, Y., Zhu, H., Ozkan, H. E., Derksen, R. C. and Krause, C. R. (2013b), Spray drift and off-target loss reductions with a precision air-assisted sprayer, *Transactions of the ASABE*, 56(6), 1273–81.

Cool, S. R., Pieters, J. G., Seatovic, D., Mertens, K. C., Nuyttens, D., et al. (2017), Development of a stereovision-based technique to measure the spread patterns of granular fertilizer spreaders, *Sensors*, 17, 1396. doi:10.3390/s17061396.

Corwin, D. L. and Lesch, S. M. (2005), Apparent soil electrical conductivity measurements in agriculture, *Computers and Electronics in Agriculture*, 46, 11–43.

Crozier, C. R. and Roberson, G. T. (2014), Granular fertilizer spreader swath uniformity achieved at varying target application rates, *Crop Management*. doi:10.2134/CM-2013-0103-RS.

Erickson, B. and Widmar, D. A. (2015), 2015 precision agricultural services dealership survey results, West Lafayette, IN: Purdue University, http://agribusiness.purdue.edu/files/resources/2015-crop-life-purdue-precision-dealer-survey.pdf (accessed 21 February 2018).

Ess, D. R. and Morgan M. T. (2010), *The Precision Farming Guide for Agriculturalists*, Moline, IL: John Deere Publishing.

Felizardo, K. R., Mercaldi, H. V., Cruvinel, P. E., Olivera, V. A. and Steward, B. L. (2016), Modeling and model validation of a chemical injection sprayer system, *Applied Engineering in Agriculture*, 32(3): 285–97.

Fulton, J. P., Shearer, S. A., Higgins, S. F., Darr, M. J. and Stombaugh, T. S. (2005a), Rate response assessment from various granular VRT applicators, *Transactions of the ASAE*, 48(6), 2095–103.

Fulton, J. P., Shearer, S. A., Higgins, S. F., Hancock, D. W. and Stombaugh, T. S. (2005b), Distribution pattern variability of granular VRT applicators, *Transactions of the ASAE*, 48(6), 2053–64.

Fulton, J. P., Shearer, S. A., Higgins, S. F. and McDonald, T. P. (2013), A method to generate and use as-applied surfaces to evaluate variable-rate fertilizer application, *Precision Agriculture*, 14, 184–200.

Gebhardt, M. R., Day, C. L., Goering, C. E. and Bode, L. E. (1974), Automatic sprayer control system, *Transactions of the ASAE*, 17(6), 1043–7.

Giles, D. K. and Comino, J. A. (1990), Droplet size and spray pattern characteristics of an electronic flow controller for spray nozzles, *Journal of Agricultural Engineering Research*, 47, 249–67.

Godwin, R. J., Richards, T. E., Wood, G. A., Welsh, J. P. and Knight, S. M. (2003), An economic analysis of the potential for precision farming in UK cereal production, *Biosystems Engineering*, 84(4), 533–45.

Grisso, R., Alley, M., Thomason, W., Holshouser, D. and Roberson, G. T. (2011), Precision farming tools: Variable-rate application, Virginia Cooperative Extension Publication 442-505, Blacksburg, VA: Virginia Tech University, https://pubs.ext.vt.edu/content/dam/pubs_ext_vt_edu/442/442-505/442-505_PDF.pdf (accessed 2 July 2018).

Gundy, G. J., Dille, J. A. and Asebedo, A. R. (2017), Efficacy of variable rate soil-applied herbicides based on soil electrical conductivity and organic matter differences, *Advances in Animal Biosciences*, 8(2), 277–82.

Han, Y. J., Bode, L. E. and Hummel, J. W. (1986), Controlling chemical application rate with bypass spray nozzles, *Transactions of the ASAE*, 29(5), 1221–7.

Han, S., Hendrickson, L. L., Ni, B. and Zhang, Q. (2001), Modification and testing of a commercial sprayer with PWM solenoids for precision spraying, *Applied Engineering in Agriculture*, 17(5), 591–4.

Hartkamp, A. D., White, J. W. and Hoogenboom, G. (1999), Interfacing geographic information systems with agronomic modeling: A review, *Agronomy Journal*, 91, 761–72.

Heege, H. J., Reusch, S. and Thiessen, E. (2008), Prospects and results for optical systems for site-specific on-the-go control of nitrogen-top-dressing in Germany, *Precision Agriculture*, 9, 115–31.

Holland, K. H., Lamb, D. W. and Schepers, J. S. (2012), Radiometry of proximal active optical sensors (AOS) for agricultural sensing, *IEEE Journal of Selected Topics in Applied Earth Observations and Remote Sensing*, 5, 1793–1802.

Hughes, K. L. and Frost, A. R. (1985), A review of agricultural spray metering, *Journal of Agricultural Engineering Research*, 32, 197–207.

Kang, F., Pierce, F. J., Walsh, D. B., Zhang, Q. and Wang, S. (2011), An automated trailer sprayer system for targeted control of cutworm in vineyards, *Transactions of the ASABE*, 54(4), 1511–19.

Karkee, M., Steward, B. and Kruckeberg, J. (2014), Automation of pesticide application systems, *in* Q. Zhang and F. J. Pierce (Eds), *Agricultural Automation: Fundamentals and Practices* (pp. 263–93), Boca Raton, FL: CRC Press.

Kitchen, N. R., Sudduth, K. A., Drummond, S. T., Scharf, P. C., Palm, H. L., et al. (2010), Ground-based canopy reflectance sensing for variable-rate nitrogen corn fertilization, *Agronomy Journal*, 102, 71–84.

Krebs, M., Rautmann, D., Nordmeyer, H. and Wegener, J. K. (2015), Development of a direct injection system without time lag for application of plant protection products, *Landtechnik*, 70(6), 238–52.

Kuo, B. C. and Golnaraghi, F. (2002), *Automatic Control Systems*, Hoboken, NJ: John Wiley & Sons.

Kweon, G. and Grift, T. E. (2006), Feed gate adaptation of a spinner spreader for uniformity control, *Biosystems Engineering*, 95(1), 19–34.

Liu, H. and Zhu, H. (2016), Evaluation of a laser scanning sensor on detection of complex shaped targets for variable-rate sprayer development, *Transactions of the ASABE*, 59(5), 1181–92.

Liu, H., Zhu, H., Shen, Y., Chen, Y. and Ozkan, H. E. (2014), Development of digital flow control system for multi-channel variable-rate sprayers, *Transactions of the ASABE*, 57(1), 273–81.

Luck, J. D., Pitla, S. K., Shearer, S. A., Mueller, T. G., Dillon, C. R., et al. (2010), Potential for pesticide and nutrient savings via map-based automatic boom section control of spray nozzles, *Computers and Electronics in Agriculture*, 70, 19–26.

Luck, J. D., Pitla, S. K., Sama, M. P. and Shearer, S. A. (2015), Flow, spray pattern, and droplet spectra characteristics of an electronically actuated variable-orifice nozzle, *Transactions of the ASABE*, 58(2), 261–9.

Miller, P. C. (2003), Patch spraying: future role of electronics in limiting pesticide use, *Pest Management Science*, 59, 566–74.

Miller, M. S. and Smith, D. B. (1992), A direct nozzle injection controlled rate spray boom, *Transactions of the ASAE*, 35(3), 781–5.

Morris, D. K., Ess, D. R., Hawkins, S. E. and Parsons, S. D. (1999), Development of a site-specific application system for liquid animal manures, *Applied Engineering in Agriculture*, 15(6), 633–8.

Nawar, S., Corstanje, R., Halcro, G., Mulla, D. and Mouazen, A. M. (2017), Delineation of soil management zones for variable-rate fertilization: A review, *Advances in Agronomy*, 143, 175–244.

Oerke, E.-C., Gerhards, R., Menz, G. and Sikora, R. A. (2010), *Precision Crop Protection – the Challenge and Use of Heterogeneity*, Dordrecht, the Netherlands: Springer.

Oliveira, L. F., Scharf, P. C., Vories, E. D., Drummond, S. T., Dunn, D., et al. (2013), Calibrating canopy reflectance sensors to predict optimal mid-season nitrogen rate for cotton, *Soil Science Society of America Journal*, 77, 173–83.

Ortiz, B. V., Perry, C., Sullivan, D., Lu, P., Kemerait, R., et al. (2012), Variable rate application of nematicides on cotton fields: A promising site-specific management strategy, *Journal of Nematology*, 44(1), 31–9.

Peteinatos, G. G, Weis, M., Andújar, D., Ayala, V. R. and Gerhards, R. (2014), Potential use of a ground-based sensor technologies for weed detection, *Pest Management Science*, 70, 190–9.

Raun, W. R., Solie, J. B., Johnson, G. V., Stone, M. L., Mullen, R. W., et al. (2002), Improving nitrogen use efficiency in cereal grain production with optical sensing and variable rate application, *Agronomy Journal*, 94, 815–20.

Raun, W. R., Solie, J. B., Stone, M. L., Martin, K. L., Freeman, K. W., et al. (2005), Optical sensor-based algorithm for crop nitrogen fertilization, *Communications in Soil Science and Plant Analysis*, 36, 2759–81.

Reichard, D. L. and Ladd, T. L. (1983), Pesticide injection and transfer system for field sprayers, *Transactions of the ASAE*, 25(2), 258–60.

Řeznik, T., Lukas, V., Charvát, K., Charvát Jr., K., Horáková, Š., et al. (2016), FOODIE data models for precision agriculture, *Proceedings of the 13th International Conference on Precision Agriculture*, https://ispag.org/proceedings/?action=abstract&id=1983 (accessed 1 May 2018).

Roberts, D. F., Kitchen, N. R., Scharf, P. C. and Sudduth, K. A. (2010), Will variable-rate nitrogen fertilization using corn canopy reflectance sensing deliver environmental benefits? *Agronomy Journal*, 102, 85–95.

Samborski, S. M., Tremblay, N., and Fallon, E. (2009), Strategies to make use of plant sensors-based diagnostic information for nitrogen recommendations, *Agronomy Journal*, 101, 800–16.

Scharf, P. C., Shannon, D. K., Palm, H. L., Sudduth, K. A., Drummond, S. T., et al. (2011), Sensor-based nitrogen application out-performed producer-chosen rates for corn in on-farm demonstrations, *Agronomy Journal*, 103(6), 1683–91.

Sharda, A., Fulton, J. P., McDonald, T. P., Zech, W. C., Darr, M. J., et al. (2010), Real-time pressure and flow dynamics due to boom section and individual nozzle control on agricultural sprayers, *Transactions of the ASABE*, 53(5), 1363–71.

Sharda, A., Luck, J. D., Fulton, J. P., McDonald, T. P. and Shearer, S. A. (2013), Field application uniformity and accuracy of two rate control systems with automatic section capabilities on agricultural sprayers, *Precision Agriculture*, 14, 307–22.

Sharda, A., Fulton, J. P. and Taylor, R. K. (2016), Performance of variable-orifice nozzles for liquid fertilizer applications, *Applied Engineering in Agriculture*, 32(3), 347–52.

Sharda, A., Franzen, A., Clay, D. E. and Luck, J. (2018), Precision variable equipment, *in* D. K. Shannon, D. E. Clay and N. R. Kitchen (Eds), *Precision Agriculture Basics* (pp. 155–68), Madison, WI: ASA, CSSA and SSSA.

Shen, Y., Zhu, H., Liu, H., Chen, Y. and Ozkan, E. (2017), Development of a laser-guided embedded-computer-controlled air-assisted precision sprayer, *Transactions of the ASABE*, 60(6), 1827–38.

Solari, F., Shanahan, J., Ferguson, R., Schepers, J. and Gitelson, A. (2008), Active sensor reflectance measurements of corn nitrogen status and yield potential, *Agronomy Journal*, 100, 571–9.

Souza, E. G., Scharf, P. C. and Sudduth, K. A. (2010), The influence of sun position and clouds on reflectance and vegetation indices of greenhouse-grown corn, *Agronomy Journal*, 102, 734–44.

Steward, B. L. and Humburg, D. S. (2000), Modeling the Raven SCS-700 chemical injection system with carrier control with sprayer simulation, *Transactions of the ASAE*, 43(2), 231–45.

Stone, M. L., Solie, J. B., Raun, W. R., Whitney, R. W., Taylor, S. L., et al. (1996), Use of spectral radiance for correcting in-season fertilizer nitrogen deficiencies in winter wheat, *Transactions of the ASAE*, 39, 1623–31.

Stone, M. L., McKee, K. D., Formwalt, C. W. and Benneweis, R. K. (1999), ISO 11783: An electronic communications protocol for agricultural equipment, ASAE Distinguished Lecture # 23, St. Joseph, MI: American Society of Agricultural and Biological Engineers.

Stone, M. L., Benneweis, R. K. and Bergeijk, J. V. (2008), Evolution of electronics for mobile agricultural equipment, *Transactions of the ASABE*, 51(2), 385–90.

Sudduth, K. A., Drummond, S. T. and Kitchen, N. R. (2015), Operational characteristics of commercial crop canopy sensors for nitrogen application in maize, *in* J. V. Stafford (Ed.), *Precision Agriculture 15 – Proceedings of the 10th European Conf. on Precision Agriculture* (pp. 51–8), Wageningen, the Netherlands: Wageningen Academic Publishers.

Swinton, S. M. (2005), Economics of site-specific weed management, *Weed Science*, 53(2), 259–63.

Tompkins, F. D., Howard, K. D., Mote, C. R. and Freeland, R. S. (1990), Boom flow characteristics with direct chemical injection, *Transactions of the ASAE*, 33(3), 737–43.

Tona, E., Calcante, A. and Oberti, R. (2018), The profitability of precision spraying on specialty crops: A technical-economic analysis of protection equipment at increasing technological levels, *Precision Agriculture*, 19(4), 606–29. doi:10.1007/s11119-017-9543-4.

Tremblay, N., Bouroubi, M. Y., Bélec, C., Mullen, R., Kitchen, N. R., et al. (2012), Corn response to nitrogen is influenced by soil texture and weather, *Agronomy Journal*, 104, 1658–71.

Vidrine, C. G., Goering, C. E., Day, C. L., Gebhardt, M. R. and Smith, D. B. (1975), A constant pesticide application rate sprayer model, *Transactions of the ASABE*, 18(3), 439–43.

Vondricka, J. and Schulze Lammers, P. (2009a), Real-time controlled direct injection system for precision farming, *Precision Agriculture*, 10, 421–30.

Vondricka, J. and Schulze Lammers, P. (2009b), Measurement of mixture homogeneity in direct injection systems, *Transactions of the ASABE*, 52(1), 61–6.

Vondricka, J. and Schulze Lammers, P. (2009c), Evaluation of a carrier control valve for a direct nozzle injection system, *Biosystems Engineering*, 103, 43–8.

Womac, A. R. and Bui, Q. D. (2002), Design and tests of a variable-flow fan nozzle, *Transactions of the ASAE*, 45(2), 287–95.

Yost, M. A., Kitchen, N. R., Sudduth, K. A., Sadler, E. J., Baffaut, C., et al. (2016), Long-term impacts of cropping systems and landscape positions on claypan-soil grain crop production, *Agronomy Journal*, 108, 713–25.

Yost, M. A., Kitchen, N. R., Sudduth, K. A., Sadler, E. J., Drummond, S. T., et al. (2017), Long-term impact of a precision agriculture system on grain crop production, *Precision Agriculture*, 18(5), 823–42.

Zhu, H., Ozkan, H. E., Fox, R. D., Brazee, R. D. and Derksen, R. C. (1998), Mixture uniformity in supply lines and spray patterns of a laboratory injection sprayer, *Applied Engineering in Agriculture*, 14(3), 223–30.

Zhu, H., Rosetta, R., Reding, M. E., Zondag, R. H., Ranger, C. M., et al. (2017), Validation of laser-guided variable-rate sprayer for managing insects in ornamental nurseries, *Transactions of the ASABE*, 60(2), 337–45.

Chapter 8

Spray technologies in precision agriculture

Paul Miller, Silsoe Spray Applications Unit Ltd, UK

1 Introduction

The application of plant protection products plays a key role in the production of most crops. Methods of application depend on the type of crop being treated, with the main ones being:

- Field crops – mainly treated with boom sprayers;
- Bush and tree crops – treated with air-assisted or tunnel-type sprayers;
- Glasshouse and specialist food crops – treated with purpose-designed and -built equipment – often at small scale, manually operated and with handheld delivery systems.

This chapter is concerned only with equipment designed to operate with field crops although many of the principles discussed will relate to other application systems. For field crops grown in Northern European conditions, the costs of plant protection products typically account for between 30% and 50% of the direct costs of production, and so there are important financial incentives in ensuring that such products are applied in a manner that maximises their effectiveness and minimises the quantities applied. Many such products have a toxicological profile that means that it is important to avoid exposure to non-target surfaces and organisms and their use is regulated based partly on the effect that such substances may have on non-target systems. An important performance parameter for any application system therefore relates to the extent to which off-target exposure can be minimised.

http://dx.doi.org/10.19103/AS.2017.0032.09

When considering the role that spray applications can play within precision agriculture approaches for field crops, there are three scales that are important, namely:

- A whole field/plot area scale where it is important to ensure that the applied dose is effectively controlled and delivered in a manner that will maximise effectiveness and minimise losses to non-target systems;
- A patch spraying scale that involves treating parts of a field area differently. Each patch can be subject to separate operations determined from sensing systems used at the time of application or from a treatment map generated from data collected at other times;
- A plant scale treatment where applications are directed at single plants or small groups of plants with a spot treatment applied to the detected plant(s). Such approaches are usually operated in real time with a sensing system to identify plant(s) to be treated and a delivery system that applies product only to the detected targets and with the minimum contamination of surrounding plants and areas.

Each of the above scales will involve particular characteristics of the application system, but there will be many features that are common to all scales. In this chapter, some general features of spraying equipment for treating field crops that are relevant to precision agriculture are considered before addressing features that are relevant to each of the above scales and reviewing some case studies.

It has been recognised that computer-based control systems are likely to be at the centre of any spray application system operating as part of a precision agriculture environment with key components shown in Fig. 1 concerning:

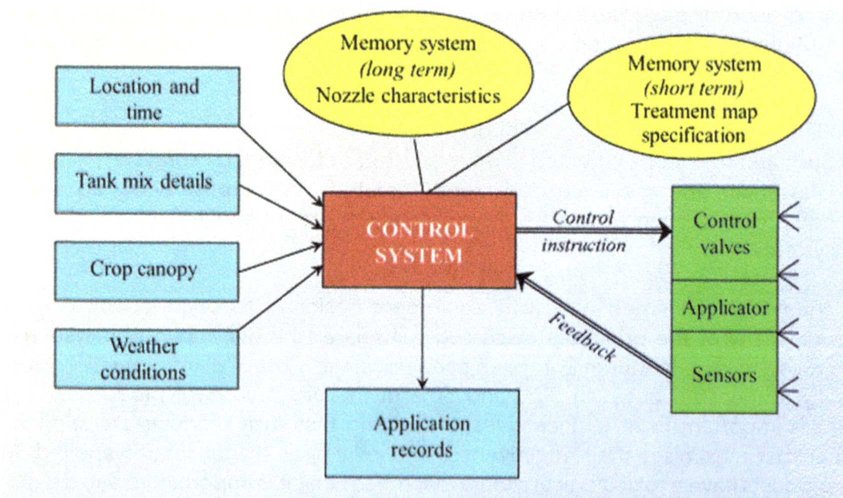

Figure 1 Typical structure of a sensing and control system for a crop sprayer operating as part of a precision agriculture approach. Source: Adapted from Miller (2003a).

- Inputs relating to location and time, the physical properties of the liquid to be delivered, condition of the crop and target pest/weed/disease and weather conditions;
- An on-board memory containing details of the machine characteristics including the specifications of the delivery nozzle that will be updated for each batch of treatments requiring a given set-up;
- A loadable memory containing information that will include a treatment map for individual fields;
- A control algorithm for the delivery system including sensors to provide feedback information;
- A means to automatically monitor and record the treatments applied.

2 Features of field crop sprayers for precision agriculture

2.1 Control of delivered dose

For most conventional crop sprayers, the delivered dose is controlled by selecting a nozzle size to deliver a defined flow rate that, at a given operating pressure and nominal forward speed, will apply a set volume of spray liquid per unit area. This is verified in a calibration routine and the machine then operates to apply the set volume of liquid containing a fixed concentration of active product(s) loaded into the machine. Changes in applied dose that would result from changes in the nominal forward speed are accommodated by changes in pressure at the nozzle, and the machine will typically monitor pressure and/or flow rates to give information to the operator as well as a basis for the pressure-based dose control system. However, since flow rate through an orifice (the nozzle) is a function of the square root of pressure, relatively large changes in pressure are required for a given change in flow rate. In practice, this limits the turn-down ratio that can be achieved with conventional dose control systems to some 1.0:1.25 when fitted with conventional nozzles and 1.0:1.40 when fitted with 'extended range/variable pressure' nozzles. This ratio is mainly a function of the spray volume distribution pattern achieved by such nozzles although it has been recognised that changes to the droplet size distribution with changing pressure may further limit this ratio (Paice et al., 1996). Combellack and Miller (1999) proposed that the definition of turn-down ratio for spatially variable application should be based on a constant droplet size distribution with a tolerance in the volume median diameter (v.m.d.) of ±10%, but there is no evidence that this has been accepted or taken up commercially. Most of the turn-down ratio associated with conventional and extended range flat fan nozzles will probably be used, accommodating variations in spraying speeds such that there is little scope to use pressure control systems to change the applied dose in a precision spraying system. Paice et al. (1996) indicated that a turn-down ratio of 3:1 would be desirable to accommodate forward speed variations, but in practice the use of a set calibrated speed and a defined field layout will mean that turn-down ratios in the order of 1.5:1 are adequate to compensate for variations in forward speed in many situations. However, other approaches have been used to change applied dose levels, and some of these have also been used to extend the speed range over which effective control of the delivered dose can be maintained as well as facilitating variable dose applications. Most dose control systems currently available for precision application change applied dose by changing the volume of a dilute mixture of plant protection products and water such that changes in dose will also result in changes in the applied volume of spray liquid.

2.1.1 On/off control

On/off control is probably the simplest form of dose control that can be used on crop sprayers and this can be implemented using solenoid valves having relatively rapid response times (<10.0 ms – Rietz et al., 1997; Miller and Watt, 1980), giving good spatial resolution in the direction of travel. Most conventional boom sprayers control the output to sections of nozzles on a boom such that the simplest control would involve switching boom sections on or off. Spatial resolution in the direction along the boom is then a function of section width (typically 3.0–6.0 m) although the current trend towards individual nozzle control will improve this resolution, typically down to 0.5 m – see Section 2.2. With a single spray line on a boom, on/off control only gives two levels of delivered dose – full and zero. This then requires a very high level of performance from any detection and mapping approaches. Computer modelling studies reported by Paice et al. (1998) suggested that such an approach would not give stable long-term control of grass weeds in cereal crops. The use of multiple boom lines with on/off control on both lines can increase the number of dose levels that can be achieved but will incur additional cost, complexity and increased difficulties when flushing and cleaning the sprayer.

On/off controls have been used in conjunction with other control strategies in the design of systems for the spatially variable application of plant protection products – for example Gerhards and Oebel (2006) and Paice et al. (1996).

2.1.2 Nozzle clusters (multiple nozzle systems)

Multiple nozzle holders that mount up to four nozzles in a single holder have been developed commercially (Sökefeld, 2010) and used to accommodate changes in forward speed and the requirements for spatially variable application. Nozzles can be switched individually or simultaneously to obtain a wide range of outputs such that turn-down ratios of more than 10:1 can be achieved but with some changes in the droplet size distribution/spray quality. Miller et al. (1997) described the performance of such a system in which a three-nozzle arrangement could achieve a turn-down ratio of 5:1, with the nozzles in each assembly being actuated pneumatically. The sequencing of the nozzle on/off control in each cluster allowed for the time taken to switch off a spray and establish a spray following the opening of the control valves such that gaps in the output delivery were avoided.

The arrangement of using multiple nozzles in a single holder can also be used to adjust the physical characteristics of a delivered spray to match spatially variable requirements. For example, when spraying close to a field boundary, there may be a requirement to minimise the risk of drift even if this increases the risk of reduced efficacy in the performance of the applied products. Miller et al. (1997) suggested that a switching system based on two nozzles would provide an adequate range of dose delivery to accommodate forward speed changes. Therefore, if a four-nozzle assembly were to be fitted with two conventional flat nozzles and two air-induction nozzles, then the conventional nozzles could be used to treat the main part of a field and the air-induction nozzles used close to the field boundary where high levels of drift control are required with the switching done automatically from a field map.

2.1.3 Twin-fluid nozzles and spinning discs

Twin-fluid (or gaseous energy) nozzles use both pressurised air and liquid supplies into the nozzle body and, for a given nozzle configuration, the resulting flow rate and droplet size distribution is a function of the two pressures (Matthews et al., 2014). Studies by Western et al. (1989) confirmed that nozzle output and the droplet size distribution could be controlled independently for relatively low flow rates and that the drift risk with such nozzles was lower than that for conventional nozzles operating at comparable flow rates. Sökefeld (2010) suggested that early commercial designs of twin-fluid nozzle ('Airtec' by Cleanacres Machinery Ltd, Hazleton, Cheltenham, Gloucestershire, GL54 4DX, UK and 'AirJet' by TeeJet Technologies, Wheaton Facility, 200 W North Avenue, Glendale Heights, Illinois, 60139, USA) were capable of turn-down ratios of 2:1 whilst retaining the same droplet size distribution/spray quality.

A subsequent design of twin-fluid nozzle specifically aimed at use in precision agriculture described by Combellack and Miller (1998, 1999) achieved turn-down ratios of up to 5.5:1 whilst retaining an almost constant droplet size and with an air input requirement that was some 50% lower than that of earlier designs. Wind tunnel studies with this nozzle design also showed that it could operate with levels of drift in the order of 25% of those of conventional nozzles at comparable flow rates.

For spinning disc and cage systems, the droplet size produced is a function of both liquid flow rate and rotational speed, and therefore a spray with a given mean droplet size can be generated over a range of flow rates by adjusting rotational speed. Such a characteristic could be valuable when aiming to vary the dose of chemical delivered but with a defined droplet size distribution/spray quality. Frost (1981) showed that the range of flow rates that could be used to produce a given droplet size increased as the droplet size increased. Heijne (1978) showed that a commercial spinning disc system could achieve a turn-down ratio of 7:1 when mean droplet sizes ranged from 100 to 350 μm. Parkin and Siddique (1990) reported results obtained with a rotary cage atomiser, showing a 10:1 turn-down ratio (flow rates from 2.0 to 20.0 L/min) for mean droplet sizes in the range of 100–180 μm.

2.1.4 Pulse width modulation

Pulse width modulation is a common method used for controlling electrically powered devices by turning the power supply on and off very quickly. Giles and Comino (1990) described a system in which a rapid response solenoid valve was closely coupled to a spray nozzle, with the valve being operated at frequencies in the order of 10 Hz. Flow through the assembly was controlled by the valve duty cycle – the proportion of the time that the valve was held open for each cycle. Results from studies with this system showed that when operating with flat fan nozzles, the system could achieve a turn-down ratio of 3.21:1, with the droplet size increasing as the flow rate was reduced such that v.m.d. increased by some 8.1% across the flow rate range. Giles et al. (1999) used this control approach to develop a system for the precision application of agricultural plant protection products. Trials with this system showed that following a 3:1 step increase in flow rate, the system overshot the set point by approximately 20% but was within 5% of the set point in 5 seconds. Since the system enabled operating pressures to be reduced and flow rate maintained, this enabled the small droplet component within a spray to be reduced – a strategy that could then be used to control drift when operating in sensitive areas such as those close to field boundaries. Recent commercial developments based on this principle have integrated the automatic control of nozzle pressure and duty cycle into a unit that is then compatible with conventional spray rate controllers.

2.1.5 Variable orifice nozzles

Variable orifice nozzle designs have been developed based on different operating principles. Some designs have used materials or a collection of elements arranged to form an orifice configured such that as pressure is fed into the nozzle, the size of the output orifice increases by elastic deformation. The performance of such systems tends to be variable and unpredictable, and there are few performance data relating to such systems in the published literature. An adaptation of this principle for applying liquid fertilisers has been described by Brady et al. (2014) in which a pre-orifice in a stream jet nozzle was covered by a flexible sheath that expands as pressure is applied to the nozzle inlet. Results from tests with the design showed that turn-down ratios in the order of 5:1 could be achieved compared with a value of 1.7:1 for a conventional stream jet nozzle.

A number of alternative designs typically use a spring and/or diaphragm arrangement to adjust the size of a metering orifice within a nozzle by moving a conical shaped plunger positioned above the orifice. As the supply pressure is increased, the conical plunger lifts out of the orifice, increasing the effective orifice area and thus giving a larger increase in flow rate than what would have been achieved through a fixed orifice. Womac and Bui (2002) described and analysed the performance of such an arrangement that used a diaphragm and a control liquid pressure to adjust the position of the metering plunger. They found that a turn-down ratio of 13:1 could be achieved in output flow rates, with the sensitivity to input pressure being related to the control pressure on the diaphragm as expected. For a given control pressure, droplet size increased almost linearly with flow rate (and hence input pressure), with smaller droplet sizes measured at the higher pressure settings (control and line pressures) as expected. The performance of a commercial design of variable orifice nozzle (the 'Varitarget' nozzle) was studied by Daggupati (2007) who found that the design could achieve a turn-down ratio in flow rate of around 3.5:1 when operating at pressures of between 300 and 700 kPa, with commercial data suggesting that turn-down ratios as high as 10:1 can be achieved over a pressure range of 100–600 kPa and depending on the nozzle tip fitted to the assembly but without close control of the droplet size distribution produced. The results presented by Daggupati (2007) confirmed that mean droplet sizes tended to increase with increasing flow rate through this design of nozzle system.

An important feature of the performance of nozzle systems involving moving parts, such as the 'Varitarget' nozzle, is hysteresis – the difference between a flow rate setting when flow is increasing and that when flow is decreasing. Unpublished data for this type of nozzle design suggest that hysteresis can result in flow variations of up to ±30% of a nominal or mean value – see Fig. 2.

2.1.6 Injection metering systems

Injection metering systems are a potentially attractive method for controlling the dose delivery of a crop sprayer since keeping the diluent (normally water) separate from the

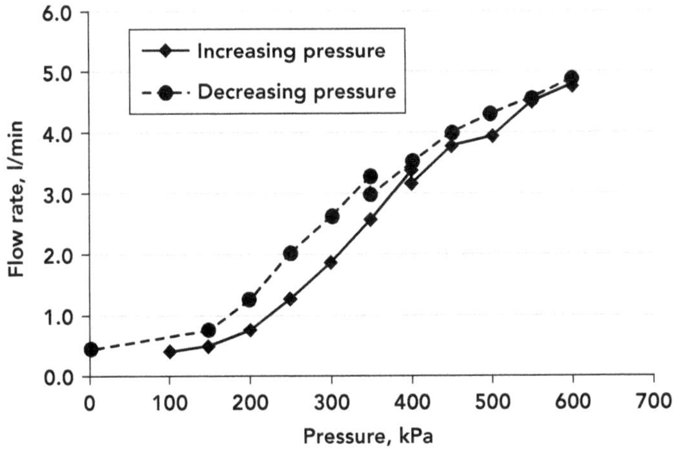

Figure 2 Measured characteristics of a commercial design of variable orifice nozzle showing the wide range of flow rate and the effect of hysteresis.

formulated products until the time of application can reduce the machine contamination and hence the wash-out and flushing requirements as well as enabling different active ingredients at different doses to be applied in a spatially variable manner. Injection metering also enables the dose of chemical formulation(s) to be varied independently of the applied volume and to target different formulations in a spatially variable manner.

There are two main arrangements for injection metering systems on crop sprayers (Miller, 2003b):

1 With the metering pump(s) positioned upstream of the main system pump. In this case, the metering pump is operating against a relatively low pressure head, and therefore, a wide range of pump types can be used including the peristaltic design. However, since the output capacity of the main pump is greater than the maximum flow rate through all the nozzles, there needs to be a re-circulation loop such that unused dilute spray liquid is returned to the input side of the main pump. This in turn means that when the system responds to a step change in the demanded output, the response has a ramped characteristic as the effects of the recirculating mixture are accounted for. Because dilute plant protection products are passing through the main pump, there is greater level of machine contamination than is the case when injection takes place downstream of the main pump – see (2) below. Pipework carrying dilute chemicals tends to be relatively long, and there is no opportunity to deliver different mixtures or concentrations to different parts of the boom.
2 With the metering pump(s) positioned downstream of the main pump system. With this arrangement, the metering pump is working to deliver formulated product into spray lines that are at full spraying pressure, and therefore, the metering pump needs a characteristic that is largely independent of the pressure that it must deliver into. This means that piston pumps are commonly used for such applications, but the main pump is operating only with the diluent (water). Injection can take place at any point between the main pump and the nozzles and, although some allowance needs

to be made for mixing, experiments have shown that this can generally be achieved in substantially less than 1.0 m length of pipe. Miller and Smith (1992) proposed a system that directly injected formulated products into the nozzle, but it is likely that such an approach would require a metering pump for each nozzle since dividing a metered flow could give inaccurate application if, for example, a nozzle feed line became blocked. This makes such a system relatively expensive and complex. Having long lengths of feed line carrying concentrated formulated products under pressure can give problems with respect to the risk of failure (leakage) and with cleaning and flushing procedures.

Any injection metering system can operate in two possible modes:

1 With variable spray concentration, such that changes in dose are delivered by changing the concentration of formulation(s) in the spray liquid. Changes to the formulations included in the spray liquid can also be accommodated with this approach. This mode of operation has the potential advantage that the nozzle used can be selected to give spray characteristics that closely match target requirements (see Section 2.3) and be used for the range of dose deliveries. However, system response times can then be relatively slow as changes in concentration move through the pipework in the sprayer to the nozzles. Antuniassi et al. (1997) examined the response characteristics of some injection metering systems and indicated that response times to a step change in demand may be greater than 15 s particularly where injection is on the input side of the main pump. Frost (1990) discussed how plumbing components can be used in practical designs and Paice et al. (1995) showed that response times could be substantially reduced by using small bore pipework and accounting for pressure drops that this may incur via the system control unit.
2 With a constant spray concentration, such that changes in dose are delivered by changing the volume of spray liquid using one of the approaches outlined above. The main use of the injection metering system operating in this mode is to keep the active formulated products separate from the water and hence facilitate reduced residual spray liquid and the need for flushing and cleaning, rather than the control of delivered dose.

All injection metering pumps must be able to operate with the full range of plant protection product formulations. This will involve the ability to resist chemical attack and corrosion and to accommodate a wide range of flow characteristics including viscosity which may, for example, vary with temperature. Frost (1990) described a system in which formulated products are delivered from a metering cylinder by pumping water into the base of the cylinder and with the formulated product separated from the water by a piston. In this arrangement, the metering pump needs only to overcome friction within the system, and the use of a calibrated gear pump acted as both pump and flow meter in a closed loop control arrangement. This system was used in an experimental patch sprayer described by Paice et al. (1995) but has not been developed beyond the commercial prototype stage.

While a number of injection metering systems have been developed commercially, the uptake of such systems has been limited probably due to the following factors identified by a number of authors including Sudduth et al. (1995):

1 The relatively high cost and complexity of such systems.
2 The difficulty in accommodating a wide range of formulations. Most injection metering systems are aimed at using liquid formulations, although systems for granules (but not powders) have been developed, and the wide range of concentrate volumes that constitute a single dose that makes the design of a system that will accommodate the full range very difficult. Metering pumps will typically operate over a 10:1 range of flow rates, while formulations may have full dose volumes ranging from 250 mL/ha to 5.0 L/ha. Many systems therefore have different metering pump size options to accommodate the range of formulations – again adding to the expense and complexity.
3 The interface with the formulation container, the need in some cases to return any unused formulation back into its original container and problems associated with the cleaning, flushing and priming of delivery lines carrying concentrated formulations.

2.1.7 Other systems

Bjugstad et al. (2012) described a system for control of dose delivery using two tanks – one loaded with dilute formulations at the maximum concentration to be used and the other with clean water. Each tank was connected to a conventional diaphragm pump and the flows from the two tanks were mixed to change the delivered concentration depending on spatial specification (a treatment map) and the speed of the sprayer. Results of tests with an experimental machine showed that the system delivered the required dose within ±5.0% for 95% of the treated area and had the advantage of being relatively insensitive to the physical properties of the formulations.

2.2 Spatial resolution

The resolution at which spatially selective spraying is operated is likely to be a major factor in determining its cost-effectiveness. Two considerations are relevant:

- the spatial distributions of the weed/disease/pest to be treated together with the timescales over which these might change; and
- the characteristics of the delivery system that will influence both the lateral and longitudinal resolution that can be achieved.

Much of the analysis examining the required spatial resolution has focused on the application of herbicides to weed patches, with fewer studies directed at fungicides or insecticides. Grass weed patches in arable fields are relatively stable and persistent, and therefore, mapping such patches as a basis for deriving a treatment map is feasible. Rew et al. (1996, 1997) analysed the effects of spatial resolution when treating patches of grass weeds in arable crops and, as expected, found that the savings in herbicide use reduced as spatial resolution was increased. For example, using a typical strategy for controlling common couch (*Elymus repens*) and increasing the spatial resolution from 2.0 to 4.0 m reduced the herbicide saving from 70% down to 63%. Paice et al. (1996) concluded that, considering the likely increased cost of mapping and treatment at the finer scale, the practical optimum resolution was likely to be at least 4.0 m. Miller and Lutman (2008) also recognised the higher costs associated with operating at a finer resolution particularly when the mapping of weed patches was based on manual surveying. It is possible that the potential development of automated weed detection methods may enable mapping and treatment at a small resolution in an economically viable way (Murdoch et al., 2014).

While fungal diseases may develop in patches (Dammer, 2010), currently such disease patches would need to be mapped manually and treated quickly after mapping since, in certain conditions, the disease could spread rapidly. As with weed patches, sensors to automatically detect crop diseases would change the options for applying spatially variable treatments, but such sensors are not yet commercially available. This suggests that a resolution equal to the width of a boom would be adequate for such treatments. Similarly, the spatial resolution to treat pest infestations is unlikely to be at less than the width of the boom.

Most of the options for controlling the dose delivered by crop sprayers outlined in Section 2.1 operate across the full width of a boom. The current trend for increasing boom widths on commercial sprayer designs for operating in field crops (now up to 48 m) means that if spatially variable application were to be implemented using current equipment, the lateral scale of resolution would be very high. While it is technically feasible to envisage separate pressure control or injection metering systems for different boom sections, this substantially increases the cost and complexity of the complete machine, and this has not been attempted commercially to date. On/off control with either single or multiple spray lines can be readily implemented on conventional sprayer designs at a scale corresponding to boom section widths – typically 3.0–6.0 m. Commercial sprayers are now available with individual nozzle control via Controller Area Network (CAN) (e.g. Househam Sprayers Ltd, Woodhall Spa, Lincolnshire, LN10 6YQ, UK), and this approach would facilitate a finer lateral resolution for on/off control and potentially for pulse width modulation systems. Nozzles are commonly spaced at 0.5 m intervals along a boom but arranged to give sprays that fully overlap so that any control strategy at the resolution of the nozzle spacing will not have a sharp cut-off edge.

Longitudinal resolution is a function of the response time of the sprayer and the spraying speed. Systems based only on the use of solenoid valves are therefore likely to have a longitudinal resolution approximately equivalent to boom section widths (circa 1.0 s at 4.0 m/s). Those with pressure control systems would be approximately equivalent to the boom length, and some injection metering systems would be much longer but predictable so that a control system could look ahead with map-based systems.

2.3 Matching physical characteristics of sprays to target requirements

While it is likely that most precision spray applications would involve variations in dose, it has been shown that different target geometries retain different proportions of an emitted spray depending on the spray quality/droplet size distribution generated by the nozzle. Results from laboratory studies measuring deposits on idealised 1.0 mm stainless steel rods mounted vertically showed large differences between different nozzle designs producing different droplet size distributions – see Fig. 3.

The results in Fig. 3 have been linked with results from field trials that show higher levels of control of grass weeds in cereals crops when treated with finer sprays having a smaller mean droplet size distribution (Butler Ellis et al., 2008). However, the use of finer

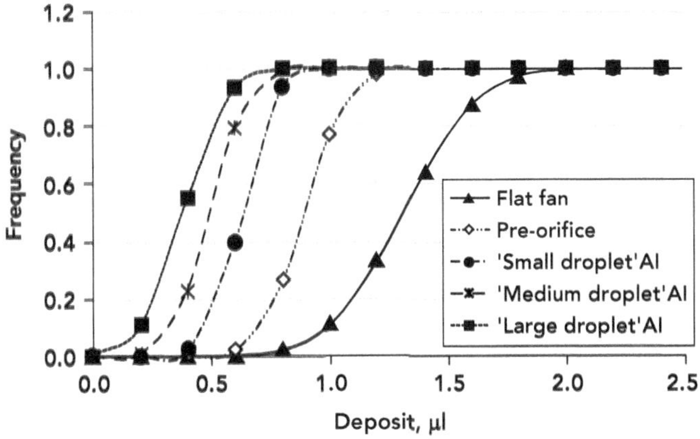

Figure 3 Deposit distributions measured on vertical 1.0-mm diameter rods sprayed with different nozzle designs, all having a flow rate of 0.8 L/min at a pressure of 300 kPa. Source: Adapted from Miller et al. (2010).

sprays will incur a greater risk of drift (see Section 2.4) such that it may be appropriate to use a finer spray only when treating a patch of grass weeds. There is therefore scope to consider spatially variable applications using nozzles with characteristics that are directly matched to the spray target particularly when using multiple nozzle systems as discussed in Section 2.1.2 although such approaches have not been widely developed to date.

2.4 Minimising drift and exposure of systems outside the treatment area

The control of spray drift is now an important aspect relating to the operation of field crop sprayers with implications for managing the exposure of non-target organisms and surfaces. Using buffer zones, both within the cropped area and at the edge of the crop, is now an established method for controlling the exposure to pesticides of non-target systems such as surface water and non-target plants in many countries. Concerns relating to toxicity levels and the sensitivity of non-target systems to pesticide exposure in recent years have led to proposals to increase the widths of such buffer zones, enabling the approval of products where the degree of risk reduction offered by a smaller buffer zone would not be sufficient (Bryon and Hamey, 2008). The use of relatively large buffer zones is not popular with farmers. One way of enabling the widths of buffer zones to be reduced is to use nozzles/spraying arrangements that have been shown to reduce drift when operating close to field boundaries – for example as in the LERAP scheme in the UK (Anon., 2001). The widths of buffer zones, the extent by which they can be reduced and the widths of field area close to the boundary that must be sprayed with the drift-reducing method depend on the details of the scheme and the product(s) to be applied. Farmers may be reluctant to use the drift-reducing application arrangements over the whole field area because of a risk to efficacy. Mapping the regions close to field boundaries with the specifications relating to buffer zone implementation allows compliance to be achieved automatically with an added advantage that such compliance can be recorded.

3 Case study 1: designing and developing a system for spot treatment of volunteer potatoes

The control of volunteer potatoes in crops such as onions, leeks and carrots is an important issue influencing the yield, quality and ease of harvesting of such crops. Traditionally the control of such weeds has been by repeated overall spraying of selective herbicides, but the revision of European Regulations EU 91/414 relating to the registration of active substances for plant protection has meant that many of the formulated products used for such applications are no longer available. The options to use selective herbicides for the control of weeds such as volunteer potatoes are now very limited. Spot treatment with a total herbicide was identified as a possible alternative to overall spraying based on the technology available (e.g. Hague and Tillett, 2001) and the fit with the agronomic requirements for the vegetable crops considered. The work to develop the spot spraying system involved:

1 The development of a weed detection algorithm (not discussed in detail in this chapter);
2 The development of an application system that would give adequate deposits on the target weed with the minimum potential contamination of surrounding crop plants;
3 The design and construction of an experimental prototype machine;
4 Laboratory experiments and field trials to assess the performance of system components and the complete machine.

The weed detection algorithm identified the plan outline of a detected weed and positioned this with respect to a spray boom and nozzles that would apply the total herbicide (Miller et al., 2012a) – see Fig. 4. The process of identifying the weed outlines and associated spray nozzle schedules was repeated at a rate of 30 Hz in successive images as they passed through the field of view of the camera system (Fig. 5) to refine the calculation of the area to be treated. Provision was made to spray a proportion of this plan area as a variable that could be adjusted in the field depending on the relative risks of damage to neighbouring crop plants and the control of the weed. This was determined by assessing whether nozzles at the edge of the area to be sprayed could remain off while still achieving the required percentage of the treatment area. If this was the case, then the switching on and off for nozzles passing over the central part of the target was eroded equally until the preset proportion of weed plan area to be treated was reached.

The specification for the nozzles concerned the uniform application of herbicide sprays based on existing formulation types, to areas less than 100 mm square when travelling at a speed of 5.0 km/h with the minimum of drift and splash outside of the treated area. This involved the rapid establishment and cut-off for a spray in a total cycle time of 0.025 s. Nozzle spacing was to be matched to crop rows but would be in the order of 100 mm. A review of potential nozzle systems meeting this specification identified

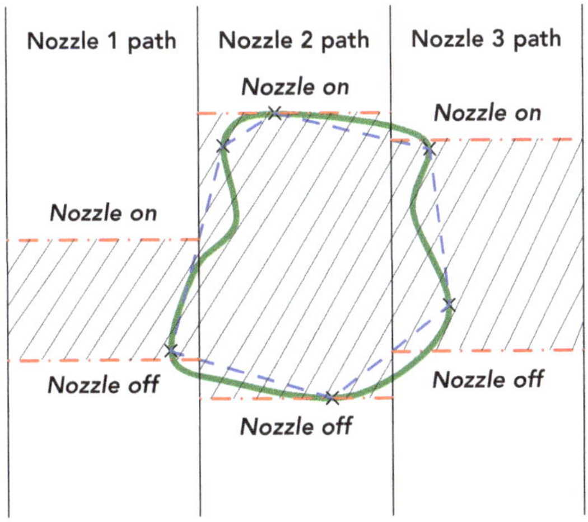

Figure 4 Diagram showing how a detected weed shape was mapped to the control of spray nozzles.

Figure 5 An example of an analysed image showing the centre of a carrot crop bed (blue crosses), crop rows (as green lines) and detected volunteer potatoes bounded by polygons (in red).

two candidate designs: a fluidic type that generated an oscillating liquid stream from the nozzle by a passive arrangement of feed channels within the nozzle body and an oscillating hypodermic syringe nozzle in which the needle was mounted on a vibrating arm and excited by an electrical vibrator. An evaluation of these two options was conducted (Miller et al., 2012b), and while both were shown to technically meet the specification, the fluidic design was selected for a prototype machine based on commercial considerations. Further development of this nozzle design was then undertaken in conjunction with Hypro EU Ltd, Station Road, Longstanton, Cambridge, CB24 3DS, UK, to produce a final design (the 'Alternator' nozzle – see Fig. 6a) that incorporated a small latching solenoid matched to the nozzle body and the ability to change spray tips within the nozzle body using a cartridge system. The performance of the nozzle assembly was assessed in terms of the

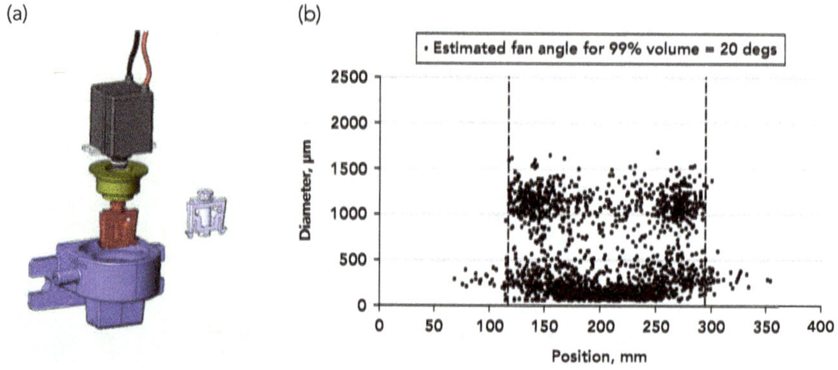

Figure 6 (a) Left: nozzle and solenoid valve cartridge assembly showing two options for the fitted spray tip. (b) Right: droplet size and spatial distribution pattern for the nozzle operating at 75 kPa pressure as measured with an imaging system (see Tuck et al., 1997) showing large droplets with a bi-modal size distribution and a relatively sharp cut-off at the edge of the pattern.

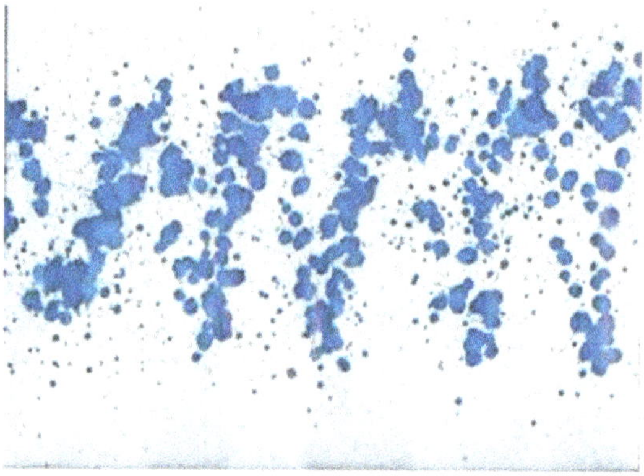

Figure 7 Pattern achieved from the fluidic nozzle supplied with water and tracer dye at 0.75 bar running right to left at a height of 0.5 m and a speed of 2.10 m/s.

droplet size (Fig. 6b), velocity and spray volume distribution patterns (Fig. 7) generated both with a static nozzle in still air and with a moving nozzle in wind tunnel conditions (Miller et al., 2012b).

The performance of the unit was evaluated in a range of crop and field conditions, assessing the levels of weed control achieved, damage to crop plants and the deposition of sprays both on target weeds and the surrounding crop plants – see Fig. 8. Detailed performance assessments were conducted when spraying mugwort (*Artemisia vulgaris*) in a crop of leeks with two machine settings: one spraying the central 40% of a weed image (conservative strategy aimed at minimising the risk of crop damage) and the other spraying 80% of a weed image (riskier strategy aimed at a high weed kill). Results showed that deposits on targeted weeds were more than two orders of magnitude greater than

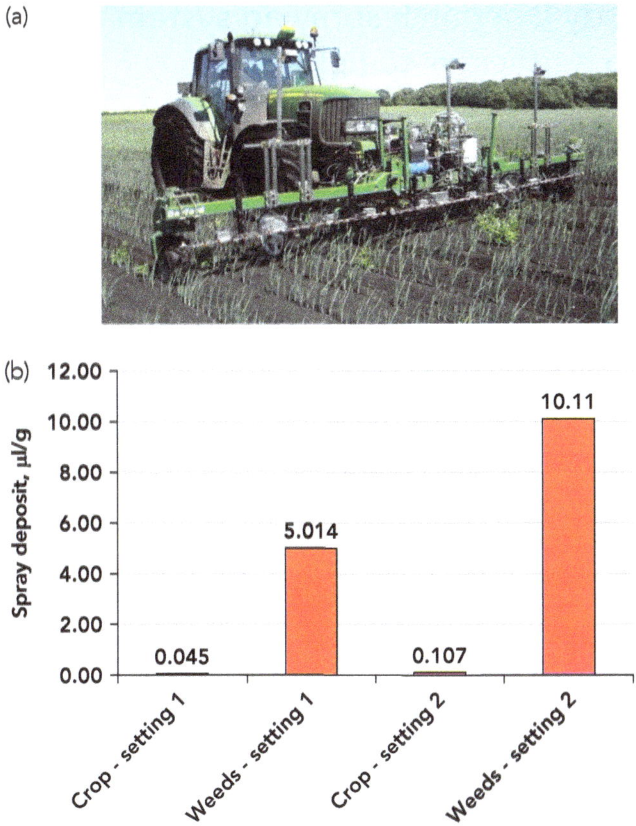

Figure 8 (a) The final prototype version of the spot spraying system working in a crop of leeks and (b) the results of measurements of spray deposits on target weeds and crop plants within 150 mm radius of a treated plant for two settings of the machine.

those on surrounding crop plants (Fig. 8b), with deposits for the two settings directly as expected. Overall results from a series of field trials in a range of crop, weed and weather conditions showed that the unit could achieve levels of weed control that were at least competitive with all alternative approaches and with commercially acceptable levels of crop damage (Miller et al., 2012a).

It was recognised that a key factor leading to the high levels of weed kill and low levels of crop damage recorded with the system was that the large weeds targeted retained the relatively large droplets effectively. Provision was made with the nozzle cartridge assemblies to fit nozzles, giving much finer spray qualities for treating patches of small grass weeds in a range of crop situations. As a result of the experimental work and prototype development, a commercial machine was produced and offered for sale by Garford Farm Machinery Ltd, Hards Lane, Frognall, Deeping St. James, Peterborough, PE6 8RP, UK, who partnered the project work throughout.

4 Case study 2: a patch spraying system for applying herbicides to field crops

When considering the design and development of a patch spraying system that would provide a first-level step for the commercial introduction of such a system, the approaches as demonstrated and described by Paice et al. (1995) based on a combination of injection metering, multiple spray lines, small bore delivery pipes and on/off control were regarded as too complex and expensive for such an introductory system. It was therefore decided to use a multiple nozzle system (see Section 2.1.2) using three conventional nozzles to deliver flow rates over a 5:1 range at a defined spray quality. An important objective for the development was the treatment of grass weed patches in cereal crops and therefore the ability to maintain a fine/medium spray quality was important – see Section 2.3.

A summary specification for the system was defined as follows (Miller, 2003b; Miller et al., 1997):

- An accuracy of delivered dose to better than ±5.0% of target;
- A spatial resolution in the transverse direction to be 4.0 m or greater with a response time of the dose control system of less than 1.0 s such that resolution in the longitudinal direction was of a similar magnitude (depending on spraying speed);
- An ability to operate with a wide range of commercially available formulations.

The control system for the mounted sprayer was linked to a unit in the tractor that supported a field location system based on satellite navigation (GPS) and which also carried the treatment map specifying the doses to be applied in different parts of the field. Weed patch mapping was by manual observations recorded using the same basic unit as mounted on the tractor either using an all-terrain scouting vehicle or with the unit mounted in a combine harvester and weed presence/absence recorded using a push button system. Recorded weed patch maps (Fig. 9a) were transformed into treatment maps (Fig. 9b) using software that added buffers around identified weed patches to account for:

- seed movement since visual mapping was undertaken in one season to provide treatment maps that would be used in the following season;
- any positioning errors during either mapping or treatment;
- the response time characteristics of the dose control system.

The spray delivery system initially used three nozzles at each position along the boom with the switching of nozzles and control of spraying pressure implemented using a compressed air arrangement powered with an air compressor driven from the pump shaft on the sprayer. Experience from field trials with the unit indicated that for many applications, an adequate turn-down ratio could be achieved using a two-nozzle arrangement and this configuration was used in the later trials with the system.

The results from full-scale field trials with the system (Fig. 10) showed that the sprayer could deliver the dose to the specification, but that the development of treatment maps based on visual mapping of weed patches was time-consuming, expensive and prone to errors.

(a)
(b)

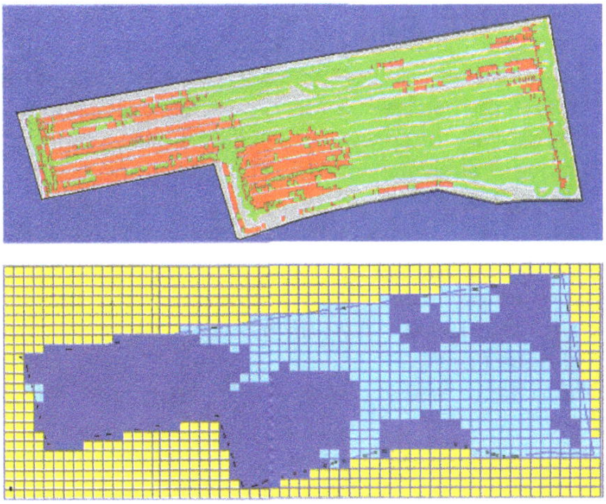

Figure 9 (a) Weed patches as mapped visually from a combine harvester; (b) treatment map derived from the weed patch map.

Figure 10 Experimental patch sprayer operating in a crop of peas.

5 Conclusion

Much of the discussion in this chapter has been set against the background of treating crops grown on relatively wide areas including the example of the spot treatment of large weeds in vegetable crops grown in rows, and this has strongly influenced the technologies that have been discussed. There is a considerable body of work that has examined the application of chemical and non-chemical treatments to individual plants using versions of autonomous robots, and such systems are likely to have applications in high-value crops grown in small areas. The effectiveness of all plant protection products is very influenced by the timeliness of application, and hence, work rates are important. Some schedules for applying such products recommend that treatments are to be applied in less than 3 days from the time of assessing the need for the treatment. The types of system included in this review are generally capable of operating in such timescales.

Many of the developments of targeted spraying systems took place in the late 1990s or early 2000s with a focus for much of the work directed at the use of herbicides for weed control because of the evidence of patchiness and the speed of propagation of weeds. The performance of systems has been shown to meet specifications particularly as far as the spatially variable delivery of sprays is concerned. However, the commercial uptake of such systems has been relatively slow. In 2008, Miller and Lutman cited the following reasons for this limited uptake:

- The economic case for adoption is not clear with the costs of equipment and mapping not readily recovered particularly for high dose/low dose strategies;
- Patch spraying is regarded as risky with high risks particularly associated with errors in weed (patch) detection;
- A lack of field data to show that the risks associated with patch spraying are manageable;
- A reluctance by farmers and agronomists to spend time and effort manually mapping weed patches recognising that research investigating automated systems for weed detection could change this in the future;
- No current regulatory incentive to reduce herbicide use even though Government (and European) policy is to minimise use.

The current situation is much as it was in 2008. Some more recent studies have aimed at quantifying the performance and advantages of systems based on the principles outlined in this chapter. For example, work by Zaman et al. (2011) considered the development of a spot sprayer using conventional narrow spray angled nozzles coupled to solenoid valves as part of a system based on the real-time detection of weeds using height discrimination and ultrasonic detectors. Results from this work demonstrated the feasibility of spot application in blueberry crops. González-de-Soto et al. (2016) evaluated the performance of a robotised patch sprayer that used an injection metering system and conventional nozzles as the basis for spray delivery control and concluded that such a system could achieve treatment and control of more than 99.5% of detected weeds with substantial savings in herbicide use. This study involved assumptions relating to idealised weed detection and evaluated the performance of the spray delivery in both laboratory and field conditions. A technical-economic analysis by Tona et al. (2017) suggested that while on-off and canopy adjusted application systems could be cost-effective over conventional approaches, more complex approaches could probably not be justified assuming the likely costs of units required to deliver such treatments. There have been commercial developments over the last decade that have refined the design of a number of the systems described in this chapter particularly to reduce costs and improve performance and reliability. There are now indications that some of these systems are being involved in developments with realistic commercial objectives and potential outcomes.

Work to automatically detect weed patches has made progress (see e.g. Murdoch et al., 2014; Weis and Sökefeld, 2010), and is considered elsewhere in this book, but there is currently no commercially available system operating to do so. Problems relating to herbicide resistance for some grass weed species have become more serious, and now some farmers are mapping the heaviest infestations of blackgrass patches in cereal crops with a view to applying a total herbicide to the patch destroying both weed and crop but preventing seed return to the soil in the patch.

Many sprayers designed for operation in field crops now have a computer-based control system with links to a field location system using satellite navigation methods and a map-based recording system. Such systems are already delivering improved precision in spray application and will provide the platform for further improvements in the future both with and without links to spatial variability.

6 Future trends

While current crop spraying systems can get data relating to the field in which they are operating from downloaded maps or using field location and mapping systems, all of the data relating to the products that are being applied are currently input manually from the product label or use settings/configurations that the operator derives from information on the product label. If the sprayer control system could have information relating to the product to be applied, then this would enable decisions relating to optimised application to be made automatically particularly if inputs from other sensors relating to the crop and weather conditions are also available – see Fig. 1. The potential for reading product labels on the sprayer and automatically recording the quantities and details of formulations loaded into the machine has been investigated and demonstrated (Watts et al., 2003; Peets et al., 2009), but, to date, these systems have not been developed commercially. The particular advantages of such an approach would relate to:

- An improved automated management of drift risk accounting for local rules and conditions at the time of spraying;
- The potential to account for both crop and weather conditions at the time of spraying – sensors are already available to characterise the crop canopy, but further research is required to relate crop canopy conditions to application parameters to optimise responses;
- The ability to automatically generate records that fully meet existing and possible future requirements.

It is also likely that new sensing systems for both weeds and diseases will be developed in the future (as discussed elsewhere in this book), and this will have an important influence on the potential for both spatial and temporally variable applications. Work on distinguishing weeds from crop continues to be a major technical subject for study and has been reviewed by a number of authors (e.g. Weis and Sökefeld, 2010). The detection of grass weed patches in cereal crops has recently made progress and is now close to commercial exploitation (see Murdoch et al., 2014 and commercial literature relating to 'eyeweed'). The situation relating to both diseases and insect pests is more challenging. Even with improved sensor technologies (West et al., 2010; Martinelli et al., 2015), it is likely that there will be few patchy diseases that can be detected in time for spatially variable spray treatments to be effective (West et al., 2010).

While spatial variability is an important component of precision spray application, it is also important to deliver the appropriate dose to the appropriate target to maximise biological control. The interface between application systems, decision support tools and improved sensing systems is likely to be the basis for key developments in the future.

There continues to be considerable research and commercial development interest in the platforms from which precision sprayers might operate. Autonomous vehicles have been developed and used experimentally but are not yet a commercial reality. Applications from unmanned aerial platforms (drones) are also being investigated for specialist conditions where the alternative would be manual application and where the limited payload of such aerial systems can be managed effectively.

7 Where to look for further information

Further information on the design and performance of agricultural spraying systems can be found in journals such as *Biosystems Engineering*, *Crop Protection* and *Precision Agriculture*. Details of nozzle design and performance are reported in *Atomization and Sprays*. The key conference relating to Pesticide Application is now that run biannually by the Association of Applied Biologists as 'International Advances in Pesticide Application' and with proceedings published in *Aspects of Applied Biology*. Key conference papers may also be reported in the proceedings of The American Society for Agriculture and Biological Engineers (ASABE).

8 References

Anon. (2001). Local Environmental Risk Assessment for Pesticides (LERAP). Horizontal Boom Sprayers, A step-by-step guide to reducing aquatic buffer zones in the arable sector. Pesticides Safety Directorate, Department for Environment, Food and Rural Affairs, London, UK.

Antuniassi, U. R., Miller, P. C. H. and Paice, M. E. R. (1997). Dynamic and steady-state dose response of some chemical injection metering systems. *Proceedings of the Brighton Crop Protection Conference – Weeds*, 17–20 November 1997. The British Crop Protection Council, Surrey, UK, pp. 587–692.

Bjugstad, N., Ensby, T., Holth, E. H., Moen, O. K. and Thorstensen, H. F. (2012). FlexiDose – a spraying system developed for variable pesticide dosage without using pesticide injection methods. *Aspects of Applied Biology, International Advances in Pesticide Application*, 144, 181–8.

Brady, M. J., Goddard, R. S. and Kately, S. J. (2014). A novel device for the variable rate application of liquid fertilisers 'Trident'. *Aspects of Applied Biology, International Advances in Pesticide Application*, 122, 411–13.

Butler Ellis, M. C., Miller, P. C. H. and Orson, J. H. (2008). Minimising drift while maintaining efficacy – the role of air-induction nozzles. *Aspects of Applied Biology, International Advances in Pesticide Application*, 84, 59–66.

Byron, N. and Hamey, P. (2008). Setting unsprayed buffer zones in the UK. *Aspects of Applied Biology, International Advances in Pesticide Application*, 84, 123–30.

Combellack, J. H. and Miller, P. C. H. (1998). Does the technology exist to efficiently and effectively patch spray weeds? In: *Precision Weed Management in Crops and Pastures*. R. W. Medd and J. E. Prately (Eds). Proceedings of a workshop 5–6 May 1990, Wagga Wagga. Published for CRC for Weed Management Systems, University of Adelaide, South Australia, Australia, pp. 55–61.

Combellack, J. H. and Miller, P. C. H. (1999). The development of a twin-fluid nozzle for precision agriculture. *Proceedings of the 1999 Brighton Conference – Weeds*, 15–18 November 1999. The British Crop Protection Council, Surrey, UK, pp. 473–80.

Daggupati, N. P. (2007). Assessment of the Varitarget nozzle for variable rate application of liquid crop protection products. Master of Science Thesis, Kansas State University, Manhattan, KS.

Dammer, K.-H. (2010). Variable rate application of fungicides. In: *Precision Crop Protection – The Challenge and the Use of Heterogeneity*. E.-C. Oerke, R. Gerhards, G. Menz and R. A. Sikora (Eds). Springer, Dordrecht, Heidelberg, London and New York, pp. 349–63.

Frost, A. R. (1981). Rotary atomisation in the ligament formation mode. *Journal of Agricultural Engineering Research*, 26, 63–78.

Frost, A. R. (1990). A pesticide injection metering system for use on agricultural spraying machines. *Journal of Agricultural Engineering Research*, 46, 55–70.

Gerhards, R. and Oebel, H. (2006). Practical experiences with a system for site-specific weed control in arable crops using real-time image analysis and GPS-controlled patch spraying. *Weed Research*, 46, 185–93.

Giles, D. K. and Comino, J. A. (1990). Droplet size and spray pattern characteristics of an electronic flow controller for spray nozzles. *Journal of Agricultural Engineering Research*, 47, 249–67.

Giles, D. K., Stone, M. L. and Dieball, K. (1999). Distributed network system for control of spray droplet size and application rate for precision chemical application. *Precision Agriculture'99*. J. V. Stafford (Ed.). Proceedings of the 2nd European Conference on Precision Agriculture, Sheffield Academic Press, Sheffield, UK, pp. 857–66.

González-de-Soto, M., Emmi, L., Perez-Ruiz, M., Aguera, J. and Gonzalez-de-Santos, P. (2016). Autonomous systems for precise spraying – Evaluation of a robotised patch sprayer. *Biosystems Engineering*, 146, 165–82.

Hague, T. and Tillett, N. D. (2001). A bandpass filter approach to crop row location and tracking. *Mechatronics*, 11(1), 1–12.

Heijne, C. G. (1978). A study of the effects of disc speed and flow rate on the performance of the Micron 'Battleship'. *Proceedings of the British Crop Protection Conference – Weeds*, 20–23 November 1978. The British Crop Protection Council, Surrey, UK, pp. 20–3.

Martinelli, F., Scalenghe, R., Davino, S., Panno, S., Scuderi, G., Ruisi, P., Villa, P., Stroppiana, D., Boschetti, M., Goulart, L. R., Davis, C. E. and Dandekar, A. M. (2015). Advanced methods of plant disease detection. A review. *Agronomy for Sustainable Development*, 35(1), 1–25.

Matthews, G. A., Bateman, R. and Miller, P. C. H. (2014). *Pesticide Application Methods*. 4th edn. John Wiley and Sons Ltd, West Sussex, UK.

Miller, P. C. H. (2003a). The current and future role of application in improving pesticide use. *Proceedings of the BCPC International Congress Crop Science and Technology*, 10–12 November 2003. The British Crop Protection Council, Surrey, UK, pp. 247–54.

Miller, P. C. H. (2003b). Handling and dose control. In: *Optimising Pesticide Use*. M. Wilson (Ed.). John Wiley and Son Ltd, West Sussex, UK, pp. 45–73.

Miller, P. C. H. and Combellack, J. H. (1997). The performance of an air/liquid nozzle system suitable for applying herbicides in a spatially selective manner. In: *Precision Agriculture'97*. J. V. Stafford (Ed.). Proceedings of the First European Conference on Precision Agriculture, Volume 1. BIOS Scientific Publishers, Oxford, UK, pp. 651–60.

Miller, P. C. H. and Lutman, P. J. W. (2008). A review of the factors influencing the technical feasibility and potential commercial uptake of the patch spraying of herbicides in arable crops. *Aspects of Applied Biology, International Advances in Pesticide Application*, 84, 265–72.

Miller, M. S. and Smith, D. B. (1992). A direct nozzle injection controlled spray boom. *Transactions of the ASAE*, 35(3), 781–5.

Miller, P. C. H. and Watt, B. A. (1980). The use of a high speed photographic technique to examine the response time characteristics of a solenoid spray valve. *Journal of Agricultural Engineering Research*, 25, 217–20.

Miller, P. C. H., Paice, M. E. R. and Ganderton, A. D. (1997). Methods of controlling sprayer output for spatially variable herbicide application. *Proceedings, Brighton Crop Protection Conference – Weeds*. The British Crop Protection Council, Surrey, UK, pp. 641–4.

Miller, P. C. H., Butler Ellis, M. C., Bateman, R., Lane, A. G., O'Sullivan, C. M., Tuck, C. R. and Robinson, T. H. (2010). Deposit distributions on targets with different geometries and treated with a range of spray characteristics. *Aspects of Applied Biology, International Advances in Pesticide Application*, 99, 241–8.

Miller, P. C. H., Tillett, N. D., Hague, A. and Lane, A. G. (2012a). The development and field evaluation of a system for the spot treatment of volunteer potatoes in vegetable crops. *Aspects of Applied Biology, International Advances in Pesticide Application*, 114, 113–20.

Miller, P. C. H., Tillett, N. D., Swan, T., Tuck, C. R. and Lane, A. G. (2012b). The development and evaluation of nozzle systems for use in targeted spot spraying applications. *Aspects of Applied Biology, International Advances in Pesticide Application*, 114, 159–66.

Murdoch, A., Flint, C., Pilgrim, R., De La Warr, P., Camp J, Knight, B., Lutman, P., Magri, B., Miller, P., Robinson, T., Sanford, S. and Walters, N. (2014). eyeWeed: Automating mapping of black-grass (*Alopecurus myosuroides*) for more precise applications of pre- and post-emergence herbicides and detecting potential herbicide resistance. *Aspects of Applied Biology, Crop Production in Southern Britain: Precision Decisions for Profitable Cropping*, 127, 151–8.

Paice, M. E. R., Miller, P. C. H. and Bodle, J. (1995). An experimental machine for evaluating spatially selective herbicide application. *Journal of Agricultural Engineering Research*, 60, 107–16.

Paice, M. E. R., Miller, P. C. H. and Day, W. (1996). Control requirements for spatially selective herbicide sprayers. *Computers and Electronics in Agriculture*, 14, 163–77.

Paice, M. E. R., Day, W., Rew, L. J. and Howard, A. (1998). A stochastic simulation model for evaluating the concept of patch spraying. *Weed Research*, 38, 373–8.

Parkin, C. S. and Siddique, H. A. (1990). Measurements of the drop spectra from rotary cage aerial atomizers. *Crop Protection*, 9, 33–8.

Peets, S., Gasparin, C. P., Blackburn, D. W. K. and Godwin, R. J. (2009). RFID tags for identifying and verifying agrochemicals in food traceability systems. *Precision Agriculture*, 10(5), 382–94.

Rew, L. J., Cussans, G. W., Mugglestone, M. A. and Miller, P. C. H. (1996). A technique for surveying spatial distribution of weeds and the potential reduction in herbicide use from patch spraying (*Elymus repens* L.) common couch, in cereal fields. *Weed Research*, 36, 283–92.

Rew, L. J., Miller, P. C. H. and Paice, M. E. R. (1997). The importance of patch mapping resolution for sprayer control. *Aspects of Applied Biology, Optimising Pesticide Applications*, 48, 49–56.

Rietz, S., Palyi, B., Ganzelmeier, H. and Laszlo, A. (1997). Performance of electronic controls for field sprayers. *Journal of Agricultural Engineering Research*, 68, 399–407.

Sökefeld, M. (2010). Variable rate technology for herbicide application. In: *Precision Crop Protection – the challenge and the use of Heterogeneity*. E.-C. Oerke, R. Gerhards, G. Menz and R. A. Sikora (Eds). Springer, Dordrecht, Heidelberg, London and New York, pp. 335–47.

Sudduth, K., Borgelt, S. C. and Hou, J. (1995). Performance of a chemical injection sprayer system. *Transactions of the ASAE*, 11(3), 343–8.

Tona, E., Calcante, A. and Oberti, R. (2017). The profitability of precision spraying on speciality crops: A technical-economic analysis of protection equipment at increasing technical levels. *Precision Agriculture* online (*in Press*).

Tuck, C. R., Butler Ellis, M. C. and Miller, P. C. H. (1997). Techniques for measuring droplet size and velocity distributions in agricultural sprays. *Crop Protection*, 16(7), 619–28.

Watts, A. J., Miller, P. C. H. and Godwin, R. J. (2003). Automatically recording sprayer inputs to improve traceability and control. *Proceedings, The BCPC International Congress – Crop Science and Technology*, 1. The British Crop Protection Council, Surrey, UK, pp. 323–8.

Weis, M. and Sökefeld, M. (2010). Detection and identification of weeds. In: *Precision Crop Protection – The Challenge and the Use of Heterogeneity*. E.-C. Oerke, R. Gerhards, G. Menz and R. A. Sikora (Eds). Springer, Dordrecht, Heidelberg, London and New York, pp. 119–34.

West, J. S., Bravo, C., Oberti, R., Moshou, D., Ramon, H. and McCartney, H. A. (2010). Detection of fungal diseases optically and pathogen inoculum by air sampling. In: *Precision Crop Protection – The Challenge and the Use of Heterogeneity*. E.-C. Oerke, R. Gerhards, G. Menz and R. A. Sikora (Eds). Springer, Dordrecht, Heidelberg, London and New York, pp. 135–49.

Western, N. M., Hislop, E. C., Herrington, P. J. and Jones, E. I. (1989). Comparative drift measurements for BCPC reference hydraulic nozzles and for an Airtec Twin-Fluid nozzle under controlled conditions. *Proceedings, The BCPC Conference* – Weeds, 12–15 November 1989. The British Crop Protection Council, Surrey, UK, pp. 641–8.

Womac, A. R. and Bui, Q. D. (2002). Design and tests of a variable flow fan nozzle. *Transactions of the ASAE*, 45(2), 287–95.

Zaman, Q. U., Esau, T. J., Schumann, A. W., Percival, D. C., Chang, Y. K., Read, S. M. and Farooque, A. A. (2011). Development of a prototype automated variable rate sprayer for real-time spot-application of agrochemicals in wild blueberry fields. *Computers and Electronics in Agriculture*, 76, 175–82.

Chapter 9

Intelligent machinery for precision agriculture

Qin Zhang, Washington State University, USA; Joseph Dvorak, University of Kentucky, USA; and Timo Oksanen, Aalto University, Finland

1 Introduction

Invented in the early 1980s in the United States, precision crop production technology divides a large field into a number of smaller management zones. Field operations in each zone are performed in response to the inter- and intra-field variability in soils or crops to achieve optimized returns on inputs while preserving resources. Performing field operations precisely to achieve better crop growth is not a new concept. For thousands of years, our ancestors manually exercised very small area-based, if not plant-based, precise farming practices in response to actual crop growth at each location. The innovative core of modern precision agriculture is the use of agricultural machinery to perform precise responsive field operations in mechanized precision crop production.

One feature of mechanized agriculture is making the operations uniform to achieve high efficiency. To enable responsive field operations with uniformly implemented machinery requires the machinery to have additional control capabilities. For example, to apply agro-chemicals precisely according to actual needs requires that the machinery not only controls the application rate on the go, but also is guided to follow the crop rows or other desired paths accurately. A few additional capabilities that are required to perform such variable-rate application include an auto-steer system, a map- or sensor-based application rate controller and a section control. Integration of such automated functions will furnish

http://dx.doi.org/10.19103/AS.2017.0032.10

agricultural machinery with some capabilities that are normally performed by a human and make it 'intelligent'. Such 'intelligent' capabilities include the perception, reasoning and controlling operations in the fields with variability to achieve the predetermined operational goals, with or without human supervision. While this definition could set a lower threshold for a machine to be considered intelligent, probably almost all automated machinery used in agriculture would fit this definition because of the high uncertainty and variability associated with its operations. The current level of state-of-the-art intelligent agricultural machinery is somewhere between automatically implementing repeated tasks to decrease human intervention and near autonomously performing predefined specific tasks repeatedly (Han et al., 2015).

2 Automated guidance systems: overview and global navigation satellite system (GNSS)-based systems

2.1 Overview

One fundamental function of intelligent agricultural machinery is to automatically steer the machine accurately along desired paths, often within a couple of centimetres range, to either avoid row overlap/skip or minimize the crop damage in operations. One of the earliest auto-steer systems guided a tractor to follow furrows in the field (Willrodt, 1924), and another early successful example was the use of wires carrying low-frequency current signals to guide a tractor travelling on desired pathways (Schafer and Young, 1979). In the 1980s, the development of computer vision technology provided the opportunities for developing vision-based tractor auto-steer systems (Reid and Searcy, 1987). While the steering control mechanisms on those auto-steer systems could be fairly different, one common and indispensable element is the navigation sensor, which provides either the position or the posture, or both, for the vehicle to follow the desired path (Tillett, 1991). There are many types of navigation sensors such as GNSS (Bell, 2000), machine vision (Reid and Searcy, 1987; Rovira-Más et al., 2005), radar (Reina et al., 2011) and laser (Barawid et al., 2007).

The navigation sensors can rely on either a) global positioning (GNSS satellite-based systems, of which GPS was one of the earliest implementations) or b) relative local positioning sensors that measure the distance to the previous swath (like row feelers) or marks left from previous operation (like tramlines in crop canopy). Any global positioning system besides GNSS is rare these days, even if in the early years, beacon systems were proposed. However, relevant local sensors may be based on various technologies, for example, vision systems, laser range finders, radar, mechanical row feelers or ultrasonic distance sensors. Both global and local sensors have their specific advantages and disadvantages, and therefore hybrid systems have been created either by using sensor fusion algorithms or by using different technologies in different phases of operation, for example, one for the main area and the other for the headland.

2.2 GNSS-based guidance systems

The potential for auto-steering agricultural machinery has been enriched by GNSS and the advent of precision agriculture technologies. As a result, these new technologies not only provide location information to steer the machinery, but also position the machinery to perform location-specific tasks (Reid et al., 2000). The use of automated guidance systems for tractors and other agricultural machineries has enabled optimal coverage of operations in arable farming and reduced the overlap of parallel swaths by a few percentage points depending on the operation and working width (Griffin et al., 2005). Auto-steering of agricultural machinery can be accurately operated on the preferred course, performing precise operations from land preparation to harvesting. According to Deere & Co. (2013), a high-accuracy guidance system can help to reduce overlap by up to 90% through accurate and repeatable guidance in both curved and straight tracks in the field. This helps farmers gain financial benefits through increased operational efficiency, reduced resource inputs and pass overlap, and variable rate implementation, from seeding to water and/or fertilizer application (Shockley et al., 2011). Auto-steer technology has often been praised by farmers as the most effective precision agriculture technology.

Although a regular GNSS can provide metre-level positioning accuracy, it is insufficient to provide guidance for the decimetre-level accuracy required for most field operations in agricultural production. GNSS receivers require some kinds of correction signal to compensate for noise created by the atmosphere and other errors along the signal path for automated guidance purposes. One type of correction signal is the regional satellite-based augmentation system which is an umbrella for various satellite-based broadcast services (such as WAAS, representing Wide Area Augmentation System, in North America; and EGNOS, representing European Geostationary Navigation Overlay Service, in Europe). Another correction system includes the use of various commercial correction signals broadcasted by either satellites, that is, OmniSTAR, or regional providers over cellular networks. To get an even more accurate positioning signal, it is common for agribusinesses and farmers to either build real-time kinetic (RTK) correction stations or make use of corrective satellite differential receivers and transmitters (Brimery, 2005; Rovira-Más et al., 2010).

Besides signal correction, the other crucial component to improve accuracy of antenna-based positioning systems on mobile machinery applications is attitude estimation, that is, the roll and pitch angles of tilt plus heading orientation if the navigation sub-system for guidance requires more information besides the direction of travel. Two approaches are commonly used for attitude estimation: a) using inertial (accelerometer and gyroscope) and magnetic sensors (Earth's magnetic field) to create an inertial measurement unit (IMU) or b) using multiple GNSS receivers. Three axis accelerometers not only measure acceleration of the vehicle, but also sense gravity. Fused to three axis gyroscope data measuring angular velocities, both tilt and angular velocity of the vehicle can be calculated. The IMU drift error is normally corrected based on a Kalman filter or other statistics-based methods. Later, the IMU data can be used to correct GNSS position and course. Acceleration and angular speed can also be integrated into the same filter where triangulation of GNSS satellite signals can be solved. On the other hand, using two GNSS receivers, sharing the same correction signal, can compute their position in a global framework and if the known locations are, for example, 2 m apart, the heading of the vehicle is always known in spite of earlier movement history without the need for any calibration. Generally, RTK GPS has an accuracy of ±2 cm, while an IMU composed of fibre-optic gyroscopes has an accuracy of 0.2°.

GNSS-based navigation systems guide the machinery by following the set of predetermined work paths, which are defined by two points, A and B. These A and B points are used as the reference points to create a navigation map using commercialized software in actual field operations. The navigation sensors are used to obtain the vehicle posture (roll, pitch and yaw angles) and position with respect to the selected projected coordinate system, such as Universal Transverse Mercator coordinates. The navigation system then adjusts the steering of the machine to maintain travel along the A–B line. Without automated guidance, the wider the tool, the harder it is to keep it side by side with the previous swath as the human driver is sitting far from the control point.

3 Automated guidance systems: vision-based systems

3.1 Overview

Visual sensing is one of the effective relative local positioning methods. Guidance systems based on vision systems are beneficial in operations that are carried out in fields that contain some landmarks (where to find features, e.g. rows of plants or trees). It offers significant advantages when compared to other sensors providing similar information – high-resolution images provide rich information about the scene ahead of moving machinery. In addition, it makes available instantaneous updates of what occurs in the worksite, allowing the perception of variety of object properties such as position, dimensions, texture or colour (Rovira-Más et al., 2010). However, the use of vision systems in plain field operations, that is, in harrowing, is hard due to lack of clear features or because of optical disturbances (Kannas et al., 2008).

Visual sensors are electronic devices that use images to capture the surrounding scenario, that is, CCD cameras, thermographic scanners and stereo vision rigs. On intelligent agricultural machinery, vision systems can be based on monocular or stereo cameras. Monocular camera systems are sufficient for detecting objects with easily recognizable colour differences from the background under visible light (such as seedlings on base soil in daylight) or with distinguishable differences in some specific non-visible bands of light. In comparison to monocular camera systems, stereo camera systems bring additional value when the colour-based information is not sufficient – to record the shape of interested objects in a challenging environment, such as green plants on green turf. Another technical challenge for vision-based guidance system is to be adaptive to environmental lighting conditions, which is normally solved either by adaptive algorithms or by using an auxiliary lighting system to stabilize lighting conditions.

Use of visual sensing is essential in finding the path for local positioning of the moving machinery with respect to surrounding objects for precise operation of the automatically guided vehicle. Even a minimum degree of automation will require reliable perception to provide awareness of space around the machinery. The systematic arrangement of field crops in equally spaced rows provides useful features for a vision system to position a vehicle with respect to the rows in order to provide guidance directly. This idea has been in use for over 20 years (Reid and Searcy, 1987; Rovira-Más et al., 2005), and is still in use today due to the difficulties of building a robust system capable of adapting to any lighting situation. Monocular cameras are commonly used as visual perception sensors to find the path in a field with row crops. The processing of images in this application involves the identification of the relative position between the moving machinery and the crop rows.

3.2 Case study for vision-based guidance systems

While many monocular cameras can perceive colour in field scenes, most of the algorithms developed for navigation are programmed to work on monochrome images to minimize the computing power requirement. Regardless of the method used, a monochrome image is typically presented in a greyscale image with 256 levels of intensity between black (level 0) and white (level 255). A binarization operation is then used to map all the pixels in the image having values between 0 and 255, to either black (0) or white (255), according to their membership to class 'object of interest' or 'background'. In navigation applications, the purpose is to separate the 'crop rows' and the 'soil' to achieve a strong reduction in information with regard to grey levels from 256 levels to 2 levels (Fig. 1a). To guide the machinery in tracking crop rows, there is no need to define the entire geometry of the rows; rather, the centrelines of rows could more accurately provide such information. Thus, a midpoint encoding method (Rovira-Más et al., 2005) is often used to extract a thin centreline holding the critical information for tracking the rows. Two methods, regression analysis and the Hough transform, are commonly used to find the desired path (in terms of a navigation target point as illustrated in Fig. 1b) after the centrelines for at least two rows of crops have been identified. Regression analysis is a popular statistical technique to estimate the best fit for a line given a scattered set of points, but it is very sensitive to the presence of outliers which are not unusual in digital images acquired for navigational purposes. In comparison to regression analysis, the Hough transform is a mathematical transformation that computationally enhances the alignment of points and increases efficiency due to its low sensitivity to outliers, and therefore is more suitable for navigation applications (Rovira-Más et al., 2010).

While monocular cameras acquire the key information for navigating agricultural machinery performing fieldwork, the adoption of stereoscopic vision improves the monocular vision by adding the third dimension, that is, the distances between scene objects and the camera. In contrast to monocular cameras creating planar images where every pixel is the result of the two-dimensional projection of the 3D world, stereovision adds the third coordinate, or range, that completes the full location of any point in a 3D Cartesian frame. The core of a stereovision navigation system is a stereovision-based navigation algorithm to identify the paths from the disparity images obtained from the field scenes and extracting the appropriate steering angles based on the position errors of the moving machinery with respect to the identified pathway. This can be accomplished

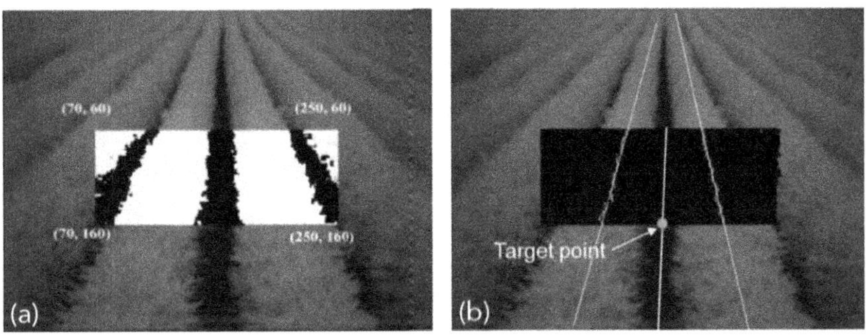

Figure 1 Row crop field path finding operation on a monochrome image (see text for further explanation).

in three steps: first compute 3D disparity points of the detected objects, followed by creating an elevation map of the scene and then determining a navigation point in this elevation map (Kise et al., 2005). Creating an elevation map based on acquiring a pair images of the scene (Fig. 2a illustrates one field image of an acquired pair) is the core in this three-step process. An 'elevation map' is the 3D spatial field map that describes crop height at each point on the field (Fig. 2b).

One method of finding the navigation point in an elevation map is cross correlation, a standard statistical technique for estimating the degree to which two series are correlated (Tsai et al., 2003), to fit the 3D crop rows. By this approach, the centre of the elevation map is defined as the viewpoint for navigation and the middle point between two crop rows being followed is defined as the navigation point, similar to the way a human determines the target while driving a tractor. This method uses a general crop profile function *g(i)*, an ideal cosine function, for fitting a cross section of crop rows:

$$g(i) = \cos\left(\frac{2\pi}{w/d} i\right), \left(0 \le i \le 3\frac{w}{d}\right) \tag{1}$$

where *w* is the inter-row spacing and *d* is the resolution of an elevation map. Normally, a crop profile fitting function containing three periods of cosine curve, as shown in Fig. 2b, will be sufficient for this purpose.

The period of *g(i)* equals the inter-row spacing, which makes this function depend only on the inter-row spacing, regardless of the type of crop or its stage. The navigation point from the stereo field scene image can be determined by aligning the actual crop row profiles obtained from a crop elevation map to the ideal profile represented by the general crop profile function using the following equation:

$$R(\tau) = \frac{\sum_{i=0}^{m-1}\left(f(i+\tau) - \bar{f}\right)\left(g(i) - \bar{g}\right)}{\sqrt{\sum_{i=0}^{m-1}\left(f(i+\tau) - \bar{f}\right)\sum_{i=0}^{m-1}\left(g(i) - \bar{g}\right)}} \tag{2}$$

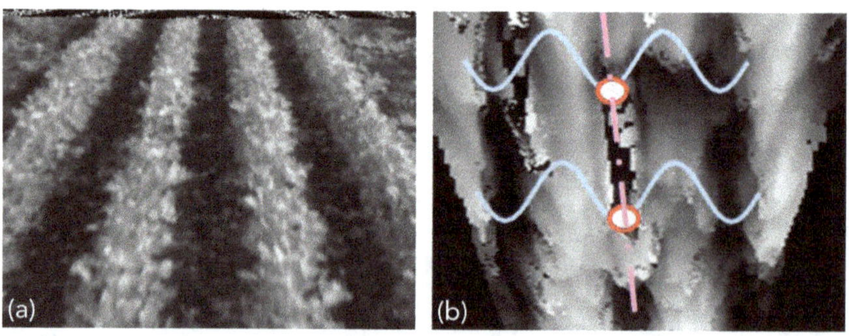

Figure 2 Row crop field path finding operation on a stereovision image: (a) one field image of an acquired pair and (b) the elevation map extracted from the acquired pair.

where $R(\tau)$ is the cross correlation coefficient; τ is a shift between the two signals; $f(i)$ is the data series of actual crop profile obtained from the elevation map; $g(i)$ is the ideal crop profile provided by the general crop profile function; $\bar{f}$ and $\bar{g}$ are the means of corresponding series, respectively; m is the number of data points in g series.

By repeatedly computing $R(\tau)$ by changing τ from 0 to one complete cycle, τ that results in the maximum $R(\tau)$ value is selected as the best match point for the actual crop profile and the ideal crop profile in this cross correlation analysis, and then it is used to realign the general crop profile function. After being realigned at the optimal τ point obtained, the middle point of the general crop profile function is defined as the navigation point. Figure 2b shows the navigation points identified on two cross sections in the elevation map. This approach can accurately identify a navigation point based on the crop elevation map. It is recommended to limit the navigation point search zone close to the longitudinal centre of the elevation map for achieving more stable and accurate guidance. When the crop is at an early stage or the crop size is very thin and the rows are sparse, some other crop profile functions could be more suitable, such as rectangular wave function (Kise et al., 2005).

4 Path planning

4.1 Overview

For any mechanized field operation, several steps are often required in managing the machinery to perform the task. Normally, the first step is to plan the working paths of the machinery to ensure complete coverage of the field. Often this process is optimized to minimize the fieldwork time, but other agronomic and farm management considerations like soil compaction and field access may be more important (Bochtis et al., 2012).

Path plan generation has several considerations: one of the most important is the general fieldwork pattern to be used. Common fieldwork patterns include headland patterns, circular patterns, contour patterns to match terrain or combinations of these (Bochtis et al., 2007; Jin and Tang, 2011). Often field geometry dictates that dividing the field into separate sub-regions is more efficient, so field decomposition must also be considered at this point (Oksanen and Visala, 2009). For headland patterns, the direction of the paths must also be considered. Historically, the pattern has been set by machinery limitations, terrain type, regional practices and crops. Crops grown under centre pivot irrigation use a circular pattern. Producers growing corn and soybeans in the American Midwest use a headland pattern with the direction often based on the longest edge of the field. Producers in hilly and terraced terrain frequently place paths along the contours of the terraces.

Once the field has been divided into path patterns, these paths must be allocated to the available machinery. As with pattern type and path direction, producers have developed several different methods of path allocation. One common method is leader-follower. With leader-follower allocation, the leader starts a path and the follower then works the path adjacent to the leader. Subsequent followers can then continue working on the next available adjacent paths. When the leader finishes a path, it then travels to a new path while skipping the paths that are already being worked by its followers. The leader-follower method can make navigation more complicated because all the machinery is in the same area of the field, which can lead to potential collision. It can also create more non-working time as paths must constantly be skipped.

An allocation method that addresses these issues is the work zone approach. With work zones, the field is divided into regions that are assigned to each vehicle. At its simplest, this could be starting machines at opposite ends of the field. The machines work towards each other until they meet in the middle. Work zones become more complicated when the field has to be divided into more than two zones, as accurately assessing completion times of each zone can be difficult beforehand. Machinery with different operating widths, travel speeds and engine power will be able to cover varying areas of land in a given time period. Complex field geometries can also make creating work zones difficult. Finally, if a breakdown or adverse field conditions stops or delays one machine, even perfectly assigned work zones will not be completed as intended. In the event of these issues, the efficiency gains of the work zone method will disappear as other machines will have to wait or take over completion of unfinished zones.

Path ordering is the last stage of fieldwork planning. Each field machine must determine the order in which the paths assigned to it are worked. With a single machine in a basic rectangular field, this can be as simple as selecting the next closest path every time it finishes one path. However, machinery dynamics and implement widths may mean that it is more efficient to make turns if adjacent passes are skipped.

Although each fieldwork planning stage (path creation, allocation and ordering) is a different process, they are often optimized or performed together since each step affects the others. Producers have traditionally focused on ease of implementation in selecting methods. Automated guidance technologies have had a major impact on this process. Guiding from A-B lines rather than just following the previous row has meant that offsets between passes must be equal so implements of different widths cannot be used in the same field at the same time. In addition, if multiple machines are being used in the same field, the A-B lines must be shared between machines. Without A–B line sharing, machines must work in non-contiguous regions of a field or even completely different fields. More recently developed guidance equipment is capable of handling the complex curvature from paths that follow terrain contours, but early equipment forced producers to adopt straight A-B lines. Improved informatics will cause even more changes as coordination between machines and real-time optimization become possible. They also enable the adoption of previously impossible but more efficient methods to cover fields that are completely based on optimization algorithms. Future adoption of completely autonomous machines will require these steps to be efficiently handled by the machine control system in order to be successful.

Researchers have developed many methods to automate and optimize these fieldwork planning steps. Some have focused on simply optimizing the path direction in a headland pattern to minimize overlap or turning time (Zandonadi et al., 2011). Others have focused on path ordering or combinations of creation and ordering (Bochtis and Sørensen, 2009; Hameed et al., 2011). The basic problem of how to use a given set of field machinery to cover a particular area is very challenging. The problem is similar to classic computer science optimization problems such as the travelling salesperson problem or the vehicle routing problem (VRP). These types of problems are computationally intractable and without defined solution techniques. Instead, advanced metaheuristics like genetic algorithms (Alba and Dorronsoro, 2004), simulated annealing (Osman, 1993) and Tabu search (Toth and Vigo, 2002) are employed to provide approximate solutions. Agronomic considerations also place unique constraints in the optimization performed. Driving over-planted crops often harms yields and must be avoided, but unrestricted travel poses no issue in pre-plant operations. This means that the optimization used for a pre-plant spraying operation must be different from that used for post-plant spraying even though the same equipment may be used for both. Harvesting operations face similar complications, but the regions where navigation is flexible change as harvest progresses through a field.

Although the steps in fieldwork planning have been optimized in different ways, most projects share a similar procedure. The initial step is the creation of a model that describes the field and the travel options of the vehicle. This model is often constructed to match the format and constraints of common computer science optimization problems so that decades of developments into those problems can be used. For transformations into the travelling salesperson problem or the VRP, this requires defining various points in the field as nodes and developing a network graph that describes the connections and travel costs between those points. An optimization technique is then used to attempt to minimize the costs associated with the fieldwork. Finally, the routing solution must then be transformed into a format, such as waypoints, that can be programmed into the automated guidance system or shown on an in-cab navigation display.

4.2 Path planning case study

Node assignment, which is the first step in turning a physical field into a mathematical VRP representation, can be accomplished using different methods. Initially, the field is covered by work paths defined by their endpoints (Fig. 3a). Some approaches have assigned two nodes per path by using the endpoints as nodes (Bochtis and Sørensen, 2009), while others

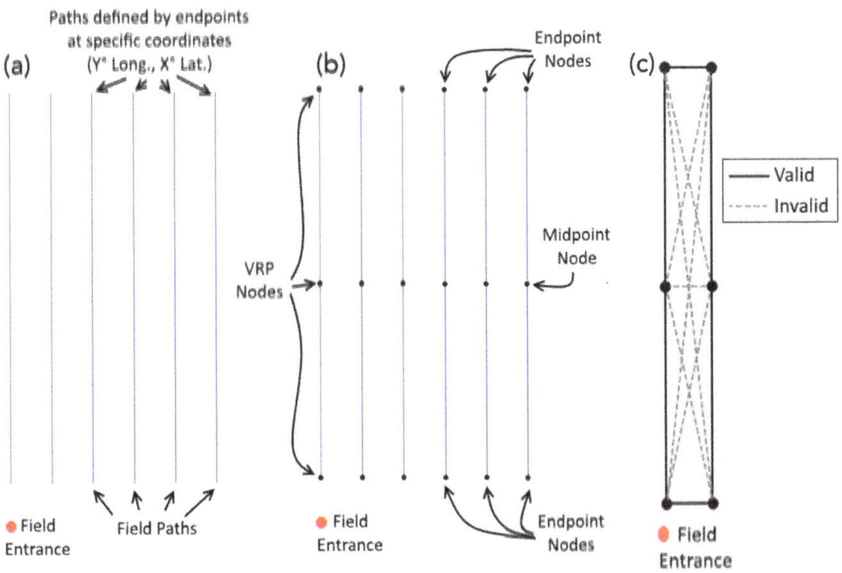

Figure 3 Field with (a) paths to be worked by a vehicle, (b) with the VRP nodes assigned to those paths and (c) invalid and valid arcs for travel (represented for only two paths).

have used three nodes per path (the two endpoints and the midpoint) (Seyyedhasani and Dvorak, 2017), as shown in Fig. 3b.

Costs must then be assigned to the arcs that connect the nodes. This cost assignment varies depending on what aspects of fieldwork the model is expected to optimize. Costs have been based on travel time (Seyyedhasani and Dvorak, 2017), non-working time (Bochtis et al., 2009) and even soil compaction potential (Bochtis et al., 2012). The arcs that connect certain nodes do not represent valid travel paths (Fig. 3c), and very large costs are assigned to these invalid arcs to discourage the selection of these arcs in any solution.

The solution to a VRP is a route for each available vehicle. These routes are ordered lists of nodes that describe the individual travel paths of each vehicle. Finally, a fitness function must be created that mathematically evaluates the ability of a given solution (set of vehicle routes) to achieve the optimization criteria. Options for fitness functions include minimization of total non-working time, minimization of soil compaction, minimization of total time for all vehicles, minimization of the time to complete the fieldwork and weighted functions of several factors (Seyyedhasani and Dvorak, 2017).

Once the task has been properly expressed as a VRP, one of the many techniques available to produce solutions to the VRP can be used to generate the vehicle routes. These include traditional heuristics like Clarke-Wright (Bochtis et al., 2013; Spekken and de Bruin, 2013), and more complex techniques such as genetic algorithms (Hameed et al., 2011) and Tabu search (Seyyedhasani and Dvorak, 2017). Each solution method has trade-offs in terms of ability to produce more optimal solutions, computation time and risk of producing invalid or extremely poorly optimized solutions (Toth and Vigo, 2002).

Figure 4a shows a field from path generation research with its optimal path direction shown (Hameed et al., 2011). These paths were used with a VRP designed to minimize the time to complete the field using five machines. Tabu search was utilized to provide the solution to the VRP, which resulted in the routings shown in Fig. 4b.

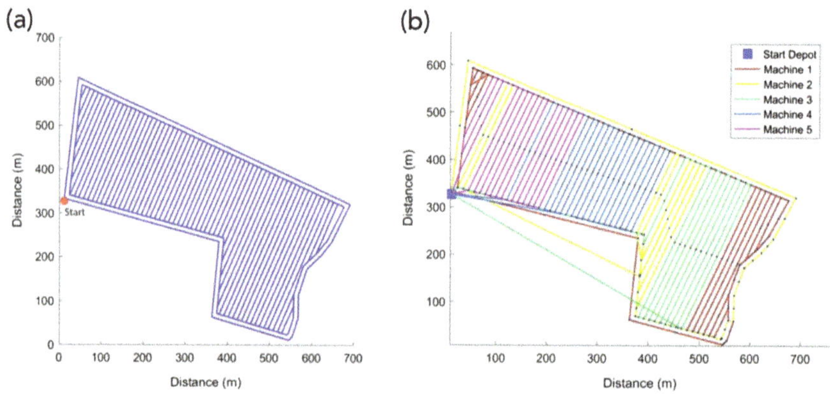

Figure 4 An example of field routing optimization for coordinating five machines working effectively in the same field (see text for further explanation).

5 Automated actuation systems

5.1 Overview

Research indicates that a tractor can operate autonomously when the auto-steer system on the tractor is integrated with a pre-programmed site-specific implementation plan. Figure 5 presents a prototype of an autonomous tractor developed by a research team from the University of Illinois at Urbana-Champaign in late 1990s. Equipped with an auto-steer system and supported by a map-based implementation plan, the tractor could navigate itself and move from a garage to a designated field to perform the pre-programmed field operations autonomously (Reid et al., 2000). To provide the capability of automation, the tractor used a set of navigation sensors providing the machine position and heading information for guiding the machine. The navigation sensors included a motion model for determining the current motion status accurately and robustly, a navigation planner for creating steering signals in real time and a steering controller (supported by a set of feedback sensors) for keeping the machine on the desired course while performing various field operations.

As work on autonomous steering methods progressed, it became apparent that a robust method for assessing guidance accuracy was required. For most agricultural tasks, the tractor is driven in parallel paths across the field. The task of the guidance system is to ensure the tractor is located at precise locations on these paths at certain times. Positional deviations parallel to the paths merely manifest as fluctuations in speed. However, the deviations perpendicular to each individual desired vehicle path represent significant positional errors. These perpendicular deviations are called cross-track errors.

Figure 5 An autonomous tractor developed at the University of Illinois at Urbana-Champaign in the late 1990s. This autonomous tractor could navigate itself transporting from garage to field performing pre-programmed operations autonomously (Reid et al., 2000).

The guidance accuracy of a vehicle can be assessed relatively simply by producing statistics that describe the distribution of cross-track errors.

One drawback of cross-track error is that it requires the exact location of the intended vehicle path. The path must be determined prior to testing; its position must be accurately described and this information must be provided to the guidance system. In actual agriculture environments, these tasks are often difficult or impossible. Therefore, the International Organization for Standardization (ISO) standard describing the testing of agriculture machinery systems, ISO 12188-2 (ISO/ASABE, 2012), uses relative cross-track error as its error measure. Relative cross-track error measures the difference between two passes of a vehicle when it is guided along the same path. Because relative cross-track error is based on relative differences, it is not able to detect systematic bias errors that occur identically in both passes. However, this is not a big drawback, as producers are generally more concerned with relative accuracy between passes rather than absolute accuracy. This is the case whether producers are using autonomous guidance to minimize implement overlap between passes in a single field operation or guide machinery on the same tracks in different operations for controlled traffic farming. Since the guidance systems in agriculture vehicles are most often based on GNSS location systems, which exhibit a random walk error pattern, these errors are divided into pass-to-pass errors (revisit time less than 15 min) and long-term guidance errors (revisit times greater than 1 h). The pass-to-pass error describes the guidance system's performance when used to minimize overlap between consecutive passes in a field. The long-term guidance error, which is generally worse than pass-to-pass error, is most useful when the vehicle is driven along the same path in different field operations.

ISO 12188-2 (ISO/ASABE, 2012) provides details on conducting the test and analysing the data it produces. Following the test procedure requires autonomous travel down and back on a straight, level test course at least 100 m long. During this travel, the location of the vehicle is sampled at a rate no less than 10 Hz. The test is to be performed at three different speeds, 'fast' (5.0 m s^{-1}), 'medium' (2.5 m s^{-1}) and 'slow' (0.1 m s^{-1}). The standard also requires the test sequence to be repeated at least three times with different GNSS constellations over two consecutive days. In calculating error, a positive value indicates that the later pass was to the right of the track from the previous pass (based on the travel direction of the later pass). Results reported by an ISO 12188-2 test include the average error which represents directional bias, the variability and the cumulative distribution of errors as represented by the 50%, 95% and maximum absolute values of the relative cross-track error. The 95% value of the cumulative distribution is the key quality indicator from this test.

5.2 Case study for automated actuation systems

Figure 6a displays outbound and return paths of a field machine performing a test run specified by ISO 12188-2, and Fig. 6b shows the relative cross-track error from this test run graphically (Rounsaville et al., 2016). The accuracy statistics for this single pass test are listed in Table 1: The negative value of the average relative cross-track error indicates that the guidance system of this machine has a bias towards driving to the left of the A–B line. The relatively large standard deviation indicates that the steering was constantly corrected by the auto-guidance system but with a relatively low tracking accuracy throughout the test. Finally, the cumulative distribution values indicated that the maximum expected cross-track error could reach 24 cm.

Table 1 Relative cross-track error statistics for the sample test run

Mean (cm)		−9.81
Standard deviation (cm)		5.71
Cumulative distribution values	50%	9.43
	95%	19.55
	Maximum	24.28

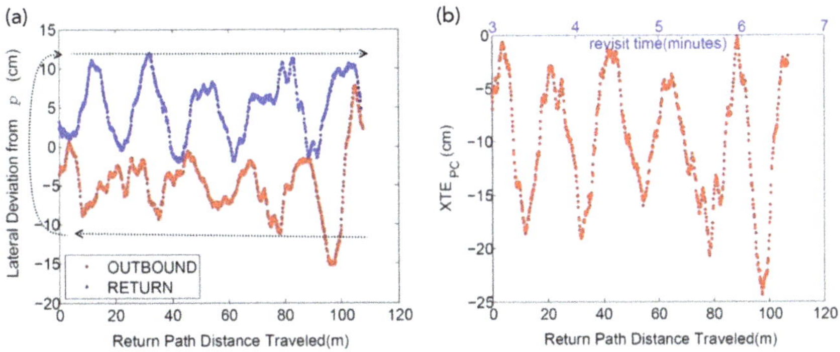

Figure 6 An example of field machine path tracking recorded during a single test run following the ISO 12188-2 procedures. The black dotted line (not to scale) in (a) shows the machine travel directions and (b) shows the relative cross-track error.

6 Implement controls

Automatic guidance of a vehicle requires some actuators. Traditionally, a human driver has used the steering actuator of the tractor to keep the tool on the desired path and in the first era of automated guidance systems the same concept was utilized. For automation systems, control of multiple actuators simultaneously is not a problem, at least if the controlled system is not modular. To increase accuracy and to compensate for disturbances that are hard to compensate with tractor steering, so-called implement steering gives an additional degree of freedom for control. Furthermore, implement guidance is a system that consists of an implement steering actuator plus guidance control to keep the control point of implement on desired path.

Implement steering can be realized in alternative mechanical ways: side shift in hitch, controlled joint in drawbar, steering support wheels, steering counters or side shift of tools mounted to implement frame, just to mention a few options. Active control of steering actuators requires power and control of the actuators. Currently, usually the steering is realized by taking hydraulic power from tractor and using electrohydraulic valves to control hydraulic cylinders, but electric drives using high voltage from the tractor are foreseen in the future.

Implement guidance is considered a stand-alone system that utilizes implement steering but is not connected to the tractor guidance. Therefore, the tractor can be controlled by other means, either by human or by another stand-alone system. Connecting two systems together has been studied (Backman et al., 2013; Werner et al., 2013). The specific challenge is related to modularity of agricultural machinery: a tractor made by one manufacturer must be compatible with an implement made by another company and the guidance system may be of another brand. The data communication between various parts of a multi-brand system had been addressed by ISO, creating ISO 11783 standard series, commercially known as ISOBUS (ISO, 2010). Currently, ISO 11783 standard series does not specify any means to control implement steering. However, this has already been proposed (Oksanen and Backman, 2016).

7 Future trends: increasingly autonomous systems and their potential impacts

7.1 Towards more autonomous systems

Fully autonomous agricultural machinery will require greater situational awareness than that provided by the navigation sensors that support auto-steer. In addition to navigation, human operators are also expected to monitor the progress of the field operation and ensure the safety of the machine as well as people and property in the area. Many sensors and systems have been developed that support these activities in order to increase productivity. For an autonomous field machine, all of these systems must be integrated so that the entire system operates safely and with the accuracy, throughput and abilities of traditional manned machines.

Field condition-detecting sensors have been used to ensure or improve machine productivity. For example, automatic intra-row weeding has been enabled using visual sensing systems to differentiate weeds from crops (Fennimore et al., 2016). 3D LIDAR sensors have been used to map individual plants in a field (Weiss and Biber, 2011). Due to its robustness and availability at a lower price (Andújar et al., 2011), ultrasonic sensors are being used in agriculture for a variety of applications, such as to detect the canopies for automatic control of spraying in orchards and vineyards (Zaman et al., 2005; Gil et al., 2013). Ultrasonic sensors have also been used to measure sward height and forage mass in forage production (Fricke et al., 2011), and are widely deployed to control the boom height in modern agricultural sprayers (Herbst et al., 2015).

Different types of sensing systems have also been developed to identify humans or other objects in the worksite to ensure or improve safety. For example, stereo vision systems have proven capable of identifying human presence and locating their position around moving tractors in agricultural fields (Yang and Noguchi, 2012; Wei et al., 2005). To detect humans that approach running PTO shafts, microwave, passive infrared and radar-type sensors have been used on tractors (Shutske et al., 2001; Venem et al., 2006). Ultrasonic sensors can detect humans near agricultural machinery provided they are not wearing soft fur coats, but they only respond to the closest object so obstacles hidden in tall crops could be missed (Dvorak et al., 2016). Thermal cameras have proven capable of identifying animals hidden in forage crops (Steen et al., 2012). While detecting human and animal presence is important for safe machine operation, it is also necessary to identify other objects in a machine operating environment. LIDAR and a combination system including ultrasonic and visual light camera have been used to detect and classify objects near machinery (Doerr et al., 2013; Kise et al., 2005; Leidenfrost et al., 2013).

While GNSS-based devices are widely utilized for location in agricultural applications, mobile robotic systems for use in other industries have not been able to rely on the robust location provided by GNSS for a variety of reasons. This lack of acceptable location sensors has led to the development of simultaneous localization and mapping or SLAM, which is a framework that enables a mobile robot to determine its location and map its surroundings while navigating through an unknown environment (Dissanayake et al., 2001). The SLAM concept is a framework and can be implemented with a variety of different algorithms; however, the basic idea is to use a sensor that provides information on the environment surrounding the vehicle. This sensor data is analysed to find features or landmarks, which represent a map of the environment. By comparing differences between subsequent readings from the sensor, the system updates its estimate of its location within the map. Identification and tracking of landmarks and estimation of vehicle location are all probabilistic in nature, so analysis techniques such as Kalman filters are used in this step. In applications where the high-precision location services provided by GNSS devices are not available, SLAM has enabled autonomy that would not be practical otherwise. While SLAM is of limited use in traditional outdoor agricultural settings where location determination is not an issue with GNSS, certain applications such as in greenhouses (González et al., 2009), barns or dense orchards (Zhang et al., 2013) can still benefit from SLAM techniques. The probabilistic mapping component of SLAM can also be beneficial in building up field maps even if location is provided directly from other sensors (Biber et al., 2012; Ruckelshausen et al., 2009).

7.2 Future impacts of autonomous systems

As technologies become mature, it is foreseeable that the adoption of intelligent machinery in agricultural production will quickly gear up as farmers are exposed to the benefits of using such technologies.

The use of intelligent machinery can also help increase productivity in crop production. As the supply of skilled seasonal tractor drivers becomes a challenge, farmers are looking at the option of adopting machines with more automatic or even autonomous functions to reduce the skill required to perform the work efficiently. Automation could improve productivity by reducing mistakes caused by operator fatigue, by being able to guide machines operating on the same lines year after year and by being able to extend field operation hours to after dark through using auto-steered machines. Over the past two decades, the agricultural industry has taken some sequential steps to adopt various automation technologies. It started with yield mapping technology to collect harvest yield data. Then auto-guidance technology was adopted – initially using a light bar and later a full RTK GPS-based auto-steer for tillage and for chemical applications. This was followed by adopting prescription application (auto-steered) for seeds, fertilizers and other agrochemicals based on collected data including soil type, soil test data and previous crop and yield. The benefits that farmers could get from those more precise operations include (1) seed cost saving by not planting on waterways scattered through the farm, by no double planting into the end rows and by reduced seeding rate in low productivity areas; and (2) a dramatic reduction of over-application of chemicals and fertilizers by reduced swath overlap, end row and waterway application through auto-steered operations and by prescription applications.

While autonomous systems present new opportunities to crop producers, they also introduce new risks such as damage to property or people. Already, deaths have been attributed to on-road automotive vehicles operating under the control of an automated navigation system. Although such a tragic outcome has not been reported with autonomous agricultural vehicles, the large size and mass of traditional agriculture equipment create potential safety risks. The agricultural industry has a long history of developing a wide variety of sensors and systems to detect dangerous situations and to protect people and property. Fully autonomous systems just represent the next level of development in these systems, as an autonomous vehicle cannot rely on a human operator as the final safety check. Some factors in agricultural production make developing these safety systems easier than the similar development underway for on-road vehicles. For example, agricultural fields are generally controlled by a single entity that determines exactly what people and vehicles are present at any one time. On the other hand, agricultural fields are less structured than roads, and the wide variety of agricultural field operations means that the set of acceptable actions for a vehicle is not constant. For example, a tractor tilling a field should stop its operation if its sensors detect that it is about to drive into a neighbouring field of almost ripe corn. However, during silage harvest, the autonomous guidance system should ensure the machine is driving through that same stand. Development of appropriate safety systems (sensors, logic and actions) for the wide variety of agricultural tasks will continue to be an area of active advancement for many years.

Autonomous machines also provide opportunities in agricultural production that have been impractical before. Transporting a human operator through the field establishes a minimum bound on vehicle size. Human operator time is also relatively expensive which has been a primary factor encouraging the development of larger and faster machines. The relentless push for larger and faster machinery is causing issues with transportability and soil compaction (Shearer et al., 2010). In a future with autonomous machines, efficiencies of scale and traction requirements for certain field operations may still call for the creation of massive machines, but now operational considerations rather than the costs of human labour will drive these decisions. While it is too early in autonomous vehicle development to predict the eventual outcome, researchers have proposed a variety of machinery systems. These range from swarms of many small battery-powered machines as envisioned by the MARS project between AGCO's Fendt subsidiary (Blender et al., 2016) to autonomous versions of current high-horsepower tractors (optionally with cabs for operator flexibility) represented by CNH Industrial's concept machines (CNH, 2016). The proponents of using multiple small vehicles have touted their reduced potential for damage to property or people, the redundancy in the case of equipment failure and the ability for small machines to work small and oddly shaped fields efficiently (Blackmore et al., 2002). For larger machines, the benefits include requiring that producers purchase fewer sets of the equipment with potentially expensive autonomy equipment and less adjustment since operation will be similar to current practices. Researchers are currently investigating the potential of various machinery models and future developments in autonomy and agriculture in general will ultimately determine the direction of agricultural machinery.

Another noticeable impact of adopting automation technology is the much lower dependency on low- to semi-skilled human labourers in performing fieldwork, especially for some high-value speciality crop production. An unpublished study on comparing labour requirements for weed control in organic vegetable production at an Eastern Washington farm shows that the use of an auto-guided tractor could increase the efficiency by 30 times. This resulted in a dramatic change to seasonal labour requirements for vegetable production, which is a challenge in US organic vegetable industry. The adoption of automation technologies caused a revolutionary change in the skill requirements for achieving more precise operations. It is a commonly accepted opinion today that automated or robotic farming is going to occur within a generation. This will result in some major changes in agricultural workforce structure away from a heavy dependency on sufficient low- to semi-skilled seasonal fieldworkers to an increased dependency on well-educated and highly trained professionals that provide commercial services. This change in the agricultural workforce would have a big impact on rural area demographic profiles and therefore the rural economy.

8 Where to look for further information

The research and development of intelligent machinery suitable for precision agriculture applications being very active both at fundamental technology and at commercial product development levels. Readers could find more up-to-date information either from relevant scientific journals, such as *Computers and Electronics in Agriculture, Biosystems Engineering, Precision Agriculture* and *Transactions of the ASABE*, if interested in fundamental technologies, or from equipment manufacturers' websites, such as Deere & Company and CNH, if interested in new products.

9 References

Alba, E. and Dorronsoro, B. 2004. Solving the vehicle routing problem by using cellular genetic algorithms. In J. Gottlieb and G. Raidl (Eds), *Evolutionary Computation in Combinatorial Optimization* (Vol. 3004, pp. 11–20). Berlin Heidelberg: Springer.

Andújar, D., Escolà, A., Dorado, J. and Fernández-Quintanilla, C. 2011. Weed discrimination using ultrasonic sensors. *Weed Research*, 51: 543–7.

Backman, J., Oksanen, T. and Visala, A. 2013. Applicability of the ISO 11783 network in a distributed combined guidance system for agricultural machines. *Biosystems Engineering*, 114: 306–17.

Barawid, O. C., Mizushima, A., Ishii, K. and Noguchi, N. 2007. Development of an autonomous navigation system using a two-dimensional laser scanner in an orchard application. *Biosystems Engineering*, 96(2): 139–49.

Bell, T. 2000. Automatic tractor guidance using carrier-phase differential GPS. *Computers and Electronics in Agriculture*, 25(1–2): 53–66.

Biber, P., Weiss, U., Dorna, M. and Albert, A. 2012. Navigation system of the autonomous agricultural robot BoniRob. In *Workshop on Agricultural Robotics: Enabling Safe, Efficient, and Affordable Robots for Food Production (Collocated with IROS 2012)*. IEEE Robotics & Automation Society. IEEE, https://www.ieee.org/.

Blackmore, S., Have, H. and Fountas, S. 2002. *A Specification of Behavioural Requirements for an Autonomous Tractor*. Paper presented at the 6th International Symposium on Fruit, Nut and Vegetable Production Engineering conference, Potsdam – Bornim, Germany.

Blender, T., Buchner, T., Fernandez, B., Pichlmaier, B. and Schlegel, C. 2016. Managing a mobile agricultural robot swarm for a seeding task. In *IECON 2016–42nd Annual Conference of the IEEE Industrial Electronics Society*. IEEE, https://www.ieee.org/.

Bochtis, D., Vougioukas, S., Ampatzidis, Y. and Tsatsarelis, C. 2007. Field operations planning for agricultural vehicles: A hierarchical modeling framework. *Agricultural Engineering International:*

CIGR Journal. International Commission of Agricultural and Biosystems Engineering (CIGR), http://cigr.org/.

Bochtis, D. D., Vougioukas, S. G. and Griepentrog, H. W. 2009. A mission planner for an autonomous tractor. *Transactions of the ASABE*, 52(5): 1429–40. International Commission of Agricultural and Biosystems Engineering (CIGR), http://cigr.org/.

Bochtis, D. D. and Sørensen, C. G. 2009. The vehicle routing problem in field logistics, part I. *Biosystems Engineering*, 104: 447–57.

Bochtis, D. D., Sørensen, C. G. and Green, O. 2012. A DSS for planning of soil-sensitive field operations. *Decision Support Systems*, 53: 66–75.

Bochtis, D. D., Sørensen, C. G., Busato, P. and Berruto, R. 2013. Benefits from optimal route planning based on B-patterns. *Biosystems Engineering*, 115(4): 389–95.

Brimery, J. 2005. Nothing runs like a precision farming system. *Progressive Engineer*. March 2005 Issue.

CNH Industrial. 2016. *CNH Industrial Brands Reveal Concept Autonomous Tractor Development: Driverless Technology to Boost Precision and Productivity*. Available at: http://www.prnewswire.com/news-releases/cnh-industrial-brands-reveal-concept-autonomous-tractor-development-driverless-technology-to-boost-precision-and-productivity-300317452.html, Accessed on 3 May 2017.

Deere & Co. 2013. *John Deere Guidance Systems*. Available at https://www.deere.com/common/docs/products/equipment/agricultural_management_solutions/guidance_systems/brochure/en_GB_yy1114823_e.pdf, Accessed on 3 May 2017.

Dissanayake, M., Newman, P., Clark, S., Durrant-Whyte, H. and Csorba, M. 2001. A solution to the simultaneous localization and map building (SLAM) problem. *IEEE Transactions on Robotics and Automation*, 17: 229–41.

Doerr, Z., Mohsenimanesh, A., Laguë, C. and McLaughlin, N. B. 2013. Application of the LIDAR technology for obstacle detection during the operation of agricultural vehicles. *Canadian Biosystems Engineering*, 55: 2.9–2.16.

Dvorak, J. S., Stone, M. L. and Self, K. P. 2016. Object detection for agricultural and construction environments using an ultrasonic sensor. *Journal of Agricultural Safety and Health*, 22: 107–19.

Fennimore, S. A., Slaughter, D. C., Siemens, M. C., Leon, R. G. and Saber, M. N. 2016. Technology for automation of weed control in specialty crops. *Weed Technology*, 30: 823–37.

Fricke, T., Richter, F. and Wachendorf, M. 2011. Assessment of forage mass from grassland swards by height measurement using an ultrasonic sensor. *Computers and Electronics in Agriculture*, 79: 142–52.

Gil, E., Llorens, J., Llop, J., Fàbregas, X., Escolà, A. and Rosell-Polo, J. R. 2013. Variable rate sprayer. Part 2 – Vineyard prototype: Design, implementation, and validation. *Computers and Electronics in Agriculture*, 95: 136–50.

Giles, D. K., Delwiche, M. J. and Dodd, R. B. 1989. Sprayer control by sensing orchard crop characteristics: Orchard architecture and spray liquid savings. *Journal of Agricultural Engineering Research*, 43: 271–89.

González, R., Rodríguez, F., Sánchez-Hermosilla, J. and Donaire, J. G. 2009. Navigation techniques for mobile robots in greenhouses. *Applied Engineering in Agriculture*, 25: 153–65.

Griffin, T., Lambert, D. and Lowenberg-DeBoer, J. 2005. Economics of lightbar and auto-guidance GPS navigation technologies. In J. V. Stafford (Ed.) *Precision Agriculture '05*. 5th European Conference on Precision Agriculture, Uppsala, Sweden, pp. 581–7.

Hameed, I. A., Bochtis, D. D. and Sørensen, C. G. 2011. Driving angle and track sequence optimization for operational path planning using genetic algorithms. *Applied Engineering in Agriculture*, 27: 1077–86.

Han, S., Steward, B. L. and Tang, L. 2015. Chapter 5. Intelligent agricultural machinery and field robots. In Zhang, Q. (Ed.), *Precision Agriculture Technology for Crop Farming*. Boca Raton, FL: CRC Press.

Herbst, A., Osteroth, H.-J., Fleer, W. and Stendel, H. 2015. *A Method for Testing Automatic Spray Boom Height Control Systems.* Presented at the 2015 ASABE Annual International Meeting, New Orleans, Louisiana.

ISO. 2010. *11783: Tractors and Machinery for Agriculture and Forestry – Serial Control and Communications Data Network* (Vol. 11783, p. 94). Switzerland: ISO.

ISO/ASABE. 2012. Tractors and machinery for agriculture and forestry – Test procedures for positioning and guidance systems in agriculture – Part 2: Testing of satellite-based auto-guidance systems during straight and level travel, vol. 12188-2, p. 6.

Jin, J. and Tang, L. 2011. Coverage path planning on three-dimensional terrain for arable farming. *Journal of Field Robotics*, 28: 424–40.

Kannas, K., Oksanen, T. and Visala, A. 2008. Machine vision based parallel navigation for tractor-harrow. In *Proc. of CIGR International Conference of Agricultural Engineering*. International Commission of Agricultural and Biosystems Engineering (CIGR), http://cigr.org/.

Keicher, R. and Seufert, H. 2000. Automatic guidance for agricultural vehicles in Europe. *Computers and Electronics in Agriculture*, 25: 169–94.

Kise, M., Zhang, Q. and Noguchi, N. 2005. An obstacle identification algorithm for a laser range finder-based obstacle detector. *Transactions of the ASABE*, 48: 1269–78.

Leidenfrost, H. T., Tate, T. T., Canning, J. R., Anderson, M. J., Soule, T., Edwards, D. B. and Frenzel, J. F. 2013. Autonomous navigation of forest trails by an industrial-size robot. *Transactions of the ASABE*, 56: 1273–90.

Oksanen, T. and Visala, A. 2009. Coverage path planning algorithms for agricultural field machines. *Journal of Field Robotics*, 26: 651–68.

Oksanen, T. and Backman, J. 2016. Implement guidance model for ISO 11783 standard. In *Proc. of 5th IFAC AgriControl conference 2016*. Seattle, USA. International Federation of Automatic Control (IFAC), https://www.ifac-control.org/.

Osman, I. 1993. Metastrategy simulated annealing and tabu search algorithms for the vehicle routing problem. *Annals of Operations Research*, 41(4): 421–51.

Reid, J. F. and Searcy, S. W. 1987. Vision-based guidance of an agricultural tractor. *IEEE Control Systems*, 7(12): 39–43.

Reid, J. F., Zhang, Q., Noguchi, N. and Dickson, M. 2000. Agricultural automatic guidance research in North America. *Computers and Electronics in Agriculture*, 25(1): 155–67.

Reina, G, Underwood, G., Brooker, G. and Durrant-Whyte, H. 2011. Radar-based perception for autonomous outdoor vehicles. *Journal of Field Robotics*, 28(6): 894–913.

Rounsaville, J. D., Dvorak, J. S. and Stombaugh, T. S. 2016. Methods for Calculating Relative Cross-Track Error for ASABE/ISO Standard 12188-2 from Discrete Measurements. *Transactions of the ASABE*, 59(6), 1609–16.

Rovira-Más, F., Zhang, Q. and Hansen, A. C. 2010. *Mechatronics and Intelligent Systems for Off-road Vehicles*. London, UK: Springer-Verlag.

Rovira-Más, F., Zhang, Q., Reid, J. F. and Will, J. D. 2005. Visual crop row detection algorithm for automated guidance. *Journal of Automobile Engineering*, 219, 999–1010.

Ruckelshausen, A., Biber, P., Dorna, M., Gremmes, H., Klose, R., Linz, A., Rahe, R., Resch, R., Thiel, M., Trautz, D. and Trautz, D. 2009. *BoniRob – An Autonomous Field Robot Platform for Individual Plant Phenotyping*. Paper presented at the Proceedings of the 7th European Conference on Precision Agriculture, Wageningen: Academic Publishers.

Seyyedhasani, H. and Dvorak, J. S. 2017. Using the vehicle routing problem to reduce field completion times with multiple machines. *Computers and Electronics in Agriculture*, 134: 142–50.

Schafer, R. L. and Young, R. E. 1979. An automatic guidance system for tractors. *Transactions of the ASAE*, 22(1): 46–9, 56.

Shearer, S. A., Pitla, S. K. and Luck, J. D. 2010. *Trends in the Automation of Agricultural Field Machinery*. Paper presented at the Club of Bologna, Bologna, Italy.

Shockley, J. M., Dillon, C. R. and Stombaugh, T. S. 2011. A whole farm analysis of the influence of auto-steer navigation on net returns, risk, and production practices. *Journal of Agricultural and Applied Economics*, 43(1): 57–75.

Shutske, J. M., Gilbert, W. and Chaplin, J. 2001. Evaluation of a microwave and infrared human-presence sensing system for agricultural equipment. *Journal of Agricultural Safety and Health*, 7: 253–64.

Spekken, M. and de Bruin, S. 2013. Optimized routing on agricultural fields by minimizing maneuvering and servicing time. *Precision Agriculture*, 14(2): 224–44.

Steen, K. A., Villa-Henriksen, A., Therkildsen, O. R. and Green, O. 2012. Automatic detection of animals in mowing operations using thermal cameras. *Sensors*, 12: 7587–97.

Tillett, N. D. 1991. Automation guidance sensors for agricultural field machines: A Review. *J. Agri. Eng. Res.*, 50: 167–87.

Toth, P. and Vigo, D. 2002. *The Vehicle Routing Problem*. Philadelphia, Pennsylvania: Society for Industrial and Applied Mathematics.

Tsai, D., Lin, C. and Chen, J. 2003. The evaluation of normalized cross correlations for defect detection. *Pattern Recognition Letters*, 24(15): 2525–35.

Venem, M. T., Shutske, J. M. and Gilbert, W. J. 2006. Testing and creation of a safety system to disengage the PTO of a tractor. *Applied Engineering in Agriculture*, 22: 5–12.

Wei, J., Rovira-Más, F., Reid, J. F. and Han, S. 2005. Obstacle detection using stereo vision to enhance safety of autonomous machines. *Transactions of the ASABE*, 48: 2389–97.

Weiss, U. and Biber, P. 2011. Plant detection and mapping for agricultural robots using a 3D LIDAR sensor. *Robotics and Autonomous Systems*, 59: 265–73.

Werner, R., Kormann, G. and Mueller, S. 2013. Systematic model based path tracking control of actively steered implement in simulation and experiment. In *Proc. of 4th IFAC Agricontrol conference*. Laxenburg, Austria: IFAC.

Willrodt, F. L. 1924. Steering attachment for tractors. *US Patent No. 1506706*.

Yang, L. and Noguchi, N. 2012. Human detection for a robot tractor using omni-directional stereo vision. *Computers and Electronics in Agriculture*, 89: 116–25.

Zaman, Q. U., Schumann, A. W. and Miller, W. M. 2005. Variable rate nitrogen application in Florida citrus based on ultrasonically-sensed tree size. *Applied Engineering in Agriculture*, 21: 331–5.

Zandonadi, R. S., Luck, J. D., Stombaugh, T. S., Sama, M. P. and Shearer, S. A. 2011. A computational tool for estimating off-target application areas in agricultural fields. *Transactions of the ASABE*, 54: 41–9.

Zhang, J., Chambers, A., Maeta, S., Bergerman, M. and Singh, S. 2013. 3D perception for accurate row following: Methodology and results. In *IEEE/RSJ International Conference on Intelligent Robots and Systems*, pp. 5306–13. IEEE, https://www.ieee.org/.

Chapter 10

Controlled traffic farming in precision agriculture

Diogenes L. Antille, National Centre for Engineering in Agriculture, University of Southern Queensland, Australia; Tim Chamen, Controlled Traffic Farming Europe Ltd, UK; Jeff N. Tullberg, National Centre for Engineering in Agriculture, University of Southern Queensland, Australia; Bindi Isbister, Department of Primary Industries and Regional Development, Agriculture and Food, Australia; Troy A. Jensen, Guangnan Chen and Craig P. Baillie, National Centre for Engineering in Agriculture, University of Southern Queensland, Australia; and John K. Schueller, Department of Mechanical and Aerospace Engineering, University of Florida-Gainesville, USA

1 Introduction

In the past few decades, there has been a continuous drive towards the development and adoption of larger, and more powerful, agricultural machinery (Kutzbach, 2000; Jørgensen, 2012). Larger machinery is often related with timeliness, higher work rates and lower labour requirements, which has led to significant improvements both in efficiency and productivity (Vermeulen et al., 2010). A drawback of this trend has been the associated increase in machinery weight, which has, to some extent, offset advances made by the industry in developing improved running gear, such as in tyre (e.g. radial ply tyres) and track technology (e.g. rubber belts) to reduce contact pressures (Ansorge and Godwin, 2008; Antille et al., 2013; Misiewicz et al., 2015). The progressive increase in axle loads, as observed for example with harvesting equipment (e.g. Ansorge and Godwin, 2007; Bennett et al., 2015), means that soil stresses have also continued to increase, extending deeper into the subsoil (e.g. $\geq$0.3 MPa at 400 mm deep) and exceeding historic values, such as those resulting from in-furrow ploughing (Koolen et al., 1992; Chamen, 2015).

http://dx.doi.org/10.19103/AS.2017.0032.11

1.1 Traffic-induced soil compaction

Traffic compaction[1] has adverse effects on the physical, chemical and biological properties of soils, thus affecting important soil processes and functions, and crop productivity (Soane and van Ouwerkerk, 1995; Wu et al., 1995; Lipiec and Hatano, 2003). As such, compaction compromises water infiltration, increases the frequency and duration of waterlogged conditions, reduces gaseous exchange between soil and the atmosphere, and restricts root penetration and therefore exploitation of water and nutrients in the subsoil (O'Sullivan and Simota, 1995). These effects can, in turn, lead to increased risk of erosion and runoff, and therefore nutrient and sediment transport to watercourses (diffuse pollution), and increased risk of flooding (Rickson, 2014; Graves et al., 2015; Alaoui et al., 2018). There is also an elevated risk of greenhouse gas emissions from soils affected by compaction due to poor aeration (Ruser et al., 1998, 2006; Antille et al., 2015a). The effects of soil compaction are often persistent, particularly in the subsoil, and they are intensified with repeated passes (Raghavan et al., 1976, 1977). Alleviation of compaction through natural processes varies widely depending on soil type, the extent of wetting and drying, and freezing and thawing cycles (Pollard and Webster 1978; Dexter, 1991). For example, heavy clay soils with shrinking–swelling properties may recover from the effect of compaction to a greater extent than typically medium- and lighter-textured soils with no shrinking–swelling capacity (Radford et al., 2007). However, in all soils, the rate of amelioration of such compaction decreases with an increase in soil depth (Kay, 1990; McHugh et al., 2009), which reinforces the need to avoid compaction, particularly in the subsoil. In intensively managed arable cropping systems, the frequency of traffic with farm machinery does not normally allow for natural soil alleviation; therefore, tillage repair treatments are often required between crop cycles (Alakukku, 1999; Spoor, 2006). Remediation of compaction through tillage is energy-demanding as draught increases exponentially with an increase in soil depth (Godwin, 2007; Godwin and O'Dogherty, 2007). Consequently, remediation of deep soil compaction can prove impractical and often uneconomic at depths greater than approximately 400 mm (Spoor and Voorhees, 1986; Håkansson and Reeder, 1994; Tullberg, 2000). Deep loosening of soil carries the risk of recompaction at such depth with subsequent traffic leading to a recurrent cycle of compaction–(deep) tillage–recompaction, which aggravates the problem (Spoor et al., 2003; Chamen, 2011). Therefore, effective technical solutions must lie in 'traffic management' rather than 'tillage management'.

Concerns over the long-term sustainability of arable cropping associated with progressive deterioration of the soil resource have contributed to the development of more efficient field traffic management strategies, namely, controlled traffic farming (CTF) systems. These systems have evolved in response to evidence of significant soil damage from compaction caused by field traffic. Reviews by Tullberg et al. (2007) and Chamen (2015) provide in-depth analyses of the factors that motivated the development of CTF systems using purposely modified commercially available machinery. Research (e.g. Taylor, 1983; Tullberg, 2010) and on-farm practice (e.g. Walsh, 2008; Newall, 2013) have shown that CTF systems are effective means of managing compaction by confining all load-bearing wheels[2] to the least possible area of permanent traffic lanes. Controlled traffic systems have fundamental advantages in maintaining 'good' soil structural conditions

1. The term 'compaction' refers here to that originating from traffic of farm machinery under relatively 'soft' soil conditions where the natural load-carrying capacity of the soil is exceeded. For some vehicles, compaction impacts have been somewhat offset by higher operating speeds.
2. The term 'wheels' is used here synonymously with rubber belt traction systems.

of non-trafficked crop beds, with improved trafficability and timeliness compared with conventional traffic systems (e.g. McHugh et al., 2009; Millington et al., 2017).

1.2 Objectives and scope

It is helpful to examine the concept of CTF, highlight the requirements of such a mechanisation system, and provide a synthesis of the potential benefits which may be realised through careful integration of CTF with precision agriculture (PA) technologies. These benefits are explored in terms of overall 'system' efficiency and, in particular, the sustainability dimensions of CTF.

2 Controlled traffic farming systems: definition and requirements

The concept of CTF was mentioned in an early study by Hylton-Jolliffe and Halkett (1858). Later, CTF was trialled in the United States after the 1950s as a strategy to mitigate soil compaction impacts on crop yield (e.g. Cooper et al., 1969; Dumas et al., 1973, 1975; Morling, 1982; Taylor, 1983; Carter et al., 1991) and in the United Kingdom in the 1980s (Chamen et al., 1990, 1992; Chamen and Longstaff, 1995), but it was not adopted in commercial-scale farming until the 1990s, initially in Australia (Tullberg et al., 2007), and subsequently in northern and central Europe (e.g. Chamen, 2015; Galambošová et al., 2017), and Canada (CTFA, 2017), respectively. Modification of commercially available machinery and development of precision (±0.02 m accuracy) guidance systems (RTK-DGNSS, Real-Time Kinematic-Differential Global Navigation Satellite System) greatly facilitated on-farm adoption of CTF (Dijksterhuis et al., 1998; Vermeulen and Chamen, 2010).

The Australian Controlled Traffic Farming Association Inc. (ACTFA, http://actfa.net/) defines CTF as a system in which:

- All machinery has the same or modular working and track width so that field traffic can be confined to the least possible area of permanent traffic lanes.
- All machinery is capable of precise guidance along permanent traffic lanes, and
- The layout of permanent traffic lanes is designed to optimise surface drainage and operational logistics.

These elements are essential components of a CTF system (Vermeulen et al., 2010; Tullberg, 2014). The following practices are also facilitated by and in agreement with CTF (after ACTFA, 2018):

- Minimal soil disturbance (e.g. shallow tillage and no-tillage). Strategic tillage may be occasionally required in long-term no-tillage systems for mechanical control of herbicide-resistant weeds (e.g. Melland et al., 2017; Dang et al., 2018) and restore optimal growing conditions on some soils. In sandy soils with hardpans, deep tillage (e.g. 500 mm) with topsoil slotting (soil inclusion plates attached to the rear of deep ripper tines) is an effective technique to prolong the de-compaction response in those soils (Blackwell et al., 2016).

- Increased frequency of rainfed cropping and cover crops to maximise biomass production, residue returned to soil and therefore greater opportunities for carbon sequestration (Antille et al., 2015a). In Australia, typical cropping frequencies in CTF systems under no-tillage vary between 1.0 and 1.2, which compares to about 0.7 in non-CTF with conventional tillage.
- Precise management such as inter-row seeding and accurate placement of agrochemicals and fertilisers.
- Spatial (subfield-scale) monitoring, mapping and management at progressively finer resolution within a defined spatial framework, which distinguishes between permanent traffic lanes and crop beds. This is assisted by continuous regeneration of soil structure in crop beds (e.g. Hulugalle et al., 2017) and elimination of traffic compaction-induced soil variability.
- Accurate and repeatable on-farm research trials based on a defined spatial framework and enabled by PA technologies, GPS guidance and yield mapping.

Conversion from a conventional system, with unmatched machinery and track gauge widths, to CTF should consider the steps listed in Table 1 (after Isbister et al., 2013, 2018).

In well-designed CTF systems, such as those commonly used in grain cropping in Australia, permanent traffic lanes typically occupy less than 15% of the total cultivated field area, provided that the section width of heavy load-bearing tyres is 500 mm or less (Tullberg et al., 2007; Tullberg, 2014). Without CTF, varying equipment and track gauge widths often translate into disorganised or 'random' traffic patterns, which can cover more than 85% of the total cultivated area each time a crop is produced. In non-CTF systems, even when no-tillage is practised, the field area affected by traffic can be as high as 45% (Kroulík et al., 2009). Alternative CTF systems to the single track width (known as 'Com-Trac') have been also developed, and many of these are more readily adoptable within European farming systems (e.g. Galambošová et al., 2017). For 'Com-Trac' systems, the width of the track matches that of the vehicle that is most costly or difficult to modify, typically the harvesting equipment. For combine harvesters in which the front axle has single tyres configuration, the standard track width is 3 m; hence, all other equipment needs to have equal wheel spacing, as shown in Fig. 2.

Figure 3 shows the use of Trackman® (Newell et al., 1998) to calculate and illustrate the relative footprint of mechanisation systems representing conventional tillage[3] and zero-tillage with unmatched machinery and track gauge widths, and CTF with 3 m wheel spacing. Trackman® is a wheel track mapping software, which was developed to determine both the extent and location of wheeling within a field given the mechanisation system available at the farm. The software can be used for decision-making to investigate 'what-if' scenarios when considering conversion to CTF. This is possible by making repetitive adjustments to equipment widths and wheel spacing options until the mechanisation system is optimised to suit the farm-specific application. By providing information on the machinery modifications required, an economic evaluation can then be undertaken to assist decision-making. Other decision support tools are also available, for example, the 'CTF Calculator' (http://www.ctfcalculator.org/). This latter tool extends Trackman® and can be used to provide estimates of the area of a field affected by traffic and assess compaction management options, including CTF and deep tillage (WAAA, 2018).

3. 'Conventional tillage' involves primary and secondary tillage operations for crop establishment (ASABE, 2005, 2009).

Table 1 Components of controlled traffic farming (CTF) systems (after Isbister et al., 2013, 2018)

Component	Description
Guidance system	• Requirement for accurate guidance systems that enable machinery to return consistently and precisely to the same (permanent) traffic lanes. This is provided by global navigation satellite systems with a real-time kinematic (RTK) ±0.02 m correction, • Service from continuous operation reference station networks is available, which reduces capital investment, • High-quality guidance delivers auto-steer, which provides the required accuracy (reduced overlapping), particularly for large cropping fields. Electronic guidance systems may allow for savings of up to 10% due to reduced overlapping (McCallum and Long, 2005), • Mechanical guidance systems and centre guide rows in the crop may also be an option (see Webb et al., 2004).
Machinery matching	• When establishing CTF, the following machinery-specific elements need to be considered: ◦ Deciding on imperial or metric system: a problem commonly encountered when dealing with different machinery brands or models. Sometimes, spreaders and sprayers may be run in metric, and planters and combine harvesters in imperial allowing for some overlap of the cutter bar, but ensuring no crop is missed. ◦ Choosing the operating width and match in multiples: a convenient operating width ratio is 3:1, although other combinations (Fig. 1) are possible (e.g. 2:1). Unloading of grain from the combine to a chaser bin 'on-the-go' may require extension of the combine's auger or lateral extension of the bin or both. ◦ Matching wheel track spacing: combine harvesters typically have 3 m wheel spacing, hence, other equipment (e.g. 2.2 m wide) may be modified accordingly to match such width. Hydraulically adjustable axles (e.g. sprayers) and axle extensions or cotton reels may also be used (e.g. tractors, sprayers and chaser bins).
Permanent traffic lanes' design	• Traffic lanes may be cropped or left bare, depending on the system, which needs to be assessed against susceptibility to erosion, weed pressure and shallow soil cultivation for pre-emergence herbicides (Rainbow, 2005). • Seeding equipment: tines or disc units may be removed for bare traffic lanes. Tines on traffic lanes may be operated at shallower depths than those in the crop bed to achieve the required soil cover without loosening the soil at depth, and to maintain trafficability conditions of consolidated soil. Discs typically operate at shallower depths and offer improved depth control than tines (Aikins et al., 2017). • Yield penalties (up to 40%) may occur in seeded traffic lanes, but this is compensated for by that achieved in crop beds; additional benefits are provided by maintaining a crop/residue cover (Neale, 2011; Galambošová et al., 2017).
Layout considerations	• Tools available for planning layout: aerial digital photogrammetry, digital elevation models, soil type maps, historic yield maps, site-specific knowledge. • Layout planning should consider potential adoption of variable rate technology (VRT).

(*Continued*)

Table 1 (*Continued*)

Component	Description
	• Risk of erosion associated with CTF layout needs careful consideration. Such risk may be pre-assessed using digital terrain models coupled with a multi-direction flow path prediction model (Poulton et al., 2015; Vacher et al., 2016). This technique enables high-resolution contoured surfaces and flow path prediction maps to be produced, which can be used to (1) identify 'hot spots' for erosion, (2) determine likely changes in surface water flows caused by traffic lanes (length, orientation) and contour banks, and (3) pre-assess the effectiveness of potential corrective measures for control of excess surface water. • Up/downslope orientation of traffic lanes is preferred to across slope due to reduced potential for soil loss, particularly in high rainfall areas (Tullberg et al., 2001; Titmarsh et al., 2003). Such orientation reduces the risk of flow concentration and therefore safe discharge of runoff. • Raised beds or subsurface drainage (e.g. mole drainage) provide an option to mitigate the risk of waterlogging (e.g. Godwin et al., 2015).

Different colours in Fig. 3 denote different pieces of farm equipment. In the conventional tillage scenario, the same tractor (red), a front wheel assist (FWA) fitted with dual tyres, is associated with the seeder, scarifier and blade plough. In the CTF and zero-tillage scenarios, the tractor (also in red) is displayed as a FWA, but fitted with single instead of duals tyres. The combine harvester (blue) is consistent across all situations. The 'footprint' displayed at the bottom of each section represents the accumulation of all wheel tracks for the associated pieces of equipment and for each scenario (note the footprint colours are consistent with the colours displayed in the figure). When the 'footprints' align as shown in the CTF example, the area of the field affected by compaction is significantly reduced. However, where random traffic occurs with multiple passes of equipment, the 'footprint' or compacted area is significant. In this latter situation, some areas of the field have up to seven passes of load-bearing wheels during a single cropping cycle.

Alternatives to the common CTF system, in which all vehicles have 3 m wheel spacing, include the following arrangements, which are based on the relative widths of farm equipment tracks (after Chamen, 2006; Vermeulen and Chamen, 2010):

- 'Half-Trac': a system with two track widths, one exactly half the width of the other. Implement widths are a direct multiple of one or other of the track widths.
- 'Twin-Trac': a system that uses two track widths; the wider track straddles adjacent passes of the narrower track, and implement width is the addition of the two track widths or a direct multiple of it.
- 'Out-Trac': a system that uses a single common standard track width, but allows the widest vehicle (usually the harvester) to track 'outwith' the narrower tracks while centred on them. Implements may be any common width or direct multiple.
- 'Ad-Trac': a system with two track widths, the narrower using one track of the wider, resulting in an additional track. Implements may be any common width or direct multiple.

Vermeulen and Chamen (2010) also provide schematic examples of the above-cited options for controlled traffic. For these alternative systems, CTF Europe Ltd. (http://www.controlledtrafficfarming.com/Home/Default.aspx) has proposed a tier approach that aims

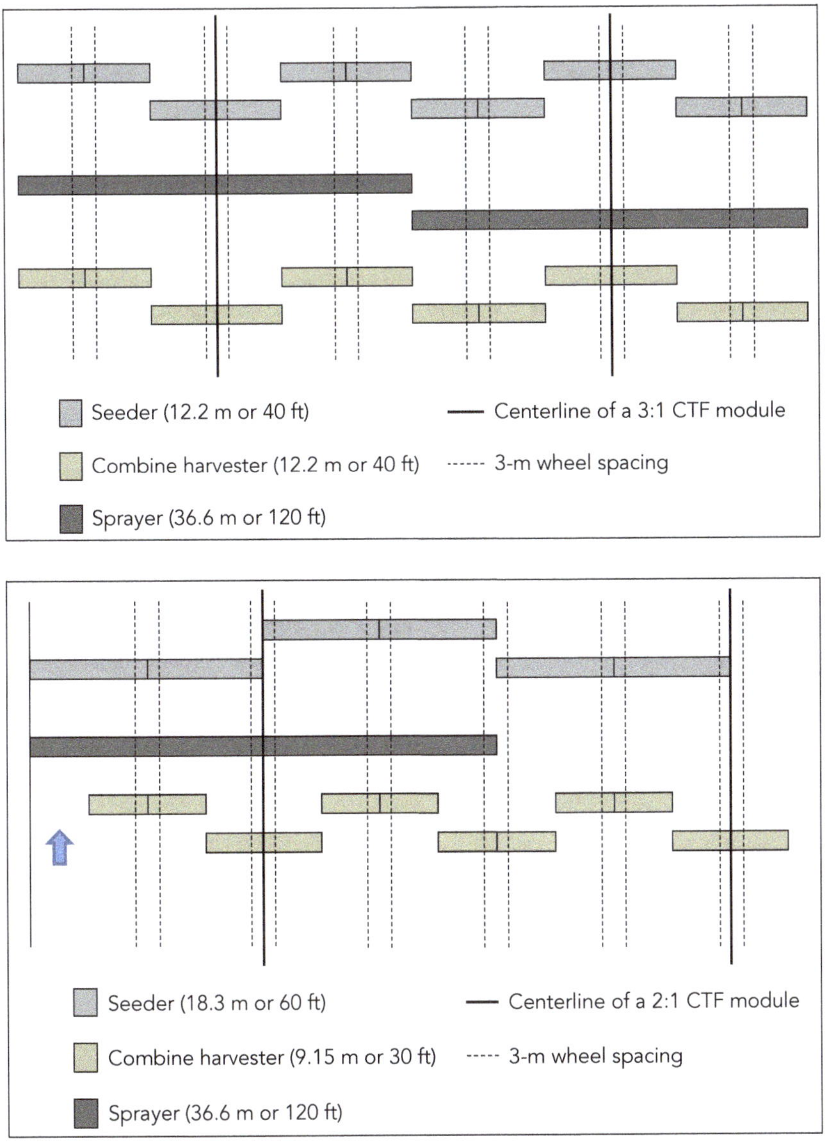

Figure 1 Schematic of two different CTF system setups. Top: a 3:1 seeder and combine harvester-to-sprayer matching ratio. Bottom: a 2:1 seeder-to-sprayer matching ratio in which the combine harvester's operating width is half and one-quarter the width of the seeder and sprayer, respectively (re-drawn from Isbister et al., 2018). The blue arrow in the bottom figure indicates that the combine harvester runs the starting lap half-empty. For all machinery, the wheel spacing is 3 m.

to encourage farmers to progressively reduce the area of the field subject to vehicular traffic through improvements in the design of the CTF system. This tier system includes the following (as per cent of tracked area): 30–40% (tier 1), 20–30% (tier 2), 10–20% (tier 3)

Figure 2 An 8400-series John Deere CTF-compatible tractor (left) and close-up of front (centre) and rear (right) axles extended to 3 m. Left: note the tractor positioned on the permanent traffic lanes of a 9 m module 'Com-Trac' CTF system combined with no-tillage (Felton, Queensland, Australia). The machinery width ratio used in this example is 3:1, that is, 9 m front combine harvester and planter bar, and 27 m boom width sprayer. Photo: courtesy of Rob McCreath (Felton, QLD, Australia).

and 10% or less (tier 4). Tier 4 may be only achievable with the use of gantry systems (e.g. Chamen et al., 1992, 1994; Taylor, 1994), which have been used in horticulture with satisfactory results (e.g. Pedersen et al., 2015, 2016). Seasonal controlled traffic systems have also been adopted and are designed to confine most field operations, usually with the exception of harvesting, to semi-permanent traffic lanes (Vermeulen and Mosquera, 2009). These systems represent a technical solution for the vegetable, sugar and cotton industries, for example, where incompatibilities between harvesting equipment used with different crops in the rotation are common (Braunack and McGarry, 2006; McPhee and Aird, 2013; Antille et al., 2016a). The CTF technical manual developed by Isbister et al. (2013) provides practical considerations to match machinery for CTF. Information is also

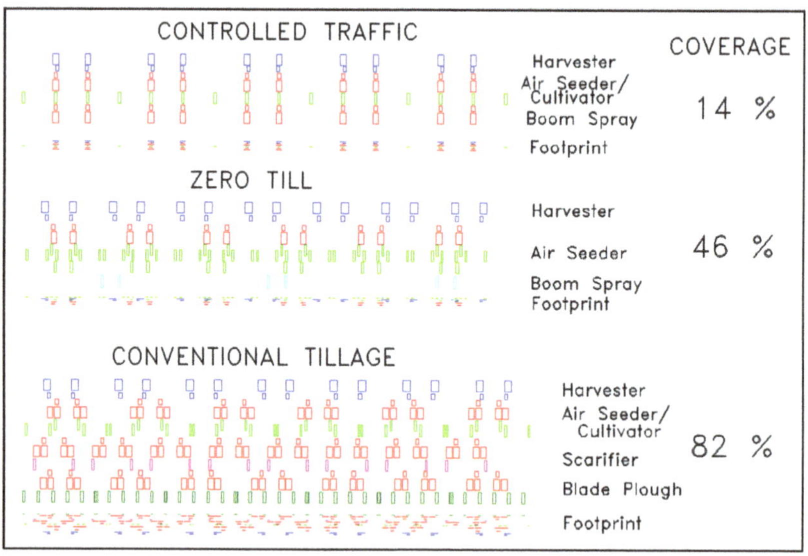

Figure 3 Typical wheeled areas for controlled traffic (3 m centre, 9 m module, zero-tillage) and unmatched machinery in zero-tillage and conventional tillage systems, respectively, determined with the use of Trackman® (Newell et al., 1998).

available from Isbister et al. (2018), and the websites of ACTFA and CTF Europe Ltd., which appear at the end of this chapter.

2.1 Advanced tyre technologies

Developments in tyre technology have enabled tyres to operate at significantly lower inflation pressures while supporting the same load and speed compared with conventional radial tyres (MICHELIN™, 2018). For example, 'increased (also 'improved') flexion' (IF) and 'very high flexion' (VF) tyres deflect relatively more (from 24% to 32%) at the same inflation pressure compared to standard radial tyres (from 18% to 22%), which provides increased contact patch length[4] and thus area (e.g. Smith et al., 2014). An IF tyre allows up to approximately 20% more load at the same inflation pressure (or 20% lower inflation pressure to support the same load) whereas a VF tyre allows up to about 40% more load (or 40% lower inflation pressure to support the same load). When cyclic field operations (CFO) or cyclic harvesting operations are combined with IF tyres, they allow for an additional 50% load capacity operating under 10 km h^{-1} (Harris, 2017). Low-pressure tyres are commercially available for a relatively wide range of farm equipment, including high-power tractors, combine harvesters, sprayers and crop conditioning equipment, and chaser bins. Research conducted at Harper Adams University, UK (Godwin et al., 2015, 2017), has shown that low (ground) pressure tyre technology can be considered as a cost-effective alternative to CTF, particularly when full conversion to this system is cumbersome due to incompatibilities between tyre configuration, track gauge widths and crop row configuration, and because of road transport-related regulations. For some equipment, tyres may also offer more versatility compared with rubber tracks but they do have their limits. For a VF tyre to carry a load of 10 Mg with a cyclic load, the inflation pressure would be 1.6 bar and the tyre around 1 m wide depending on the tyre diameter. On a combine harvester, this would make the machine about 4 m wide, assuming it could actually be fitted to the combine.

3 Sustainability of controlled traffic farming

Controlled traffic farming (CTF) can have potential benefits on (1) overall soil health, (2) crop performance, (3) fertiliser and water (rainfall and irrigation) use efficiency and (4) greenhouse gas emissions. Research and farm practice have shown that the overall improvements in efficiency that can be achieved with CTF usually translate into increased system resilience and therefore profit margins.

3.1 Effects on soil functions

The adverse effects of traffic-induced compaction on overall soil health are well documented (e.g. van Ouwerkerk and Soane, 1995; Soane and van Ouwerkerk, 1995). These effects are cumulative and can lead to a progressive decline in crop productivity (Alakukku, 1996; Shah et al., 2017). Soil compaction affects soil permeability to air and water movement through the profile (Vomocil and Flocker, 1961; Lipiec and Hatano,

4. Extra length rather than width is particularly valuable for CTF systems because these tyres reduce contact pressures without increasing tracked area.

2003), which is caused by reduced pore size and disruption of pore connectivity, with the associated reduction in surface water infiltration and hydraulic conductivity (Meek et al., 1992; Mossadeghi-Björklund et al., 2016). Hussein et al. (2017) found that differences in terminal infiltration rates in Red Ferrosols were up to ten-fold higher in the crop beds of a CTF system compared with the same soil managed under conventional traffic (non-CTF). The effect of traffic compaction on soil hydraulic properties is exemplified in Fig. 4 (Antille et al., 2016b). This example shows that disruption in the connectivity of soil pores caused by compaction occurs mainly between larger, vertically oriented drainage pores, and that compaction has relatively little impact on pores holding water at potentials greater than 1000 kPa; this effect was also shown by Connolly et al. (2001).

Reduced soil (macro)-porosity impairs internal drainage (Vero et al., 2014), which can set conditions for increased runoff and erosion (Rickson, 2014), increase the frequency and duration of waterlogged conditions (Alaoui et al., 2018) and enhance nitrogen (N) losses through denitrification (Torbert and Wood, 1992; Tullberg et al., 2018). Modelling work by Owens et al. (2016) showed that up to 90% reduction in soil erosion rates from the Great Barrier Reefs catchments, with the associated reduction in sediment, nutrient and pesticide discharge to surface waters, may be achieved when changing management practices from conventional tillage with no traffic control to no-tillage and CTF (Fig. 5). These results agree with those from earlier studies by Tullberg et al. (2001) and Li et al. (2009), which suggested that tillage and traffic effects on runoff and sediment yield appear to be cumulative. Owens et al. (2016)'s analysis assumed that 15% of the cultivated field area of the CTF system was affected by traffic (permanent traffic lanes)

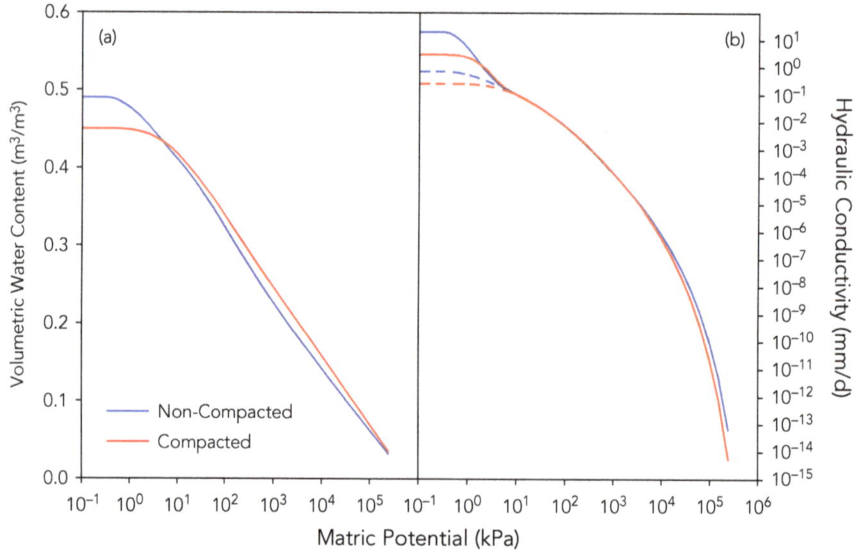

Figure 4 An example of the effect of compaction on soil water retention (a) and hydraulic conductivity (b) estimated by the SWIM3 soil water model (Huth et al., 2012) for a Vertisol (after Antille et al., 2016b). Note the relatively small contribution of larger pores on water retention near saturation in (a) and the relative difference between total hydraulic conductivity (solid lines) and micropore conductivity (dotted lines) in (b).

and that recommended management practices for erosion control (e.g. contour banks) were also in place.

Soil structural development in non-trafficked crop beds is also observed when the soil is tilled annually (e.g. Chamen et al., 1990; Vermeulen and Klooster, 1992), but faster rates of natural amelioration are expected when CTF is coupled with no-tillage (Bullock et al., 1985; McHugh et al., 2009). Good soil structural conditions are necessary for adequate functioning of soil organisms (Rabot et al., 2018), and these contribute to aggregate formation and development of biopores, which are largely implicated in soil water retention and soil internal drainage (Leeds-Harrison et al., 1986; Kautz, 2015). A study by Pangnakorn et al. (2003) in rainfed cropping in southern Queensland showed that earthworm counts under conventional (random) compared with controlled traffic

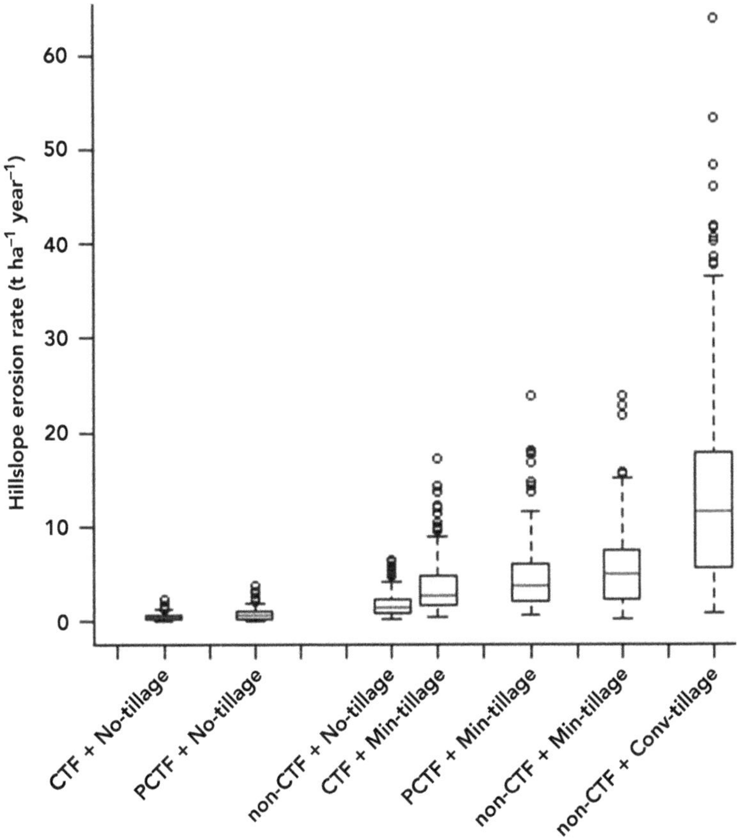

Figure 5 Average annual soil erosion in the Fitzroy Basin (Queensland, Australia) for soil management scenarios that include controlled (CTF) and partially controlled (PCTF) traffic farming, no-tillage, minimum (Min-) and conventional (Conv-) tillage, respectively (after Owens et al., 2016). Variation for each management scenario denotes inter-annual variation (1970–2015) for soils and climates. Box plots show Min, Q_1, Med, Q_3 and Max, respectively. Points numerically distant from the rest of the data are outliers.

no-till were approximately three and eight times higher, respectively, than under annually wheeled, tilled soil. Wheel traffic tended to reduce earthworm count, but wheeling followed by tillage had significantly less effect. Soil compression during wheeling did not appear to be the main cause of earthworm death, but rather restricted movement and oxygen depletion (Pangnakorn et al., 2003).

3.2 Effects on crop performance and fertiliser use efficiency

A distinction is made between direct and indirect effects of soil compaction on crop productivity (Chamen, 2006). Direct effects refer to the extent to which crop growth and physiological development are restricted by processes such as water and nutrient uptake, which are adversely affected by soil mechanical impedance (e.g. root expansion and exploration, biomass accumulation and partitioning). By contrast, indirect effects are those that relate to timeliness. This includes the time required to fix pre-existing compaction to enable satisfactory establishment of the next crop in due time, and the ability to complete in-season field operations (e.g. sowing, spraying and fertilisation). In most circumstances, the opportunities to conduct such operations occur within a relatively narrow window as it is determined by the crop (e.g. physiological stage), weather and soil conditions being appropriate (field access, trafficability and workability) for traffic with farm machinery (Earl, 1997; De Toro, 2005; Gut et al., 2015). Failure to conduct in-season operations at the right time will impact crop productivity (e.g. mismatch between fertiliser application and nutrient demand by the crop). Impacts on crop quality may also occur, for example, as a result of delayed spraying for crop protection purposes or late harvest (Bednarz et al., 2002); in all circumstances with the associated impact on crop profit margins. In crops established on soil with 'residual' compaction, such as that originated during harvest of the previous crop, the impact on yield is commonly explained by reduced plant stand and greater plant stand variability. For example, in cotton a 20% reduction in plant stand below a target of about 10–12 plants m^{-2} due to compaction can result in up to 15% reduction in yield or more if the effect is due to variability in plant stand (Hadas et al., 1985). For Australian broadacre grain production, the daily loss of crop value caused by delays in sowing (outside the optimum) was estimated at approximately 1.5% per day of the total crop value (Tullberg, 2014)[5].

Table 2 summarises reported information on impacts of soil compaction on crop yield. Optimum soil bulk density values are dependent on crop and soil type, and seasonal effects of weather (McKyes et al., 1979). Therefore, optimum values for a given crop vary within a range, and these may be narrower or wider depending on year-specific conditions.

Because yield is affected by compaction, the amount of fertiliser recovered in crop (agronomic efficiency) is concurrently reduced (Wolkowski, 1990; Hussein et al., 2018), which has financial implications for growers and adverse effects on the environment (Soane and van Ouwerkerk, 1995). The mechanisms by which fertiliser use efficiency is affected by compaction are discussed in several studies (e.g. Lipiec and Stępniewski, 1995; Arvidsson, 1999; Antille et al., 2016a), and the effect on plant uptake and subsequent recovery in crop is nutrient-specific as it depends on the complex interactions between root growth, soil water and aeration status, and degree of compactness. Generally, factors that restrict

5. Tullberg's (2018) analysis was conducted to determine the sowing capacity required to establish a crop within the optimum window for that crop, without significant financial losses. The same concept is applied here to show that delays in 'accessing' the field due to unsuitable trafficability conditions can have financial impacts of the same magnitude. Permanent traffic lanes in CTF largely eliminate this problem.

Table 2 Yield of common crops achieved in the absence of field traffic relative to yield obtained when the crop was grown under traffic intensities that are typical of the cropping system (100% means no difference)

Crop type	Relative yield (or range)	Soil type(s)	References
Winter cereals	112%	Red ferrosol	Hussein et al. (2017)
Winter cereals	104–135%	Loam, sandy loam	Hamilton et al. (2003), Galambošová et al. (2017) and Godwin et al. (2017)
Winter cereals	114–127%	Heavy clay	Radford et al. (2001), Tullberg et al. (2001) and Neale (2011)
Grain legumes	110–143%	Clay, clay loam	Khalilian et al. (1991) and Botta et al. (2007)
Summer cereals	100–175%	Loam, clay	Ngunjiri and Siemens (1995), Jensen et al. (2000), Radford et al. (2001) and Hussein et al. (2018)
Cotton	106–128%	Silt loam, sandy clay loam, clay	Hadas et al. (1985), Akinci et al. (2004), Kulkarni et al. (2010) and Hulugalle et al. (2017)
Oilseeds	128–195%	Clay loam, clay	Bayhan et al. (2002) and Chan et al. (2006)

Winter cereals: wheat, barley. Summer cereals: grain sorghum, maize. Grain legumes: soybeans. Oilseeds: oilseed rape, sunflower. Cotton: includes lint or lint + seed.

water movement within the soil profile (e.g. pore connectivity) and from the soil matrix to the roots will affect nitrogen (N) uptake since the main mechanism implicated in N transport to plant roots is mass flow (Barber et al., 1963; Kirkby et al., 2009). Nitrogen recovered in cereal (grain) and grass (biomass) correlates well with soil water availability or rather with water used by the crop during the crop cycle (Melaj et al., 2003; Antille et al., 2015b, 2017). Factors that affect root elongation, such as increased soil mechanical strength, will restrict nutrient absorption through root interception, and this mechanism is particularly important for phosphorus, because of its relatively low mobility within the soil (Wiersum, 1962; Prummel, 1975). Threshold values of soil strength above which root elongation stops vary depending upon the crop, but these typically range between 2 and 2.5 MPa for most arable crops (Taylor and Ratliff, 1969a,b). Traffic with harvesting equipment (overall load about 35 Mg) on Vertisols (moisture content: 20–25%, w/w) was reported to increase cone index in the 0–500 mm depth range from about 1 MPa prior to traffic to more than 3 MPa after traffic (Braunack and Johnston, 2014). Observations from that work were in close agreement with those reported by Ansorge and Godwin (2007) and Antille et al. (2013), albeit on medium-textured soil.

Hussein et al. (2017) showed differences in N-fertiliser use efficiency from wheat established under CTF and non-CTF systems of up to 40% (Fig. 6). Furthermore, differences in N recovered in grain between traffic treatments were irrespective of fertiliser-N formulation. Hussein et al. (2017)'s work agreed closely with Gregorich et al. (2014) demonstrating that significant improvements in fertiliser-N recoveries may not be realised with enhanced N formulations alone (e.g. slow and controlled-released fertilisers) and that avoidance of (random) traffic compaction is a pre-requisite for improved fertiliser use efficiency. This is an important practical consideration for N management as much research effort is being spent on the role of novel fertiliser formulations in improving nutrient use efficiency (e.g. Watts et al., 2014), without necessarily accounting for the fact that this is largely explained by overall soil condition.

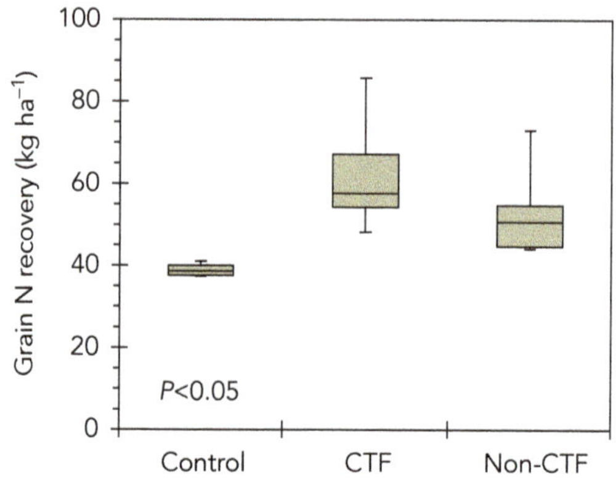

Figure 6 Controlled (CTF) and non-controlled (non-CTF) traffic effects on nitrogen (N) fertiliser use efficiency in wheat (*Triticum aestivum* L.). Box plots show: Min, Q_1, Med, Q_3 and Max, respectively, for N rates between 100 and 300 kg ha^{-1}, except for control ($N = 0$). 'Control' is the zero-N treatment from both CTF and non-CTF traffic systems, respectively (after Hussein et al., 2017).

3.3 Effects on energy use and greenhouse gas emissions

Crop production systems that reduce the need for tillage will significantly reduce on-farm energy use and associated emissions. These include direct emissions from fuel use during tillage operations (e.g. Burt et al., 1994) and carbon dioxide (CO_2) that is released to the atmosphere through oxidation of soil organic matter (e.g. Chatskikh and Olesen, 2007). Oxidation of soil organic matter is enhanced by tillage, particularly in warm and moist environmental conditions (Ding et al., 2002). There exists a relationship between traffic-induced soil compaction and tillage whereby increased traffic intensity leads to increased need for tillage repair treatments (Arndt and Rose, 1966). Consequently, energy requirements of primary and secondary tillage could increase by factors of up to 2.5 and 3.25, respectively, depending on the depth of the operation and the specific soil conditions under which such operations are conducted (Williford, 1980; Hadas et al., 1986; Chamen et al., 1996). In CTF systems, Lamers et al. (1986) reported energy savings of up to approximately 50% due to lower rolling resistance and wheel slip on permanent traffic lanes compared with trafficking over cultivated and relatively soft soil, and similar observations were made by Luhaib et al. (2017). Tullberg (2000) showed that the effect of wheeling on the draft of tillage implements (chisel and sweep tines positioned behind a 90 HP tractor in a black Vertisol, 250 mm depth) increased total draft by 30% compared with the same implement operated in non-wheeled soil. Tullberg used these observations to demonstrate that approximately 50% of a tractor's power output can be squandered in the process of creating and subsequently disrupting its own wheel compaction in tillage operations.

Anecdotal evidence from growers who practise CTF and no-tillage indicates up to 50% reduction in tractor fuel use for field operations. Of the total area used for arable cropping in England (about 1.8 M ha), approximately 0.5 M ha were established using reduced or no-tillage (DEFRA, 2009). For cereal cropping, where reduced tillage is not practised, average diesel fuel use was approximately 95 L ha^{-1} (see also Taylor et al., 1993). If CTF and no-tillage allow for a reduction of 50% in fuel use (of 95 L ha^{-1}) by avoiding compaction across the arable cropping area in England, this would save approximately 82×10^6 L of diesel fuel per year, equivalent to 220 Mg CO_2e or 125 kg ha^{-1} of CO_2e per year (assume 1 L of diesel ≈ 2.7 kg CO_2e). Another study by Tullberg (2014) in Australian grain production systems showed that tillage-based systems[6] required 52 L ha^{-1} of diesel for field operations whereas the most efficient non-CTF system under no-tillage required no less than 25 L ha^{-1}. When energy losses due to rolling resistance were minimised by driving over consolidated soil (permanent traffic lanes) of a CTF system, diesel use was further reduced to approximately 13 L ha^{-1} (Tullberg, 2014).

For arable systems that rely on tillage, reductions in energy use where traffic is controlled are mainly due to lower soil-specific resistance in the absence of compaction, tillage operations conducted at shallower depths when remediation of deep compaction is not required, fewer tillage operations and reduced power loss in tractive efficiency because of lower motion resistance and reduced wheel slip (Dickson and Ritchie, 1996; Tullberg, 2000).

Life cycle analyses conducted by Gasso et al. (2014) suggested that potential emissions reductions from adoption of CTF in intensively managed arable cropping systems would be between 20% and 45% for N_2O, and between 370% and 2100% for CH_4, and that direct emissions from field operations could be reduced by at least 20% compared with non-CTF systems. Such emissions reduction potential agreed with measured data available in the literature, as compiled by Antille et al. (2015a,c). Higher N-fertiliser use efficiencies (NUE) typically achieved in CTF systems (Fig. 6) are consistent with the above observations, and suggest disproportionally lower NUE at high N application rates (e.g. ≥200 kg ha^{-1} N) in non-CTF systems because of the non-linear response relationship between N rate and (direct) N_2O emissions (Millar et al., 2010; Scheer et al., 2016). Figure 7 shows short-term (45 days) emissions of N_2O after injection of anhydrous ammonia (82% N) at a rate of 80 kg ha^{-1} N to a black Vertisol established to wheat in southern Queensland (Australia). Mean N_2O emissions from simulated 'random' traffic were similar to those from permanent traffic lanes of the CTF system, and significantly higher than those from permanent crop beds (Tullberg et al., 2011). A negative sum of CH_4 fluxes (absorption) was observed in permanent crop beds whereas the sum of fluxes from both wheeled soils was positive (emission). Hence, overall traffic treatment effects were significant ($p < 0.05$). Total emissions of N_2O and CH_4 over the measured period post-seeding (45 days) were 58 (permanent crop bed), 325 (permanent traffic lanes) and 370 (random traffic) kg ha^{-1} CO_2e, respectively. This indicates a 45-day post-seeding total CO_2e emission from the CTF system used in the study of 90 kg ha^{-1}, that is, 39 kg ha^{-1} from the 12% cropped area occupied by permanent traffic lanes and 51 kg ha^{-1} from the remaining 88% permanent crop beds. Such losses represent approximately 40% of the emissions of 214 kg ha^{-1} likely from a randomly trafficked soil where 50% of the cropped field area is wheeled. Relatively higher N_2O fluxes from the random traffic treatment compared with permanent traffic lanes simply reflects the combined N-fertiliser × traffic effect on emissions. This effect was not observed in permanent traffic lanes because fertiliser is placed at or prior to planting next to the plant row, and in the system used in Tullberg's study, traffic lanes were not cultivated. Further studies by Tullberg et al. (2018) supported those initial observations, and confirmed that adoption of CTF could reduce total soil emissions by 30–50%, which was consistent across a wide range of soil types, crop rotations and environmental conditions.

6. In Australia, grain production systems that rely on tillage typically practise minimum ('conservation or non-inverting') shallow tillage with stubble retention.

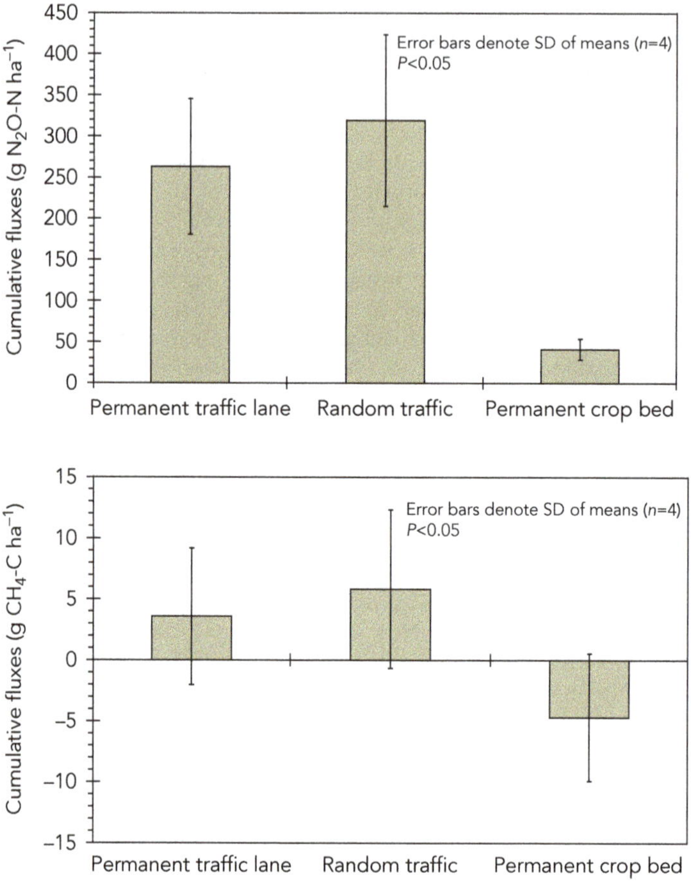

Figure 7 Short-term (45 days) cumulative fluxes of nitrous oxide (N_2O) and methane (CH_4) following establishment of wheat on a black Vertisol in southern Queensland (Australia) showing the effect of traffic compaction on emissions (after Tullberg et al., 2011).

4 Coupling controlled traffic farming with precision agriculture

The beneficial effects of CTF are often associated with reduced soil compaction and related mechanisms, which have a direct impact on crop yield, but the absence of widespread traffic-induced soil compaction does not always translate automatically into increased yield. A time lag generally occurs before tangible agronomic benefits can be realised following adoption of CTF, which largely depends on the interaction between the natural environment (e.g. soil type, rainfall) and the overall management of the production system. Inputs (e.g. fertiliser, irrigation), crop husbandry (e.g. crop rotation, crop protection) and management of soil-related constraints (e.g. drainage, hardsetting soil, removal of natural compaction, sodicity/salinity) need to be concurrently optimised to enable those agronomic benefits to be delivered. An important practical consideration that will assist farmers with such an optimisation process is the adoption of PA technologies to enable management of spatial variability at the sub-field scale (Robertson et al., 2007). Adoption of variable rate technology (VRT) should be preceded by adoption of CTF, as this will eliminate soil variability induced by traffic, particularly compaction, and the soil processes and functions that may be impaired as a result.

The interaction of controlled traffic and PA can be seen from the very beginnings of PA. Figure 8 shows what is believed to be the first publishing of a yield map automatically generated during grain combining (Searcy et al., 1989). The straight, aligned nature of two narrow low-yielding regions in 1986 indicated that they were probably man-made. Discussion with the farm manager brought to light the fact that farm workers had crossed the field with heavy equipment to access another field. The yield map allowed the manager to determine how much yield those actions had cost them and motivated the prohibition of such traffic. PA provides an opportunity to monitor the effects of field traffic and to demonstrate how controlled traffic can minimise such affected areas.

Table 3 provides a summary of CTF factors, which are at the interface with PA and highlights the importance of CTF in enabling greater uniformity and precision in the relationship between soil/crop and machine.

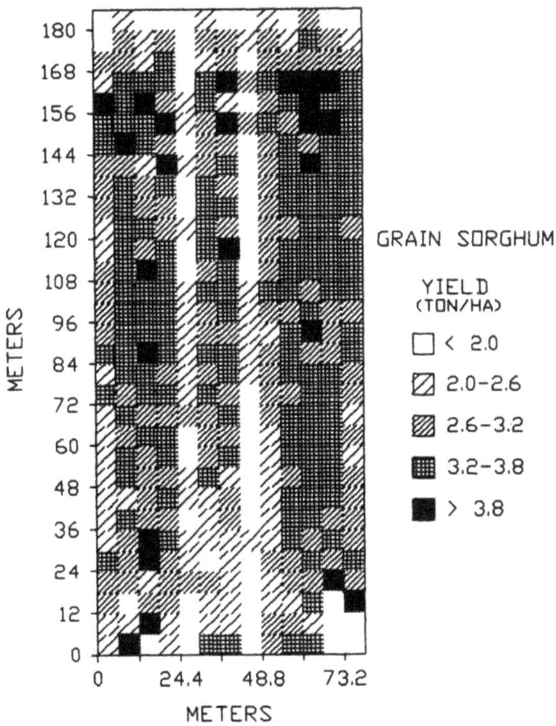

Figure 8 Sorghum combining yield map from 1986 showing effects of pre-planting traffic (from Searcy et al., 1989).

Table 3 CTF factors at the interface with PA, which facilitate greater precision in the relationship between soil/crop and machine

Factor	Description	References
Timeliness, trafficability and field efficiency	• Improved field access for all field operations: well-designed permanent traffic lanes provide a means for safe disposal of excess surface water. Therefore, the risk of waterlogging can be rapidly overcome, which ensures that traffic lanes are trafficable sooner after rain. This is critical when timeliness is so important to the outcome of crop establishment as well as in-crop (increased gap between potential and attainable yield) and harvest operations (loss of actual yield and quality), and optimal timing is usually influenced by rainfall events. • Improved load-carrying capacity in permanent traffic lanes also increases the window of opportunities for field operations. Improved tractive efficiency on compacted traffic lanes: reduced rolling resistance and wheel slip, and improved fuel use efficiency. • Layout of permanent traffic lanes enables optimisation of in-field logistics routing, minimising manoeuvring and in-field servicing time, including time associated with supporting operations during harvest (e.g. loading/unloading of grain bins). • Increased in-field efficiency is achieved through reduced overlapping between adjacent swaths (e.g. tillage, fertiliser applications, spraying). Decision support systems (DSS) for optimisation of in-field machinery management (fleet trajectory planning) are available, including DSS that estimate energy savings from such optimisation processes, and route planning for autonomous vehicles.	Parvin et al. (2005), Bochtis et al. (2010a,b, 2014), Spekken and de Bruin (2013), Vero et al. (2014), Gut et al. (2015), Conesa-Muñoz et al. (2016), Spekken et al. (2016), Rodias et al. (2017), Couto et al. (2018) and Seyyedhasani and Dvorak (2018).
Improved uniformity and precision	• RTK-GPS auto-steer systems achieve precision of ±0.02 m approximately 70% of the time, but this level of precision can be significantly affected by, for example, transverse movement allowed by tractor tyres. Transverse displacement of implements relative to tractors can be even greater. CTF does not eliminate such effects, but they are reduced and largely offset by the use of consistent field patterns and/or implement auto-steer. • Matched implement widths and consistent cultivation patterns also provide a more consistent machine/crop relationship, for example, in mechanically controlling herbicide-resistant weeds with minimal crop damage or inter-row application of herbicide (shield and band spraying). • Inter-row seeding: commonly in the centre of the inter-row, but sometimes greater precision is needed for seed placement nearer the row of previous or standing crop.[a] Other examples include stubble handling, fertiliser placement. Precise management is also required for surface drainage and runoff control.	Bromet (2006), Yule (2007), Werth et al. (2008) and Dang et al. (2018).

(Continued)

Table 3 (*Continued*)

Factor	Description	References
Reduced crop and soil variability	• Greater uniformity of crop development facilitates management decisions (e.g. irrigation, spraying/defoliation, fertiliser application, harvest operation). Reduced crop variability is associated with reduced traffic-induced variability in soil hydraulic and mechanical properties influencing crop growth and nutrition. • The effectiveness of selective herbicides, which have narrow windows for application, is improved when targeted weeds show similar developmental stages. • Fertiliser use efficiency can be improved by synchronising fertiliser supply (rate and timing of application) with crop demand for nutrients. Variable rate technology allows for fertiliser management at the sub-field scale, but better synchronisation between demand and supply can be achieved when the crop in different management zones within a field has the same physiological stage. Fertiliser rates can be adjusted to meet actual nutrient demand (at the sub-field scale), but the (optimum) timing of application needs to be the same for the entire crop in the field. • Reducing variability is important, and management zones need to be defined based on understanding causes as they relate to soil type and landscape, and managed by variable rate inputs. Pre-defined management zones need to be seasonally consistent more than 70% of the time for VRT to be effective and profitable.	Wall and Friesen (1991), Godwin et al. (2003), Yule (2007), Tillett et al. (2008), Lawes et al. (2009), Barik et al. (2014), Norton (2014), Alesso et al. (2016) and Snyder (2017).

[a] Reduced water repellence and improved water-holding capacity is observed in some soils (e.g. Doerr et al., 2000) closer to previous crop rows; hence, greater level of precision is required for seed placement near-crop rows in these circumstances.

Yield maps may be used for visual examination of a crop uniformity from side by side CTF and non-CTF fields, respectively. As discussed in Table 3, greater uniformity is observed in CTF compared with non-CTF crops, and Yule (2007) attributed this improvement to optimised soil–plant–water relations (more uniform plant available water) and reduced waterlogging when layout of permanent traffic lanes has been designed for effective surface drainage. The small-scale variability that may be lost from yield maps by kriging is visible, together with field-scale variability, in high-quality satellite imagery, as further evidence of the fact that much of the field-scale variability is simply the result of management activity and not a true reflection of soil type (Yule, 2007).

Generally, the input optimisation processes of PA do not acknowledge that the basic pixel of yield maps is derived from the cutting width of the combine harvester (e.g. 15 m) and a travel distance (e.g. 10–20 m) determined by grain processing delay. Hence, yield maps do not recognise smaller-scale soil state variability, such as that produced by vehicle traffic. Similarly, it cannot distinguish yield variability that results from non-uniform seed distribution during planting (e.g. Aikins et al., 2017) or fertiliser spreading (e.g. Jensen and Pesek, 1962; Antille et al., 2015d). For this latter operation, a coefficient of variation (CV) of 15% or less is considered as 'good' relative to a target CV of typically 10% (Miller et al., 2009). However, this implies that the best area deciles receive 67% more inputs than the worst (Tullberg, 2018). Small spatial-scale yield variability of similar or greater magnitude to the field-scale variability dealt with by VRT is not seen in kriged yield maps, unlike high-resolution (e.g. less than 4 m^2 pixel scale) satellite or UAV images. Such smaller-scale resolution is required to depict spatial variability induced by management, but can also be observed by sprayer-mounted NDVI sensors (Tullberg, 2018). When CTF is designed to reduce within-field variability (e.g. waterlogging, erosion and compaction), crop establishment and growth are more uniform, and PA tools (e.g. variable rate inputs of water and fertiliser) are able to deliver greater efficiencies. Controlled traffic farming eliminates variability created by non-organised traffic patterns and the application of PA in CTF allows for in-depth analysis of 'system' performance, the identification of causes of variability and management options. This is more effective at small pixel scale.

Precision agriculture is often perceived as the ultimate solution for improved productivity despite the fact that it does not always account for variability at the sub-machine width spatial scale. The effects of traffic compaction have significant economic impacts, and these are relatively well-known (e.g. Walsh, 2002; Chamen et al., 2015; Graves et al., 2015), and the same is true for variability caused by non-uniform machine performance (e.g. Horrell et al., 1999; Richards and Hobson, 2013). These unwanted effects can be addressed by CTF and proper machinery calibration (Tullberg, 2018).

5 Future trends and conclusion

There is both circumstantial and direct evidence which demonstrates the significant productivity and sustainability benefits associated with adoption of controlled traffic farming (CTF). These benefits may be fully realised when CTF is jointly practised with no-tillage and assisted by the range of precision agriculture (PA) technologies available. Farmers often recognise the synergistic effect of integrating CTF with no-tillage and PA; hence, producing outcomes that are greater than the sum of the parts. Important contributing factors are those associated with improved trafficability and field access (timeliness), and those associated with greater precision and uniformity. Effective integration of CTF and PA requires an understanding of factors that limit productivity, the nature and extent of variability and the management options available to farmers. Adoption of CTF, and its allied technologies, is therefore encouraged as a technically and economically viable option to improve productivity, resource-use efficiency and all dimensions of sustainability.

Studies on CTF economics have consistently shown that it is a profitable technological innovation for arable land-use (e.g. Kingwell and Fuchsbichler, 2011; Chamen et al., 2015; Galambošová et al., 2017) and grass production systems (e.g. Peets et al., 2017). Despite these benefits, large-scale adoption of CTF has been low,[7] with the exception of grain production systems in Australia where approximately 30% of the total grain production area is now managed under CTF. The main barriers for adoption of CTF have been equipment incompatibilities and the need to modify machinery to suit a specific system design, often at the own farmers' risk of loss of product warranty. Other barriers include reliance on contracting operations and land tenure systems. These structural barriers may be overcome with forward planning when conversion to CTF is built into the machinery

7. The extent of adoption in the United Kingdom is unknown but it could now be significant and can only be revealed by a national survey.

replacement programme, and organisations such as ACTFA and CTF Europe have developed suitable schemes to assist farmers in such a process.

6 Disclaimer

Mention of trade names or commercial products in this chapter is solely for the purpose of providing accurate information and does not imply recommendation, endorsement or otherwise by the authors or their institutions. Data presented in figures and tables is fully credited to the original sources, and corresponding citations are provided.

7 Where to look for further information

Further information can be retrieved from the following sources:

- Australian Controlled Traffic Farming Association Inc.: http://www.actfa.net/
- Controlled Traffic Farming Alberta: http://www.controlledtrafficfarming.org/
- Controlled Traffic Farming Europe Ltd.: http://www.controlledtrafficfarming.com/Home/Default.aspx
- Controlled Traffic Farming Technical Manual: https://www.nacc.com.au/wp-content/uploads/2015/05/NACC_Controlled_Traffic_Farming_Technical_Manual.pdf
- CTF Solutions: http://www.ctfsolutions.com.au/easyweb3/WEBID-132275-ep_code-index.html/
- Harper Adams University (Shropshire, UK), Controlled Traffic Farming and Soil Tillage Research: https://www.harper-adams.ac.uk/research/project.cfm?id=4
- Precision Agriculture, Australia: http://www.precisionagriculture.com.au/
- The National Centre for Precision Farming, Harper Adams University (Shropshire, UK): https://www.harper-adams.ac.uk/research/ncpf/
- The Soil and Water Management Centre: http://www.soilandwater.org.uk/
- WideTract: https://widetract.com/

8 References

ACTFA (2018). Controlled traffic farming defined. *Australian Controlled Traffic Farming Association Inc.* Retrieved: http://actfa.net/ (Accessed on 16 January 2018).

Aikins, K. A., Antille, D. L., Jensen, T. A. and Blackwell, J. (2017). Advances in residue management mechanisms of zero-tillage planters. ASABE Paper No.: 1700449. St Joseph, MI: ASABE. DOI:10.13031/aim.201700449.

Akinci, I., Cakir, E., Topakci, M., Canakci, M. and Inan, O. (2004). The effect of subsoiling on soil resistance and cotton yield. *Soil and Tillage Research*, 77(2): 203–10. DOI:10.1016/j.still.2003.1012.1006.

Alakukku, L. (1996). Persistence of soil compaction due to high axle load traffic. II. Long-term effects on the properties of fine-textured and organic soils. *Soil and Tillage Research*, 37(4): 223–38. DOI:10.1016/0167-1987(96)01017-3.

Alakukku, L. (1999). Subsoil compaction due to wheel traffic. *Agricultural and Food Science in Finland*, 8(4–5): 333–51.

Alaoui, A., Rogger, M., Peth, S. and Blöschl, G. (2018). Does soil compaction increase floods? A review. *Journal of Hydrology*, 557: 631–42. DOI:10.1016/j.jhydrol.2017.12.052.

Alesso, C. A., Cipriotti, P. A., Linguetti, C. G., Masola, M. J., Imhoff, S. C. and Antille, D. L. (2016). Spatial distribution of soil mechanical strength within a controlled traffic farming system as determined by cone index and geostatistical techniques. ASABE Paper No.: 16-2456741. St Joseph, MI: ASABE. DOI:10.13031/aim.20162456741.

Ansorge, D. and Godwin, R. J. (2007). The effect of tyres and a rubber track at high axle loads on soil compaction, Part 1: Single axle-studies. *Biosystems Engineering*, 98(1): 115–26. DOI:10.1016/j.biosystemseng.2007.06.005.

Ansorge, D. and Godwin, R. J. (2008). The effect of tyres and a rubber track at high axle loads on soil compaction-Part 2: Multi-axle machine studies. *Biosystems Engineering*, 99(3): 338–47. DOI:10.1016/j.biosystemseng.2007.11.014.

Antille, D. L., Ansorge, D., Dresser, M. L. and Godwin, R. J. (2013). Soil displacement and soil bulk density changes as affected by tire size. *Transactions of the ASABE*, 56(5): 1683–93. DOI:10.13031/trans.56.9886.

Antille, D. L., Chamen, W. C. T., Tullberg, J. N. and Lal, R. (2015a). The potential of controlled traffic farming to mitigate greenhouse gas emissions and enhance carbon sequestration in arable land: A critical review. *Transactions of the ASABE*, 58(3): 707–31. DOI:10.13031/trans.58.11049.

Antille, D. L., Hoekstra, N. J. and Lalor, S. T. J. (2015b). Field-scale evaluation of calcium ammonium nitrate, urea, and urea treated with N-(N-butyl) thiophosphoric triamide applied to grassland in Ireland. *Communications in Soil Science and Plant Analysis*, 46(11): 1345–61. DOI:10.1080/00103624.2015.1033540.

Antille, D. L., Imhoff, S. C., Alesso, C. A., Chamen, W. C. T. and Tullberg, J. N. (2015c). Potential to increase productivity and sustainability in Argentinean agriculture with controlled traffic farming: A short discussion. *Acta Technologica Agriculturae*, 18(3): 83–7. Special Issue: Controlled Traffic Farming. DOI:10.1515/ata-2015-0016.

Antille, D. L., Gallar, L., Miller, P. C. H. and Godwin, R. J. (2015d). An investigation into the fertilizer particle dynamics off-the-disc. *Applied Engineering in Agriculture*, 31(1): 49–60. DOI:10.13031/aea.31.10729.

Antille, D. L., Bennett, J. M. and Jensen, T. A. (2016a). Soil compaction and controlled traffic considerations in Australian cotton-farming systems. *Crop and Pasture Science*, 67(1): 1–28. DOI:10.1071/cp15097.

Antille, D. L., Huth, N. I., Eberhard, J., Marinoni, O., Cocks, B., Poulton, P. L., et al. (2016b). The effects of coal seam gas infrastructure development on arable land in Southern Queensland, Australia: Field investigations and modeling. *Transactions of the ASABE*, 59(4): 879–901. DOI:10.13031/trans.59.11547.

Antille, D. L., Godwin, R. J., Sakrabani, R., Seneweera, S., Tyrrel, S. F. and Johnston, A. E. (2017). Field-scale evaluation of biosolids-derived organomineral fertilizers applied to winter wheat in England. *Agronomy Journal*, 109(2): 654–74. DOI:10.2134/agronj2016.09.0495.

Arndt, W. and Rose, C. W. (1966). I. – The distribution of traffic in tillage operations. *Journal of Agricultural Engineering Research*, 11(3), 170–5. DOI:10.1016/S0021-8634(66)80016-1.

Arvidsson, J. (1999). Nutrient uptake and growth of barley as affected by soil compaction. *Plant and Soil*, 208(1): 9–19. DOI:10.1023/a:1004484518652.

ASABE (2005). ASAE EP291.3 FEB2005 (R2013): Terminology and definitions for soil tillage and soil-tool relationships. St Joseph, MI: American Society of Agricultural and Biological Engineers.

ASABE (2009). Standard ASAE S414.2 MAR2009 (R2013): Terminology and definitions for agricultural tillage implements. St Joseph, MI: American Society of Agricultural and Biological Engineers.

Barber, S. A., Walker, J. M. and Vasey, E. H. (1963). Mechanisms for movement of plant nutrients from soil and fertilizer to plant root. *Journal of Agricultural and Food Chemistry*, 11(3): 204–7. DOI:10.1021/jf60127a017.

Barik, K., Aksakal, E. L., Islam, K. R., Sari, S. and Angin, I. (2014). Spatial variability in soil compaction properties associated with field traffic operations. *Catena*, 120: 122–33. DOI:10.1016/j.catena.2014.04.013.

Bayhan, Y., Kayisoglu, B. and Gonulol, E. (2002). Effect of soil compaction on sunflower growth. *Soil and Tillage Research*, 68(1): 31–8. DOI:10.1016/S0167-1987(02)00078-8.

Bednarz, C. W., Shurley, W. D. and Anthony, W. S. (2002). Losses in yield, quality, and profitability of cotton from improper harvest timing. *Agronomy Journal*, 94(5): 1004–11. DOI:10.2134/agronj2002.1004.

Bennett, J. McL., Woodhouse, N. P., Keller, T., Jensen, T. A. and Antille, D. L. (2015). Advances in cotton harvesting technology: A review and implications for the John Deere round baler cotton picker. *Journal of Cotton Science*, 19(2): 225–49.

Blackwell, P., Isbister, B., Riethmuller, G., Barrett-Lennard, E., Hall, D., Lemon, J., et al. (2016). Deeper ripping and topsoil slotting to overcome subsoil compaction and other constraints more economically: Way to go! In: *Proceedings of the Grains Research Update, Session 5: Soils & Nutrition*. Perth, WA, Australia: Grains Research and Development Corporation. Retrieved: http://www.giwa.org.au/2016researchupdates (Accessed on 18 April 2018).

Bochtis, D. D., Sørensen, C. G., Busato, P., Hameed, I. A., Rodias, E., Green, O., et al. (2010a). Tramline establishment in controlled traffic farming based on operational machinery cost. *Biosystems Engineering*, 107(3): 221–31. DOI:10.1016/j.biosystemseng.2010.08.004.

Bochtis, D. D., Sørensen, C. G., Green, O., Moshou, D. and Olesen, J. (2010b). Effect of controlled traffic on field efficiency. *Biosystems Engineering*, 106(1): 14–25. DOI:10.1016/j.biosystemseng.2009.10.009.

Bochtis, D. D., Sørensen, C. G. C. and Busato, P. (2014). Advances in agricultural machinery management: A review. *Biosystems Engineering*, 126: 69–81. DOI:10.1016/j.biosystemseng.2014.07.012.

Botta, G. F., Pozzolo, O., Bomben, M., Rosatto, H., Rivero, D., Ressia, M., et al. (2007). Traffic alternatives for harvesting soybean (*Glycine max* L.): Effect on yields and soil under a direct sowing system. *Soil and Tillage Research*, 96(1–2): 145–54. DOI:10.1016/j.still.2007.05.003.

Braunack, M. V. and Johnston, D. B. (2014). Changes in soil cone resistance due to cotton picker traffic during harvest on Australian cotton soils. *Soil and Tillage Research*, 140: 29–39. DOI:10.1016/j.still.2014.1002.1007.

Braunack, M. V. and McGarry, D. (2006). Traffic control and tillage strategies for harvesting and planting of sugarcane (*Saccharum officinarum*) in Australia. *Soil and Tillage Research*, 89(1): 86–102. DOI:10.1016/j.still.2005.07.002.

Bromet, N. (2006). Weed management using GPS guidance. PhD Thesis. Brisbane, QLD, Australia: School of Agronomy in Horticulture, University of Queensland.

Bullock, P., Newman, A. C. D. and Thomasson, A. J. (1985). Porosity aspects of the regeneration of soil structure after compaction. *Soil and Tillage Research*, 5(4): 325–41. DOI:10.1016/s0167-1987(85)80001-5.

Burt, E. C., Reeves, D. W. and Raper, R. L. (1994). Energy utilization as affected by traffic in a conventional and conservation tillage system. *Transactions of the ASAE*, 37(3): 759–62. DOI:10.13031/2013.28137.

Carter, L. M., Meek, B. D. and Rechel, E. A. (1991). Zone production system for cotton: Soil response. *Transactions of the ASAE*, 34(2): 354–60. DOI: 310.13031/12013.31668.

Chamen, W. C. T. (2006). Controlled traffic farming: Literature review and appraisal of potential use in the U.K. Home-Grown Cereals Authority Research Review No.: 59. Kenilworth, UK: Agriculture and Horticulture Development Board.

Chamen, W. C. T. (2011). The effects of low and controlled traffic systems on soil physical properties, yields and the profitability of cereal crops on a range of soil types. PhD Thesis. Bedford, UK: School of Applied Science, Cranfield University.

Chamen, T. (2015). Controlled traffic farming – From worldwide research to adoption in Europe and its future prospects. *Acta Technologica Agriculturae*, 18(3): 64–73. DOI:10.1515/ata-2015-0014.

Chamen, W. C. T. and Longstaff, D. J. (1995). Traffic and tillage effects on soil conditions and crop growth on a swelling clay soil. *Soil Use and Management*, 11(4): 168–76. DOI:10.1111/j.1475-2743.1995.tb00951.x.

Chamen, W. C. T., Chittey, E. T., Leede, P. R., Goss, M. J. and Howse, K. R. (1990). The effect of tyre/soil contact pressure and zero traffic on soil and crop responses when growing winter wheat. *Journal of Agricultural Engineering Research*, 47: 1–21. DOI:10.1016/0021-8634(90)80026-Q.

Chamen, W. C. T., Watts, C. W., Leede, P. R. and Longstaff, D. J. (1992). Assessment of a wide span vehicle (gantry), and soil and cereal crop responses to its use in a zero traffic regime. *Soil and Tillage Research*, 24(4): 359–80. DOI:10.1016/0167-1987(92)90119-V.

Chamen, W. C. T., Dowler, D., Leede, P. R. and Longstaff, D. J. (1994). Design, operation and performance of a gantry system: Experience in arable cropping. *Journal of Agricultural Engineering Research*, 59(1): 45–60. DOI:10.1006/jaer.1994.1063.

Chamen, W. C. T., Cope, R. E., Longstaff, D. J., Patterson, D. E. and Richardson, C. D. (1996). The energy efficiency of seedbed preparation following mouldboard ploughing. *Soil and Tillage Research*, 39(1–2): 13–30. DOI:10.1016/S0167-1987(96)01051-3.

Chamen, W. C. T., Moxey, A. P., Towers, W., Balana, B. and Hallett, P. D. (2015). Mitigating arable soil compaction: A review and analysis of available cost and benefit data. *Soil and Tillage Research*, 146: 10–25. DOI:10.1016/j.still.2014.09.011.

Chan, K. Y., Oates, A., Swan, A. D., Hayes, R. C., Dear, B. S. and Peoples, M. B. (2006). Agronomic consequences of tractor wheel compaction on a clay soil. *Soil and Tillage Research*, 89(1): 13–21. DOI:10.1016/j.still.2005.06.007.

Chatskikh, D. and Olesen, J. E. (2007). Soil tillage enhanced CO_2 and N_2O emissions from loamy sand soil under spring barley. *Soil and Tillage Research*, 97(1): 5–18. DOI:10.1016/j.still.2007.08.004.

Conesa-Muñoz, J., Bengochea-Guevara, J. M., Andujar, D. and Ribeiro, A. (2016). Route planning for agricultural tasks: A general approach for fleets of autonomous vehicles in site-specific herbicide applications. *Computers and Electronics in Agriculture*, 127: 204–20. DOI:10.1016/j.compag.2016.06.012.

Connolly, R. D., Freebairn, D. M., Bell, M. J. and Thomas, G. (2001). Effects of rundown in soil hydraulic condition on crop productivity in south-eastern Queensland – A simulation study. *Australian Journal of Soil Research*, 39(5): 1111–29. DOI:10.1071/sr00089.

Cooper, A. W., Trouse, A. C. and Dumas, W. T. (1969). Controlled traffic in row crop production. In: *Proceedings of the 7th International Congress 'Agricultural Engineering'* (pp. 1–6). Organised by The International Commission of Agricultural and Biosystems Engineering. Section III. Baden-Baden, Germany: CIGR.

Couto, L. D., Tran-Jørgensen, P. W. V. and Edwards, G. T. C. (2018). Model-Based Development of a multi-algorithm harvest planning system. *Advances in Intelligent Systems and Computing*, 676: 19–33. DOI:10.1007/978-3-319-69832-8_2.

CTFA (2017). CTFA Project Final Report 2014–2017. Controlled Traffic Farming Alberta, Canada. Retrieved: http://www.controlledtrafficfarming.org/images/CTFA%20Project%20Final%20Report%202014-17%20May%2014.pdf (Accessed on 03 February 2018).

Dang, Y. P., Balzer, A., Crawford, M., Rincon-Florez, V., Hongwei, L., Melland, A. R., et al. (2018). Strategic tillage in conservation agricultural systems of north-eastern Australia: Why, where, when and how? *Environmental Science and Pollution Research*, 25(2): 1000–15. DOI:10.1007/s11356-017-8937-1.

Department for Environment, Food and Rural Affairs (DEFRA) (2009). Energy use on farm: Results from the farm business survey 2007/2008. London, UK: Department for Environment, Food and Rural Affairs.

De Toro, A. (2005). Influences on timeliness costs and their variability on arable farms. *Biosystems Engineering*, 92(1): 1–13. DOI:10.1016/j.biosystemseng.2005.06.007.

Dexter, A. R. (1991). Amelioration of soil by natural processes. *Soil and Tillage Research*, 20(1): 87–100. DOI:10.1016/0167-1987(91)90127-j.

Dickson, J. W. and Ritchie, R. M. (1996). Zero and reduced ground pressure traffic systems in an arable rotation. 1. Cultivation power requirement. *Soil and Tillage Research*, 38(1–2): 71–88. DOI:10.1016/0167-1987(96)01026-4.

Dijksterhuis, H. L., Van Willigenburg, L. G. and Van Zuydam, R. P. (1998). Centimetre-precision guidance of moving implements in the open field: A simulation based on GPS measurements. *Computers and Electronics in Agriculture*, 20(3): 185–97. DOI:10.1016/S0168-1699(98)00016-7.

Ding, G., Novak, J. M., Amarasiriwardena, D., Hunt, P. G. and Xing, B. (2002). Soil organic matter characteristics as affected by tillage management. *Soil Science Society of America Journal*, 66(2): 421–9. DOI:10.2136/sssaj2002.4210.

Doerr, S. H., Shakesby, R. A. and Walsh, R. P. D. (2000). Soil water repellency: Its causes, characteristics and hydro-geomorphological significance. *Earth Science Reviews*, 51(1–4): 33–65. DOI:10.1016/s0012-8252(00)00011-8.

Dumas, W. T., Trouse, A. C., Smith, L. A., Kummer, F. A. and Gill, W. R. (1973). Development and evaluation of tillage and other cultural practices in a controlled traffic system for cotton in the Southern Coastal Plains. *Transactions of the ASAE*, 16(5): 872–5. DOI:10.13031/2013.37648.

Dumas, W. T., Trouse, A. C., Smith Jr., L. A., Kummer, F. A. and Gill, W. R. (1975). Traffic control as a means of increasing cotton yields by reducing soil compaction. ASAE Paper No.: 75-1050. St Joseph, MI: ASAE.

Earl, R. (1997). Prediction of trafficability and workability from soil moisture deficit. *Soil and Tillage Research*, 40(3–4): 155–68. DOI:10.1016/S0167-1987(96)01072-0.

Galambošová, J., Macák, M., Rataj, V., Antille, D. L., Godwin, R. J., Chamen, W. C. T., et al. (2017). Field evaluation of controlled traffic farming in Central Europe using commercially available machinery. *Transactions of the ASABE*, 60(3): 657–69. DOI:10.13031/trans.11833.

Gasso, V., Oudshoorn, F. W., Sørensen, C. A. G. and Pedersen, H. H. (2014). An environmental life cycle assessment of controlled traffic farming. *Journal of Cleaner Production*, 73: 175–82. DOI:10.1016/j.jclepro.2013.10.044.

Godwin, R. J. (2007). A review of the effect of implement geometry on soil failure and implement forces. *Soil and Tillage Research*, 97(2): 331–40. DOI:10.1016/j.still.2006.06.010.

Godwin, R. J. and O'Dogherty, M. J. (2007). Integrated soil tillage force prediction models. *Journal of Terramechanics*, 44(1): 3–14. DOI:10.1016/j.jterra.2006.01.001.

Godwin, R. J., Wood, G. A., Taylor, J. C., Knight, S. M. and Welsh, J. P. (2003). Precision farming of cereal crops: A review of a six year experiment to develop management guidelines. *Biosystems Engineering*, 84(4): 375–91. DOI:10.1016/S1537-5110(03)00031-X.

Godwin, R., Misiewicz, P., White, D., Smith, E., Chamen, T., Galambošová, J., et al. (2015). Results from recent traffic systems research and the implications for future work. *Acta Technologica Agriculturae*, 18(3): 57–63. DOI:10.1515/ata-2015-0013.

Godwin, R. J., Misiewicz, P. A., Smith, E. K., Millington, W. A. J., White, D. R., Dickin, E. T., et al. (2017). Summary of the effects of three tillage and three traffic systems on cereal yields over a four-year rotation. *Aspects of Applied Biology*, 134: 233–41.

Graves, A. R., Morris, J., Deeks, L. K., Rickson, R. J., Kibblewhite, M. G., Harris, J. A., et al. (2015). The total costs of soil degradation in England and Wales. *Ecological Economics*, 119: 399–413. DOI:10.1016/j.ecolecon.2015.07.026.

Gregorich, E. G., McLaughlin, N. B., Lapen, D. R., Ma, B. L. and Rochette, P. (2014). Soil compaction, both an environmental and agronomic culprit: Increased nitrous oxide emissions and reduced plant nitrogen uptake. *Soil Science Society of America Journal*, 78(6): 1913–23. DOI:10.2136/sssaj2014.03.0117.

Gut, S., Chervet, A., Stettler, M., Weisskopf, P., Sturny, W. G., Lamandé, M., et al. (2015). Seasonal dynamics in wheel load-carrying capacity of a loam soil in the Swiss Plateau. *Soil Use and Management*, 31(1): 132–41. DOI:10.1111/sum.12148.

Hadas, A., Wolf, D. and Rawitz, E. (1985). Residual compaction effects on cotton stand and yields. *Transactions of the ASAE*, 28(3): 691–6. DOI: 610.13031/12013.32321.

Hadas, A., Wolf, D. and Rawitz, E. (1986). Prediction of draft requirement – Soil compaction relations in tilling moist soils. *Soil and Tillage Research*, 8: 51–64. DOI:10.1016/0167-1987(86)90322-3.

Håkansson, I. and Reeder, R. C. (1994). Subsoil compaction by vehicles with high axle load-extent, persistence and crop response. *Soil and Tillage Research*, 29(2–3): 277–304. DOI:10.1016/0167-1987(94)90065-5.

Hamilton, G., Bakker, D., Houlbrooke, D., Hetherington, R. and Spann, C. (2003). Permanent raised beds prevent waterlogging and increase the productivity of dryland farming areas in Western Australia. In: *Proceedings of the 16th Triennial Conference of the International Soil and Tillage Research Organisation* (pp. 524–30). Brisbane, Australia: ISTRO.

Harris, B. J. (2017). Increased deflection agricultural radial tires following the Tire and Rim Association. IF, VF, and IF/CFO: Load and inflation standards. ASABE Paper No.: 913C0117, Distinguished Lecture Series No.: 39. St Joseph, MI: ASABE. DOI:10.13031/913C0117.

Horrell, R., Metherell, A. K., Ford, S. and Doscher, C. (1999). Fertiliser evenness – losses and costs: A study on the economic benefits of uniform applications of fertiliser. *Proceedings of the New Zealand Grassland Association*, 61: 215–20.

Hulugalle, N. R., Weaver, T. B. and Finlay, L. A. (2017). Soil structure in permanent beds under irrigated cotton-based cropping systems in a Vertisol. *Soil and Tillage Research*, 165: 107–12. DOI:10.1016/j.still.2016.08.002.

Hussein, M. A., Antille, D. L., Chen, G., Luhaib, A. A. A., Kodur, S. and Tullberg, J. N. (2017). Agronomic performance of wheat (*Triticum aestivum* L.) and fertiliser use efficiency as affected by controlled and non-controlled traffic of farm machinery. ASABE Paper No.: 1700586. St Joseph, MI: ASABE. DOI:10.13031/aim.201700586.

Hussein, M. A. H., Antille, D. L., Chen, G., Kodur, S. and Tullberg, J. N. (2018). Agronomic performance of sorghum (*Sorghum bicolor* (L.) Moench) and fertilizer use efficiency as affected by controlled and non-controlled traffic of farm machinery. ASABE Paper No.: 1800250. St Joseph, MI: ASABE. DOI:10.13031/aim.201800250.

Huth, N. I., Bristow, K. L. and Verburg, K. (2012). SWIM3: Model use, calibration, and validation. *Transactions of the ASABE*, 55(4): 1303–13. DOI: 1310.13031/12013.42243.

Hylton-Jolliffe, W. J. and Halkett, P. A. (1858). On guide way agriculture: Being a system enabling all the operations of the farm to be performed by steam-power. *Journal of the Society for Arts*, 7(316): 41–58.

Isbister, B., Blackwell, P., Riethmuller, G., Davies, S., Whitlock, A. and Neale, T. (2013). Controlled traffic farming technical manual. Geraldton, WA, Australia: Northern Agricultural Catchments Council. Retrieved: http://nacc.com.au/wp-content/uploads/2015/05/NACC_Controlled_Traffic_Farming_Technical_Manual.pdf (Accessed on 12 December 2017).

Isbister, B., Metz, N. and Coles, J. (2018). Setting common runlines for machinery in a controlled traffic farming (CTF) system. Government of Western Australia, Department of Primary Industries and Regional Development (Agriculture and Food). Retrieved: https://www.agric.wa.gov.au/soil-compaction/setting-common-runlines-machinery-controlled-traffic-farming-ctf-system (Accessed on 21 January 2018).

Jensen, D. and Pesek, J. (1962). Inefficiency of fertilizer use resulting from nonuniform spatial distribution: II. Yield losses under selected distribution patterns. *Soil Science Society of America Journal*, 26(2): 174–8. DOI:10.2136/sssaj1962.03615995002600020023x.

Jensen, T., Powell, G. and Neale, T. (2000). Yield and protein variation within a controlled traffic system. In: *Proceedings of 10th Agronomy Conference 'Science and Technology: Delivering Results for Agriculture?'* Hobart, TAS, Australia: The Regional Institute. Retrieved: http://www.regional.org.au/au/asa/2001/1/a/neale.htm (Accessed on 20 July 2015).

Jørgensen, M. H. (2012). Agricultural field machinery for the future – From an engineering perspective. *Agronomy Research*, 10(1): 109–13.

Kautz, T. (2015). Research on subsoil biopores and their functions in organically managed soils: A review. *Renewable Agriculture and Food Systems*, 30(4): 318–27. DOI:10.1017/s1742170513000549.

Kay, B. D. (1990). Rates of change of soil structure under different cropping systems. *Advances in Soil Science*, 12: 1–52. DOI:10.1007/978-1-4612-3316-9_1.

Khalilian, A., Hood, C. E., Palmer, J. H., Garner, T. H. and Bathke, G. R. (1991). Soil compaction and crop response to wheat/soybean interseeding. *Transactions of the ASAE*, 34(6): 2299–303. DOI:10.13031/2013.31871.

Kingwell, R. and Fuchsbichler, A. (2011). The whole-farm benefits of controlled traffic farming: An Australian appraisal. *Agricultural Systems*, 104(7): 513–21. DOI:10.1016/j.agsy.2011.04.001.

Kirkby, E. A., Le Bot, J., Adamowicz, S. and Roemheld, V. (2009). Nitrogen in physiology: An agronomic perspective and implications for the use of different nitrogen forms. Proceeding No.: 653. York, UK: The International Fertiliser Society.

Koolen, A. J., Lerink, P., Kurstjens, D. A. G., van den Akker, J. J. H. and Arts, W. B. M. (1992). Prediction of aspects of soil-wheel systems. *Soil and Tillage Research*, 24(4): 381–96. DOI:10.1016/0167-1987(92)90120-z.

Kroulík, M., Kumhála, F., Hůla, J. and Honzík, I. (2009). The evaluation of agricultural machines field trafficking intensity for different soil tillage technologies. *Soil and Tillage Research*, 105(1): 171–5. DOI:10.1016/j.still.2009.07.004.

Kulkarni, S. S., Bajwa, S. G. and Huitink, G. (2010). Investigation of the effects of soil compaction in cotton. *Transactions of the ASABE*, 53(3): 667–74. DOI:10.13031/2013.30058.

Kutzbach, H. D. (2000). Trends in power and machinery. *Journal of Agricultural Engineering Research*, 76(3): 237–47. DOI:10.1006/jaer.2000.0574.

Lamers, J. G., Perdok, U. D., Lumkes, L. M. and Klooster, J. J. (1986). Controlled traffic farming systems in the Netherlands. *Soil and Tillage Research*, 8: 65–76. DOI:10.1016/0167-1987(86)90323-5.

Lawes, R. A., Oliver, Y. M. and Robertson, M. J. (2009). Capturing the in-field spatial-temporal dynamic of yield variation. *Crop and Pasture Science*, 60(9): 834–43. DOI:10.1071/cp08346.

Leeds-Harrison, P. B., Shipway, C. J. P., Jarvis, N. J. and Youngs, E. G. (1986). The influence of soil macroporosity on water retention, transmission and drainage in a clay soil. *Soil Use and Management*, 2(2): 47–50. DOI:10.1111/j.1475-2743.1986.tb00678.x.

Li, Y. X., Tullberg, J. N., Freebairn, D. M. and Li, H. W. (2009). Functional relationships between soil water infiltration and wheeling and rainfall energy. *Soil and Tillage Research*, 104(1): 156–63. DOI:10.1016/j.still.2008.10.023.

Lipiec, J. and Hatano, R. (2003). Quantification of compaction effects on soil physical properties and crop growth. *Geoderma*, 116(1), 107–36. DOI:10.1016/S0016-7061(03)00097-1.

Lipiec, J. and Stępniewski, W. (1995). Effects of soil compaction and tillage systems on uptake and losses of nutrients. *Soil and Tillage Research*, 35(1–2): 37–52. DOI:10.1016/0167-1987(95)00474-7.

Luhaib, A. A. A., Antille, D. L., Tullberg, J. N., Hussein, M. A. and Chen, G. (2017). Effect of controlled traffic farming on energy saving in Australian grain cropping systems. ASABE Paper No.: 1700583. St Joseph, MI: ASABE. DOI:10.13031/aim.201700583.

McCallum, M. and Long, B. (2005). Update on CTF and GPS uptake by farmers in SA. In: *Proceedings of the 3rd Australian Controlled Traffic Farming Conference*. Gatton, QLD, Australia: Australian Controlled Traffic Farming Association Inc. Retrieved: https://www.actfa.net/actfa-conferences/2005-ctf-conference/ (Accessed on 16 December 2017).

McHugh, A. D., Tullberg, J. N. and Freebairn, D. M. (2009). Controlled traffic farming restores soil structure. *Soil and Tillage Research*, 104(1): 164–72. DOI:10.1016/j.still.2008.10.010.

McKyes, E., Negi, S., Douglas, E., Taylor, F. and Raghavan, V. (1979). The effect of machinery traffic and tillage operations on the physical properties of a clay and on yield of silage corn. *Journal of Agricultural Engineering Research*, 24(2): 143–8. DOI:10.1016/0021-8634(79)90048-9.

McPhee, J. E. and Aird, P. L. (2013). Controlled traffic for vegetable production: Part 1. Machinery challenges and options in a diversified vegetable industry. *Biosystems Engineering*, 116(2): 144–54. DOI:10.1016/j.biosystemseng.2013.06.001.

Meek, B. D., Rechel, E. R., Carter, L. M., DeTar, W. R. and Urie, A. L. (1992). Infiltration rate of a sandy loam soil: Effects of traffic, tillage, and plant roots. *Soil Science Society of America Journal*, 56(3): 908–13. DOI:10.2136/sssaj1992.03615995005600030038x.

Melaj, M. A., Echeverría, H. E., López, S. C., Studdert, G., Andrade, F. and Bárbaro, N. O. (2003). Timing of nitrogen fertilization in wheat under conventional and no-tillage system. *Agronomy Journal*, 95(6): 1525–31. DOI:10.2134/agronj2003.1525.

Melland, A. R., Antille, D. L. and Dang, Y. P. (2017). Effects of strategic tillage on short-term erosion, nutrient loss in runoff and greenhouse gas emissions. *Soil Research*, 55(3): 201–14. DOI:10.1071/SR16136.

MICHELIN (2018). Michelin Ultraflex® Technologies. Retrieved: http://agricultural.michelinman.com/us/Michelin-innovation/Michelin-Ultraflex-Technologies (Accessed on 7 December 2017).

Millar, N., Robertson, G. P., Grace, P. R., Gehl, R. J. and Hoben, J. P. (2010). Nitrogen fertilizer management for nitrous oxide (N_2O) mitigation in intensive corn (Maize) production: An emissions reduction protocol for US Midwest agriculture. *Mitigation and Adaptation Strategies for Global Change*, 15(2): 185–204. DOI:10.1007/s11027-010-9212-7.

Miller, P. C. H., Audsley, E. and Richards, I. R. (2009). Costs and effects of uneven spreading of nitrogen fertilisers. Proceeding No.: 659. York, UK: The International Fertiliser Society.

Millington, W. A. J., Misiewicz, P. A., White, D. R., Dickin, E. T., Mooney, S. J. and Godwin, R. J. (2017). An investigation into the effect of traffic and tillage on soil properties using X-ray computed tomography. ASABE Paper No.: 1700380, St Joseph, MI: ASABE. DOI:10.13031/aim.201700380.

Misiewicz, P. A., Blackburn, K., Richards, T. E., Brighton, J. L. and Godwin, R. J. (2015). The evaluation and calibration of pressure mapping system for the measurement of the pressure distribution of agricultural tyres. *Biosystems Engineering*, 130: 81–91. DOI:10.1016/j.biosystemseng.2014.12.006.

Morling, R. W. (1982). Pros and cons of 'controlled traffic farming'. ASAE Paper No.: 82-1043. St Joseph, MI: ASAE. ISSN: 0145-0166.

Mossadeghi-Björklund, M., Arvidsson, J., Keller, T., Koestel, J., Lamandé, M., Larsbo, M., et al. (2016). Effects of subsoil compaction on hydraulic properties and preferential flow in a Swedish clay soil. *Soil and Tillage Research*, 156: 91–8. DOI:10.1016/j.still.2015.09.013.

Neale, T. (2011). The cost of grain harvester compaction. Toowoomba, QLD, Australia: Precision Agriculture. Retrieved: http://www.precisionagriculture.com.au/news-and-articles.php (Accessed on 20 July 2015).

Newall, A. (2013). Why controlled traffic: A consultant's perspective: Integrating agronomy into CTF. In: *Proceedings of the 1st International Controlled Traffic Farming Conference*. Toowoomba, Queensland, Australia: Australian Controlled Traffic Farming Association Inc. Retrieved: https://www.actfa.net/actfa-conferences/international-ctf-conference-2013/ (Accessed on 20 July 2015).

Newell, G., Jensen, T. and Walsh, P. (1998). A wheel track computer program to assist in controlled traffic farming practices. In: *Proceedings of the 1998 International Conference on Engineering in Agriculture*. Paper No.: SEAg 98/079. Barton, ACT, Australia: Society for Engineering in Agriculture.

Ngunjiri, G. M. N. and Siemens, J. C. (1995). Wheel traffic effects on corn growth. *Transactions of the ASAE*, 38(3): 691–9. DOI:10.13031/2013.27882.

Norton, R. (2014). Combating climate change through improved agronomic practices and input-use efficiency. *Journal of Crop Improvement*, 28(5): 575–618. DOI:10.1080/15427528.2014.924331.

O'Sullivan, M. F. and Simota, C. (1995). Modelling the environmental impacts of soil compaction: A review. *Soil and Tillage Research*, 35(1): 69–84. DOI:10.1016/0167-1987(95)00478-B.

Owens, J., Shaw, M. and Silburn, M. (2016). Modelling reductions of pollutant loads due to improved management practices in the Great Barrier Reef catchments: Grains cropping modelling, Technical Report (p. 111). Brisbane, QLD, Australia: Queensland Department of Natural Resources and Mines, Queensland Government.

Pangnakorn, U., George, D. L., Tullberg, J. N. and Gupta, M. L. (2003). Effect of tillage and traffic on earthworm populations in a Vertosol in South-East Queensland. In: *Proceedings of the 16th International Soil and Tillage Research Organisation Conference* (pp. 881–5). Brisbane, QLD, Australia: ISTRO.

Parvin, D. W., Martin, S. W., Cooke Jr., F. and Freeland Jr., B. B. (2005). Economics and marketing: Effect of harvest season rainfall on cotton yield. *Journal of Cotton Science*, 9(3): 115–20.

Pedersen, H. H., Sørensen, C. G., Oudshoorn, F. W., Krogsgård, P. and Munkholm, L. J. (2015). Evaluation of onion production on sandy soils by use of reduced tillage and controlled traffic farming with wide span tractors. *Acta Technologica Agriculturae*, 18(3): 74–82. DOI:10.1515/ata-2015-0015.

Pedersen, H. H., Oudshoorn, F. W., McPhee, J. E. and Chamen, W. C. T. (2016). Wide span – Re-mechanising vegetable production. *Acta Horticulturae*, 1130: 551–7. DOI:10.17660/ActaHortic.2016.1130.83.

Peets, S., Chamen, W. C. T., Godwin, R. J., White, D. R., Misiewicz, P. A. and Hargreaves, P. R. (2017). System design and the economics for controlled traffic farming (CTF) in grass silage production. ASABE Paper No.: 1700145. St Joseph, MI: ASABE. DOI:10.13031/aim.201700145.

Pollard, F. and Webster, R. (1978). The persistence of the effects of simulated tractor wheeling on sandy loam subsoil. *Journal of Agricultural Engineering Research*, 23(2): 217–20. DOI:10.1016/0021-8634(78)90050-1.

Poulton, P. L., Huth, N. I., Caccetta, P., Cocks, B., Wallace, J. and Wu, X. (2015). Monitoring impacts from broad-scale resource development on surface water flows in agricultural landscapes. In: *Proceedings of the 17th Australian Society of Agronomy Conference 'Building Productive, Diverse and Sustainable Landscapes'*. Hobart, TAS, Australia: Australian Society of Agronomy Inc. Retrieved: http://agronomyaustraliaproceedings.org/images/sampledata/2015_Conference/pdf/agronomy2015final00163.pdf (Accessed on 18 April 2018).

Prummel, J. (1975). Effect of soil structure on phosphate nutrition of crop plants. *Netherland Journal of Agricultural Science*, 23: 62–8.

Rabot, E., Wiesmeier, M., Schlüter, S. and Vogel, H. J. (2018). Soil structure as an indicator of soil functions: A review. *Geoderma*, 314: 122–37. DOI:10.1016/j.geoderma.2017.11.009.

Radford, B. J., Yule, D. F., McGarry, D. and Playford, C. (2001). Crop responses to applied soil compaction and to compaction repair treatments. *Soil and Tillage Research*, 61: 157–66. DOI:10.1016/S0167-1987(01)00194-5.

Radford, B. J., Yule, D. F., McGarry, D. and Playford, C. (2007). Amelioration of soil compaction can take 5 years on a Vertisol under no till in the semi-arid subtropics. *Soil and Tillage Research*, 97(2): 249–55. DOI:10.1016/j.still.2006.01.005.

Raghavan, G. S. V., McKyes, E. and Chassé, M. (1976). Soil compaction patterns caused by off-road vehicles in eastern Canadian agricultural soils. *Journal of Terramechanics*, 13(2): 107–15. DOI:10.1016/0022-4898(76)90017-3.

Raghavan, G. S. V., McKyes, E., Stremshorn, E., Beaulieu, B. and Gray, A. (1977). Vehicle compaction patterns in clay soil. *Transactions of the ASAE*, 20(2): 218–20, 225. DOI:10.13031/2013.35527.

Rainbow, R. W. (2005). Use of controlled traffic systems with auto-steer to enhance inter-row cropping and opportunities for introduction of non-chemical weed control systems. In: *Proceedings of the 3rd Australian Controlled Traffic Farming Conference*. Gatton, QLD, Australia: ACTFA. Retrieved: https://www.actfa.net/wp-content/uploads/2014/01/steer-to-enhance-interrow-cropping-and-opportunities-for-introduction-of-non-chemical-weed-control-systems-Rohan-Rainbow.pdf (Accessed on 27 March 2018).

Richards, I. R. and Hobson, R. D. (2013). Method of calculating effects of uneven spreading of fertiliser nitrogen. Proceeding No.: 734. York, UK: The International Fertiliser Society. ISBN: 978-0-85310-371-4.

Rickson, R. J. (2014). Can control of soil erosion mitigate water pollution by sediments? *Science of the Total Environment*, 468–469: 1187–97. DOI:10.1016/j.scitotenv.2013.05.057.

Robertson, M., Isbister, B., Maling, I., Oliver, Y., Wong, M., Adams, M., et al. (2007). Opportunities and constraints for managing within-field spatial variability in Western Australian grain production. *Field Crops Research*, 104(1–3): 60–7. DOI:10.1016/j.fcr.2006.12.013.

Rodias, E., Berruto, R., Busato, P., Bochtis, D., Sørensen, C. G. and Zhou, K. (2017). Energy savings from optimised in-field route planning for agricultural machinery. *Sustainability (Switzerland)*, 9(11), Article No.: 1956. DOI:10.3390/su9111956.

Ruser, R., Schilling, R., Steindl, H., Flessa, H. and Beese, F. (1998). Soil compaction and fertilization effects on nitrous oxide and methane fluxes in potato fields. *Soil Science Society of America Journal*, 62(6): 1587–95. DOI:10.2136/sssaj1998.03615995006200060016x.

Ruser, R., Flessa, H., Russow, R., Schmidt, G., Buegger, F. and Munch, J. C. (2006). Emission of N_2O, N_2 and CO_2 from soil fertilized with nitrate: Effect of compaction, soil moisture and rewetting. *Soil Biology and Biochemistry*, 38(2): 263–74. DOI:10.1016/j.soilbio.2005.05.005.

Scheer, C., Rowlings, D. W. and Grace, P. R. (2016). Non-linear response of soil N_2O emissions to nitrogen fertiliser in a cotton-fallow rotation in sub-tropical Australia. *Soil Research*, 54(5): 494–9. DOI:10.1071/sr14328.

Searcy, S. W., Schueller, J. K., Bae, Y. H., Borgelt, S. C. and Stout, B. A. (1989). Mapping of spatially variable yield during grain combining. *Transactions of the ASAE*, 32(3): 826–9. DOI:10.13031/2013.31077.

Seyyedhasani, H. and Dvorak, J. S. (2018). Reducing field work time using fleet routing optimization. *Biosystems Engineering*, 169: 1–10. DOI:10.1016/j.biosystemseng.2018.01.006.

Shah, A. N., Tanveer, M., Shahzad, B., Yang, G., Fahad, S., Ali, S., et al. (2017). Soil compaction effects on soil health and crop productivity: An overview. *Environmental Science and Pollution Research*, 24(11): 10056–67. DOI:10.1007/s11356-017-8421-y.

Smith, E. K., Misiewicz, P.A., Chaney, K., White, D.R. and Godwin, R.J. (2014). Effect of tracks and tyres on soil physical properties in a sandy loam soil. ASABE Paper No.: 141912659. St Joseph, MI: ASABE. DOI:10.13031/aim.20141912659.

Snyder, C. S. (2017). Enhanced nitrogen fertiliser technologies support the '4*R*' concept to optimise crop production and minimise environmental losses. *Soil Research*, 55(6): 463–72. DOI:10.1071/SR16335.

Soane, B. D. and van Ouwerkerk, C. (1995). Implications of soil compaction in crop production for the quality of the environment. *Soil and Tillage Research*, 35(1–2): 5–22. DOI:10.1016/0167-1987(95)00475-8.

Spekken, M. and de Bruin, S. (2013). Optimized routing on agricultural fields by minimizing maneuvering and servicing time. *Precision Agriculture*, 14(2): 224–44. DOI:10.1007/s11119-012-9290-5.

Spekken, M., de Bruin, S., Molin, J. P. and Sparovek, G. (2016). Planning machine paths and row crop patterns on steep surfaces to minimize soil erosion. *Computers and Electronics in Agriculture*, 124: 194–210. DOI:10.1016/j.compag.2016.03.013.

Spoor, G. (2006). Alleviation of soil compaction: Requirements, equipment and techniques. *Soil Use and Management*, 22(2): 113–22. DOI:10.1111/j.1475-2743.2006.00015.x.

Spoor, G. and Voorhees, W. B. (1986). Deep tillage. *Soil and Tillage Research*, 8: 317–19. DOI:10.1016/0167-1987(86)90345-4.

Spoor, G., Tijink, F. G. J. and Weisskopf, P. (2003). Subsoil compaction: Risk, avoidance, identification and alleviation. *Soil and Tillage Research*, 73(1–2): 175–82. DOI:10.1016/S0167-1987(03)00109-0.

Taylor, J. H. (1983). Benefits of permanent traffic lanes in a controlled traffic crop production system. *Soil and Tillage Research*, 3(4): 385–95. DOI:10.1016/0167-1987(83)90040-5.

Taylor, J. H. (1994). Chapter 22 – Development and benefits of vehicle gantries and controlled traffic systems. In: Soane, B. D. and van Ouwerkerk, C. (Eds), *Developments in Agricultural Engineering*, volume 11 (pp. 521–37). Amsterdam, the Netherlands: Elsevier. DOI:10.1016/B978-0-444-88286-8.50030-1.

Taylor, H. M. and Ratliff, L. F. (1969a). Root elongation rates of cotton and peanuts as a function of soil strength and soil water content. *Soil Science*, 108(2): 113–19. DOI:10.1097/00010694-196908000-00006.

Taylor, H. M. and Ratliff, L. F. (1969b). Root growth pressures of cotton, peas, and peanuts. *Agronomy Journal*, 61(3): 398–402. DOI:10.2134/agronj1969.00021962006100030020x.

Taylor, A. E. B., O'Callaghan, P. W. and Probert, S. D. (1993). Energy audit of an English farm. *Applied Energy*, 44(4): 315–35. DOI:10.1016/0306-2619(93)90022-h.

Tillett, N. D., Hague, T., Grundy, A. C. and Dedousis, A. P. (2008). Mechanical within-row weed control for transplanted crops using computer vision. *Biosystems Engineering*, 99(2): 171–8. DOI:10.1016/j.biosystemseng.2007.09.026.

Titmarsh, G., Waters, D., Wilson, A., Wockner, G., Burgis, M. and Connolly, R. (2003). Experiences in runoff and erosion of controlled traffic farming systems in southern Queensland. In: *Proceedings of the 16th Triennial Conference of the International Soil and Tillage Research Organisation* (pp. 1235–40). Brisbane, QLD, Australia: ISTRO.

Torbert, H. A. and Wood, C. W. (1992). Effects of soil compaction and water-filled pore space on soil microbial activity and N losses. *Communications in Soil Science and Plant Analysis*, 23(11–12): 1321–31. DOI:10.1080/00103629209368668.

Tullberg, J. N. (2000). Wheel traffic effects on tillage draught. *Journal of Agricultural Engineering Research*, 75(4): 375–82. DOI:10.1006/jaer.1999.0516.

Tullberg, J. (2010). Tillage, traffic and sustainability – A challenge for ISTRO. *Soil and Tillage Research*, 111(1): 26–32. DOI:10.1016/j.still.2010.08.008.

Tullberg, J. N. (2014). Chapter 3: Energy in crop production systems. In: Bundschuh, J. and Chen, G. (Eds), *Sustainable Energy Solutions in Agriculture* (pp. 53–76). London, UK: Taylor & Francis Group (CRC Press).

Tullberg, J. N. (2018). Chapter 2: Developments in mechanization technology: Controlled traffic farming. In: Chen, G. (Ed.), *Advances in Agricultural Machinery and Technologies* (pp. 27–47). London, UK: Taylor & Francis Group (CRC Press).

Tullberg, J. N., Ziebarth, P. J. and Li, Y. (2001). Tillage and traffic effects on runoff. *Australian Journal of Soil Research*, 39(2): 249–57. DOI:10.1071/sr00019.

Tullberg, J. N., Yule, D. F. and McGarry, D. (2007). Controlled traffic farming – From research to adoption in Australia. *Soil and Tillage Research*, 97(2): 272–81. DOI:10.1016/j.still.2007.09.007.

Tullberg, J., McHugh, A. D. J., Khabbaz, B. G., Scheer, C. and Grace, P. (2011). Controlled traffic/permanent bed farming reduces GHG emissions. In: *Proceedings of the 5th World Congress of Conservation Agriculture 2011 'Resilient Food Systems for a Changing World'*. Brisbane, QLD, Australia: Australian Centre for International Agricultural Research. Retrieved: http://aciar.gov.au/files/node/13991/controlled_traffic_permanent_bed_farming_reduces_g_10248.pdf (Accessed on 18 April 2018).

Tullberg, J. N., Antille, D. L., Bluett, C., Eberhard, J. and Scheer, C. (2018). Controlled traffic farming effects on soil emissions of nitrous oxide and methane. *Soil and Tillage Research*, 176: 18–25. DOI:10.1016/j.still.2017.09.014.

Vacher, C. A., Antille, D. L., Huth, N. I. and Raine, S. R. (2016). Assessing erosion processes associated with establishment of coal seam gas pipeline infrastructure in Queensland, Australia. ASABE Paper No.: 16-2461210. St Joseph, MI: ASABE. DOI: 10.13031/aim.20162461210.

van Ouwerkerk, C. and Soane, B. D. (1995). ISTRO workshop on 'The effects of soil compaction on physical, chemical and biological factors in the environment', 25 August 1993, Melitopol, Ukraine. [Editorial]. *Soil and Tillage Research*, 35(1–2), 1–4. DOI:10.1016/0167-1987(95)00480-G.

Vermeulen, G. D. and Chamen, W. C. T. (2010). Controlled traffic farming to improve soil structure and crop productivity. Proceeding No.: 678. York, UK: The International Fertiliser Society. ISBN: 978-0-85310-315-8.

Vermeulen, G. D. and Klooster, J. J. (1992). The potential of a low ground pressure traffic system to reduce soil compaction on a clayey loam soil. *Soil and Tillage Research*, 24(4): 337–58. DOI:10.1016/0167-1987(92)90118-U.

Vermeulen, G. D. and Mosquera, J. (2009). Soil, crop and emission responses to seasonal-controlled traffic in organic vegetable farming on loam soil. *Soil and Tillage Research*, 102(1): 126–34. DOI:10.1016/j.still.2008.08.008.

Vermeulen G. D., Tullberg J. N. and Chamen W. C. T. (2010). Chapter 8: Controlled traffic farming. In: Dedousis, A. and Bartzanas, T. (Eds), *Soil Engineering*. Soil Biology Series, Volume 20 (pp. 101–20). Berlin Heidelberg, Germany: Springer-Verlag. DOI:10.1007/978-3-642-03681-1_8.

Vero, S. E., Antille, D. L., Lalor, S. T. J. and Holden, N. M. (2014). Field evaluation of soil moisture deficit thresholds for limits to trafficability with slurry spreading equipment on grassland. *Soil Use and Management*, 30(1): 69-77. DOI:10.1111/sum.12093.

Vomocil, J. A. and Flocker, W. J. (1961). Effect of soil compaction on storage and movement of soil air and water. *Transactions of the ASAE*, 4(2): 242–6. DOI:10.13031/2013.41066.

WAAA (2018). CTF Calculator – Calculating the trafficked area of your paddock: Soil compaction management. Geraldton, WA, Australia: Western Australian Agriculture Authority. Available at: http://www.ctfcalculator.org/.

Wall, D. A. and Friesen, G. H. (1991). Effectiveness of selective herbicides for the control of annual weed species in lathyrus (*Lathyrus sativus* L.). *Crop Protection*, 10(6): 496–500. DOI:10.1016/S0261-2194(91)80139-7.

Walsh, P. (2002). New method yields a worm's eye view. *Farming Ahead*, 132: 16–18.

Walsh, P. (2008). CTF farmer, my experiences, grains. In: *Proceedings of the 6th Australian Controlled Traffic Farming Conference*. Dubbo, NSW, Australia: Australian Controlled Traffic Farming Association Inc. Retrieved: http://www.actfa.net/wp-content/uploads/2014/02/CTF-Farmer-My-Experiences-Grains.pdf (Accessed on 18 April 2018).

Watts, D. B., Runion, G. B., Smith Nannenga, K. W. and Torbert, H. A. (2014). Enhanced-efficiency fertilizer effects on cotton yield and quality in the coastal plains. *Agronomy Journal*, 106(2): 745–52. DOI:10.2134/agronj13.0216.

Webb, B., Blackwell, P., Riethmuller, G. and Lemon, J. (2004). Tramline farming systems technical manual. Bulletin 4607. Perth, WA, Australia: Department of Agriculture and Food of Western Australia. Available at: https://researchlibrary.agric.wa.gov.au/cgi/viewcontent.cgi?article=1192&context=bulletins (Accessed on 22 February 2018).

Werth, J. A., Preston, C., Taylor, I. N., Charles, G. W., Roberts, G. N. and Baker, J. (2008). Managing the risk of glyphosate resistance in Australian glyphosate-resistant cotton production systems. *Pest Management Science*, 64(4): 417–21. DOI:10.1002/ps.1508.

Wiersum, L. K. (1962). Uptake of nitrogen and phosphorus in relation to soil structure and nutrient mobility. *Plant and Soil*, 16(1): 62–70. DOI:10.1007/bf01378158.

Williford, J. R. (1980). A controlled-traffic system for cotton production. *Transactions of the ASAE*, 23(1): 65–70. DOI:10.13031/2013.34526.

Wolkowski, R. P. (1990). Relationship between wheel-traffic-induced soil compaction, nutrient availability, and crop growth: A review. *Journal of Production Agriculture*, 3(4): 460–9. DOI:10.2134/jpa1990.0460.

Wu, L., Swan, J. B., Allmaras, R. R. and Logsdon, S. D. (1995). Tillage and traffic influences on water and solute transport in corn-soybean systems. *Soil Science Society of America Journal*, 59(1): 185–91. DOI:10.2136/sssaj1995.03615995005900010028x.

Yule, D. (2007). CTF and PA Tools – the perfect match. In: *Proceedings of the 5th Australian Controlled Traffic Farming Conference*. Perth, WA, Australia: Australian Controlled Traffic Farming Association Inc. Retrieved: https://www.actfa.net/actfa-conferences/ctf07-conference/ (Accessed on 18 April 2018).

Part 3

Applications

Chapter 11

Precision tillage systems

Pedro Andrade-Sanchez, University of Arizona, USA; and Shrinivasa K. Upadhyaya, University of California-Davis, USA

1 Introduction

Modern agriculture is being transformed, primarily by the increasing number of information inputs used in every aspect of the farming enterprise. With the advent of the personal computer in the early 1980s, the use of digital communications and information technology in agriculture has been part of the public narrative. In a paper published in 1981, Prof William Chancellor (1981) predicted that crop production would become more efficient as information made available through the wide adoption of microcomputers supplanted the need to expend so much energy. Prof Chancellor laid out a conceptual framework for optimizing existing farm operations, as well as 'novel, high-information production processes'. For example, the concept of using 'tractor transmission control' to optimize gear selection based on torque-RPM characteristics is now a reality on modern tractors equipped with continuously variable transmissions (CVTs). Another technology that has been realized is 'traction wheel slip control', a concept originally presented by Young and Schafer (1977) for manipulating weight transfer, implement depth and tyre pressure based on sensing of the drive wheel slip, tyre sinkage and deflection, and soil strength during tillage operations. In 1984, Schafer et al. (1984) introduced the concept of prescribed tillage capable of providing adequate soil conditions for the growth of crops in view of 'future goals of productivity, energy efficiency, and resource conservation'. As represented in Fig. 1, the authors conceptualized the composition and function of a control system to implement spatially prescribed tillage operations with integrated electronics to generate

http://dx.doi.org/10.19103/AS.2017.0032.12

Figure 1 A concept drawing of a computer-controlled tillage system with satellite communication for achieving custom prescribed tillage (NTML Photo No. P10 352). Image published in Schafer et al. (1984).

information loops for monitoring and control of soil-engaging tools aided by satellite communications, which we know today as global navigation satellite systems (GNSSs).

More recently, the commercial development of information-intensive precision agriculture technologies has given rise to a number of applications aimed at improving agricultural production through site-specific management. Tillage management fits well in this new paradigm for several reasons: (1) because fossil fuel consumption during primary tillage operations accounts for well over half of the energy use on US farms (Cromwell et al., 1995) with an overall diesel fuel consumption for customary passes of heavy tillage tools ranging from 30 to 40 L/ha to produce adequate seed bed conditions (Griffith and Parsons, 1983; Cromwell et al., 1995; Hitaj and Suttles, 2016), the wide range of energy use scenarios and the magnitude of energy use present opportunities to develop more energy-efficient systems; (2) the soil mechanical state (i.e. density and compaction) in most agricultural systems exhibits significant in-field variation in both horizontal and soil profile depth location and (3) new developments in tractor/implement guidance as well as depth control of soil-engaging tools provide a platform for the implementation of precision, prescription or variable-depth tillage. This chapter provides a literature review of past developments in on-the-go soil strength sensing and the state of the art in sensor technology applied to precision tillage. Past work and present trends are reviewed, along with a review of commercially available products and needed future developments in tillage technology.

2 Depth-prescribed tillage to control subsoil compaction

The most common cause of soil compaction in modern mechanized agriculture is vehicular traffic from heavy machinery, which imposes large loads that end up being supported by the soil (Upadhyaya, 1992; Hakanssan and Medvedev, 1995; Lipiec et al., 2012). This mechanical loading on agricultural fields frequently takes place when the soil is moist and mechanical stresses reach farther into the soil profile, thus increasing the extent of compaction. Soil water content is the single most important factor that contributes to soil compaction in farmland (Chancellor, 1977; Raper and Kirby, 2006). In addition to the increase in tillage energy requirements from compacted soil, plant root development and growth are severely affected by subsoil compaction, which most probably will cause a reduction in the yield potential of crops as the extent of detrimental compaction conditions becomes more pronounced (Chancellor, 1977; Soane, 1990; Radford et al., 2001; Servadio et al., 2016). Research studies have confirmed the presence of within-field hardpan spatial variability detected with soil penetration resistance methods (Kollárová et al., 2014; Raper et al., 2005, Mohawesh et al., 2008). Mzuku et al. (2005) used an analytical approach based on geostatistics and reported variations in the spatial autocorrelation of several soil properties, including bulk density. Moran's I statistic values suggested a moderate degree of clustering in soil compaction values, consistent with in-field variability.

Field research aimed at determining the extent and location of subsoil hardpans relies extensively on the use of cone penetrometer devices. These devices quantify levels of compaction as measurements of force over area (i.e. pressure) adjusted to the ASAE (2004) Cone Index (CI) standard. Raper et al. (2005) found that site-specific subsoiling produced corn yields equivalent to those produced by uniform deep subsoiling while reducing draft forces, drawbar power and fuel use. While CI is an effective measure of soil strength by depth, the method is based on point measurements with obvious limitations when it comes to characterizing the soil strength conditions of large areas or whole fields. Because of this constraint on its practical use at large scale, newer developments in soil strength sensing have been directed towards continuous measurements of instrumented soil-engaging tools and frames. Adamchuk et al. (2004) and Hemmat and Adamchuk (2008) have provided comprehensive reviews on the use of instrumented soil-engaging tools for continuous measurements of soil properties/characteristics and data analysis approaches.

Using GNSSs for geographic position monitoring allows researchers to geo-reference and map depth-dependent soil strength information and enables the development of prescribed precision tillage maps (Mouazen and Roman, 2006). Geo-referenced measurements of soil profile compaction are actionable data that will enable the development of variable-depth tillage management strategies to alleviate the problems caused by hardened soil layers (Basso, 2003). Information-intensive precision tillage serves a variety of objectives in agricultural resource conservation. This chapter will deal with the application of soil profile compaction sensors to the implementation of prescription tillage. As variable-depth tillage systems develop even further, conservation tillage systems that selectively treat the soil and reduce the overall energy use will continue to evolve.

3 Soil sensing to enable precision tillage

Perhaps the prime mover driving the development of sensing systems that can characterize the mechanical state of soil is the desire to quantify the extent of soil compaction that adversely affects soil health at different soil profile depths. The rationale is that a sensor providing information feedback on the magnitude of soil compaction and the depth location of such conditions can be integrated into a tillage depth control system to take remedial action by targeting the corresponding depth of operation to break the hardened soil mechanically. One of the earliest research studies in this area was reported by Stafford and Hendricks (1988), who tested two configurations of subsoiler tines instrumented with customized ring force transducers. The authors found that the forces registered in a rear-facing blade in contact with undisturbed soil were sensitive to soil strength variations with depth and provided information on the vertical location of hard layers. Glancey et al. (1989) built a prototype to measure the force distribution due to soil cutting along a rigid tine. Strain gauges mounted on the front and rear faces of the tine measured the bending moments as the blade cut soil of various textures and conditions. Using a high-order polynomial regression approach with constrained optimization to represent the depth of soft tilled soil, the authors modelled force distributions over the operating depth.

Research efforts like the two cases above, along with increasing concerns about soil compaction, stimulated further developments in soil-strength-sensing systems. In the early 2000s, a number of papers were published on using instrumented tillage tools and tines to detect the location of hardened layers in the soil profile using direct methods of force measurement. Typically, these sensors generated soil strength conditions at depths up to 45 cm. Some examples of depth-profile soil strength sensors include the developments of Adamchuck et al. (2001), who used an array of strain gauges attached to a rigid shank. Chung et al. (2006) reported on the construction and field testing of a sensor instrumented with horizontal cones sticking out of a shank and directly connected to load cells. Khalilian et al. (2002) and Andrade-Sanchez et al. (2007) followed a similar approach of using force transducers housed inside a tillage tool and directly connected to cutting edges. One advantage of the direct measurement approach is that it somewhat simplifies the analytical work.

The sensors described above are only a subset of existing sensor prototypes. These and other developments cover a wide range of mechanical approaches to using force and pressure transducers to generate a signal proportional to the magnitude of cutting resistance at multiple depths in the soil profile. Figure 2 shows an example of an instrumented soil-engaging tool fabricated with five active cutting elements connected to custom-designed orthogonal ring load cells (Andrade-Sanchez et al., 2007). With the addition of GNSS-based position information, these soil profile compaction sensors proved capable of providing geo-referenced information to pinpoint the location of compacted layers in the field. Mapping soil cutting force data provides an interesting insight into soil strength variability and the possibility of improving soil conditions with the potential implementation of smart precision tillage management. To illustrate this point, Fig. 3 provides a longitudinal transect depth view of soil strength conditions characterized

Figure 2 Instrumented tillage tool for measurement of soil cutting forces in the soil profile. Rendering on the left shows the active components (i.e. octagonal load cells and cutting elements). The image on the right shows the soil sensor attached to a three-point hitch frame and mounted onto an agricultural tractor.

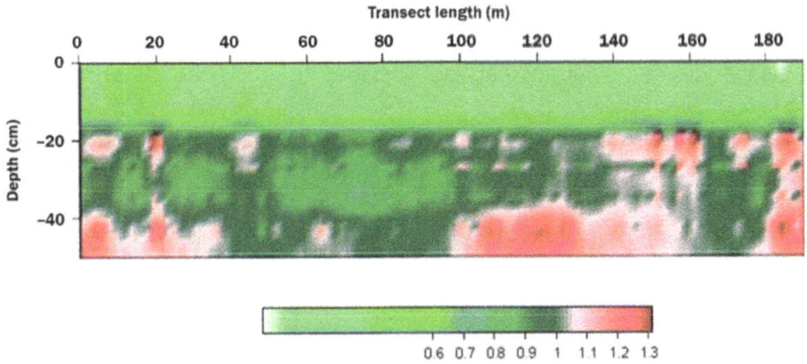

Figure 3 Map of soil cutting forces transformed to Cone Index Equivalent (CIE) quantities down to a depth of 45 cm. The colour code illustrates the soil profile conditions: a colour range from green to red is used to indicate the range of compaction-free and compacted soil conditions, respectively, in a field in Maricopa, AZ, USA.

by a soil profile sensor. This transect map shows the presence of compaction-free zones as well as portions of the transect with soil strength values that exceed the equivalent of 2.0 MPa of CI (ASAE, 2004), a threshold value commonly accepted as a soil condition that negatively affects subsoil compaction (Duiker, 2002).

Many of the above referenced papers reported the results of empirical statistical analyses of tool geometry, soil conditions and operational factors such as depth and speed that affected the variability of forces along the cutting edge of the instrumented shank. The depth location of the sensing force proved to be a first-order factor that affected the magnitude of the cutting force. The quantification of increasing forces with increasing depth provides a basis for the interest in the development of depth control systems that could operate with the information generated by soil compaction sensors to reduce the energy consumption of subsoil tillage by targeting the optimal depth of operation of the tillage tool.

4 Extracting information from soil sensors

The approach used by Andrade-Sanchez et al. (2007, 2008) was empirical in nature. Field tests included deployment of the soil profile sensor and measurements of a standard cone penetrometer over a wide range of soil types and conditions. Multiple linear regression analysis showed very high correlations between the CI, the unit pressure acting on the sensor cutting edge [i.e. the Cone Index Equivalent (CIE)] and the interaction between the CI and the depth of the cutting edge. Very high coefficients of multiple determination were obtained across all soil types and conditions tested. The existence of such a unique relationship for the data obtained for various soil types and conditions in California and the Midwestern United States was quite surprising. A recent theoretical work based on a finite element method analysis performed by Zhang et al. (2017) attempted to explain the analytical basis for the relationship between the CIE and CI. Simulations included 25 distinct soil types and conditions. Their results were consistent with the observations obtained by Andrade-Sanchez et al. (2007, 2008), suggesting the existence of a strong theoretical basis to the empirical relationships that they observed.

Soil strength sensors provide a powerful insight into the mechanics of soil fracturing when forces are applied to break the soil, such as in the case of tillage. New analytical tools such as non-linear dynamics can be adopted to study the deterministic nature of soil cutting forces, such as was done in the recent paper by Sakai et al. (2017). As stated by Sakai et al. (2017), 'Soil tillage is an essential operation in farming. Soil structures formed through soil failure during tillage are very important in maintaining productive and sustainable arable lands'. Future gains in our ability to model soil cutting processes during tillage will further enable the development of technology at the producer level.

5 Implementing depth control

Implementing an active tillage system with position control brings additional requirements, such as the ability to measure the displacement of the implement to determine the depth of operation of the soil-engaging tool and use this information as feedback for a depth control system. Research and development efforts on tool depth monitoring/control date back many years. Dyck (1975) developed a system using a gauge wheel for depth reference, along with mechanical linkage and limit switches that energized an electro-hydraulic valve to control the operational depth of the disc. McLaughlin et al. (1976) used the concept of measuring airflow through a spring as an indicator of the angular position of a reference shoe and proposed using this transducer for tillage depth control applications. Sonar displacement transducers have been used extensively as non-contact measuring devices in co-ordination with actuators to provide vertical position information and control of tillage tools (Dyck, 1985; Yasin et al., 1992). One common problem with this approach is the increase in signal noise produced at the surface level by the presence of crop residue, rocks, large soil aggregates or other objects that disperse the sound waves in multiple directions. Other developments for tool depth metering include the use of inclinometers and linear/rotary position sensors that measure the position of linkage arms (arm position sensors) on a tractor and implement system (Lee et al., 1998, 2000). Use of these devices facilitates the construction of electronic interfaces for fast communication and enhanced control on tractor-mounted (Lee et al., 2000) or pull-behind (Suomi and Oksanen, 2015) implements. The implementation of arm position sensors for depth control functions has been reported in field cultivators, seed planters and drill coulter applications (Nielsen et al., 2016) with promising advances in tractor/implement CAN bus connectivity (Suomi and Oksanen, 2015).

Vertical tillage frames are becoming popular solutions on large farms spanning thousands of hectares. These units typically weigh 13–15 t; thus, they have significant power requirements of approximately 30 kW per metre width. A reduction in the number of passes in the field translates to significant improvements in efficiency and timeliness for seedbed preparation. A conventional approach to provide depth adjustment to tillage tool frames is the use of single-point depth control, which combines mechanical settings with hydraulic cylinder actuators. Manual adjustments to the desired depth are selected based on the soil conditions and desired outcome. Recent advances in tillage tool designs show an increased use of multi-gang, vertical tillage tools equipped with electronic components that enable automatic tool depth monitoring and control functions. It is increasingly common for new tillage systems to be compatible with the tractor communication infrastructure to enable extra functionality such as independent section and in-field variable-depth control by following programmed instructions set in electronic prescription files.

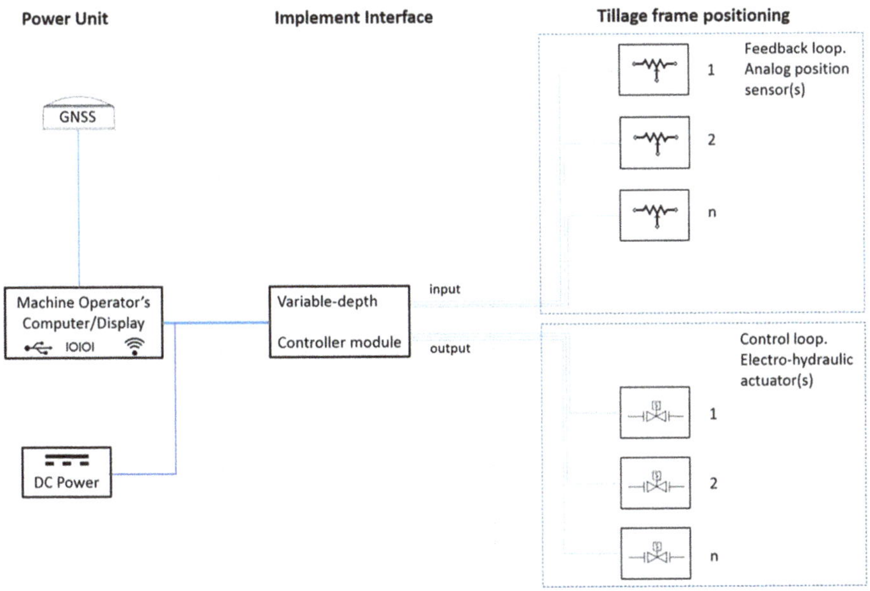

Figure 4 Conceptual schematics of a precision variable-depth control system for soil-engaging tools mounted on tillage frames.

Figure 4 illustrates the conceptual framework of an advanced tillage system with automatic depth control of the tillage frame and soil-engaging tools. Position information is generated by rotary/linear potentiometers and feedback signals that are fed into the function-specific module that controls vertical position. A variable-depth controller processes the input and incorporates the geometrical parameters of the frame and support wheels to compute vertical position. With the vertical position input in place, the controller sends signals to electro-hydraulic valves that control the directional flow of hydraulic oil to adjust the extension or contraction of hydraulic cylinder actuators. The variable-depth controller is directly connected to a terminal installed within reach of the operator who selects desired operational parameters or loads prescribed instructions for desired tillage depth through serial ports or wireless access points. GNSS messages feed the machine operator display to enable mapping of the tillage depth information.

6 Tractor/implement guidance technology

Generally speaking, the efficiency of tillage operations increases with operational improvements in the power unit and the frame supporting the soil-engaging tools. Tractor and implement guidance are examples of such technologies that bring benefits to tillage efficiency by avoiding overlaps and skips in operation. As noted by Mousazadeh (2013), fuel efficiency improvements during tillage are even more evident when advanced navigation is coupled with tractor automatic functions like engine throttle control, CVT and three-point hitch position control. Under a similar rationale, mission planning to optimize tractor field passes has a positive effect on tillage efficiency (Bochtis et al., 2009). In terms of implement guidance, active and passive systems are commercially available options (Fulton et al., 2010). These systems provide frame stability to prevent side-drifting caused by variations in side slope in the field or by the effect of directional tillage tooling like chisels. Active implement guidance systems consist of hydraulic cylinders or disc coulter actuators to steer the implement independently of the power unit. Active systems require a second GNSS receiver and steering controller for the implement, along with a hydraulic valve. In the case of passive systems, the power unit adjusts position in order for the trailing implement to maintain the desired position and heading. Passive systems are more economical because they use a simpler configuration consisting of dual GNSS receivers to keep track of the implement and tractor position simultaneously. Computer algorithms loaded in the tractor display make the necessary adjustments automatically. Advances in GNSS-based machine navigation, robotics and autonomous vehicles suggest that tillage operations will become even more efficient in the future.

7 Conclusion

For many decades now, soil dynamics research has provided a solid platform for the study of energy inputs used during tillage to transform the soil condition and make it a suitable medium for the high rate of growth and healthy development of crops as required in modern agriculture. The development of technology to measure soil strength has allowed scientists and practitioners of diverse backgrounds to connect information on the mechanical state of soil with engineering principles of machine use efficiency and productivity. Moreover, measurement of soil compaction is a powerful tool to explore soil–plant relationships that point to improvements in tillage management for crop production. On a par with the development of soil sensor technology, advanced electronic systems have evolved to enable real-time monitoring and control functions in agricultural machines. Soil state information and the user knowledge of field conditions and production functions find their way into electronic prescriptions as bytes of information in a flow to control implement performance parameters such as tillage depth. These information-intensive systems reflect the era of digital agriculture we live in and demonstrate a sustained trend towards increasing energy substitution for information.

8 Where to look for further information

The American Society of Agricultural and Biological Engineers (ASABE) has a long-standing tradition in advancing the study of soil dynamics and tillage. The book series 'Advances in Soil Dynamics' is a fundamental source of literature in this topic. Moreover, annual events and *ad hoc* meetings on topics such as precision agriculture, smart-farming, robotics, mechanization and so forth are carefully planned and promoted in the Society's web forums. Publications that result from these meetings are available on-line in the technical library (https://elibrary.asabe.org/). Scientific journals such as Elsevier's *Soil and Tillage Research* (https://www.journals.elsevier.com/soil-and-tillage-research) and *Computer and Electronics in Agriculture* (https://www.journals.elsevier.com/computers-and-electronics-in-agriculture) provide a key platform for the dissemination of information on precision tillage.

Precision tillage is a concept currently under development with private sector investment in R&D. New commercially available products for tillage operations are constantly promoted using active methods of information spread. In North America, farm equipment manufacturers have an on-line presence with news releases of new products and with direct face-to-face interaction with the public during farm shows across the United States and Canada. Portals of large equipment manufacturers are very dynamic sites connecting interested parties with their products. A large number of small- to medium-scale manufacturers have a collective presence in associations like the Farm Equipment Manufacturers Association (https://www.farmequip.org/). Information on new, advanced precision tillage technology will always find an outlet developed by the private sector to satisfy their commercial interests. When this technology is available to the public, land-grant institutions in the United States play an important role in carrying out unbiased research and informing stakeholders of new technological options through programming in Cooperative Extension Programs (https://nifa.usda.gov/cooperative-extension-system).

9 References

Adamchuk, V. I., Morgan, M. T. and Sumali, H. (2001). Application of a strain gauge array to estimate soil mechanical impedance on-the-go. *Transactions of the ASAE*, 44(6): 1377–83.

Adamchuk, V. I., Hummel, J. W., Morgan, M. T. and Upadhyaya, S. K. (2004). On-the-go soil sensors for precision agriculture. *Computers and Electronics in Agriculture*, 44: 71–91.

Andrade-Sanchez, P., Upadhyaya, S. K. and Jenkins, B. M. (2007). Development, construction, and field evaluation of a soil compaction profile sensor. *Transactions of the ASABE*, 50(3): 719–25.

Andrade-Sanchez, P., Upadhyaya, S. K., Plouffe, C. and Poutré, B. (2008). Development and field evaluation of a field-ready soil compaction profile sensor for real-time applications. *Applied Engineering in Agriculture*, 24(6): 743–50.

ASAE Standards. (2004). *S313.3. Soil Cone Penetrometer*. 50th ed., St. Joseph, MI: ASAE.

Basso, B. (2003). Perspectives of precision agriculture in conservation agriculture. In: Garcia Torres, L., Benites, J., Martínez-Vilela, A. and Holgado-Cabrera, A. (Eds), *Conservation Agriculture*. Dordrecht, the Netherlands: Springer, pp. 281–8.

Bochtis, D. D., Vougioukas, S. G. and Griepentrog H. W. (2009). A mission planner for an autonomous tractor. *Transactions of the ASABE*, 52(5): 1429–40.

Chancellor, W. J. (1977). Compaction of soils by agricultural equipment. Bulletin No. 1881. Davis, CA: Division of Agricultural Sciences, University of California Davis, 53p.

Chancellor, W. J. (1981). Substituting information for energy in agriculture. *Transactions of the ASAE*, 24(4): 802–13.

Chung, S. O., Sudduth, K. A. and Hummel, J. W. (2006). Design and validation of an on-the-go soil strength profile sensor. *Transactions of the ASABE*, 49(1): 5–14.

Cromwell, R. P., Stanley, J. M., Gallaher, R. N. and Wright, D. L. (1995). Fuel consumption and power requirement for tillage operation. Institute of Food and Agricultural Sciences, Gainesville, FL: University of Florida.

Duiker, S. W. (2002). Diagnosing soil compaction using a penetrometer (soil compaction tester). Agronomy Facts 63. State College, PA: Penn State Cooperative Extension.

Dyck, F. B. (1975). Automatic depth control for a discer. *Canadian Agricultural Engineering*, 17: 47–9.

Dyck, F. B., Wu, W. K. and Lesko, R. (1985). Automatic depth control for cultivators and air seeders developed under the AERD Program. *Proceedings of the ASAE Agri-Mation Conference & Exposition*. St Joseph, MI: ASAE, pp. 265–76.

Fulton, J., Brooke, A., Mullenix, D. and Winstead, A. (2010). *Guidance Systems: Implement Guidance Solutions*. Auburn, AL: Alabama Cooperative Extension System.

Glancey, J. L., Upadhyaya, S. K., Chancellor, W. J. and Rumsey, J. W. (1989). An instrumented chisel for the study of soil-tillage dynamics. *Soil Tillage Research*, 14: 1–24.
Griffith, D. R. and Parsons, S. D. (1983). Energy requirements for various tillage-planting systems. Bulletin NCR-202-W. West Lafayette, IN: Purdue University, Cooperative Extension Service.
Hakanssan, I. and Medvedev, V. W. (1995). Protection of soils from mechanical overloading by establishing limits for stresses caused by heavy vehicles. *Soil and Tillage Research*, 35: 85–97.
Hemmat, A. and Adamchuk, V. I. (2008). Sensor systems for measuring soil compaction: Review and analysis. *Computers and Electronics in Agriculture*, 63: 89–103.
Hitaj, C. and Suttles, S. (2016). Trends in U.S. agriculture's consumption and production of energy: Renewable power, shale energy, and cellulosic biomass. Economic Information Bulletin No. 159. Washington, DC: United States Department of Agriculture, Economic Research Service.
Khalilian, A., Han, Y. J., Dodd, R. B., Sullivan, M. J., Gorucu, S. and Keskin, M. (2002). A control system for variable depth tillage. ASAE Paper No. 021209. St. Joseph, MI: ASABE.
Kollárová, K., Žitňák, M. and Korenko, M. (2014). Hardpan location and crop yield: From information to precision tillage map. *Advanced Materials Research*, 1059: 27–33.
Lee, J., Yamazaki, M., Oida, A., Nakashima, H. and Shimizu, H. (1998). Electro-hydraulic tillage depth control system for rotary implements mounted on agricultural tractor: Design and response experiments of control system. *Journal of Terramechanics*, 35: 229–38.
Lee, J., Yamazaki, M., Oida, A., Nakashima, H. and Shimizu, H. (2000). Field performance of proposed foresight tillage depth control system for rotary implements mounted on an agricultural tractor. *Journal of Terramechanics*, 37: 99–111.
Lipiec, J., Horn, R., Pielrusiewicz, J. and Siczek, A. (2012). Effect of soil compaction on root elongation and anatomy of different cereal plant species. *Soil and Tillage Research*, 121: 74–81.
McLaughlin, N. B., Bigsby, F. W., Wilson, J. N. and Zoerb, G. C. (1976). A fluidic transducer for a tillage depth control system. *Canadian Agricultural Engineering*, 18: 10–13.
Mohawesh, O., Ishida, T., Fukumura, K. and Yoshino, K. (2008). Assessment of spatial variability of penetration resistance and hardpan characteristics in a cassava field. *Australian Journal of Soil Research*, 46: 210–18.
Mouazen, A. M. and Roman, H. (2006). Development of on-line measurement system of bulk density based on on-line measured draught, depth and soil moisture. *Soil and Tillage Research*, 86: 218–29.
Mousazadeh, H. (2013). A technical review on navigation systems of agricultural autonomous off-road vehicles. *Journal of Terramechanics*, 50: 211–32.
Mzuku, M., Khosla, R., Reich, R., Inman, D., Smith, F. and MacDonald, L. (2005). Spatial variability of measured soil properties across site-specific management zones. *Soil Science Society of America Journal*, 69: 1572–9.
Nielsen, S. K., Nørremark, M. and Green, O. (2016). Sensor and control for consistent seed drill coulter depth. *Computers and Electronics in Agriculture*, 127: 690–8.
Radford, B. J., Yule, D. F., McGarry, D. and Playford, C. (2001). Crop responses to applied soil compaction and compaction repair treatments. *Soil and Tillage Research*, 61: 157–66.
Raper, R. L. and Kirby, J. M. (2006). Soil compaction: How to do it, undo it, or avoid doing it. ASABE distinguished lecture #30. ASABE publication No. 913C0106. St. Joseph, MI: ASABE, 44p.
Raper, R. L., Reeves, D. W., Shaw, J. N., van Santen E. and Mask, P. L. (2005). Using site-specific subsoiling to minimize draft and optimize corn yields. *Transactions of the ASABE*, 48(6): 2047–52.
Sakai, K., Upadhyaya, S. K., Andrade-Sanchez, P. and Sviridova, N. V. (2017). Chaos emerging in soil failure patterns observed during tillage: Normalized deterministic nonlinear prediction (NDNP) and its application. *Chaos* 27(3): 033115.
Schafer, R. L., Young, S. C., Hendrick, J. G and Johnson, C. E. (1984). Control concepts for tillage systems. *Soil and Tillage Research*, 4: 313–20.
Servadio, P., Bergonzoli, S. and Beni, C. (2016). Soil tillage systems and wheat yield under climate change scenarios. *Agronomy*, 6: 43.
Soane, B. D. (1990). The role of organic matter in soil compactibility: A review of some practical aspects. *Soil and Tillage Research*, 16: 179–201.

Stafford, J. V. and Hendrick, J. G. (1988). Dynamic sensing of soil pans. *Transactions of the ASAE*. 31(1): 9–13.

Suomi, P. and Oksanen, T. (2015). Automatic working depth control for seed drill using ISO 11783 remote control messages. *Computers and Electronics in Agriculture*, 116: 30–5.

Upadhyaya, S. K. (1992). Deterring compaction of soils by heavy machinery. *California Agriculture*, 45(1): 19–20.

Yasin, M., Grisso, R. D. and Lackas, G. M. (1992). Non-contact system for measuring tillage depth. *Computers and Electronics in Agriculture*, 7: 133–47.

Young, R. E. and Schafer, R. L. (1977). Can automation improve traction? *Agricultural Engineering*, 58(2): 15–18.

Zhang, Q. S., Upadhyaya, S. K., Liao, Q. X. and Andrade-Sanchez, P. (2017). Finite element modeling and simulations to investigate the relationship between the Cone Index profile and draft requirements of a compaction profile sensor with depth. *Transactions of the ASABE* (accepted, in press).

Chapter 12

Variable-rate seeding systems for precision agriculture

John Fulton, The Ohio State University, USA

1 Introduction

Variable-rate seeding (VRS) has drawn interest with the early precision agriculture pioneers of the mid-1990s. The premise of VRS was that subfield variability in soil, terrain and other attributes exists in many arable farming areas of the world. Therefore, yield potential, crop emergence and crop development can vary across the field in these areas, so VRS can serve as a way to better place seed in a site-specific manner to maximize yield and profit. VRS has been implemented by some farmers as a practice to better manage farm risks through focusing or investing in areas of higher return potential and not investing as much in areas of low profitability.

In most regions of the world, VRS has not been readily adopted until recently. In the United States, the generation of seeding prescription (RX) maps remained relatively level at around 4–6% until 2008 but dramatically increased to 24% in 2013 and 50% in 2015 with projections of 59% by 2018 (Erickson and Widmar, 2015). This same US study indicated that the percentage of market for VRS prescriptions was 14% in 2015 but growing to 29% by 2018. One reason for this interest is that variable-rate technology (VRT) on planters has advanced to a point where it is embedded into agriculture machinery, making it simpler to implement. Highly accurate global navigation satellite system (GNSS) technology with innovations in VRTs has enabled farmers to create and implement VRS prescriptions to optimize soybean yields and seed placement (Hoeft et al., 2000). Figure 1 illustrates a 16-row planter used typically to plant maize and soybeans, but today this planter represents 16, 1-row planters with individual meters that singulate individual seeds and has fertilizer technology controlled independently to place fertilizer near the seed.

http://dx.doi.org/10.19103/AS.2017.0032.13

Figure 1 Modern row-crop planter with individual electric drive, dual meters to singulate individual seeds along with pulsed-width modulation (PWM) technology to independently meter fertilizer row-by-row.

Simply, VRS represents the site-specific practice of adjusting the seeding rate or amount across a field typically based on site-specific data layers such as soil type, landscape position, yield maps, remote-sensed imagery or other data that can identify yield-limiting factors. A crucial element to VRS implementation is the establishment of those factors impacting yield, site-specifically, to create seeding zones and assigning rates, thus generating the seeding prescription (RX) map. Figure 2 presents remote-sensed imagery illustrating the existence of soil variability.

Varying rates of crop inputs are typically applied using what is termed a prescription (RX) map indicating the allocation of the input across the field for defined management zones. Prescription application involves the application of two or more product rates within the same field using VRT. These technologies have grown substantially with over 28% of cropland areas in the United States having used some form of VRT in 2010 (Schimmelpfennig, 2016). Specifically, in the state of Ohio, USA, use of VRTs in the Western Lake Erie Basin increased from 4% to 14% from 2006 to 2012, respectively (USDA NRCS, 2016) (Fig. 3).

A difficulty with VRS has been assigning a seeding rate or variety to a management zone that will maximize yield and profit. Management zone development has been somewhat of an art mixed with science and individual knowledge of local crop production. Different methods exist for seeding zone development depending upon country and region of the country. While VRS requires data for a field along with time upfront to create seeding zones, there exists opportunity for return on investment (ROI). Farmers and researchers have collected a wealth of raw data, but a shortage of techniques and agronomic knowledge has prevented successful utilization of this data for advancement of prescriptive agriculture (Bullock et al., 2007) such as seeding.

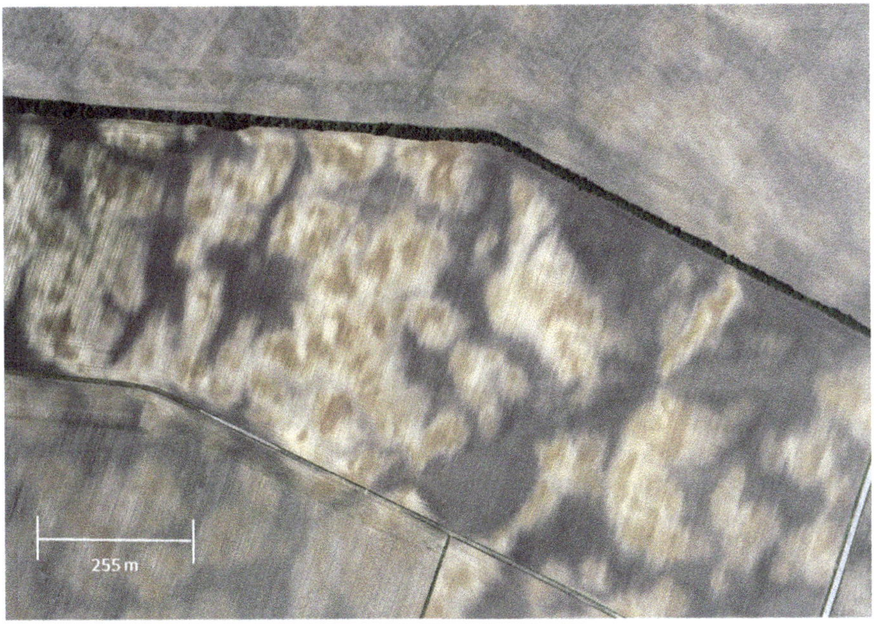

Figure 2 Visible (i.e. RGB) remote-sensed image illustrating field variability in terms of soil texture and organic matter that impact crop yields. Due to the inherent soil and terrain variability, an opportunity for variable-rate seeding was identified and implemented profitably in this field.

More recently, VRS has expanded with new planter technologies to include the ability to place two different varieties within a field commonly referred to as multi-hybrid planting. This dual placement technology also allows for differently treated seeds or two

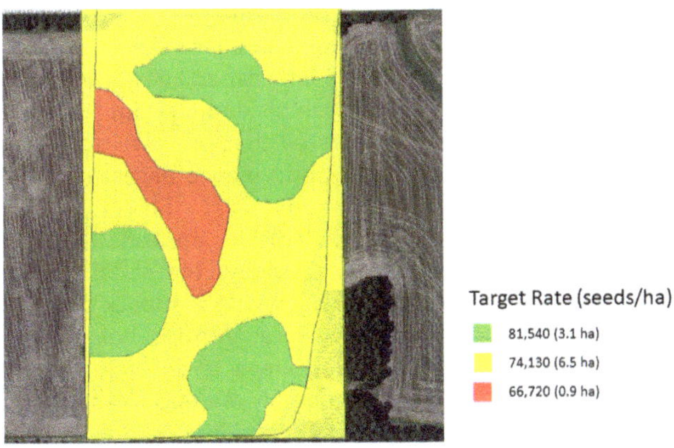

Figure 3 Illustration of a seeding prescription map for dryland maize with three different seeding rates assigned to each management zone. This type of seeding prescription is common in North and South America for those implementing VRS in maize.

different cultivars to be planted across fields or two different cultivars within the same pass. Further, technology has advanced to simplify data collection at the farm level with telemetry and wireless enabling the collection of planter, fertilizer and yield data to be captured and stored, thereby making this data more accessible to the farmer and those providing VRS prescription (RX) services. Key technologies available to implement VRS strategies include:

1 Population or seeding amount,
2 Dual meters commonly referred to as multi-hybrid,
3 Downforce placed on row units,
4 Depth, or
5 Combination of the above.

While this chapter covers VRS systems, the development of precision agriculture technologies outpaces the scientific knowledge on how best to implement the systems by farmers. The success of VRS rests on identification, and understanding and management of spatial, temporal and agronomic knowledge. Identification and management can be attained with the use of precision agriculture, but 'scientific' understanding, though limited, is growing with the wealth of data being collected and research being conducted. Precision agriculture technologies can improve field execution of operations such as planting. Mistakes during field execution can be costly especially considering the cost of seed. As an example, 2017 production costs estimate that nearly 29% of the total variable costs for maize are associated with seed costs (OSU, 2017). Therefore, accuracy of seed placement is important.

2 Variable-rate seeding technology

Two important requirements for implementing a VRS system are to (1) have the right technology and (2) identify the process and people to create prescription maps. Various technologies exist today that can improve the placement of seed and give the ability to vary population and variety (Fig. 4). Key VRS technologies include a GNSS receiver, in-cab display (Fig. 5a) and a planter with variable-rate capabilities. Variable-rate capabilities include a hydraulic drive that allows for variation of seeding population across the planter (Fig. 5e). A quality seeder or planter is required that provides accurate seed metering capabilities in order for VRT to provide the performance necessary to execute seeding prescriptions.

Figure 5 presents various technologies available for implementing precision seeding strategies. These include an in-cab display providing the ability to upload RX maps, execute RX maps, collect as-planted data and provide feedback to the tractor operator about planter performance in real time. Other precision technologies that can improve seed placement and minimize double-seeded areas of the field are automatic row or section control technology. A technology commonly referred to as 'row clutches' enables automatic section control on seeders and planters. This technology minimizes areas of the field that can get double-seeded due to overlap by using the coverage map (e.g. GNSS-marked planted areas) to automatically turn off sections or rows of the planter in previously planted areas or areas marked as no planting (Fig. 6). Automatic section control can reduce overlap areas and thereby provide seed savings on average of 4.3% while also

Figure 4 Row-crop planter equipped with variable-rate seeding capabilities along with active downforce and cab-adjustable row cleaner technologies. Image courtesy of Ohio State University.

reducing maize harvest loss (by 8.7 times) and yield loss (17%) compared to areas that are double-planted (Runge et al., 2014). Most planter manufacturers today provide automatic section control as a standard option.

Additional technologies available commercially and becoming common components on seeders and planters include active downforce and mobile applications (i.e. apps). Active downforce replaces the mechanical springs used to keep the seeding units and ground engaging components held firmly in contact with the soil. Hydraulic actuators (Fig. 5d) and airbags (Fig. 5f) are the common mechanisms used to apply downforce onto the seeding units. The importance of this technology is that soil physical properties vary spatially and so planter performance could be improved by adjusting planter settings, specifically downforce, according to field spatial variability. This technology is termed active in that hydraulic or air pressure can be varied to maintain the force being applied to the soil by the gauge wheels or components controlling depth. For example, dry soil would require more applied downforce due to higher soil strength compared to wet soil to keep the opening discs, gauge wheels and closing wheels operating and the target seeding depth. The same premise would hold if soil texture, soil strength or moisture-holding capacity varies within a field. Seeding depth and downforce management are critical for optimization of planter performance (Hanna et al., 2010). Therefore, active downforce could play a key role in implementing precision seeding including VRS to ensure accurate seeding depth and planter performance (Table 1).

Multi-hybrid technology provides the capability of placing two varieties or seed treatments independently across fields. Rather than a single meter for each row, this technology uses two meters (Fig. 7a) that turn on and off to switch between varieties or seed treatment products. This technology requires two seed bins on the planter but does allow varying of the seeding rate as well. Currently, dual-meter technology uses electric drives (Fig. 7b) to power the meters. Figure 8 presents an example of a field with two maize varieties planted.

(a)	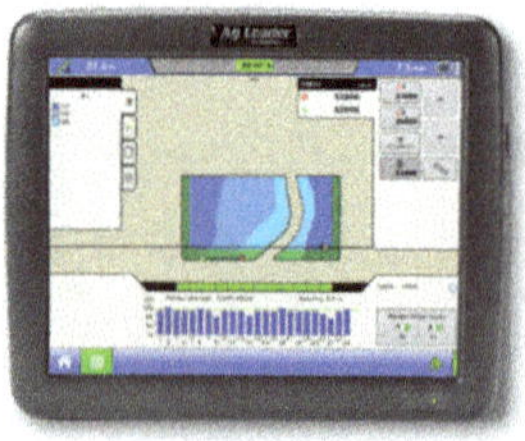	**In-cab display:** provides VRS capabilities including uploading RX maps plus collecting as-planted maps. Real-time feedback related to planter performance and regions of the field planted is provided to the tractor operator (Courtesy of Ag Leader).
(b)		**Precision meter:** uses a seed disc and double eliminator components to accurately meter individual seeds using vacuum or positive pressure (Courtesy of Precision Planting).
(c)		**Seed count sensor:** seed tube with a sensor to accurately count seeds delivered to the seed furrow. Sensor output permits a real-time estimate of variables such as seeding population and seed spacing that can be reported to the operator through the in-cab display.
(d)		**Active downforce:** hydraulic actuator used to adjust and maintain downforce of the planting units to ensure good soil–planter contact and the target seeding depth.
(e)		**Hydraulic drive:** hydraulic motor used to drive the main shaft coupled to seed metering devices enabling accurate control of seed population and providing variable-rate capabilities.
(f)	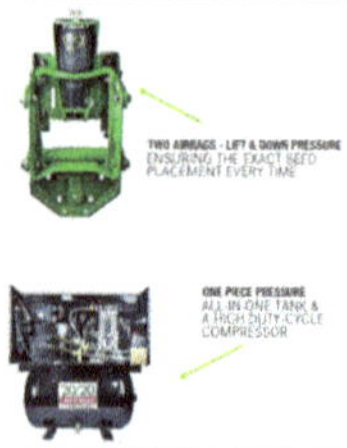	**Active downforce:** pneumatic downforce technology used to maintain downforce of the planting units to ensure good soil-planter contact and the target seeding depth (Courtesy of Precision Planting).
(g)		**Row clutch:** example row clutch that can be used on chain drive metering units to turn meters on and off for implementation of automatic row or section control technologies. The clutch shown is powered electronically but pneumatic and hydraulic options exist.

Figure 5 Example components and technologies for accurate metering of seed and implementing VRS. Images courtesy of Precision Planting and Ag Leader.

Figure 6 Illustration of automatic section control technology implemented on a maize row-crop planter. In this example, row-by-row on/off is being executed with the use of real-time kinematic (RTK) GNSS for accurate positioning information.

Important components of VRS systems are verification and evaluation to ensure seeding prescription maps are executed properly. As-planted data (Fig. 9) that is recorded and displayed by the in-cab display provide verification of variables such as planted population, seed spacing, applied downforce and vertical acceleration of individual seeding units. This information is important for verification of what actually occurred in the field and should be used in end-of-season analyses rather than assuming that the seeding prescription was executed perfectly. Further, the information can serve to identify soil and field variability as illustrated in the applied downforce in Fig. 9. In this field, ground truth data using a soil penetrometer confirmed that the high applied downforce areas (red and orange in Fig. 9) were due to soil compaction issues that resulted in emergence issues. This example illustrates the value of this type of data for understanding planter performance plus revealing other factors that can influence the crop.

Table 1 Example field data collected by Ohio State University where active downforce was implemented at two different levels with a target seeding depth of 51 mm for maize

Target seeding depth (mm)	Downforce TRTS		Mean seeding depth (mm)	CV seeding depth (%)
51	none	0 kg	47	14.5
51	Active	45.4 kg	56	5.1
51	Active	88.5 kg	58	7.8

Note the variation in seeding depth across the field especially without any downforce applied to the row units.

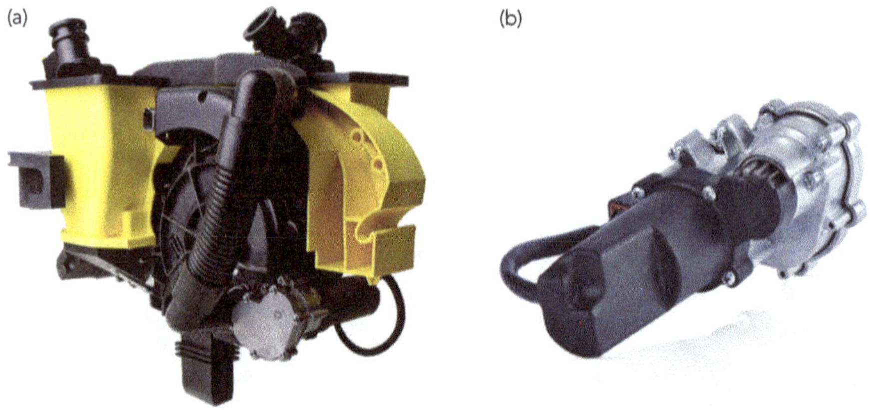

Figure 7 Example dual-meter technology (a) for a vacuum style planter that contains two seed discs with each controlled independently by electric motors (b). Image courtesy of Precision Planting.

3 Variable-rate seeding strategies

A key process for VRS is the creation of a prescription (RX) map for individual fields. Research has determined that RX creation needs to be handled on a field-by-field basis due to each field's unique variability and differences in historical management. Two main variables must be defined for VRS implementation: (1) the where (e.g. management zones) and (2) how much or what. In most cases, in designing zones, there can be around four to five variables to consider (e.g. yield, soil, terrain, etc.) paying attention to those that are yield-limiting factors.

Figure 8 Illustration of dual hybrid placement as a technology demonstration by Ohio State University using two different maize varieties to create the Ohio State University Block O. Image courtesy of the Ohio State University.

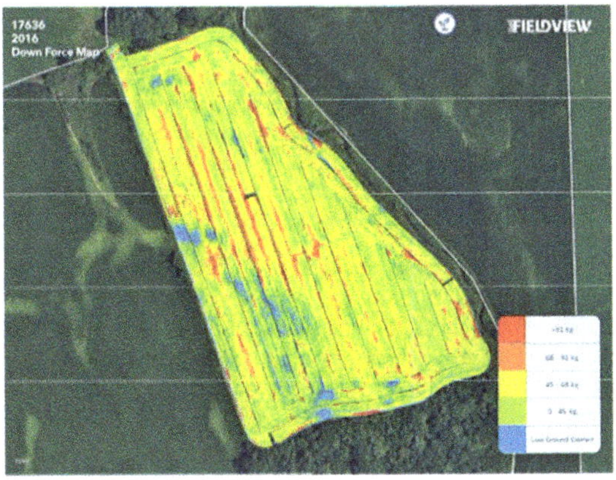

Figure 9 As-planted map showing the variation in applied downforce on a per-row basis across the field. Orange and red represent areas requiring high applied downforce (dense soil areas) with yellow moderate and green optimum for this field. Blue indicates a loss of soil to gauge wheel contact which is undesirable. This as-planted data can be valuable in verification of the planting operation and can uncover soil variations experienced during planting.

3.1 Seeding zone creation

Defining management zones offers the ability to reduce inputs in areas with less potential for production while maximizing inputs for optimal productivity within highly productive areas. The delineation of these zones can be achieved using a variety of methods. These methods can include using aerial imagery (Basnyat et al., 2004), soil type and farmer knowledge of fields (Fleming et al., 2000), soil properties (Mzuku et al., 2005) or yield history (Schepers et al., 2004). Many people creating seeding zones will use a combination of spatial data layers such as yield maps, imagery and elevation data depending upon the yield-limiting factors for a region. Table 2 outlines a few common data layers being used to create seeding zones. One question and possible issue for VRS management zone creation is what data is available for each field to develop zones.

3.2 Prescription creation

The final step in creating a seeding prescription map prior to loading the map into an in-cab display is establishing the population rate and hybrid or variety within each zone. When working with population rates (seeds/ha), one should keep rates at least 15–20% different between zones to maximize yield response to VRS. There are several philosophies that exist for connecting population rates to zones. Typically, higher populations of seed will be planted in zones with higher yield potential, while areas of lower yield potential will receive lower seed population rates (Verbeten, 2015). This approach is most prevalent in maize where high productivity areas can support higher seeding populations, producing more ears per hectare and thereby higher yield. This strategy provides an agronomic and economic advantage to the producer by maximizing yield potentials and reducing overall seed cost. For a crop like soybeans, a different strategy is typically implemented since emergence is key for maximizing yield and profit. Lower population rates are used

Table 2 Data layers commonly used to develop management zones for VRS

Data layer	Description
Water availability	Water represents the most important driver in crop yield. Water availability is impacted by rainfall over the growing season. Rainfall coupled with field soil characteristics such as soil texture, water-holding capacity, topsoil depth and organic matter content can help create management zones. Soil maps provide an initial spatial characterization that can help establish water availability
Terrain	Terrain and landscape position can significantly impact crop yield and are typically related to drainage, orientation to the sun and erosion and thereby topsoil or soil depth. Elevation data serves as the basis for creating terrain layers such as topography and can be used to determine how water drains spatially from the field, defining areas that are poor or well drained. These layers can be combined with soil, fertility and yield maps
Soil test results	Soil fertility and pH influence crop yield potential and with precision soil management can provide information that can be used for generating VRS RX maps. This soil information could reveal opportunities around determining high and low yield potential areas of a field
Yield maps	Yield maps provide critical information for prescription seeding. These maps identify high to low yielding regions of a field. Collecting yield data layers over time on a field provides important information to develop and adjust VRS management zones
Remote-sensed imagery	Remote-sensed imagery can provide both in-season information for crop development within fields along with estimates of crop-stressed areas, low to high yield regions and biomass production

For many precision agriculture practitioners, these data layers may be readily available and should be considered for developing and executing a VRS plan.

in higher productivity areas and increased within marginal soil areas. The premise is that emergence percentage will be much higher in high productivity areas and so less seed is needed to establish the optimum seeds per hectare. Soybeans are compensatory in nature and have the ability to occupy an open space or a gap, producing more branches and nodes for maximum yield. For dual hybrid placement, understanding seed traits will be important so one can determine how hybrids or varieties respond under different conditions. These agronomic strategies for maize and soybeans provide examples of how VRS has been implemented for two crops.

VRS may most likely require altering crop protection and crop nutrition strategies. It has been established that denser populations of plants can increase the possibility of disease. Fertilization may have to be implemented on a site-specific basis as well. In maize, nitrogen is a critical nutrient so varying the seeding population may require variable-rate nitrogen to maximize profit as well. Again, the premise is that different populations will have different needs from a fertility and crop protection perspective.

4 The value of variable-rate seeding

VRS has been proven to be profitable for fields with highly variable levels of productivity. Quantification of the amount of in-field variability should be considered when deciding whether to employ VRS. Nielsen et al. (2017) concluded that VRT may be of minimal value for fields with maize yields averaging from 9415 to 16 477 kg/ha. If soil types and characteristics do not vary substantially across the field, VRS may not prove to be profitable using currently available technology and agronomic knowledge.

Researchers in southern Brazil evaluated VRS using zones generated from producer knowledge of field productivity levels and also by compiling eight years of yield data to segregate into productivity zones (Hörbe et al., 2013). Low yield potential areas received a reduced population rate and experienced a 1197 kg/ha and 1900 kg/ha increase in productivity for the producer knowledge and the yield-based zones, respectively (Hörbe et al., 2013). High yield potential zones received an increased population rate and experienced an 888 kg/ha and 942 kg/ha increase in productivity for the producer knowledge and the yield-based zones, respectively (Hörbe et al., 2013).

Economic return or ROI is a primary question raised by farmers interested in VRS. Lowenberg-DeBoer (1998) evaluated the economics of VRS using yield potential zones. He found the highest ROI in areas of low yield potential soils and within fields that have a range of high- and low-, or medium- and low-level productivity soils within the Midwestern United States. Bullock et al. (1998) reported VRS of maize in the Midwestern United States could provide a ROI ranging from $0.15/ha to $12.00/ha. It has also been reported that the potential savings from VRS could be around $6.25/ha when varying seeding populations in high and low yielding areas of fields (Shanahan et al., 2004). These savings from VRS can be a means to be efficient considering the increasing cost of seed. As an example, the 2017 Ohio State University Maize Production Budgets estimate nearly 29% of the total variable costs is associated with seed costs (OSU, 2017). Similarly, as soybean seed prices continue to rise (USDA-ERS, 2017), growers are looking for ways to optimize seeding rates across their fields (Hoeft et al., 2000). Therefore, with the availability of digital tools and the amount of field-level data collection, VRS can provide an ability for farmers to explore and fully implement VRS and also gain ROI.

5 Conclusion and future trends

Precision agriculture technologies have allowed agricultural producers to utilize site-specific management of crop inputs on a subfield basis. Planting and seeding technologies have advanced to permit the implementation of VRS where population rates, hybrids or a combination can be varied over a field. Planting technologies also include in-cab displays for monitoring in-field planter or seeder performance in real time, capabilities to collect as-planted maps and changing settings from the cab. Active row-unit downforce, row cleaners and closing wheels are additional technologies being used to create an optimum seedbed while improving the accuracy of seed placement. As a result, VRS continues to increase as technology becomes more available on the farm and a database is created for individual farms to have the data in place to create sound VRS prescriptions. Interest in VRS is growing globally due to the availability of these technologies coupled with the current cost of seed. Farmers want to be efficient with the seed purchased while improving crop yields and maximizing annual profit margins with VRS providing opportunity to do this. Mistakes at planting time can be costly, so VRS not only enables matching population and hybrids to in-field variability but also helps to ensure accurate field execution to minimize planting errors.

6 Where to look for further information

- The International Society of Precision Agriculture (ISPA) provides information that includes conference proceedings; https://www.ispag.org/.
- The Society of Precision Agriculture Australia (SPAA) is a resource for precision agriculture information; https://spaa.com.au/.
- The United States Department of Agriculture (USDA) publishes reports related to precision agriculture adoption and profit. https://www.ers.usda.gov/amber-waves/2016/december/precision-agriculture-technologies-and-factors-affecting-their-adoption/
- Iowa State University in Ames, Iowa USA has conducted research and published articles discussing variable-rate seeding of maize in Iowa. An example article is 'Corn seeding rates and variable-rate seeding' available at https://crops.extension.iastate.edu/corn/production/management/planting/seedingrates.html. Additional information can be found on their Integrated Crop Management website, https://crops.extension.iastate.edu/.
- The Ohio State University provides an overview and suggestions to making variable-rate seeding profitable at http://precisionag.osu.edu.
- Pioneer maintains several publications that overview variable-rate seeding. An example would be their publication titled, 'Putting Variable-Rate Seeding to Work on Your Farm': https://www.pioneer.com/home/site/us/agronomy/library/variable-rate-seeding/.

7 References

Basnyat, P., McConkey, B., Meinert, B., Gatkze, C. and Noble, G. (2004). Agriculture field characterization using aerial photograph and satellite imagery. *IEEE Geoscience and Remote Sensing Letters*, 1(1), 7–10.

Bullock, D. G., Bullock, D. S., Nafziger, E. D., Doerge, T. A., Paszkiewicz, S. R., Carter, P. R. and Peterson, T. A. (1998). Does variable rate seeding of corn pay? *Agronomy Journal*, 90(6), 830–6.

Bullock, D. S., Kitchen, N. and Bullock, D. G. (2007). Multidisciplinary teams: A necessity for research in precision agriculture systems. *Crop Science*, 47(5), 1765–9.

Erickson, B. and Widmar, D. A. (2015). 2015 precision agricultural services dealership survey results. Purdue University Department of Agriculture Economics/Department of Agronomy, West Lafayette, IN. August 2015. http://agribusiness.purdue.edu/files/resources/2015-crop-life-purdue-precision-dealer-survey.pdf (last accessed 7 February 2018).

Fleming, K. L., Westfall, D. G., Wiens, D. W. and Brodahl, M. C. (2000). Evaluating farmer defined management zone maps for variable rate fertilizer application. *Precision Agriculture*, 2(2), 201–15.

Hanna, H. M., Steward, B. L. and Aldinger, L. (2010). Soil loading effects of planter depth-gauge wheels on early corn growth. *Applied Engineering in Agriculture*, 26(4), 551–6.

Hoeft, R. G., Aldrich, S. R., Nafziger, E. D. and Johnson, R. R. (2000). *Modern Corn and Soybean Production*, 1st Edition. MCSP Publications, Savoy, IL.

Hörbe, T. A. N., Amado, T. J. C., Ferreira, A. O. and Alba, P. J. (2013). Optimization of corn plant population according to management zones in Southern Brazil. *Precision Agriculture*, 14(4), 450–65. https://doi.org/10.1007/s11119–013-9308-7

Lowenberg-DeBoer, J. (1998). Economics of variable rate planting for corn. Staff Paper #98-2, Agricultural Economics Department, Purdue University, West Lafayette, IN.

Mzuku, M., Khosla, R., Reich, R., Inman, D., Smith, F. and MacDonald, L. (2005). Spatial variability of measured soil properties across site-specific management zones. *Soil Science Society of America Journal*, 69(5), 1572–9.
Nielsen, R. L. B., Lee, J. and Camberato, J. (2017). Yield response of corn to plant population in Indiana. Purdue University Department of Agronomy, Applied Crop Production Research Update, West Lafayette, IN. www.kingcorn.org/news/timeless/CornPopulations.pdf (last accessed 8 March 2018).
OSU. (2017). Corn production budget 2017. The Ohio State University Extension, Farm Budgets. https://farmoffice.osu.edu/farm-management-tools/farm-budgets#2017 (last accessed 9 March 2018).
Runge, M., Fulton, J. P., Griffin, T., Virk, S. and Brooke, A. 2014. ANR-2217: Automatic section control technology for row crop planters. Alabama Cooperative Extension System, Auburn University, Auburn, AL.
Schepers, A. R., Shanahan, J. F., Liebig, M. A., Schepers, J. S., Johnson, S. H. and Luchiari, A. (2004). Appropriateness of management zones for characterizing spatial variability of soil properties and irrigated corn yields across years. *Agronomy Journal*, 96(1), 195–203.
Schimmelpfennig, D. (2016). Farm profits and adoption of precision agriculture. Report Number 217, United States Department of Agriculture (USDA) and Economic Research Service (ERS).
Shanahan, J. F., Doerge, T. A., Johnson, J. J. and Vigil, M. F. (2004). Feasibility of site-specific management of corn hybrids and plant densities in the Great Plains. *Precision Agriculture*, 5(3), 207–25.
USDA NRCS. (2016). Effects of conservation practice adoption on cultivated cropland acres in Western Lake Erie Basin, 2003–06 and 2012. United States Department of Agriculture, Special Study Report March 2016.
USDA ERS. (2017). Commodity costs and returns. United States Department of Agriculture, Economic Research Service, Washington DC. Historical costs and returns: Soybeans. https://www.ers.usda.gov/data-products/commodity-costs-and-returns. (last accessed 12 March 2018).
Verbeten, B. (2015). *Varying Corn & Soybean Populations, Varieties, & Down Force*. Cornell University Cooperative Extension, Ithaca, NY.

Chapter 13

Site-specific nutrient management systems

Dan S. Long, USDA-ARS, USA

1 Introduction

Farm management has been generally defined as the science of organizing and controlling the resources of a farm such that profitability is maximized (Castle et al., 1972; Nix, 1979). Dillon (1980) succinctly defined farm management as *the process by which resources and situations are manipulated by the farm manager in trying, with less than full information, to achieve his goals*. This statement recognizes the importance of *the dynamic nature of the farm system and its environment, and the uncertain nature of the farmer's decisions, thereby implying attempted rather than sure achievement of objectives.*

Such is the case of site-specific nutrient management in which growers attempt to account for crop nutrient demand that varies in space and time within farm fields. However, ability to do so is largely limited by an incomplete understanding of the causes and magnitude of the variability underlying crop fertilizer response.

Farm management systems are created through the establishment of inputs, processes and outputs that are intended to maximize yield and profit (Fig. 1). By analogy, this concept applies to nutrient management systems as well in which inputs result from the physical environment (e.g. precipitation, temperature and growing season length) and farm management (e.g. labour, machinery, fertilizer, herbicide and other materials). Processes are actions within the farm that are directed towards the production of outputs and include but are not limited to image acquisition, field mapping, zone delineation, soil sampling and fertilizing. Outputs include grain and biomass, dollar returns and improved soil, water and air quality but they can also be negative such as greenhouse gas emissions, herbicide-resistant weeds and soil erosion.

Processes are carried out by farmers in a repeating series of steps that begins with inventory of spatial data and their recording in a geographical information system (GIS) followed by geographic analysis and planning, and application of plans through global navigation satellite systems (GNSS)-guided field implements (Fig. 2, Nielsen et al., 1996).

http://dx.doi.org/10.19103/AS.2017.0032.14
Published by Burleigh Dodds Science Publishing, 2019.

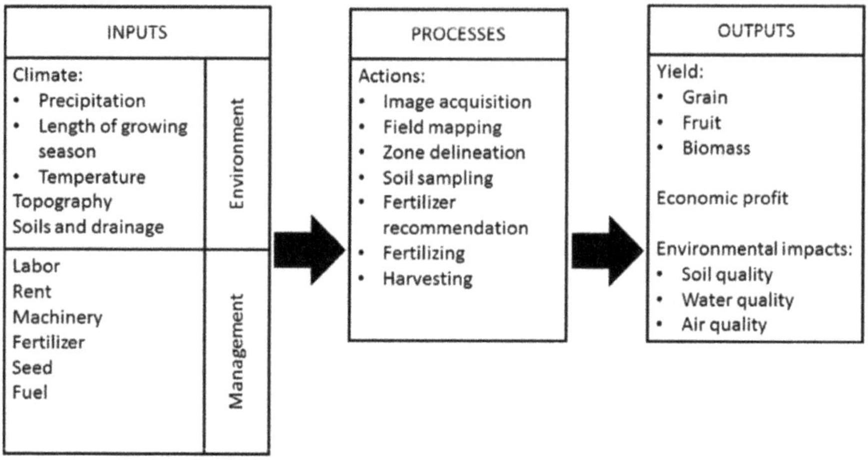

Figure 1 Components of a nutrient management system with inputs, processes and outputs.

Figure 2 Precision farming cycle represented as a repeating series of actions or processes.

Final results are monitored and compared with projected economic and environmental outcomes and recorded for use in the next cycle. As an aside, this 'precision farming cycle' is a variation of the Japanese Plan-Do-Check-Act cycle that has its origin with the Shewhart Cycle (Shewhart, 1939) and Deming Wheel (Deming, 1950) for quality control in the United States (U.S.) manufacturing sector (Fig. 3). The purpose of a cyclical series of actions is to include the development, testing and implementation of changes that will result in improvement to the system of management.

Site-specific nutrient management systems were created to improve nutrient use efficiency and balance trade-offs between profitability and environmental concerns. Fertilizers are prescribed to match crop demand for nutrients that are known to vary in space and time within farm fields. This approach embraces the 4-element 'right product, right rate, right place and right time' advanced by the International Plant Nutrition Institute (Johnston and Bruulselma, 2014). Though understanding of the environmental dynamics affecting site-specific nutrient management may be imperfect, a cornerstone of such

Published by Burleigh Dodds Science Publishing, 2019.

Figure 3 Japanese plan, do, check and action cycle.

management is knowledge evolving from the information acquired, which improves over time with each iteration of the precision farming cycle.

Environmental factors that influence crop production across the conterminous U.S. are terrain, climate, soil and plant available water (Baker and Capel, 2011). It is the combination of these four environmental factors that also determines the development of a specific nutrient management system in a certain area. Nutrient management systems differ in the information utilized and the way they operate to accomplish their common purpose. The objective of this chapter is to illustrate site-specific nutrient management with examples of systems adapted for sub-humid and semi-arid environments. After a general review of valid processes that are presently used to collect information for site-specific nutrient management, there follows examples where the achievements of different systems are described. The final section describes further developmental needs.

2 Processes to inform site-specific nutrient management

2.1 Grid sampling

In the 1990s, systematic grid point sampling of soil and laboratory analysis of resulting cores were used in the U.S. Midwest as a primary means of obtaining discrete values of soil test nitrogen (N), phosphorus (P), potassium (K) and other soil variables at geographically distributed points within fields (Wollenhaupt et al., 1994). Mathematical interpolation methods such as inverse distance weighting and kriging were then used to estimate values for rasterized maps, which were useful for spatial overlay operations in a GIS. Because of the strong role of topography on the spatial distribution of soil properties, cokriging with high resolution and readily available digital terrain data has been shown to be potentially useful for interpolation of a limited number of soil test P and K values, and to increase the efficiency of grid sampling (Kozar et al., 2002). Nonetheless, contour maps made from interpolated grid data showed that spatial variability in nutrients did not conform to soil survey maps and that nutrients were highly variable within map units. Grid sampling methods that ignored soil and landscape units were then promoted with a 1-ha cell size giving the best spatial resolution for least cost (Mallarino and Wittry, 2001).

Soil testing for immobile nutrients such as P and K is common in many production systems and site-specific management of these nutrients may involve grid sampling. For instance, the grain yield responses of corn and soybean to variable-rate and fixed-rate P fertilization were compared over 4 years in six farm fields in Iowa (Bermudez and Mallarino, 2007). Soil test P sampled on a dense grid within 6–12 ha experimental areas served as a basis for varying dry phosphate fertilizer with commercial spreading equipment. While there was no significant difference in yield between the two management systems, variable-rate P placement showed potential for reducing excess P loss to the environment as a result of reducing the P application rate to high-testing field areas. In addition, variable-rate placement applied 12.4% less P fertilizer and reduced within-field variability in soil test P.

Today, many farmers regard grid sampling as too expensive for routine collection and analysis of soil samples, but it remains common for spatial characterization of topsoil levels of P and K. In contrast, residual NO_3-N levels are highly unstable in soils due to microbial respiration and loss by leaching, and thus the frequent sampling needed for spatial characterization of fields will be cost-prohibitive. Citing greater cost-effectiveness, many growers and their advisers have now adopted the 'zone sampling' approach whereby smaller sub-regions in a field, or management zones, are delineated and sampled separately as described below.

2.2 Constructing management zones

Within-field crop variability is fundamentally the product of interaction between many ecosystem factors including topography, soils, climate, pests, animals and management. However, a cursory examination of a farm field from above is sufficient to show patterns of crop growth that tend to be grouped into different levels forming more or less definite zones. The spatial crop variability is not random, but is distributed in a pattern over the landscape. This natural phenomenon is well illustrated in traditional use of published soil surveys depicting areas of farmland that are expected to be relatively homogeneous (Mausbach and Wilding, 1991) and aerial photographs showing tonal differences in soil moisture and crop growth within fields (Milfred and Kiefer, 1976).

Management zones are defined as sub-regions of a field with homogeneous yield-limiting factors for which a single rate of a specific crop input is appropriate (Doerge, 2000) and are established based on ability to consistently capture field patterns from year to year (Mallarino and Wittry, 2001). By dividing a field into zones and sampling them individually, zone sampling captures spatial nutrient variation while reducing sampling cost. One soil testing laboratory (Midwest Laboratories, Inc., Omaha, NE, USA) recommends that the area represented by each sample be no more than 8 ha with each sample a composite of 15–25 cores within a 6.1 m radius. One of the first studies published on precision agriculture proposed the use of soil testing by soil map unit, which differed in N, P and K (Carr et al., 1991). When the Global Positioning System (GPS) allowed for mapping of soil test values, it became apparent that soil map units could be highly variable in soil nutrient values (Mallarino and Wittry, 2001). General soil maps are not recommended for delineation of N management zones because they seldom contain detailed, site-specific information (Franzen et al., 2002).

One approach to identifying zones is by measurement and mapping of the yield-determining factors at a site (King et al., 2005). Contact electrode and non-contact inductance sensors both provide a rapid, low-cost means for obtaining continuous measurements of apparent electrical conductivity (EC_a) that vary in fields with changes in clay content and mineralogy, porosity, cation exchange capacity, compaction, depth to indurated layer, water content, salinity level and other soil physical properties (Corwin and Lesch, 2003). Spatial patterns in EC_a have been correlated to yield, topsoil depth, and soil physical and chemical properties on certain soils in the U.S. Midwest (Kitchen et al., 1999, 2003). Johnson et al. (2003) showed that management zones derived from EC_a maps were useful for characterizing spatial variation of wheat and corn yields. Farmers are intimately familiar with their fields yet a map of soil EC_a was found to be more effective in identifying homogeneous sub-regions within fields than farmer interpretation of a bare soil image (Fleming et al., 2004). King et al. (2005) concluded that the method can provide useful information for the provisional delineation of soil type boundaries and crop management zones, and that expert site assessment and knowledge were necessary to confirm their existence. Kitchen et al. (2005) stated further that an EC_a map can be a good indicator of underlying soil properties; however, temporal variability must be considered when implementing this approach due to changes in crop response with environmental conditions.

A second approach to identifying zones utilizes the spatial variation in crop yield at a site. Growers are increasingly using yield monitors to acquire site-specific yield data in their fields. In the U.S. between 1996 and 2012, 5–30% of soybean, 8–34% of corn and 2–13% of spring wheat areas were mapped using a yield monitor (Schimmelpfennig, 2016). Though a single-year map may have limited value for supporting management decisions over extended time periods (Ping and Dobermann, 2005), multi-year yield data can be used to produce temporal stability maps showing high or low yielding areas that had not changed from year to year as well as unstable areas that had been highly changeable (Blackmore, 2000). Yield stability maps were used to identify homogeneous zones within fields for site-specific management of P, K and lime (Flowers et al., 2005). Taylor et al. (2001) analysed 3–7 years of yield monitor data for cornfields in Kansas and found the ability of a single year's yield map to predict yield in the following year to vary substantially. Sequential yield maps were useful for deriving a yield goal map only if the correlation among years was consistently high ($r \geq 0.70$).

Research has found that a single source document is often insufficient for guiding site-specific applications (McCann et al., 1996; Franzen et al., 1998). Multiple data layers can provide more information for subdividing fields into management zones than can be achieved with one layer of information. For example, Anderson-Cook et al. (2002) determined that soil EC_a readings and crop yield maps together could separate soil types with 90% accuracy for variable-rate fertilizer management. Hornung et al. (2006) found that a combination of bare soil imagery, topography and farmer experience more effectively defined zones of different yield productivity than a combination of bare soil imagery, soil organic matter, cation exchange capacity, soil texture and previous season's yield map. Schepers et al. (2004) used principal components analysis and unsupervised classification of principal component scores to aggregate variability in soil brightness, elevation and electrical conductivity into four management zones. Though these zones adequately characterized spatial variation in soil pH, soil EC_a, soil test P and soil organic matter, they were less consistent in characterizing variation in crop yields.

The use of a numerical technique (e.g. principle components analysis, unsupervised classification etc.) for the integration of multiple data to produce more useful information than any individual data source is known as data fusion (Liggins et al., 2001). Clustering is a data fusion technique that has been widely used to group data points into homogeneous groups. A number of different algorithms and techniques have been applied to spatial data including crop yield (Stafford et al., 1999); soil EC, elevation and slope (Fridgen et al., 2000); and multi-year cotton yield (Boydell and McBratney, 2002) for the purpose of creating management zones. Milne et al. (2012) created compact zones by applying smoothed fuzzy classification to multi-year yield data but resulting zones had a similar N response suggesting no benefit from fertilizing each zone separately. A soil map of the same field also distinguished compact zones that differed in crop N response such that it could have been more profitable to manage by soil type.

The Management Zone Analyst (MZA) software uses the fuzzy c-means unsupervised clustering algorithm to assign field observations into like classes, or management zones, and determines the appropriate number of zones (Fridgen et al., 2004). Farid et al. (2015) applied the MZA software to point values of percent sand, percent clay, elevation, soil N and soil EC sampled within a wheat field in Pakistan comprising a complex of alluvial soils. Performance indices were available to indicate that the field should be divided into four management zones that significantly differed in grain yield. Other available software for classification of management zones and automatic determination of number include ZoneMap (Zhang et al., 2010) and EZZone (Lowrance et al., 2016). Both tools are web-based and provide for interpolation of point data to a common grid. MZA does not have this feature and instead requires that data be gridded before import.

Management zones are designed to capture within-field variability in plant nutrients, crop yield or crop fertilizer response with the best technique explaining the most variance. In Montana, Long et al. (1995) found that 67% of the variance in final yield could be explained by management zones derived from a mid-season aerial photograph of ripening differences in wheat. In South Dakota, Chang et al. (2003) investigated the impact of different zone mapping techniques on soil NO_3-N and Olsen P sampling variability. Grid cell sampling (3.5 ha) yielded the greatest reduction in sampling variance below that of the whole field followed by management zone classification by GIS analysis of electrical conductivity, elevation, aspect and distance information; cluster analysis of electrical conductivity, elevation and aspect; and 1:12 000 scale soil survey maps. Variance was further reduced when areas historically impacted by homesteads or animals were sampled separately. Later, Chang et al. (2004) showed that variability in university N and P recommendations were best reduced using multiple year yield monitor data to establish yield goals followed by grid cell sampling to reduce N and P sampling error.

Principles of fertilization for zone-based nutrient management are based on sampling soils at or near planting and analysing them in the laboratory for N, P, K and S contents. For N, soil test values together with expected yield and the unit requirement for yield are then used to derive fertilizer recommendations based on university recommendation methods. Accuracy of zone-based N management to target crop N needs depends on correctness of these N fertilizer recommendations. Unfortunately, this approach provides little information on future levels of plant available N that are released over the growing season from mineralization of soil organic matter and crop residues (van Es et al., 2007). As up to 50% of crop N uptake is supplied by soil (Kramer et al., 2002), it is necessary to predict N mineralization from organic sources to synchronize the N supply with crop N demand and prevent overfertilization with N (Wade et al., 2016). Leaching can also be a major loss mechanism for NO_3-N, particularly in humid environments where infiltration exceeds runoff and nitrate levels exceed crop N demand. Furthermore, loss of NO_3-N by denitrification occurs when soils are saturated with water for prolonged periods. Consequently, economically optimum N rates (EONR) are highly variable within fields and between years at the same site (Dhital and Raun, 2016), yield goals are unpredictable (Raun et al., 2017) and they correlate poorly with recommended N rates (Scharf, 2015). Spatio-temporal variability in environmental factors affecting these N cycling processes is why responsive in-season crop sensing may be warranted.

2.3 In-season crop sensing

Beginning in the mid-late 1990s, researchers in the central Great Plains of the U.S. investigated the feasibility of active optical sensing of the crop canopy for determining crop N status for in-season N management (Stone et al., 1996; Solie et al., 1996; Sembiring et al., 1998). The rationale for in-season N management is that the information on crop N status is obtained when crop N demand is greatest and applying N at this time would improve N use efficiency. Spectral properties of the crop canopy had been found useful in predicting crop foliar N content (Curran, 1989). Work with the Minolta SPAD 502 chlorophyll meter had also indicated that analysis of differences in light reflectance from plant leaves would be useful for predicting N fertilizer needs for corn in humid regions (Schepers et al., 1992). Yield losses from N deficiency were noted to occur when chlorophyll meter readings were <95% of those for well-fertilized corn and this value was proposed as a cut-off point for applying supplemental N. Other studies with corn showed that spatial patterns in N deficiency and crop response to added N could be detected and mapped through analysis of aerial colour imagery (Blackmer et al., 1996). Deficiencies of N, seen as a lighter green colour on aerial imagery, result in leaves having less chlorophyll, less photosynthesis and increased reflectance of radiation normally absorbed as red light during photosynthesis.

The Normalized Difference Vegetation Index (NDVI), computed from radiance in the Red (630–690 nm) and NIR (760–900 nm) wavebands, is directly related to aboveground green biomass (Rouse et al., 1974). It is indirectly related to chlorophyll concentration through the negative effect of N deficiency on crop cover and crop biomass. Researchers in Oklahoma developed an in-season N management strategy that calculates N fertilizer rates as a function of an in-season estimate of grain yield and the likely yield response to additional N fertilizer (Raun et al., 2002). Crop canopy reflectance in the red and NIR converted to NDVI is continuously obtained with a downward-pointing sensor mounted to a boom on a high-wheel applicator. An in-season estimate of yield (INSEY) in an unfertilized reference area is derived by dividing NDVI by the number of days from planting to sensing. The projected N uptake for potential yield with added N is determined from the INSEY times an N-response index computed as the ratio of the NDVI of an N-rich reference and NDVI from elsewhere in the field. The fertilizer N recommendation is then computed by subtracting the projected N uptake for the potential yield in the N limiting sensed area from INSEY, and dividing by a factor (between 0.6 and 0.7) representing the expected uptake efficiency to obtain the in-season top dress N rate (Shanahan et al., 2008).

Much of the N that is applied to corn in the U.S. Midwest is lost to the environment. Part of the problem can be traced to the failure of soil-based diagnostic tools to accurately determine plant N demand. In Missouri, Scharf and Lory (2009) found that the N rate needed to economically optimize corn yield varies widely from field to field and within fields. Spatial variability in the N supplying power of soil appeared to be a major determinant of crop response to applied N and thus the optimal N rate. Field experiments showed that corn colour could explain 53–77% of the variance in corn N demand compared to potential yield or conventional soil tests that only explained <25% of variance. Therefore, the use of active light sensors was proposed for detecting corn colour based on the rationale that N-sufficient corn is darker green and taller than N-deficient corn that is lighter green and shorter. As with the chlorophyll meter, the crop's reflectance from any place in a field is compared with the reflectance from a high-N reference strip. The greater the difference in reflectance between under-fertilized corn and the N-sufficient reference, the more N that is needed. Before crop canopy colour can reliably reflect soil N fertility, corn must be more than 30 cm tall (>V6 growth stage 3 weeks after emergence, Abendroth et al., 2011) to ensure adequate root development and absorption of plant nutrients within the soil root zone.

One problem with the N-rich strip concept is that excessive levels of N can increase the potential for sulphur (S) deficiency in corn resulting in a general yellowing of new leaves, artificially low reference values and low fertilizer N recommendations. In eastern Kansas, Holland and Schepers (2010) developed a 'turn-key' approach that eliminated the need for N-rich reference strips. Their 'virtual reference strip' method relies upon an optical crop sensor with sensitivity in the 'red-edge' (680–730 nm) waveband, which is highly dependent upon chlorophyll concentration (Horler et al., 1982). Operating from a high-clearance applicator, the boom-mounted sensor acquires crop reflectance values over a full range in plant vigour from a small area of a field. Somewhere within this area there is enough N to produce full yield such that the 95th percentile of a histogram of the sensor's crop reflectance values can serve as a reference value of N sufficiency. Accordingly, field areas are identified with adequately fertilized plants, which minimizes the potential for sulphur (S) imbalances caused by excess N availability.

2.4 Grain quality sensing

Crop canopy reflectance is related to plant N through the indirect effect of N fertility on crop biomass. This concept works well where plant available water is plentiful and N is limiting but not in semi-arid environments where dryland crop growth is largely determined by water. Other problems with optical sensing include too little biomass and cover in early season, absence of variation in crop colour and excessive wetness for wheeled applicators. Fertilizer issues include poor nutrient uptake efficiency due to lack of rainfall to move the N into the root zone, stranding of broadcast granular N at the surface in dry soil and N losses due to volatilization. Rainfall is not required for late season foliar applications of liquid N fertilizer but the potential for increasing grain protein concentration (GPC) of hard wheat is difficult to predict.

As an alternative to in-season crop sensing, rugged optical spectrometers with visible-NIR sensitivity are commercially available for installation on combine harvesters to measure and map the GPC in farm fields. Thousands of protein points can be mapped across fields and are used to identify areas of N nutrition adequacy based on the causal relationship between GPC and soil N fertility (Long et al., 2005). Introduced in 2001, the ProSpectra™ Grain Analyzer (Textron Systems) was the first reflectance spectrometer developed for operational use on a combine (von Rosenberg, et al., 2000). Field tests of this instrument on an operating combine (<5.0 g protein kg^{-1} grain) were comparable with a laboratory-grade spectrometer (Long et al., 2008). Other on-combine instruments over the past 10 years included the CropScan 2000H (NIR Technology Systems, Bankstown, NSW, Australia) and AccuHarvest Grain Analyzer (Zeltex, Inc., Gaithersburg, MA, USA) but are no longer sold and have been superseded by the newer CropScan 3000H On-Header Analyzer (Next Instruments, Condell Park, NSW, Australia).

Use of GPC as a diagnostic tool in evaluating whether N was sufficient or deficient for wheat yield once a critical protein level has been identified for a particular region was first proposed by Goos et al. (1982). Critical protein levels have been found to vary by wheat class, cultivar and region. Critical protein levels can be useful for evaluating the potential yield loss to N deficiency. For example, Glenn et al. (1985) found a 9.8% yield reduction in 'Stephens' soft white winter wheat for each grain protein percentage below the critical GPC. Selles and Zentner (2001) concluded that the critical GPC as an indicator of N deficiency works well when water is not limiting but is unreliable when grain yields are negatively impacted by water stress because high protein can also occur under drought. Long et al. (2017) determined that soft white spring wheat under severe water stress with protein below the critical level did not guarantee that yields were compromised by lack of N, but rather N was sufficient above this level.

Together with yield monitor data, site-specific protein data can be used to calculate the amount of N to apply to replace that exported from the field in the harvested crop of the current year. In turn, management zones can be easily constructed to guide N fertilization of the following crop and prevent over-application that stimulates vegetative growth before the onset of drought, lowers yield and elevates protein. A map of N removal (NR) can be easily computed from maps of grain yield and protein and the following equation: $NR = (GY \times GPC) \div 0.17$, where GY is grain yield in kg ha^{-1}, GPC is expressed as a fraction and 0.17 is the fractional part of protein that is N (Engel et al., 1999; Long et al., 2000). Fertilizing with such a map would involve maintaining plant available N for the following crop by replacing only what was removed by the crop in the previous season. If yield of the harvested crop was already limited by poor nutrition, as indicated by GPC below the critical protein level, the N in deficient areas of fields can be increased to raise protein to the critical GPC level with knowledge of the amount of fertilizer N equivalent (FNE) to a unit change in protein. Current estimates of FNE vary considerably and are reported to be 14–22 kg N ha^{-1} for HRS wheat (<3000 kg ha^{-1} yield) in northern Montana (Engel et al., 1999) and 59–82 kg N ha^{-1} for HRS wheat and 54–67 kg N ha^{-1} for SWS wheat in eastern Oregon (Long et al., 2017).

3 Regional perspectives

3.1 Case studies from sub-humid U.S. Midwest

Current cropland yields in the U.S. Midwest are near their maximum production potential largely because of good soils and mild weather but an abundance of plant available nutrients and water is also responsible for the high productivity. Extended periods of drought have become infrequent since 1965 (Andresen et al., 2012) and many states within the region are now experiencing dramatic increases in the number of extreme precipitation events (Frankson et al., 2017). Sufficient N, P and K are necessary to help crops achieve their relatively high yield potential. However, leaching and other loss mechanisms not only result in nutrient runoff to the environment but also prevent nutrients from reaching plants. Therefore, a major focus of site-specific nutrient management systems in the U.S. Corn Belt is on keeping nutrients in the soil and plant available to the plant.

3.1.1 Mid-season nitrogen application to winter wheat

Responsive in-season approaches to site-specific N management started in Oklahoma with the system developed by Raun et al. (2002) that coupled plant sensing with an N fertilizer optimization algorithm that predicted yield at mid-season and the potential N removed in grain. Solie et al. (2012) improved the algorithm by introducing a generalized model that adjusts the yield calibration curve for biomass accumulation that changes with growth stage of both corn and wheat. The algorithm was further developed by Bushong et al. (2016a) who altered the proxy for crop growth (i.e. growing degree days) to count only if soil moisture was adequate for growth. In another report, Bushong et al. (2016b) evaluated current plant-sensor algorithms for predicting a difference in grain yield potential with and without added N fertilizer and concluded that they could effectively predict final yield and agronomic optimum N rate of winter wheat. Despite this progress, producers have been slow to adopt optical plant sensing. Previously, Biermacher et al. (2009) showed that advantages of in-season application of UAN are outweighed by the low cost of pre-plant application of anhydrous ammonia such that both methods are equal in economic performance. Boyer et al. (2011) found no significant difference in net returns between optical reflectance-based N management and conventional management, which clearly indicated that growers would not adopt the technology.

Few reports have been found in the literature in which the Oklahoma system was shown to improve N use efficiency and reduce pollution of ground and surface waters. Recently, however, Cao et al. (2017) documented the potential benefits of responsive, in-season N management to food security, environmental pollution and climate change in China – a country where mismanagement of N fertilizer is widespread and N recovery efficiency is poor. Impacts of precision N management were investigated in a small plot experiment in a rotation of winter wheat and summer maize. Nitrogen was applied as urea to the following six treatments:

1. Zero N fertilizer as a control
2. Conventional grower practice of applying 150 kg N ha^{-1} before planting of both crops and 150 kg ha^{-1} at stem elongation of wheat and V10 stage of maize (5 weeks after emergence, Abendroth et al., 2011)
3. University recommendation of applying 60 kg ha^{-1} at planting and 120 kg ha^{-1} at stem elongation of wheat and V10 of maize
4. 60 kg ha^{-1} at planting and topdressing of 190 kg ha^{-1} at stem elongation of wheat and 127 kg ha^{-1} at V10 of maize based on visual examination of six ramp calibration plots
5. Soil test-based application of 60 kg ha^{-1} at planting and topdressing with allowance for soil NO_3-N content at rates of 148 kg ha^{-1} at stem elongation of wheat and 187 at V10 of maize
6. Application of 60 kg ha^{-1} at planting followed by sensor-based in-season N topdress application of 123 kg N ha^{-1} at stem elongation of wheat and 115 kg ha^{-1} at V10 of maize.

The sixth treatment associated with a plant sensing-based system was consistently better for both crops than any other treatment. Compared to conventional practices, applied N was reduced by <62% thus indicating increased N use efficiencies of <123%, decreased apparent total N losses <81% and lowered emissions of greenhouse gases and N losses of <68% with no significant loss in crop yield.

3.1.2 Responsive nitrogen management in corn

The conventional practice of applying the same rate of N fertilizer to whole fields ignores inherent within-field variability in the economically optimal N fertilizer rate (EONR), known as the point where the last increment of N returns a grain yield increase large enough to pay for that N (Sawyer et al., 2006). Lory and Scharf (2003) found that yield goal-based N recommendations exceeded EONR by 90 kg ha^{-1} on average. Citing evidence that N mineralization of soil organic matter varies widely from site-year to site-year, they postulated that variability in EONR was due to spatial variation in soil N supply rather than crop yield and N demand. Leaf colour, as measured with the SPAD chlorophyll meter, is a good predictor of EONR (Scharf et al., 2006). In Missouri, this finding is the basis for crop canopy sensor-based N management and the translation of active canopy sensor data into in-season N rate decisions (Kitchen et al., 2010).

Between 2004 and 2008, Scharf et al. (2011) compared the agronomic, economic and environmental performance of sensor-based variable-rate N and producer-based constant-rate N applications in 55 cornfields across central Missouri to give a wide range in production environments. Crop reflectance sensors were installed on the front of fertilizer application equipment and positioned directly over the cornrow at a height of 50 cm above the canopy. Each of two active light sensors: Crop Circle 210 (Holland Scientific, Lincoln, Nebraska, USA) and GreenSeeker (NTech Industries, Ukiah, California, USA) was used to direct variable-rate sidedress N applications to corn at V6 to V16 growth stages. Because of a diversity of equipment in use across the farms, methods of placement and forms of N fertilizer included dribbled or injected liquid UAN, injected anhydrous ammonia and broadcast urea.

Fixed N sidedress rate chosen by the farmer and variable-rate N sidedress rate were applied in linear treatment plots across each field. During fertilizer application, the sensor-based N rate varied as a function of visible/NIR reflectance values in the field divided by visible/NIR values from a high-N reference area. After harvest, the mean yields of the two placement techniques were compared. By reducing N use and increasing yield, profitability was increased by an average of \$42 ha^{-1} over all fields. Profit was greatest when the sensor-recommended N rate was more N or substantially less N than producer-chosen rates. Sensor use reduced the total amount of applied N by 8% and reduced surplus N, or N applied in excess of that removed in grain at harvest, by 24–26%. These results confirmed that responsive, in-season N placement performed better than conventional uniform methods used by producers to establish sidedress N rates.

3.2 Case studies from semi-arid U.S. West

Plant available water is the most important factor that limits dryland crop yields in semi-arid environments. If the water supply is insufficient, plants will be incapable of utilizing all of the N and other nutrients that are available for growth and development. Therefore, site-specific nutrient management systems in the semi-arid west have largely focused on characterizing spatial patterns in fields that are associated with water-related attributes. Specialized variable-rate equipment and sampling costs are an important consideration as gross returns are often <\$200 ha^{-1} because of the relatively low yields.

3.2.1 Terrain-based nitrogen management in hard red spring wheat

In glacial till landscapes of the northern Great Plains, slope position effects on soil water-holding capacity may occur when tillage erosion causes soil loss and reduced rooting depth in upper slopes and soil deposition and increased rooting depth in lower slopes (Moulin et al., 1994). Dividing hill slopes into upper, middle and lower slopes has been used as a method to construct management zones in this region (Beckie et al., 1997; Nolan et al., 1995; Pennock and Corre, 2001). In northern Montana, Long et al. (2014) compared the hard red spring wheat yields and economic returns of variable-rate and uniform-rate N management in eight farm fields between 1994 and 2004. Digital elevation models and terrain analysis were used to divide fields into upper, south-facing middle slopes, north-facing middle slopes and lower slopes. Nitrogen recommendations for management zones were based on soil N testing and expected yields. Nitrogen fertilizer was applied as a replicated series of incremental N rates in treatment strips across each field. Fertilizer was either broadcast from a truck-mounted applicator or banded below the seed with an air seeder. Strips were cut with a production-size combine equipped with a mass flow yield monitor. At the same time, grain was hand sampled from the combine's grain bin filling auger in 20–30 s intervals during travel within each strip. These samples were later analysed for GPC in the laboratory and used to establish the dollar value of the wheat.

Mean grain protein and yield were similar between variable-rate and uniform-rate N management. Yield differences were <223 kg ha^{-1} and averaged only 18 kg ha^{-1}. Grain yield did not differ significantly among N rates within management zones. In seven of the eight sites, net returns from variable-rate N management were up to U.S.$27.97 ha^{-1} less than from uniform N management and were not profitable if Environmental Quality Incentive Program payments of U.S.$6.36 ha^{-1} were considered as part of net income. Little evidence existed that variable-rate N management based on terrain-based management zones and predetermined N recommendations improves grain yields and profits or reduces N use in water-limited, summer fallow systems of northern Montana. The reasons were traced to modest yield gains and N savings, nominal response to applied N, and inability to define management zones and optimal N rates.

3.2.2 Soil depth-based nitrogen management in soft white winter wheat

In the Pacific Northwest, soft white winter wheat is grown as an export crop to the Asian Pacific market that prefers grain with low protein concentration (<105 g kg^{-1}) for production of noodles and pastries. With help from USDA scientists, Herb March of Couse Creek Ranches (Milton-Freewater, Oregon, USA) implemented precision N practices as a means of improving the protein levels in a winter wheat–chemical fallow production system. The 2439 ha farm is situated on the foothills of the Blue Mountains with elevation rising from 334 to 973 m over 5 km. Precipitation also increases from 350 to 650 mm over this distance. Soils are derived from loess deposited over basalt bedrock (Fig. 4). The terrain is characterized by a ridge and valley pattern. Prevailing wind is from the southwest such that depth to bedrock is <60 cm for Gwin soils on windward slopes and >150 cm for Palouse soils on leeward slopes. As shallow rooted crops are more vulnerable to leaching than deeper rooted crops, the nitrate leaching potential under high rainfall is higher in

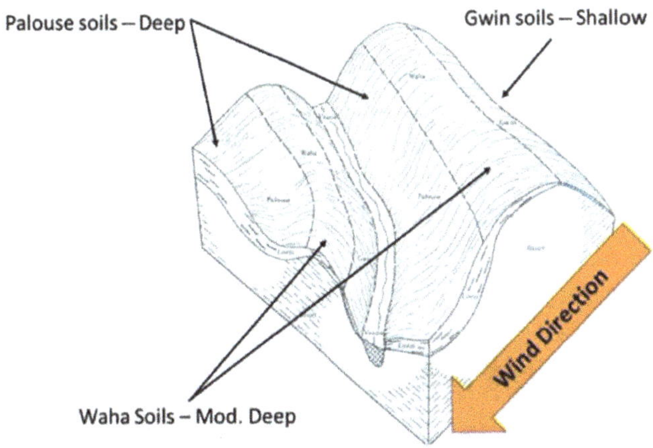

Figure 4 Ridge and valley pattern of soil map units on Herb March farm near Milton-Freewater, Oregon, USA. Block diagram adapted from USDA-SCS Soil Survey of Umatilla County Area, Oregon.

the shallow soils where soluble N cannot be readily taken up by the crop. Leaching is also important for environmental reasons as the farm is situated next to the Walla Walla River – a source of drinking water for Milton-Freewater.

Historically, a single, uniform-rate application to an entire field would give little yield on the shallow soils. Herb March would compensate by fertilizing with N at 34 kg ha^{-1} on shallow soil, 56 kg ha^{-1} on moderately deep soil and 78 kg ha^{-1} on deep soil. If it was a wet year, he would apply 34 kg N ha^{-1} to a whole field, which would give this much plus residual soil N on deep soil and compensate for what leached out of the shallow soil but would only yield about 34 kg of grain ha^{-1}. In 2010, he started using mid-season aerial NIR imagery and soil EC maps to better define the boundaries of different soils. Four management zones were developed to receive 40, 47, 72 and 112 kg N ha^{-1} in order of shallow, moderately shallow, moderately deep and deep soils. A custom-made, 9.7 m wide, hoe-type, no-till drill applied seed and liquid fertilizer in one pass. Fertilizer is applied as a mixture of urea ammonium nitrate and thiosulfate (28-00-00-5), and can be varied in 9.4 L ha^{-1} increments between 94 and 375 L ha^{-1}. Ammonium phosphate (10-34-00) and potassium chloride (00-00-60) in liquid form were also applied at constant rates of 19 and 28 L ha^{-1}.

Knowing that GPC is a useful post-harvest indicator of nitrogen nutrition sufficiency in wheat (Engel et al., 1999), the farmer equipped each of his three combine harvesters with an AgLeader mass flow yield monitor and DSquared ProSpectra Grain Quality Analyzer (Fig. 5). Resulting yield and protein maps provided information for monitoring the effectiveness of his site-specific N management system. Yield loss was avoided when the N application was reduced on shallow ground and yield was increased by putting more N on deeper ground that soil tests indicated needed more N. Varying the N rate by soil depth improved the uniformity of protein levels from field to field despite a wide range in precipitation across the farm (Table 1). Protein levels of around 110 g kg^{-1} informed Herb March that his system of N management had fulfilled its purpose. He also felt that the savings in fertilizer and increase in yield more than offset the added costs of the technology in the first year.

Figure 5 ProSpectra Grain Quality Analyzer mounted to the grain bin filling auger of Herb March's John Deere 6622 combine harvester.

Table 1 Yield and protein concentration in 18 wheat fields varying in area, elevation and rainfall across the Herb March farm near Milton-Freewater, Oregon, USA

Field	Area (ha)	Yield (kg ha^{-1})	Protein (g kg^{-1})	Elevation (m)	Rainfall (mm)
North Spence	61.7	3205	114	365	375
Bowman	41.0	5026	97	395	375
NE Garriott	80.4	4616	107	426	400
S Kelly Winter Wht	62.8	4522	108	441	400
S Kinman	74.1	4482	105	456	400
W Shumway	19.2	4824	101	456	400
S Kelly Spring Wht	62.8	1277	131	456	400
Roburn	7.2	4354	107	471	400
240	49.8	5073	107	486	375
Lawrence	232.3	4401	114	502	425
Home	67.5	4502	115	502	450
400	74.6	5254	104	532	450
Troyer (Split)	28.0	5792	102	532	450
LK (Split)	76.5	4925	110	532	475
Old Ranch	62.3	4166	112	608	475
Storey-Waldner	39.7	5066	112	638	475
West Jon	50.2	4636	108	684	525
U.K./West Wiljon	91.4	5207	111	699	525

4 Conclusions and future trends

Different site-specific nutrient management systems have developed since the mid-1980s when GPS, variable-rate equipment and other enabling technologies of precision agriculture were first introduced to farmers. In general, these systems can be divided into approaches that are either map-based or sensor-based. Map-based approaches are informed by processes that occur either in the previous season or start of the current season for example grid soil sampling or zone soil sampling. For the latter, management zones are established based on information that is gathered on certain biophysical factors that determine crop yield or simply crop yield itself. Sensor-based approaches involve optically sensing plant N sufficiency and controlling fertilizer applications in the same operation during the growing season and do not necessarily require GPS guidance.

A review of empirical studies on variable-rate fertilizer application to corn in the U.S. Midwest showed either higher or lower profitability such that economic feasibility still remains an open question (Schimmelpfennig, 2016). Economic studies in the semi-arid U.S. West are limited in number with results that are either negative (Carr et al., 1991; Long et al., 2014) or positive (Fiez et al., 1994; Beckie et al., 1997). Bongiovanni and Lowenberg-Deboer (2004) showed that site-specific N management can maintain profitability even when N is restricted to less than half the recommended uniform rate. In addition, their review of 14 studies between 1994 and 2003 revealed that variable-rate N application improved N use efficiency and reduced N leaching losses in most cases. However, many of these studies only documented reduced fertilizer use, which led to the conclusion that field studies are needed that directly measure NO_3-N losses to confirm environmental impacts of precision agriculture.

Changes in precipitation induced by climate change are projected to substantially increase N loading and threaten water resources with eutrophication across the globe (Sinha et al., 2017). The impacts are expected to be especially strong in the U.S. Corn Belt, India, China and Southeast Asia. For the U.S. Corn Belt, a 33% reduction in N inputs would be required to offset the 19% N load increase attributable to increases in precipitation that have been observed to be occurring. Site-specific nutrient management promises to help manage eutrophication and associated water quality problems but it must be capable of accounting for the impact of changing precipitation patterns on nutrient loading.

The optimization of crop production and nutrient use efficiency is a complex problem that requires a complex set of solutions to achieve improvement (Hawkesford, 2014). For example, application of N is spatially and temporally maximized when used in conjunction with germplasm optimized for traits relating to N use efficiency rather than yield alone. Other examples include (1) an increase in soluble P runoff entering Lake Erie that is an unintended consequence of no-till farming with surface applications of P fertilizer (Jarvie et al., 2017) and (2) accelerated soil acidification in the Great Plains resulting from high rates of ammonia-based fertilizers in combination with increased use of no-till (Obour et al., 2017). In these latter two examples, no-till alone may not sufficiently improve either water or soil quality without innovative cultural practices including site-specific nutrient management to avoid applying fertilizer at rates greater than expected benefits to the crop.

The traditional mass-balance approach of establishing a yield goal and applying N for expected yield with allowance for residual soil N at or near planting has proven unreliable (Raun et al., 2017). Inaccurate N recommendations for corn are the result of N losses from soil resulting from leaching and denitrification (Morris et al., 2018). In addition, the mass-balance approach fails to account for the N mineralized, which varies within the landscape with changes in soil organic carbon content (Dharmakeerthi et al., 2005). The Maximum Return to Nitrogen (MTRN) recommendation approach (available online at http://cnrc.agron.iastate.edu/, verified 24 March 2018) was developed for corn to account for N mineralization by means of updated N-response datasets derived from an extensive network of ongoing field trials (Sawyer et al., 2006). Though useful, the MTRN approach does not address year-to-year variability in temperature or rainfall (Shanahan et al., 2008). Specialized soil test procedures are available to approximate soil N supply (Bundy and Andraski, 1995; Magdoff et al., 1984; Mulvaney et al., 2001) but would be cost-prohibitive for site-specific nutrient management (Ma and Dwyer, 1999). Future N recommendation methods are needed that reliably estimate the probability of an accurate N recommendation at the field or subfield level (Morris et al., 2018).

Under semi-arid conditions, active optical sensor-based approaches are unreliable because relationships between crop spectral reflectance and crop N status are confounded by crop variation that is largely determined by plant available water. In addition, the spectral signal measured by conventional active optical sensors will be dominated by soil spectral reflectance due to limited crop cover. Combined spectral indices that isolate the vegetation signal from the integrated reflectance within the field of view of the sensor have been found to outperform single indices such as NDVI (Eitel et al., 2009) and should be incorporated into current systems. Furthermore, green scanning LiDAR (light detection and ranging) instruments emit narrowband point illumination and promise to alleviate errors as green (532 nm) point return intensity data can be used to directly quantify crop volume and derive N concentration from single leaves (Eitel et al., 2011, 2014). Mobile, multi-wavelength LiDAR sensors may be well-suited for dryland environments and represent a possible new generation of instruments for active optical crop sensing.

Yield monitors on combine harvesters make it possible to quantify spatial variability in yield across fields but do not reveal the sources of the variability. Since it is now possible to map grain protein levels in fields at the same spatial resolution as yield, studying the local relationship between yield and protein offers a way to better interpret yield maps. Furthermore, grain yield and protein maps can reveal the location and extent of areas within fields where more (or less) N was needed for yield more effectively than soil N maps derived from a limited sample number of soil test data. Adoption of this new technology will require industry support and scientific explanation. Commercial service providers will be needed to calibrate and maintain the on-combine spectroscopic instruments. Growers and their advisers will need specialized software that will process and remove errors from yield and protein data, and turn on-combine readings into task control maps for variable-rate application.

Process-oriented models offer great promise for prediction of plant available N and plant N demand under varying weather conditions. Van Alphen (2002) used a dynamic simulation model and real-time weather data to monitor soil N mineralization across a field and time the application of N fertilizer to wheat. Melkonian et al. (2008) developed the cloud-based computer simulation model, Adapt-N, that uses soil information and weather records to predict crop N uptake and optimum N application rates over time. In 2017, developers of Adapt-N were awarded the $1M Tulane University Nitrogen Reduction Challenge as being the most feasible and effective solution to reducing N and increasing use efficiency (http://www2.tulane.edu/tulaneprize/waterprize/index.cfm). It has been licensed to Yara International ASA – a large fertilizer company that manufactures a tractor-mounted, optical on-the-go system for in-season N management.

Though process-based models provide for more dynamic and locally adaptive N rate recommendations (Morris et al., 2018), input data may contain inaccuracies that can substantially affect N fertilizer predictions (Sela et al., 2017). Furthermore, when commercialized, these tools will cost farmers but will be lower than variable-rate methods that involve soil/tissue testing and sensor-based approaches (Morris et al., 2018). Similar to MTRN, Morris et al. (2018) proposed a framework for improving N fertilizer rate recommendations for corn that involves creation of a database with results from numerous field-scale strip trials evaluating N response with associated metadata about field history and N management practices. They advise that such a system must be practical and inexpensive so that farmers can use it on a routine basis.

In addition, Bayesian causal-learning approaches have been developed to deal with the large uncertainty facing environmental decision making. Utilizing a long-term database on crop N response that considers previous season's information, a spatio-temporal Bayesian updating method has achieved optimal efficiency under uncertainties of weather and markets when applied to variable-rate N application on dryland wheat in Montana (Lawrence et al., 2015). More recently, the method of Bayesian updating was applied to combining in-season plant N sensing and prior information about the response to N for a given field (McFadden et al., 2018). These Bayesian approaches may provide a tractable way to improve risk prediction and decision support that are needed in site-specific nutrient management.

Variable-rate control maps utilize polygons to represent areas such as the boundaries of management zones yet crops and soils vary continuously within fields. Managing nutrients in crop production systems when biophysical properties and processes change incrementally at both short and long distances suggests the need for representing soil and landscape variation as a continuum (N. Kitchen, pers. comm.). Since many soil properties that impact crop growth can vary at short distances, the most ideal representation of spatial patterning is through a continuum map showing gradual changes. Tools and information management systems are needed to represent the soil this way along with the management decision tools and electronic variable-rate controllers that can accept continuous data.

Geophysical sensing, involving the use of magnetometers, electrical resistance meters, ground-penetrating radar and electromagnetic conductivity meters is capable of representing soil variability as a continuum. Due to high-density surveys of large areas, mobile soil sensing can also improve overall accuracy of the measurement process, even if point accuracy is less than laboratory analysis (Sudduth et al., 1997). Various approaches for sensor data fusion of key physical and chemical soil properties are available with commercial systems (Adamchuk et al., 2004). On-combine multi-sensor systems have been described for simultaneous measurement of grain yield, GPC and straw yield that can be combined to identify areas of environmental stress in fields (Long and McCallum, 2013). The fusion of multiple sensors promises to form an integral component of precision agriculture but efficient data analysis methods are needed to handle the sheer volume of data that is generated (Mahmood et al., 2012).

A significant yet overlooked source of variability within agricultural fields arises from past land use and soil fertilizer management. In the U.S. Midwest, farm fields over time expanded in size and increased adoption of machinery that enabled farmers to do more work. Certain farmstead areas were also vacated and put into crop production. Areas where animals were confined had higher amounts of manure applied. Today, those historical field boundaries and farmsteads are reflected in grid-sampled soil fertility data. Abrupt changes in soil fertility are aligned with boundaries of these historic fields and plant available P and K are highly variable on these fields (Lerch et al., 2005). Past cultural practices and management leave residual effects and impact soil physio-chemical properties for many years. This soil nutrient variability makes it difficult to develop nutrient management plans based on randomly selected samples that would be combined for single analysis, or even if a limited number of grid soil samples were collected and analysed separately. Specialized mapping methods are needed for finding and accommodating these anomalous areas.

5 Where to look for further information

A series of Site-Specific Management Guidelines is available online at http://www.ipni.net/ssmg (International Plant Nutrition Institute) where readers can find information on many aspects of site-specific nutrient management. The 4-element 'right product, right rate, right place and right time' strategy is thoroughly described in *4R Plant Nutrition: A manual for improving the management of plant nutrition* published by the International Plant Nutrition Institute. Informative book chapters have been published on site-specific nutrient management that were written in the past by prominent individuals in the field. These include Long and Pierce (2010), Raun and Schepers (2008), Haneklaus and Schnug (2006) and Pierce and Nowak (1999). See References for full details.

Centres of expertise include:

- Nitrogen Use Efficiency Web (http://nue.okstate.edu/).
- University of Missouri, Plant Sciences Extension – Nutrient management (https://plantsciences.missouri.edu/nutrientmanagement/).
- USDA Agricultural Research Service, Nutrient use and outcome network (NUOnet) (https://www.ars.usda.gov/anrds/nuonet/nuonet-home/).

6 References

Abendroth, L. J., Elmore, R. W., Boyer, M. J. and Marley, S. K. (2011), 'Corn growth and development', PMR 1009, Iowa State University Extension, Ames, IA.

Adamchuk, V., Hummel, J., Morgan, M. and Upadhyaya, S. (2004), 'On-the-go soil sensors for precision agriculture', *Computers and Electronics in Agriculture*, 44, 71–91.

Anderson-Cook, C., Alley, M., Roygard, J., Khosla, R., Noble, R. and Doolittle, J. (2002), 'Differentiating soil types using electromagnetic conductivity and crop yield maps', *Soil Science Society of America Journal*, 66, 1562–70.

Andresen, J., Hilberg, S. and Kunkel, K. (2012), 'Historical climate and climate trends in the Midwestern USA', U.S. National Climate Assessment Midwest Technical Report, Available from the Great Lakes Integrated Sciences and Assessments Center, Available online at: http://glisa.msu.edu/media/files/NCA/MTIT_Historical.pdf (last accessed 28 May 2018).

Baker, N. and Capel, P. (2011), 'Environmental factors that influence the location of crop agriculture in the conterminous United States', Scientific Investigations Report 2011-5108, U.S. Geological Survey.

Beckie, H. J., Moulin, A. P. and Pennock, D. J. (1997), 'Strategies for variable rate N fertilization in hummocky terrain', *Canadian Journal of Soil Science*, 77, 589–95.

Published by Burleigh Dodds Science Publishing, 2019.

Bermudez, M. and Mallarino, A. (2007), 'Impacts of variable-rate phosphorus fertilization based on dense grid soil sampling on soil-test phosphorus and grain yield of corn and soybean', *Agronomy Journal*, 99, 822–32.

Biermacher, J., Epplin, F., Brorsen, B., Solie, J. and Raun, W. (2009), 'Economic feasibility of site-specific optical sensing for managing nitrogen fertilizer for growing wheat', *Precision Agriculture*, 10, 213–30.

Blackmer, T., Schepers, J., Varvel, G. and Meyer, G. (1996), 'Analysis of aerial photography for nitrogen stress within corn fields', *Agronomy Journal*, 88, 729–33.

Blackmore, S. (2000), 'The interpretation of trends from multiple yield maps', *Computers and Electronics in Agriculture*, 26, 37–51.

Bongiovanni, R. and Lowenberg-Deboer, J. (2004), 'Precision agriculture and sustainability', *Precision Agriculture*, 5, 359–87.

Boydell, B. and McBratney, A. (2002), 'Identifying potential within-field management zones from cotton-yield estimates', *Precision Agriculture*, 3, 9–23.

Boyer, C., Brorsen, B., Solie, J. and Raun, W. (2011), 'Profitability of variable rate nitrogen application in wheat production', *Precision Agriculture*, 12, 473–87.

Bundy, L. and Andraski, T. (1995), 'Soil yield potential effects on performance of soil nitrate tests', *Journal of Production Agriculture*, 8, 561–8.

Bushong, J., Mullock, J., Miller, E., Raun, W. and Arnall, D. (2016a), 'Evaluation of mid-season sensor based nitrogen fertilizer recommendations for winter wheat using different estimates of yield potential', *Precision Agriculture*, 17, 470–87.

Bushong, J., Mullock, J., Miller, E., Raun, W., Klatt, A. and Arnall, D. (2016b), 'Development of an in-season estimate of yield potential utilizing optical crop sensors and soil moisture data for wheat', *Precision Agriculture*, 17, 451–69.

Cao, Q., Miao, Y., Feng, G., Gao, X., Liu, B., Li, F., Khosla, R., Mulla, D. J. and Zhang, F. (2017), 'Improving nitrogen use efficiency with minimal environmental risks using an active canopy sensor in a wheat-maize cropping system', *Field Crops Research*, 214, 365–72.

Carr, P., Carlson, G., Jacobsen, J., Nielsen, G. and Scogley, E. (1991), 'Farming soils, not fields: A strategy for increasing fertilizer profitability', *Journal of Production Agriculture*, 4, 57–61.

Castle, E., Becker, M. and Smith, F. (1972), *Farm Business Management*, Collier-Macmillan, London, UK.

Chang, J., Clay, D., Carlson, C., Clay, S., Malo, D., Berg, R., Kleinjan, J. and Wiebold, W. (2003), 'Different techniques to identify management zones impact nitrogen and phosphorus sampling variability', *Agronomy Journal*, 95, 1550–9.

Chang, J., Clay, D., Carlson, C., Reese, C., Clay, S. and Ellsbury, M. (2004), 'Defining yield goals and management zones to minimize yield and nitrogen and phosphorus fertilizer recommendation errors', *Agronomy Journal*, 96, 825–31.

Corwin, D. and Lesch, S. (2003), 'Application of soil electrical conductivity to precision agriculture', *Agronomy Journal*, 95, 455–71.

Curran, P. (1989), 'Remote sensing of foliar chemistry', *Remote Sensing of Environment*, 30, 271–8.

Delgado, J. and Follett, R. (2010), *Advances in Nitrogen Management for Water Quality*, Soil Water Conservation Society of America, Ankeny, IA.

Deming, W. (1950), *Elementary Principles of the Statistical Control of Quality: A Series of Lectures*, Nippon Kagaku Gijutsu Remmei, Tokyo, Japan.

Dharmakeerthi, R., Kay, B. and Beauchamp, E. (2005), 'Factors contributing to changes in plant available nitrogen across a variable landscape', *Soil Science Society of America Journal*, 69, 453–62.

Dhital, S. and Raun, W. (2016), 'Variability in optimum nitrogen rates for maize', *Agronomy Journal*, 108, 2165–73.

Doerge, T. (2000), 'Management zone concepts. Site-specific management guidelines', International Plant Nutrition Institute, Available online at: http://www.ipni.net/publication/ssmg.nsf/0/C0D052F04A53E0BF852579E500761AE3/$FILE/SSMG-02.pdf (last accessed 27 March 2018).

Eitel, J., Long, D., Gessler, P., Hunt, E. and Brown, D. (2009), 'Sensitivity of ground-based remote sensing estimates of wheat chlorophyll content to variation in soil reflectance', *Soil Science Society of America Journal*, 73, 1715–23.

Eitel, J., Vierling, L., Long, D. and Hunt, E. (2011), 'Early season remote sensing of wheat nitrogen status using a green scanning laser', *Agricultural and Forest Meteorology*, 151, 1338–45.

Eitel, J., Magney, T., Vierling, L., Brown, T. and Huggins, D. (2014), 'LiDAR based biomass and crop nitrogen estimates for rapid, non-destructive assessment of wheat nitrogen status', *Field Crops Research*, 159, 21–32.

Engel, R., Long, D., Carlson, G. and Meier, C. (1999), 'Method for precision nitrogen management in spring wheat: I. Fundamental relationships', *Precision Agriculture*, 1, 327–38.

Farid, H., Bakhsh, A., Ahmad, N., Ahmad, A. and Mahmood-Khan, Z. (2015), 'Delineating site-specific management zones for precision agriculture', *Journal of Agricultural Science*, 154, 273–86.

Fiez, T., Miller, B. and Pan, W. (1994), 'Assessment of spatially variable nitrogen fertilizer management in winter wheat', *Journal of Production Agriculture*, 7, 86–93.

Fleming, K., Heerman, D. and Westfall, D. (2004), 'Evaluating soil color with farmer input and apparent soil electrical conductivity for management zone delineation', *Agronomy Journal*, 96, 1581–7.

Flowers, M., Weisz, R. and White, J. (2005), 'Yield-based management zones and grid sampling strategies: Describing soil test and nutrient variability', *Agronomy Journal*, 97, 968–82.

Frankson, R., Kunkel, K., Champion, S. and Stewart, B. (2017), 'Missouri state summary', NOAA Technical Report NESDIS 149-MO, Available online at: https://statesummaries.ncics.org/mo (last accessed 27 March 2018).

Franzen, D. W., Cihacek, L. J., Hofman, V. L. and Swenson, L. J. (1998), 'Topography-based sampling compared with grid sampling in the Northern Great Plains', *Journal of Production Agriculture*, 11, 364–70.

Franzen, D., Hopkins, D., Sweeney, M., Ulmer, M. and Halvorson, A. (2002), 'Evaluation of soil survey scale for zone development of site-specific nitrogen management', *Agronomy Journal*, 94, 381–9.

Fridgen, J.J. (2000), 'Development and evaluation of unsupervised clustering software for sub-field delineation of agricultural fields', M.S. thesis, University of Missouri, Columbia, MO.

Fridgen, J., Kitchen, N., Sudduth, K., Drummond, S., Wiebold, W. and Fraisse, C. (2004), 'Management zone analyst (MZA): Software for subfield management zone delineation', *Agronomy Journal*, 96, 100–8.

Glenn, D. M., Carey, A., Bolton, F. E. and Vavra, M. (1985), 'Effect of N fertilizer on protein content of grain, straw, and chaff tissues in soft white winter wheat', *Agronomy Journal*, 77, 229–32.

Goos, R. J., Westfall, D. G., Ludwick, A. E. and Goris, J. E. (1982), 'Grain protein as an indicator of N sufficiency for winter wheat', *Agronomy Journal*, 74, 130–3.

Haneklaus, S. and Schnug, E. (2006), 'Site-specific nutrient management: Objectives, current status, and future research needs', in Srinivasan, A. (Ed.), *Handbook of Precision Agriculture: Principles and Applications*, The Haworth Press, Inc., Binghamton, NY.

Hawkesford, M. (2014), 'Reducing the reliance on nitrogen fertilizer for wheat production', *Journal of Cereal Science*, 59, 276–83.

Holland, K. and Schepers, J. (2010), 'Derivation of a variable rate nitrogen application model for in-season fertilization of corn', *Agronomy Journal*, 102, 1415–24.

Horler, D., Dockray, M. and Barber, J. (1982), 'The red edge of plant leaf reflectance', *International Journal of Remote Sensing*, 4, 273–88.

Hornung, A., Khosla, R., Reich, R., Inman, D. and Westfall, D. (2006), 'Comparison of site-specific management zones: Soil-color-based and yield-based', *Agronomy Journal*, 98, 407–15.

Jarvie, H., Johnson, L., Sharpley, A., Smith, D., Baker, D., Bruulsema, T. and Confesor, R. (2017), 'Increased soluble phosphorus loads to Lake Erie: Unintended consequences of conservation practices?' *Journal of Environmental Quality*, 46, 123–32.

Johnson, C., Mortensen, D., Wienhold, B., Shanahan, J. and Doran, J. (2003), 'Site-specific management zones based on soil electrical conductivity in a semiarid cropping system', *Agronomy Journal*, 95, 303–15.

Johnston, A. and Bruulsema, T. (2014), '4R nutrient stewardship for improved nutrient use efficiency', *Procedia Engineering*, 83, 365–70.

King, J., Dampney, P., Lark, R., Wheeler, H., Bradley, R. and Mayr, T. (2005), 'Mapping potential crop management zones within fields: Use of yield-map series and patterns of soil physical properties identified by electromagnetic induction sensing', *Precision Agriculture*, 6, 167–81.

Kitchen, N., Sudduth, K. and Drummond, S. (1999), 'Soil electrical conductivity as a crop productivity measure for claypan soils', *Journal of Production Agriculture*, 12, 607–17.

Kitchen, N., Drummond, S., Sudduth, K. and Buchleiter, G. (2003), 'Soil electrical conductivity and topography related to yield for three contrasting soil-crop systems', *Agronomy Journal*, 95, 483–95.

Kitchen, N., Sudduth, K., Myers, D., Drummond, S. and Hong, S. (2005), 'Delineating productivity zones on claypan soils using apparent soil electrical conductivity', *Computers and Electronics in Agriculture*, 46, 285–308.

Kitchen, N., Sudduth, K., Drummond, S. and Vories, E. (2010), 'Ground-based canopy reflectance sensing for variable-rate nitrogen corn fertilization', *Agronomy Journal*, 102, 71–84.

Kozar, B., Lawrence, R. and Long, D. (2002), 'Soil phosphorus and potassium mapping using a spatial correlation model incorporating terrain slope gradient', *Precision Agriculture*, 3, 407–17.

Kramer, A. Doane, T., Horwath, W. and van Kessel, C. (2002), 'Short-term nitrogen-15 recovery vs. long-term total soil N gains in conventional and alternative cropping systems', *Soil Biology and Biochemistry*, 34, 43–50.

Lawrence, P., Rew, L. and Maxwell, B. (2015), 'A probabilistic Bayesian framework for progressively updating site-specific recommendations', *Precision Agriculture*, 16, 275–96.

Lerch, R., Kitchen, N., Kremer, R., Donald, W., Alberts, E., Sadler, J., Sudduth, K. A., Myers, D. B. and Ghidey, F. (2005), 'Development of a conservation-oriented precision agriculture system: Water and soil quality assessment', *Journal of Soil and Water Conservation*, 60, 411–21.

Liggins, M., Hall, D. and Llinas, J. (2001), *Handbook of Multisensor Data Fusion*, CRC Press, Boca Raton, FL.

Long, D. and McCallum, J. (2013), 'Mapping straw yield using on-combine light detection and ranging (lidar)', *International Journal of Remote Sensing*, 34, 6121–34.

Long, D. and Pierce, F. (2010), 'Precision farming for nitrogen management', in J. A. Delgado and R. F. Follett (Eds), *Advances in Nitrogen Management for Water Quality*, Soil Water Conservation Society of America, Ankeny, IA.

Long, D., Carlson, G. and DeGloria, S. (1995), 'Quality of field management maps', in P. C. Robert, R. H. Rust and W. E. Larson (Eds), *Proceedings of 2nd International Conference on Site-specific Management for Agricultural Systems*, ASA-CSSA-SSSA, Madison, WI, pp. 251–72.

Long, D., Engel, R. and Carlson, G. (2000) 'Method for precision nitrogen management in spring wheat: II. Implementation', *Precision Agriculture*, 2, 25–38.

Long, D., Engel, R. and Carpenter, F. (2005) 'On-combine sensing and mapping of wheat protein concentration', *Crop Management*, 4(1), Available online at https://pubag.nal.usda.gov/pubag/downloadPDF.xhtml?id=11894&content=PDF (last accessed 28 May 2018).

Long, D., Engel, R. and Siemens, M. (2008), 'Measuring grain protein concentration with in-line near infrared reflectance spectroscopy', *Agronomy Journal*, 100, 247–52.

Long, D., Whitmus, J., Engel, R. and Brester, G. (2014), 'Net returns from terrain-based variable rate nitrogen management on dryland spring wheat in northern Montana', *Agronomy Journal*, 107, 1055–67.

Long, D., McCallum, J. D., Reardon, C. L. and Engel, R. E. (2017), 'Nitrogen requirement to change protein concentration of spring wheat in Pacific Northwest', *Agronomy Journal*, 109, 675–83.

Lory, J. A. and Scharf, P. C. (2003), 'Yield goal versus delta yield for predicting fertilizer nitrogen need in corn', *Agronomy Journal*, 95, 994–9.

Lowrance, C., Fountas, S., Laikos, V. and Vellidis, G. (2016), 'EZZone – An online tool for delineating management zones', *Proceedings of 13th International Conference on Precision Agriculture*, Available online at: http://vellidis.org/wp-content/uploads/2016/07/Lowrance-Paper1931-ICPA-2016.pdf (last accessed 28 May 2018).

Ma, B. and Dwyer, L. (1999), 'Within plot variability in available soil mineral N in relation to leaf greenness and yield', *Communications in Soil Science and Plant Analysis*, 30, 1919–28.

Magdoff, F., Ross, D. and Amadon, J. (1984), 'A soil test for nitrogen availability to corn', *Soil of America Journal*, 48, 1301–4.

Mahmood, H., Hoogmoed, W. and van Henten, E. (2012), 'Sensor data fusion to predict multiple soil properties', *Precision Agriculture*, 13, 628–45.

Mallarino, A. and Wittry, D. (2001), 'Management zones soil sampling: A better alternative to grid and soil type sampling?' in *Proceedings of 13th Annual Integrated Crop Management Conference*, Iowa State University Extension, Ames, IA, pp. 159–64, Available online at http://www.agronext.iastate.edu/soilfertility/info/ICM_2001_ZoneSampling_Publ.pdf (last accessed 28 May 2018).

Mausbach, M. and Wilding, L. (1991), *Spatial Variability of Soils and Landforms*, Soil Science Society of America Spec. Pub. 28, Soil Science Society of America, Madison, WI.

McCann, B. L., Pennick, D. J., van Kessel, C. and Walley, F. L. (1996), 'The development of management units for site-specific farming', in P. C. Robert, R. H. Rust and W. E. Larson (Eds), *Proceedings of 3rd International Conference on Precision Agriculture*, ASA-CSSA-SSSA, Madison, WI, pp. 295–302.

McFadden, B., Brorsen, B. and Raun, W. (2018), 'Nitrogen fertilizer recommendations based on plant sensing and Bayesian updating', *Precision Agriculture*, 19, 79–92.

Melkonian, J., van Es, H., DeGaetano, A. and Joseph, L. (2008), 'ADAPT-N: Adaptive nitrogen management for maize using high resolution climate data and model simulations', in R. Khosla (Ed.), *Proceedings of the 9th International Conference on Precision Agriculture*, International Society of Precision Agriculture, Monticello, IL, Available online at: https://cpb-us-e1.wpmucdn.com/blogs.cornell.edu/dist/8/6785/files/2016/06/Prec-Ag-Conf-2008-Melkonian-van-Es-uhaslu.pdf (last accessed 28 May 2018).

Milfred, C. and Kiefer, R. (1976), 'Analysis of soil variability with repetitive aerial photography', *Soil Science Society of America Journal*, 40, 553–7.

Milne, A., Webster, R., Ginsburg, D. and Kindred, D. (2012), 'Spatial multivariate classification of an arable field into compact management zones based on past crop yields', *Computers and Electronics in Agriculture*, 80, 17–30.

Morris, T., Murrell, T., Beegle, D., Camberato, J., Ferguson, R., Grove, J., Ketterings, Q., Kyveryga, P. M., Laboski, C. A. M., McGrath, J. M., Meisinger, J. J., Melkonian, J., Moebius-Clune, B. N., Nafziger, E. D., Osmond, D., Sawyer, J. E., Scharf, P. C., Smith, W., Spargo, J. T., van Es, H. M. and Yang, H. (2018), 'Strengths and limitations of nitrogen rate recommendations for corn and opportunities for improvement', *Agronomy Journal*, 110, 1–37.

Moulin, A. P., Anderson, D. W. and Mellinger, M. (1994), 'Spatial variability of wheat yield, soil properties and erosion in hummocky terrain', *Canadian Journal of Soil Science*, 74, 219–28.

Mulvaney, R., Khan, S., Hoeft, R. and Brown, H. (2001), 'A soil organic nitrogen fraction that reduces the need for nitrogen fertilization', *Soil Science Society of America Journal*, 65, 1164–72.

Nielsen, G., Long, D. and Queen, L. (1996), 'Mapping potential of digitized aerial photographs and space images for site-specific crop management', *Proceedings of the SPIE 2818, Multispectral Imaging for Terrestrial Applications*, http://dx.doi.org/10.1117/12.256078

Nix, J. (1979), 'Farm management: The state of the art (or science)', *Journal of Agricultural Economics*, 39, 277–91.

Nolan, S. C., Goddard, T. W., Heaney, D. J., Penne, D. C. and McKenzie, R. C. (1995), 'Effects of fertilizer on yield at different soil landscape positions', in P. C. Robert, R. H. Rust and W. E. Larson (Eds), *Proceedings of 2nd International Conference on Precision Agriculture*, ASA-CSSA-SSSA, Madison, WI, pp. 553–8.

Obour, A., Mikha, M., Holman, J. and Stahlman, P. (2017), 'Changes in soil surface chemistry after fifty years of tillage and nitrogen fertilization', *Geoderma*, 308, 46–53.

Pennock, D. and Corre, M. (2001), 'Development and application of landform segmentation procedures', *Soil Tillage Research*, 69, 15–26.

Pierce, F. J. and Nowak, P. (1999), 'Aspects of precision agriculture', in D. L. Sparkes (Ed.), *Advances in Agronomy*, volume 67, Academic Press, San Diego, CA, pp. 1–85.

Ping, J. and Dobermann, A. (2005), 'Processing of yield map data', *Precision Agriculture*, 6, 193–212.

Raun, W. and Schepers, J. (2008), 'Nitrogen management for improved use efficiency', in J. S. Schepers and W. R. Raun (Eds), *Nitrogen in Agricultural Systems*, Agronomy Monograph 49, ASA-CSSA-SSSA, Madison, WI.

Raun, W., Solie, J., Johnson, G., Stone, M., Mullen, R., Freeman, K., Thomason, W. E. and Lukina, E. V. (2002), 'Improving nitrogen use efficiency in cereal grain production with optical sensing and variable rate application', *Agronomy Journal*, 94, 815–20.

Raun, W., Figueiredo, B., Dhillon, J., Fornath, A., Bushong, J., Zhang, H. and Taylor, R. (2017), 'Can yield goals be predicted?' *Agronomy Journal*, 109, 2389–95.

Rouse, J., Haas, R., Schell, J. and Deering, D. (1974), 'Monitoring vegetation systems in the Great Plains with ERTS', *in* S. C. Freden, E. P. Mercanti and M. Becker (Eds), *Third Earth Resources Technology Satellite–1 Symposium. Volume I: Technical Presentations*, NASA SP-351, NASA, Washington DC, pp. 309–17.

Sawyer, J., Nafziger, E., Randall, G., Bundy, L., Rehm, G. and Joern, B. (2006), 'Concepts and rationale for regional nitrogen rate guidelines for corn', PM 2015, Iowa State University Extension, Ames, IA.

Scharf, P. (2015), *Managing Nitrogen in Crop Production*, ASA-CSSA-SSSA, Madison, WI.

Scharf, P. and Lory, J. (2009), 'Calibrating reflectance measurements to predict optimal sidedress nitrogen rate for corn', *Agronomy Journal*, 101, 615–25.

Scharf, P., Brouder, S. and Hoeft. R. (2006), 'Chlorophyll meter readings can predict nitrogen need and yield response of corn in the north-central U.S.', *Agronomy Journal*, 98, 655–65.

Scharf, P. C., Shannon, D. K., Palm, H. L., Sudduth, K. A., Drummond, S. T., Kitchen, N. R., Mueller, L. J., Hubbard, V. C. and Oliveira, L. F. (2011), 'Sensor-based nitrogen applications out-performed producer-chosen rates for corn in on-farm demonstrations', *Agronomy Journal*, 103, 1683–91.

Schepers, J., Francis, D., Vigil, M. and Below, F. (1992), 'Comparison of corn leaf nitrogen concentration and chlorophyll meter readings', *Communications in Soil Science and Plant Analysis*, 23, 2173–87.

Schepers, A., Shanahan, J., Liebig, M., Schepers, J., Johnson, S. and Luchiari Jr., A. (2004), 'Appropriateness of management zones for characterizing spatial variability of soil properties and irrigated corn yields across years', *Agronomy Journal*, 96, 195–203.

Schimmelpfennig, D. (2016), 'Farm profits and adoption of precision agriculture', ERR-217, U.S. Department of Agriculture, Economic Research Service.

Sela, S., van Es, H., Moebius-Clune, B., Marjerison, R., Melkonian, J. and Moebius-Clune, D. (2017), 'Adapt-N recommendations for maize nitrogen management increase grower profits and reduce environmental losses on Northeast and Midwest USA farms', *Agronomy Journal*, 108, 1726–34.

Selles, F. and Zentner, R. P. (2001), 'Grain protein as a post-harvest index of N sufficiency for hard red spring wheat in the semiarid prairies', *Canadian Journal of Plant Science*, 81, 631–6.

Sembiring, H., Raun, W. R., Johnson, G. V., Stone, M. L., Solie, J. B. and Phillips. S. B. (1998), 'Detection of nitrogen and phosphorus nutrient status in winter wheat using spectral radiance', *Journal of Plant Nutrition*, 21, 1207–33.

Shanahan, J., Kitchen, N., Raun, W. and Schepers, J. (2008), 'Responsive in-season nitrogen management for cereals', *Computers and Electronics in Agriculture*, 61, 51–62.

Shewhart, W. (1939), *Statistical Method from the Viewpoint of Quality Control*, Graduate School of the Department of Agriculture, Washington DC.

Sinha, E., Michalak, A. and Balaji, V. (2017), 'Eutrophication will increase during the 21st century as a result of precipitation changes', *Science*, 357, 405–8.

Solie, J. B., Raun, W. R., Whitney, R. W., Stone, M. L. and Ringer, J. D. (1996), 'Optical sensor based field element size and sensing strategy for nitrogen application', *Transactions of the American Society of Agricultural Engineers*, 39, 1986–92.

Solie, J., Monroe, A., Raun, W. and Stone, M. (2012), 'Generalized algorithm for variable-rate nitrogen application in cereal grains', *Agronomy Journal*, 104, 378–87.

Stafford, J. V., Lark, R. M. and Bolam, H. C. (1999), 'Using yield maps to regionalize fields into management units', *in* P. C. Robert, R. H. Rust and W. E. Larson (Eds), *Proceedings of the 4th*

International Conference on Precision Agriculture, ASA, CSSA and SSSA, Madison, WI, pp. 225–37.

Stone, M. L., Solie, J. B., Raun, W. R., Whitney, R. W., Taylor, S. L. and Ringer, J. D. (1996), 'Use of spectral reflectance for correcting in-season fertilizer nitrogen deficiencies in winter wheat', *Transactions of the American Society of Agricultural Engineers*, 39, 1623–31.

Sudduth, K., Hummel, J., Birrell, S., (1997), 'Sensors for site-specific management', in F. J. Pierce and E. J. Sadler (Eds), *The State of Site-Specific Management for Agriculture*, ASA, CSSA and SSSA, Madison, WI.

Taylor, R., Kluitenberg, G., Schrock, M., Zhang, N., Schmidt, J. and Havlin, J. (2001), 'Using yield monitor data to determine spatial crop production potential', *Transactions of the American Society of Agricultural Engineers*, 44, 1409–14.

Van Alphen, B. (2002), 'A case study on precision nitrogen management in Dutch arable farming', *Nutrient Cycling in Agroecosystems*, 62, 151–61.

van Es, H., Kay, B., Melkonian, J. and Sogbedji, J. (2007), Nitrogen management for maize in humid regions: Case for a dynamic modeling approach, in T. Bruulsema (Ed.), *Managing Crop Nutrition for Weather*, International Plant Nutrition Institute Publication, Norcross, GA, pp. 6–13.

Von Rosenberg Jr., C. W., Abbate, A. and Drake, J. (2000), 'A rugged near-infrared spectrometer for real-time measurement of grains during harvest', *Spectroscopy*, 15, 34–8.

Wade, J., Horwath, W. and Burger, M. (2016), 'Integrating soil biological and chemical indices to predict net nitrogen mineralization across California agricultural systems', *Soil Science Society of America Journal*, 80, 1675–87.

Wollenhaupt, N., Wolkowski, R. and Clayton, M. (1994), 'Mapping soil test phosphorus and potassium for variable-rate fertilizer application', *Journal of Production Agriculture*, 7, 441–8.

Zhang, X., Shi, L., Jia, X., Seielstad, G. and Helgason, C. (2010), 'Zone mapping application for precision-farming: A decision support tool for variable rate application', *Precision Agriculture*, 11, 103–14.

Chapter 14

Site-specific irrigation systems

Amir Hagverdi, University of California-Riverside, USA; and Brian G. Leib, University of Tennessee-Knoxville, USA

1 Introduction

1.1 Site-specific irrigation management

Conventional irrigation management is based on uniform application of water across a field which may result in both over- and under-irrigation. Over-irrigation negatively impacts soil aeration, increases disease occurrence and causes leaching/movement of mobile nutrients and sediments through the root zone into groundwater and off the field towards surface water bodies. Under-irrigation adversely affects yield quantity and quality, and leaves more applied nutrient in a field that could be lost due to winter rains. *Site-specific irrigation management relies on site-specific data to understand processes governing crop water requirement and/or soil water status on a spatio-temporal scale, and utilizes variable rate technology to optimize within-field allocation of applied irrigation water.* Note that the focus of site-specific irrigation is not only on the ability to spatially vary the application of water, but also on site-specific data acquisition and processing. The ultimate goal is to maximize the economic and/or environmental value of the water applied (Kranz et al., 2012) by (i) enhancing quantity and quality of production; (ii) reducing deep percolation of water and/or run-off from irrigated fields and in turn minimizing contamination of groundwater and surface water resources through nitrate leaching, transport of sediments, nutrient and agrochemical; and (iii) alleviating the large-scale impact of irrigation on water resources sustainability by optimizing the water footprint of irrigated agriculture. The basic

http://dx.doi.org/10.19103/AS.2017.0032.15

assumption of site-specific irrigation is that crop water requirement may spatially vary due to many reasons that cause crop growth to be non-uniform across a field (Howell et al., 2012). Site-specific irrigation should emphasize the kinds of variability which can be managed by irrigation including changes in topography, soil texture/structure and drainage and salinity conditions. These are examples of parameters that impact infiltration, redistribution, storage and deep percolation of water, plant available water (PAW), leaching requirement and, consequently, characterize appropriate site-specific irrigation management strategies.

1.2 Variable rate irrigation development history

Most of the past work on site-specific variable rate irrigation (SS-VRI) has focused on pressurized mechanically moved overhead irrigation systems, mainly center pivot and lateral move. This is largely because of widespread adoption, adaptability to automation and the large area covered with a single lateral pipe (Sadler et al., 2005; Evans et al., 2013). Early developments included limited control of end guns, edge of the field stop and stop-in-slot control which allowed growers to stop the irrigation system after completion of irrigation events or at a specific position such as a field road. Later on, programmable control panels with speed control capability were developed. This feature made growers to be able to automatically apply different depths of water to a limited number of pie-shaped sectors (usually less than 10 sectors) underneath their center pivot irrigation systems, the earliest form of VRI. This method was often employed by growers when multiple crops were planted on different portions of the field (Kranz et al., 2012). More recent commercial products support greater speed control of irrigation machines resulting in more and smaller pie-shaped sectors (from <1° to 10°).

During the last two decades, a number of approaches have been developed to apply different depths of water along the moving lateral of overhead irrigation systems. This was a groundbreaking innovation in site-specific irrigation which led to the development of zone control technology for sprinklers that allowed delineation of irregularly shaped irrigation management zones (IMZs). In an early study by Roth and Gardner (1989) at the University of Arizona, a lateral move irrigation system was modified to apply different depths of water and nitrogen to a limited number of research field plots with the ultimate objective of developing water–nitrogen response surfaces. They used three spray lines each operated by separate solenoid valves while changing the travel speed of the machine. In the early 1990s, multiple site-specific irrigation research projects were initiated in the United States at Colorado (Fraisse et al., 1995a,b), University of Idaho (King et al., 1995), South Carolina (Omary et al., 1997) and Washington State University (Sadler et al., 2005).

In Idaho, King et al. (1995) studied site-specific application of water and chemicals using both lateral move and center pivot irrigation systems which consisted of a spatially referenced mapping system. In another study, King and Kincaid (2004) designed a variable flow sprinkler by changing the nozzle area using a mechanically activated pin. The main concern about this design is that changes in flow rate alter wetting patterns and water droplet size distribution of sprinklers, which in turn may create application uniformity issues (Kranz et al., 2012). In Colorado, Fraisse et al. (1995b) used the concept of pulse irrigation defined by Karmeli and Peri (1974) as a series of on (operating phase) and off (non-operative phase) irrigation cycles on a lateral move irrigation system to apply different amounts of water along the system. The main objective of Colorado State University's experiment was to perform small-plot research on conjunctive management of irrigation water and chemicals using a lateral move system. However, they also mentioned potential applications of VRI to address spatial differences such as soil type and topography. In South Carolina, VRI was designed and constructed in cooperation with the University of Georgia on a center pivot irrigation system using a multiple manifold technique (Camp et al., 1998). In Washington, researchers also used multiple manifolds with different nozzle sizes to develop a variable rate center pivot irrigation system. Further developments in the Washington State University focused on electronic controllers for the activation of solenoid valves to control a bank of sprinklers on a center pivot irrigation system (Sadler et al., 2005). Currently, commercial VRI systems usually use a combination of speed control and pulse irrigation. More recent efforts such as studies by Han et al. (2009), O'Shaughnessy et al. (2015) and Chávez et al. (2010a,b) focused on integration of VRI technology with site-specific data acquisition and processing systems using wireless technologies, geographical information system and global navigation satellite system (GNSS) applications, Internet and soil/plant monitoring systems.

The focus of this chapter is on SS-VRI using center pivot and lateral move irrigation systems which are mainly used for row-crop irrigation. However, basic concepts discussed throughout the chapter on site-specific data acquisition and mining approaches, soil mapping, zone delineation, site-specific yield estimation and irrigation scheduling should be applicable to other existing forms of site-specific irrigation technologies and/or new forms that may become available in the future for fixed sprinkler, drip and surface irrigation systems, as well as potential applications in nursery and urban irrigation settings.

2 Field-level mapping of soil variability

Mapping field-level soil heterogeneity is an essential prerequisite for almost all precision agriculture (PA) applications. Field-level spatial information of soil hydraulic attributes [mainly field capacity (FC) and permanent wilting point (PWP), which are used to calculate PAW] is required for SS-VRI management. However, direct measurement of these characteristics is challenging due to the time-consuming and labour-intensive nature of *in situ* and laboratory methods. The most widely used solution to this problem is to develop proxies of soil hydraulic attributes by collecting easily measured soil characteristics such as soil textural information (i.e. per cent of sand, silt and clay), soil bulk density (BD) and organic matter (OM) content that are typically well-correlated with soil hydraulic properties to produce pedotransfer functions (PTFs, Bouma 1989). There are many PTFs in the literature that provide non-spatial point estimation of soil water retention information. However, the literature is limited with regard to the production of high-resolution maps of soil hydraulic attributes.

The PA community predominantly focuses on both on-the-go sensors and remote sensing (RS) to map soil variability, with soil apparent electrical conductivity (ECa) being the most widely used proxy. ECa is a function of the electrical conductivity of porous media solution, the soil porosity and the cementation exponent, that is, Archie's law (Archie, 1942). When soil salinity is not a major factor, ECa may be a useful proxy of soil physical and hydraulic attributes (Sudduth et al., 2005) including depth to sand layer (Duncan, 2012), clay percentage under non-saline conditions (Saey et al., 2009) and soil texture and FC (Abdu et al., 2008). Some site-specific irrigation studies that focused on ECa to characterize soil variability have been conducted by Hedley and Yule (2009a,b), Pan et al. (2013) and Haghverdi et al. (2015a). These endeavours introduced and evaluated novel approaches to map spatial heterogeneity of soil hydraulic properties at high resolution using ECa as the main proxy.

2.1 Case study: high-resolution prediction of plant available water

This case study draws from previous work conducted and published by Haghverdi et al. (2015a) on a 73-ha field located in West Tennessee close to the Mississippi River. The soil survey indicated that the field contains Mississippi River terrace alluvial deposits that have produced Robinsonville loam, fine sandy loam, Commerce silty clay loam and Crevasse sandy loam soils. On 20 March 2014, ECa was measured at 4700 points at shallow (0–30 cm) and deep (0–90 cm) depths across the study area using a Veris 3100 machine (Veris Technologies, Salina, KS, USA) after some rainfall events when soil was assumed to be close to FC. Next, field soil sampling was conducted. On 21–22 March 2014, a truck-mounted hydraulic probe was used to sample 100 undisturbed sites at 0–100 cm depth where each soil sample was divided into four segments. Hereafter, the word 'layer' is used to distinguish among segments rather than real soil horizons. The default depth of subsamples was 25 cm, though adjustments that accounted for soil horizon transitions were made. Soil textural information (i.e. per cent of sand, silt and clay), BD and gravimetric water content were measured in the laboratory.

Ten spatial modelling scenarios that used kriging (KG), co-kriging (CKG), regression kriging (RKG), geographically weighted regression (GWR) and artificial neural network (ANN) were examined (Fig. 1). The objectives were to convert measured soil properties at point locations to continuous maps of water content at FC and PWP, and to evaluate the efficiency of the ECa as a proximal attribute in this process. FC and PWP are widely used

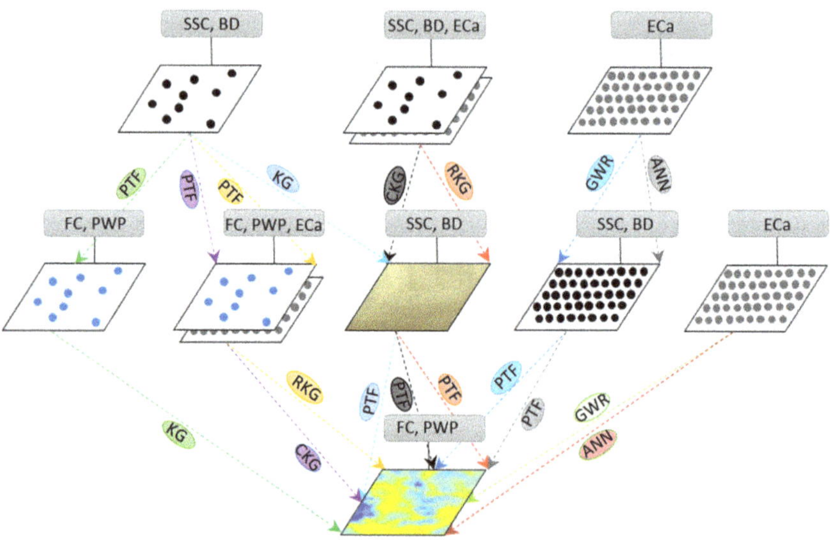

Figure 1 Ten spatial modelling scenarios using pseudo-continuous pedotransfer function (PTF, Haghverdi et al., 2012, 2014), kriging (KG), co-kriging (CKG), regression kriging (RKG), geographically weighted regression (GWR) and ANN. The inputs were per cent of sand, silt and clay (SSC), bulk density (BD, Mg m^{-3}) and soil apparent electrical conductivity (ECa, mS m^{-1}). The outputs were soil water content maps at field capacity (FC, cm^3 cm^{-3}) and permanent wilting point (PWP, cm^3 cm^{-3}) (adapted from Haghverdi et al. (2015a)).

thresholds for irrigation management. FC is a qualitative parameter that is a practical and understandable indicator of soil water holding capacity (Romano and Santini, 2002). Water retentions at −10 and −33 kPa could be considered as FC for coarse- and fine-textured soils, respectively (Rivers and Shipp, 1972). The water content at –1500 kPa is usually chosen as the PWP for all soil textures. For this study, it was assumed that at FC, matric potential was uniform across the study area (−10 kPa). PAW is obtained as the difference between water content at FC and PWP within the root zone.

2.2 Summary of findings

2.2.1 Performance of the modelling scenarios

The accuracy of soil water content maps improved when ECa was incorporated in the workflow models. Two methods showed higher performance than others: (i) ECa was required as input to establish a GWR model on discrete sampling locations which was subsequently used to generate maps of FC and PWP, and (ii) PTF was applied to predict water retention at FC and PWP on discrete sampling locations, and subsequently those predictions together with ECa data were used to generate maps of FC and PWP using RKG or CKG as the interpolation approach.

2.2.2 Application of ECa

Proximal and easily collected data such as ECa are affected by a variety of soil- and/or crop-related properties, hence should be applied and interpreted very cautiously. Unfortunately, complexity of proximal sensor measurements and in particular ECa mapping are sometimes overlooked in research studies (Corwin and Lesch, 2010) by the PA community and to a greater extent in practical applications by growers and PA professionals. The key site-specific question that should always be precisely answered is what underlying processes and soil–plant attributes affect ECa measurements. In the field of study, intensive soil sampling revealed that the spatial distribution of soil texture was the main factor governing soil water retention at all layers, and ECa measurements showed a high correlation with soil textural properties and soil water status at the time of sampling. Some studies showed that it is better to conduct ECa mapping under wet conditions because the relationship between ECa and texture is more pronounced near FC (Auerswald et al., 2001; Earl et al., 2003). Useful guidelines and insights on the application of ECa in PA have been provided by Corwin and Lesch (2003) and by Corwin and Lesch (2010).

Since sampling is time-consuming and expensive, it is desirable to minimize sampling density. However, one should note that the sampling density and scheme affect the error associated with the spatial prediction process (Herbst et al., 2006). Sampling at half of the variogram range (if there are existing variograms of the soil/crop properties) is suggested as a 'rule of thumb' to collect enough samples to make reliable prediction for PA (Kerry et al., 2010). Debaene et al. (2014) showed that 1.5 samples ha^{-1} is adequate to predict soil OM content and texture. Iqbal et al. (2005) recommended sampling intervals of <100 m in order to detect boundaries of soil hydraulic properties in an alluvial floodplain soil in the region of the Mississippi Delta. Taking 1 sample ha^{-1} is usually considered the greatest sampling density that a farmer can afford (Kerry et al., 2010). However, an additional concern of site-specific irrigation management is to collect deep enough samples to sufficiently cover the effective water uptake zone by crop roots.

Incorporation of ECa or similar easily collected dense proxies in the spatial mapping process is suggested either as a guide for directed subsequent sampling or as an input for spatial modelling. This alleviates the need for more difficult data collection (such as soil texture and BD) through sampling, which in turn reduces the associated cost of the mapping process while maintaining an adequate accuracy level.

3 Delineation of irrigation management zones

A management zone (MZ) is generally defined as a subregion of a field that is relatively homogeneous with respect to soil–landscape attributes. However, the phrase 'management zones' remains uncertain unless additional information is included to clarify the goal in subdividing the field (Kitchen et al., 2005). In PA, the focus of available studies has been mainly on delineation of yield-based productivity zones and MZs for variable rate application of fertilizer (precision nutrient management). In contrast, there are only a handful of studies on IMZ delineation (e.g. Haghverdi et al., 2015b, 2016; Mouazen et al., 2014; Boluwade et al., 2015; Fortes et al., 2015), while there is a critical need to dynamically develop approaches for IMZ delineation in an accurate and inexpensive manner (Evans et al., 2013). In order to accomplish this objective, appropriate attributes and algorithms for IMZ delineation need to be identified along with some criteria to define optimum number of IMZs for a given field.

Delineating zones for SS-VRI management is challenging. The ultimate goal should be to identify and group together those areas within a field that need identical irrigation treatment. A variety of soil-related attributes such as topography, restricted subsurface soil layers, soil type and soil salinity may be important for IMZ delineation. In practice, mainly easily collected inexpensive non-invasive proximal data collected using RS techniques or on-the-go sensors (e.g. yield maps, topography, satellite photographs, canopy images and ECa) that are representative and spatially correlated to these soil properties are used for zoning. As stated earlier in this chapter, the main drawback of proximal sensor measurements is that they are complex and affected by multiple soil-crop properties (Corwin and Lesch, 2010). Therefore, they should show similar spatial distributions to main factors of concern in irrigation decisions, in order to be considered as an appropriate input to delineate IMZs.

Applying unsupervised clustering techniques and zoning via user-defined thresholds are the two main methods to delineate MZs. Clustering techniques such as *k*-means and fuzzy *k*-means use the inherent structure and distribution of the data in order to group similar observations together into distinct classes and have been widely used to identify MZs (Córdoba et al., 2013). These methods usually produce separated irregular (freeform) shape zones across a field. Integer linear programming (ILP) can be used to make rectangular and pie-shaped zones (Cid-Garcia et al., 2013; Haghverdi et al., 2015b).

3.1 Case study: delineating management zones for variable rate irrigation on a no-tillage cotton field in West Tennessee

This case study was conducted and published by Haghverdi et al. (2015b) on the aforementioned field in western Tennessee which contained two center pivot irrigation

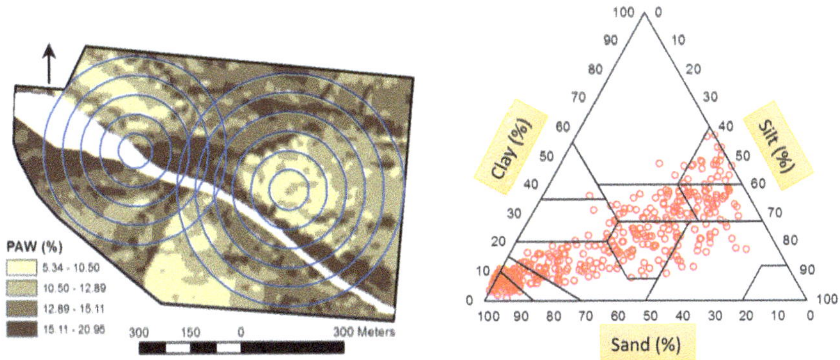

Figure 2 Left panel: Map of plant available water (i.e. PAW: FC-PWP) produced by geographically weighted regression (GWR) model for the entire sampling depth (1 m) [adapted from Haghverdi et al. (2015b)]. Right panel: The distribution of soil texture in samples from the field of study (adapted from Haghverdi et al. (2015a)).

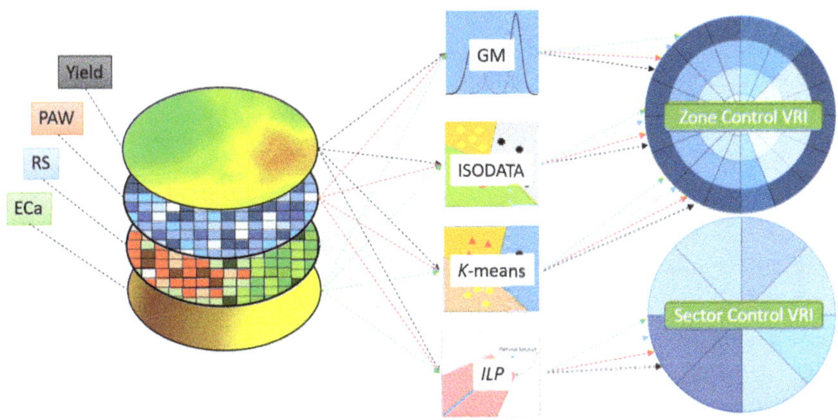

Figure 3 Multiple strategies to delineate IMZs for zone control and sector control variable rate irrigation (VRI) center pivot machines. The clustering approaches are Gaussian mixture (GM) model, ISODATA-maximum likelihood (ISODATA), *k*-means and integer linear programming (ILP). The inputs attributes are yield, plant available water (PAW, estimated as explained in part 2), space images from Landsat 7 and 8 satellites (RS) and soil apparent electrical conductivity (ECa) (adapted from Haghverdi et al. (2015b)).

systems that were used for supplemental irrigation (Fig. 2). The objectives were to evaluate multiple input attributes and algorithms for IMZ delineation, and then identify the optimum number of zones based on a particular attribute. Three unsupervised clustering techniques (i.e. *k*-means, ISODATA-maximum likelihood and Gaussian mixture model) with different combinations of input soil physical and proximal attributes (i.e. PAW, ECa, yield data and Landsat 7 and 8 satellite images) were examined (Fig.

3). Most center pivot irrigators prefer to have less than 10 MZs in a field (Evans et al., 2013). Therefore, clustering procedures were repeated to delineate from 2 to 10 IMZs. In addition, an optimization problem was formulated using the ILP technique to find the optimum number and positions of pie-shaped zones underneath the two center pivots. Pie-shaped zones are appropriate for center pivot machines with speed control capabilities (sector control VRI) where speeding the system up and slowing the system down provide less and more time for irrigation, respectively. Machines with sector control capabilities only adjust travel speed and do not provide adequate flexibility to fully manage within-field heterogeneity because sprinklers apply the same irrigation depth along the entire lateral, lacking the capability to delineate irregularly shaped zones (Kranz et al., 2012). However, this is a relatively easy-to-operate and affordable VRI strategy for those growers who possess center pivot irrigation systems because most of the available center pivots have control panels with a speed control module. The knowledge and full use of this capability provide a great opportunity to get growers acquainted with the VRI concept and bridge the knowledge and experience gap between current conventional uniform irrigation practice and cutting-edge commercially available zone control technology.

The per cent of total variance of PAW (following dividing the field into a given number of IMZs) was used as an evaluation measure to identify the optimum number of IMZs. This criterion indicates how much variability is explained as the number of IMZs increases, and was calculated following Fraisse et al. (2001) and Brock et al. (2005). First, the PAW variance was determined for each zone and weighed considering the zone area (Equation 1), then summed up to obtain the overall within-zone variance of PAW (Equation 2), and finally converted to the per cent of total variance (Equation 3):

$$\mathrm{VAR}_Z = \frac{1}{n_Z}\sum_{i=1}^{n_Z}(PAW_i - PAW_Z)^2 \tag{1}$$

$$\mathrm{VAR}_T = \sum_{i=1}^{m}\mathrm{VAR}_i \times \frac{n_Z}{n_T} \tag{2}$$

$$\%\,of\,total\,VAR = 100 \times \left(\mathrm{VAR}_T / \mathrm{VAR}_{WF}\right) \tag{3}$$

VAR_Z: weighted variance for zone Z

PAW_i: plant available water for cell i (25 m^2 in this case study)

PAW_Z: mean plant available water in zone Z

n_Z: number of cells in zone Z

VAR_T: summed and weighted within-zone variances

n_T: total number of cells across field

m: number of zones

VAR_{WF}: variance calculated for all cells across the field (represents conventional uniform irrigation with only one zone)

3.2 Summary of findings

3.2.1 Zoning algorithms

The unsupervised clustering algorithms provided similar results and good agreement among delineated zones using the clustering techniques (i.e. *k*-means, ISODATA-maximum likelihood and Gaussian mixture model) observed. The findings indicate that for a field with high spatial structure, all well-known clustering methods recognize similar patterns for IMZs.

3.2.2 Input attributes

Soil ECa followed by the satellite images showed high potential for IMZ delineation. Successful applications of ECa for zone delineation were also reported by Moral et al. (2010) and Kitchen et al. (2005). The spatial pattern of captured data by space images was affected by soil water status and crop phenology pre- and in-season, respectively. Not all space images were appropriate for IMZ delineation; therefore, it is important to screen and find reliable images. In practice, farmer's knowledge on field conditions could be extremely useful to choose the most appropriate data set and this knowledge should be incorporated in the zone delineation process. Yield data indicated less potential for zoning and showed some inconsistency mainly because different factors including physical and chemical properties of soil, irrigation regimes, pests and diseases and climate affect yield in a complex manner. Averaging yield maps across years could help find a more stable scheme of yield distribution, yet productivity zones may not necessarily match the irrigation zones.

3.2.3 Optimum number of zones

The results of this case study showed that having more than 4 to 5 zones only slightly reduced the per cent of total variance (Fig. 4), hence did not improve the ability to explain soil heterogeneity, a trend also reported by Zhang et al. (2010). Note that IMZ maps are not identical to irrigation prescription maps. Site-specific data on soil/crop water status are needed to convert MZ maps to irrigation prescription maps. It is expected that management cost (i.e. cost of instrumentation, measurements and data interpretation) increases as the number of zones increases. Therefore, it is important to find the optimum number of IMZs for each field. Having too many IMZs means making too many irrigation decisions, something that can easily overwhelm growers.

3.2.4 Dynamic in-season zoning

If in-depth vertical soil variation is significant, the spatial arrangement of zones may vary throughout the growing season considering the expansion of the effective root zone. In this case, multiple zoning schemes may be needed within a cropping season (Haghverdi et al., 2015b). In practice, the number and spatial arrangement of IMZs may be influenced by a variety of parameters including available water for irrigation, crop response to over/under-irrigation and temporal variability in seasonal climatic parameters. Continuous monitoring of soil water status and/or plant stress throughout the season helps in adjusting spatial arrangement of IMZs, if needed.

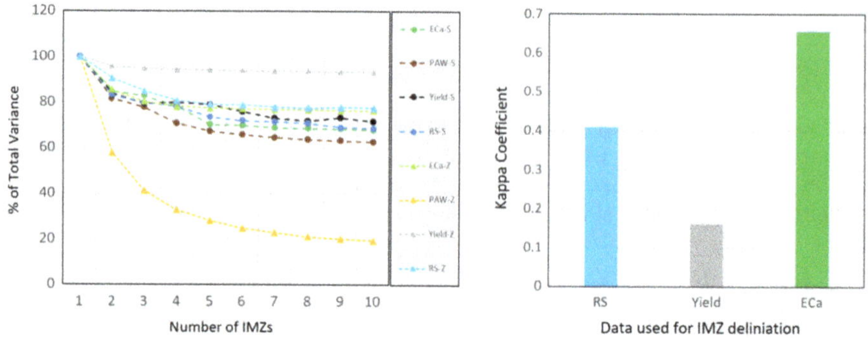

Figure 4 Performance of different input attributes and number of irrigation management zones (IMZs) averaged for the two center pivot systems in the study site. Left panel shows per cent of total variance (Equation 3) and input attributes were ECa (soil apparent electrical conductivity), PAW (plant available water, estimated as explained in part 2), yield (cotton lint yield in 2013 and 2014) and RS (reflectance values from panchromatic band taken by Landsat 7 and 8 satellites). Letter Z indicates zoning was done for zone control VRI scenario using *k*-means approach, and letter S shows zoning was conducted using integer linear programming (ILP) for sector control VRI strategy. Right panel depicts kappa coefficient (Cohen, 1960) values for the zones delineated by ILP. The kappa coefficient shows inter-classification agreement of different zoning strategies where a positive coefficient indicates that agreement exceeds chance and the magnitude of the coefficient reveals the agreement strength (adapted from Haghverdi et al. (2015b)).

4 Quantifying the potential impact of variable rate irrigation

It is important to understand how site-specific irrigation technologies and management strategies affect crop production for a given field. Currently, it is not easy to quantify the potential production benefits of SS-VRI irrigation systems over conventional uniform irrigation strategies, which contributes to growers' present uncertainty in making investment in VRI systems. There are only a few studies in the literature documenting the impact of SS-VRI on irrigation water use efficiency (WUE) and crop yield, and the reported results are not consistent among studies. Bronson et al. (2006) evaluated cotton lint yield response to applied irrigation rate for different landscape positions (i.e. bottom slope vs side slopes) and reported no difference between the two, thereby suggesting that VRI would not be beneficial at the experimental site. King et al. (2006) studied the impact of SS-VRI systems on potato yield in comparison with conventional uniform irrigation. Soil moisture sensors were used to adjust water application based on estimated evapotranspiration (ET) at each MZ, while conventional uniform irrigation was to base irrigation depth on the average irrigation requirement for the experimental units. They reported 4% greater yield under SS-VRI scenario with an increase in income of about half the cost of the technology. Some other studies (e.g. Vories et al., 2016; Sui et al., 2015) have used VRI systems, yet mainly to apply different amounts of water to irrigation research plots without investigating the impact of VRI versus uniform irrigation to address field-level soil variability.

4.1 Case study: evaluating rainfed, uniform and variable rate irrigation strategies

This case study was conducted and published by Haghverdi et al. (2016) on the aforementioned field located in West Tennessee with the mean monthly growing season temperature and precipitation equal to 21 °C and 97 mm month^{-1} from May to November, respectively. A two-year cotton irrigation experiment (2013–14) was implemented. The main objective was to estimate the potential impact of SS-VRI systems on cotton production and irrigation water use. Therefore, a variety of site-specific data were collected and used to design and evaluate some site-specific water production functions (SS-WPFs). The best SS-WPF was selected to generate yield maps for four irrigation and zoning strategies: (i) rainfed, (ii) conventional uniform supplemental irrigation, (iii) sector control VRI and (iv) a hypothetical zone control VRI capable of irrigating each cell (25 m^2) at an optimum level. For sector control VRI, the aforementioned ILP approach was adopted to find the optimum arrangement and number of IMZs, while this time the optimization objective was to maximize cotton lint yield. For preprocessing the data (i.e. four categories: soil, irrigation, fertilizer and yield), principal component analysis and *k*-means were used and, for deriving WPFs, multiple linear regression, ANN and *k*-nearest neighbour (*k*-NN) techniques were employed (Fig. 5).

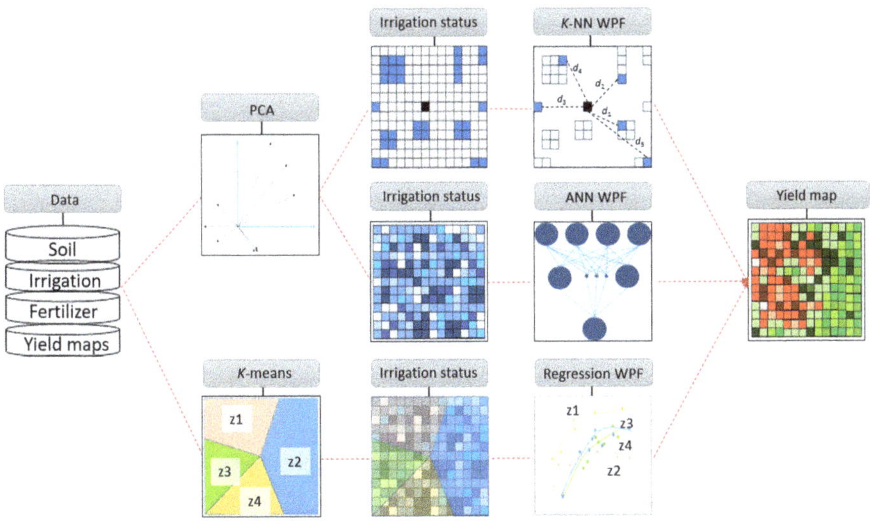

Figure 5 The three modelling scenarios used to develop site-specific water production functions (SS-WPFs) to predict cotton yields for the two center pivots in the study site. The input/output attributes in soil category: (soil apparent electrical conductivity, soil textural information, soil bulk density, soil volumetric water content at the time of sampling, at field capacity and at permanent wilting point); in crop category: (cotton lint yield in 2013 and 2014); in irrigation category (seasonal applied irrigation in 2013 and 2014); and in fertilizer category (maps of applied phosphorus and potassium by the grower). The statistical approaches used for preprocessing of data and subsequently development of WPFs were PCA: principal component analysis, *k*-means, *k*-NN: *k*-nearest neighbour, ANN: artificial neural networks and regression (adapted from Haghverdi et al. (2016)).

4.2 Summary of findings

4.2.1 Performance of the modelling approaches

K-NN WPF showed the highest performance (root-mean-square error equal to 0.131 Mg ha^{-1} and 0.194 Mg ha^{-1} in 2013 and 2014, respectively). *K*-NN is a non-parametric technique and works as a smart screening engine to select only a few relevant data points for prediction. This approach helps *k*-NN WPF to take advantage of the high number of yield data points to distinguish patterns without fitting coefficients. One should note that *k*-NN WPF introduced in this study provides discrete predictions of yield over a range of applied irrigation, hence providing only reliable estimations in the neighbourhood of measured applied irrigation depths.

Site-specific WPFs are examples of data-driven models that could be developed using farming data generated nowadays by PA. These models could be used as after-the-fact tools to evaluate different irrigation and zoning strategies and to further enhance site-specific irrigation management. When water is limited for full irrigation, SS-WPFs can be used to find the optimum deficit irrigation strategy to allocate water among IMZs. In addition, SS-WPFs are useful to identify the best irrigation strategy (which soil to follow) for zones with a mixture of soils which is likely to happen in pie-shaped zones delineated for sector control VRI systems. Note that the performance of data-driven models such as SS-WPFs may differ and should be tested for different soils, crops and climatic conditions. To derive robust SS-WPFs, all major attributes affecting crop response to irrigation should be incorporated in the modelling process.

Readers should consider that there are fundamental differences in characteristics and applications of the traditional crop WPFs and proposed SS-WPFs in this chapter. Classical WPFs are typically established using data collected from research plots through multi-year experiments in controlled conditions. The classical WPFs are developed for reuse in similar places and/or for the same place in the future. In contrast, the SS-WPFs are designed for site-specific use to convert site-specific farming data to useful information, an opportunity that has only become possible by recent advances and the spread of precision farming equipment which has led to collection of numerous site-specific data by growers.

4.2.2 Impact of site-specific irrigation strategies on cotton lint yield

The average estimated cotton lint yields in 2013 and 2014 by *k*-NN SS-WPF for the two center pivots in the study site were: rainfed (100%, 0.94 Mg ha^{-1}), uniform supplemental irrigation (139%, 1.3 Mg ha^{-1}), sector control VRI with 10 IMZs (144%, 1.35 Mg ha^{-1}) and the hypothetical high-resolution zone control VRI (155%, 1.45 Mg ha^{-1}). High in-season rainfall is typical in West Tennessee, yet some dry periods occur when supplemental irrigation is crucial to fulfilling ET demand. Over the course of this experiment, the applied supplemental irrigation was less than 20% of precipitation. The impact of site-specific irrigation strategies on crop production strategies is expected to be more pronounced in drier years. Preseason wet soil condition and colder temperatures forced the farmer to postpone planting cotton in 2013, which impacted yield pattern (Fig. 6) and crop response to irrigation mainly due to the inability of the cotton to accumulate adequate heat units.

The main objective in crop-based zoning using WPFs is to maximize production, while in soil-based zoning (explained in Section 3), the main objective is to maximize explained soil

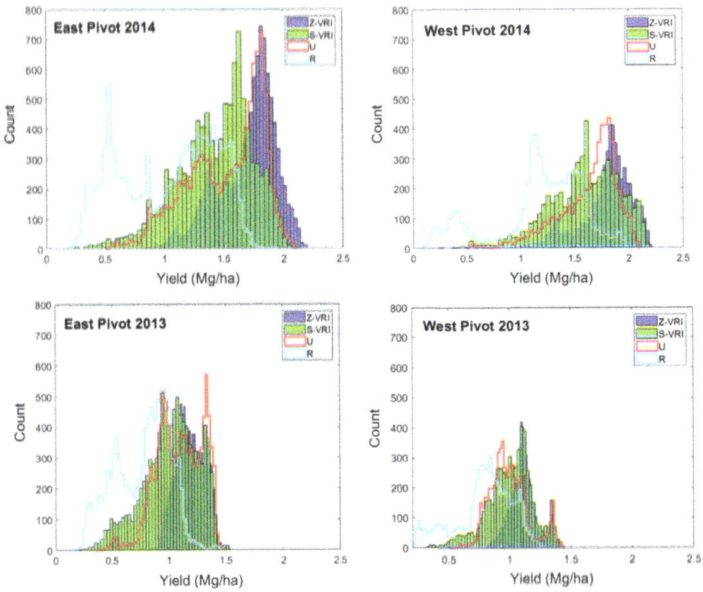

Figure 6 Histogram of cotton lint yield estimated using site-specific WPF (developed by *k*-nearest neighbour) for the center pivots in the field of study. Yield data were estimated in 2013 and 2014 cropping seasons for the four different water management strategies: R (rainfed), U (conventional uniform), S-VRI (sector control VRI with 4 IMZs) and Z-VRI (a hypothetical zone control VRI with capability to irrigate each 25 m^2 cell at the optimum level) (adapted from Haghverdi et al. (2016)).

variability. The crop-based zoning using WPFs is influenced by annual climatic conditions and farmer management practices. A long-term farming data record that captures different climatic and irrigation management scenarios is expected to lead to more stable IMZs. In practice, it is suggested to start with soil-based zoning and then analyse crop response to irrigation depths across zones using farming data and site-specific models (such as *k*-NN WPF) to improve arrangement and number of IMZs in an iterative process.

5 Site-specific irrigation management

5.1 Prescription maps for variable rate irrigation

The two major approaches for variable rate application of inputs are (i) map-based: site-specific data are collected and converted to prescription maps prior to implementation and (ii) sensor-based: real-time sensor data are collected and converted to application rate via a pre-developed application algorithm. For SS-VRI systems, ideally a combination of the two approaches is needed to first delineate IMZs, and then throughout the cropping season, create an irrigation prescription map before each irrigation event using real-time site-specific data collected via sensors. The three main methods that have been used/proposed for irrigation scheduling using SS-VRI systems are (i) soil moisture-based (Liang et al., 2016; King et al., 2006; Kim et al., 2009; O'Shaughnessy et al., 2015), (ii) ET-based (Hedley and Yule 2009b) and (iii) plant feedback-based (O'Shaughnessy et al., 2015) approaches.

Liang et al. (2016) deployed a network of soil moisture sensors to measure soil matric potential across IMZs and subsequently converted the data to equivalent volumetric water content information using the soil water retention equation of van Genuchten (1980) to create irrigation prescription maps. In another study, O'Shaughnessy et al. (2015) used a field-calibrated neutron probe to measure soil water content on a weekly basis in order to calculate the replenishment of crop water use to FC in the top 1.5 m of soil. King et al. (2006) measured soil water contents using TDR soil moisture sensors which were subsequently used to adjust water application based on estimated ET at each MZ. The main practical question for the soil moisture-based approach is the number, location and depths of soil moisture sensors which should be installed per IMZ. Hedley and Yule (2009b) conducted a simulation study to run a water balance model based on ET, to irrigate each zone when a critical soil moisture deficit is reached. O'Shaughnessy et al. (2015) introduced a plant feedback method which uses meteorological data and a wireless network of infrared thermometers (mounted on the pivot lateral and placed in the well-watered spots throughout the field). In their case study, the pivot scanned the field on even days and irrigated using an automatically created prescription map on odd days.

5.2 Uniformity and spatial resolution of variable rate systems

It is necessary to evaluate application uniformity of SS-VRI systems and assess their reliability to apply different depths of water and change the application rate as prescribed. Table 1 summarizes the reported results from several case studies by Sui and Fisher (2015), Han et al. (2009), O'Shaughnessy et al. (2013), Dukes and Perry (2006) and Yari et al. (2017) on application uniformity of various-size center pivot or lateral move systems equipped with different types of sprinkler packages. The drop hoses were spaced from 1.5 m to 3 m apart and pressure regulators were rated at 41–138 kPa. Some systems were equipped with commercially available VRI technologies, while the others were outfitted with prototype technologies developed by researchers. Application rates varied from 11% to 100% (100% means constant application with no cycling) with different equivalent water depths. The collectors were put in a variety of grid, arc-wise and transect patterns. Overall, coefficient of uniformity (CU) and low quarter distribution uniformity (DUlq) averaged about 88% (min: 73% and max: 95%) and 78% (min: 73% and max: 95%) between studies, respectively.

Han et al. (2009) and Sui and Fisher (2015) reported some reductions in CU for lower application/cycling rates. However, the overall impact of application/cycling rates on CU was not noticeable between studies. Reported results indicate that SS-VRI systems adequately applied the desired irrigation depths in most cases. Sui and Fisher (2015) reported approximately on average 13% difference between actual and desired application amounts and noticed higher error for lower application depths. O'Shaughnessy et al. (2013) observed less than 3 mm error (root-mean-square error) between actual and desired application depths. They reported that some variations in depths of applied water are expected due to wind direction and speed. Dukes and Perry (2006) observed a noticeable influence by nozzle type on the uniformity and attributed that to the distinct spray pattern of each sprinkler type.

Table 1 Summary of some reported results on uniformity and performance of SS-VRI systems

	System							Sprinkler						Test			
Ref.	Type	Make[a]	VRI	Spans	Length (m)	Press[£] (kPa)	Flow (m3/min)	Type[€]	Height (m)	Spacing (m)	Press (kPa)	WET (m)	Wind (m/s)	application	Collectors	CU	DUlq
Dukes and Perry (2006)	CP	Reinke	developed by researchers and Computronics Corp. Ltd.	4	186	262	0.9	Nelson R3000 rotator sprinkler nozzles	1.3	3.05	67		<3.8	20%, 50%, 80%, 100% at two speed rates 7% and 11% (100% @ 11% speed was 9mm)	2 outer spans- 180 collectors 6 lines (3 reps) Spacing: 3m by 15 m	89-95%	84-95%
Dukes and Perry (2006)	LM	Valley	developed by researchers	6	375	206[£]	3.8	Senninger LDN (fixed plate)	1.5	2.3	138		<5	25%, 50%, 75%, 100% at two speed rates 11% and 16% (100% @ 11% speed was 20–21mm)	2 centre-most spans- 108 collectors 6 lines (3 reps) Spacing: 3 m by 7.6m	73–83%	60–74%
								Senninger i-WOB (moving plate)	1.8	3	138		<5			84–91%	73–82%
Sui and Fisher (2015)	CP	Valley	Valley VRI	4	233	234[¥] 290	1.33	Fixed-pad Senninger LDN-UP3	1.8	2.74	103	7–13.4	3.58	30%, 50%, 70%, 100% (100% was 25.4 mm)	4 spans 210 collectors 2 lines Spacing: 2.44m with 12° between lines	78–89%	
Yari et al. (2017)	CP	Valley	Valley VRI	5	294	240		LESA, Nelson R3000 rotator	1.5	2.29	120		2.4	8.1mm, 14.2mm, 20.3mm	spans 2-5 Spacing: 3 m with 4° between lines	[¥]92–94%	
O'Shaughnessy et al. (2013)	CP	Valley	Valley VRI	3	131	172		Fixed plate Senninger LDN	1.5	1.5	41	6–9.0	4.1–9.5	30%, 50%, 70%, 80%, 100% (two travel speeds)	spans 2-3 transect, arc-wise, and grid patterns	86–90%	76–83%
	CP	Valley	Valley VRI	6	260	172		Fixed plate Senninger LDN	1.5	1.5	41	6–9.2	2–7.9	30%, 50%,70%, 80%, 100%	Spans 5-6 transect, arc-wise, and grid patterns	86–90%	76–83%

(Continued)

Table 1 (*Continued*)

	System							Sprinkler					Test				
Ref.	Type	Make[α]	VRI	Spans	Length (m)	Press[£] (kPa)	Flow (m3/min)	Type[€]	Height (m)	Spacing (m)	Press (kPa)	WET (m)	Wind (m/s)	application	Collectors	CU	DUlq
King et al. (2005)	LM		developed by researchers	3	100				1.5	3	138		<2.5	36%, 52%, 68%, 84%, 100%	multiple grid patterns	86–95%	
Han et al. (2009)	LM	Reinke	developed by researchers	2	76	172		Nelson 36 quick-change spinner nozzles	2	3	140			100%, 75%, 50%, 25% (100 % was 25 mm)	68 collectors spacing: 0.75m × 1.5 m grid	80–95%	

[£] Dukes and Perry (2006): inlet pressure was in non-variable-rate irrigation mode and at the pivot point. Sui and Fisher (2015): inlet pressure at the pivot point was 234 kPa and 290 kPa for the constant application rate and for variable application rate test, respectively.
[€] LESA: low elevation spray application; LDN: low drift nozzle; CP: center pivot; LM: lateral move; WET: Sprinkler wetted radius; Nelson Irrigation Corporation (Walla Walla, Washington); Senninger by Hunter Industries (Clermont, Florida).
[¥] Yari et al. (2017): reported results are for the experimental set-up on 4 September 2014.
[α] Reinke Manufacturing Company, Inc.: (Deshler, Nebraska, USA); Valley Irrigation by Valmont Industries, Inc.: (Omaha, Nebraska, USA).

There is a transition area between adjacent zones in which application depth changes gradually which causes edge effect between adjacent IMZs both along the irrigation system and in travel direction (Kranz et al., 2012). That is because, for a typical moving overhead irrigation system, there is an overlap between wetting areas of sprinklers such that each spot on the ground is irrigated by multiple sprinklers. The edge effect reduces application uniformity when different depths of water are applied in each zone (Sui and Fisher, 2015; O'Shaughnessy and Evett, 2015). The width and length of edge effect mainly depend on wetted diameter of sprinklers and overlap pattern, yet other factors including wind (speed and direction) and GNSS accuracy may change the location and shape of the area of influence. Chávez et al. (2010b) reported an average positioning error of 2.5 ± 1 m due to differential global positioning system reading inaccuracy. O'Shaughnessy and Evett (2015) tested a center pivot system equipped with a commercial VRI package and noticed that (i) the width of edge effects between adjacent irrigation zones ranged from 1.52 m to 6.1 m and (ii) the mean length of the edge effect was 9.1 m in the travel direction of the pivot. The spatial resolution of VRI systems and therefore the minimum size of IMZ in practice are influenced by the edge effect, a fact that should be considered for VRI implementation. In a recent case study, Higgins et al. (2016) developed a methodology to determine IMZ size and reported 23 m × 23 m as the smallest size for an independent IMZ for a nine-span center pivot equipped with a commercial VRI package.

Mechanical characteristics of center pivot and lateral move irrigation systems impact water application patterns across a field. These systems are often called continuous move systems; yet, in reality, a series of stop-advance sequences occurs independently for each tower along the lateral when these systems operate. O'Shaughnessy et al. (2013) suggested future field trials to assess the effect of intermittent movement of inner towers on performance of their 6-span center pivot irrigation system. Chávez et al. (2010c) studied and modelled the inherent lateral move water application errors and developed a methodology to use VRI systems to compensate for those errors.

5.3 Challenges and opportunities for adoption of site-specific irrigation systems

Current uses of commercial site-specific VRI technologies are on a fairly coarse scale (Kranz et al., 2012), and adoption of these machines by growers has been slow to develop and hard to rationalize considering energy and water saving at current prices along with current high implementation and management costs (Evans et al., 2013). This is also accompanied by the fear of complexity and lack of experience associated with successful implementation of VRI technologies among growers and service providers along with limited reported research-based results on the impacts of VRI technologies on crop production.

The focus of this chapter and the main potential benefits of VRI technologies are to enhance irrigation WUE and improve crop production by maintaining soil water content within the optimum range across the field while minimizing run-off and deep percolation, and making the best use of stored water within the root zone (Hedley and Yule, 2009b). Other potential advantages of site-specific VRI technologies are (i) ease of site-specific chemigation/fertigation, (ii) avoiding irrigation of non-cropped areas (e.g., roads, ponds, ditches, farm tracks, drainage pathways and water ways), (iii) enhanced mixed cropping capability and (iv) applying liquid animal waste to specific areas in a field especially when there are regulatory limitations on amount and location of application to protect wells and to reduce leaching and contamination of wildlife areas (Kranz et al., 2012; Hedley et al., 2009).

A study on adoption of variable rate technology by cotton growers in Texas, USA, showed higher adoption rates for areas with higher field-level variability and reported higher likelihood of variable rate technology adoption with younger growers, growers with big operations and those who use computers in their farming operations (Nair et al., 2012). A recent study in Brazil showed that expansion of precision farming is related to field-observed agronomic and economic benefits (Borghi et al., 2016). A study on PA technology adoption by cotton growers in the United States revealed that belief in potential improvement in environmental quality and adoption of other PA technologies were the main factors correlated with the early adoption of three PA technologies: grid soil sampling, yield monitoring and RS (Watcharaanantapong et al., 2014).

Growers and policymakers are becoming more interested in SS-VRI mainly because of the increasing limitation on irrigation water supplies and environmental issues around the world (Evans et al., 2013). In Europe, the greatest short-term enhancement in WUE is projected to come from precision irrigation adoption of overhead irrigation systems (Monaghan et al., 2013). This is because a majority of irrigated lands in Europe utilize overhead irrigation systems (including center pivots, sprinklers and high-pressure rain guns) for which the commercial VRI technology has been developed. A variety of factors such as commodity price, water price and availability, water quality, degree of field-level soil heterogeneity and climatic conditions influence expected improvement in production quality and quantity by site-specific irrigation systems. This economic improvement will not be high enough in all situations to cover the cost associated with implementation and management of these systems. However, widespread adoption of site-specific irrigation systems may be beneficial at a global scale by enhancing the water footprint and environmental impacts of irrigated agriculture through improvement in field-level WUE and reduction in nitrogen leaching out of the root zone (King et al., 2006).

6 Future trends and conclusion

6.1 Development of decision support tools

PA is progressing with emerging innovations in instrumentation and measurements, a revolution that is changing the concept of an agricultural unit from farm to subfield and down to individual plants. In the future, most farmers in developed countries will have access to precision farming equipment and be collecting numerous site-specific geospatial data. Previous VRI-related research projects have mainly focused on (i) designing irrigation systems to spatially vary irrigation rates and (ii) developing software/hardware for guiding the VRI system (Pan et al., 2013). Future research efforts should focus on the development of novel approaches for affordable data acquisition as well as robust data processing. At the same time, reliable automated decision support tools should be developed for real-time processing of data and control of irrigation systems. This is important because most research efforts have focused on providing more information to growers who have limited time to process the information (Howell et al., 2012). Recent advances in the development of smartphone applications for irrigation management (e.g. Bartlett et al., 2015; Vellidis et al., 2016) open up new horizons for availability of information to growers for real-time site-specific irrigation management.

6.2 Site-specific deficit irrigation

Site-specific irrigation implementation involves extra irrigation system hardware, labour and site-specific data to detect spatio-temporal changes in soil water status and/or crop stress in each IMZ (King et al., 2006). More studies are needed on site-specific strategies for deficit irrigation using VRI technologies specifically for regions where water allocation restrictions are prescribed for irrigation and/or water charges exist. These constraints can provide enough economic motivation for growers to make investments towards advanced irrigation technologies.

6.3 High-resolution monitoring of soil water status and ET

Development of underground soil moisture sensors (Dong et al., 2013) is a good example of emerging innovations that can substantially advance growers' ability for high-resolution monitoring of field-level soil water status. In addition, given the recent progresses in RS-based ET mapping, for example, development of Google Earth Engine Evapotranspiration Flux, Allen et al. (2015), new tools are likely to be developed in the near future to generate affordable spatial ET maps with high enough spatio-temporal resolution for SS-VRI management. Such maps can be potentially used to delineate IMZ, establish SS-WPFs and create irrigation prescription maps throughout the season.

6.4 On-farm irrigation research

Traditional irrigation/agronomic research is typically conducted in controlled environments following statistical principles on small-plot trials over multiple sites and across several years on a property allocated to research. These research trials are time-consuming and labour-intensive and hence may become unrealistic for the near future (Drummond et al., 2003). In addition, scaling and transforming findings of small-plot research trials to operational management of growers' fields are challenging. On-farm irrigation experimentation through cooperation between researchers and growers becomes more viable with more trustable results using site-specific irrigation systems which help to develop site-specific irrigation solutions.

7 List of abbreviations

ANN	Artificial neural network
BD	Bulk density
ECa	Apparent electrical conductivity
ET	Evapotranspiration
FC	Field capacity
GIS	Geographical information system
GPS	Global positioning system
GNSS	Global navigation satellite system
GWR	Geographically weighted regression
ILP	Integer linear programming
IMZ	Irrigation management zone
K-NN	K-nearest neighbour

KG Kriging
MLR Multiple linear regression
MZ Management zone
PA Precision agriculture
PAW Plant available water (also known as root zone available water holding capacity)
PTF Pedotransfer function
PWP Permanent wilting point
RKG Regression kriging
RS Remote sensing
SS-VRI Site-specific variable rate irrigation
SS-WPF Site-specific water production function
VRI Variable rate irrigation
WPF Water production function
WUE Water use efficiency

8 Where to look for further information

- *Journals: Precision Agriculture* (publisher: Springer), *Computers and Electronics in Agriculture* (publisher: Elsevier), *Transactions of the ASABE* (publisher: American Society of Agricultural and Biological Engineers), *Irrigation Science* (publisher: Springer), *Irrigation and Drainage Engineering* (publisher: American Society of Civil Engineers), *Agricultural Water Management* (publisher: Elsevier).
- *Books: Geostatistical Applications for Precision Agriculture* (publisher: Springer).
- *Events organized by:* American Society of Agricultural and Biological Engineers (website: www.asabe.org), International Society of Precision Agriculture (website: www.ispag.org), tri-societies [American Society of Agronomy (website: www.agronomy.org), Crop Science Society of America (website: www.crops.org), and Soil Science Society of America (website: www.soils.org)].

9 References

Abdu, H., Robinson, D. A., Seyfried, M. and Jones, S. B. (2008). Geophysical imaging of watershed subsurface patterns and prediction of soil texture and water holding capacity. *Water Resources Research*, 44(4). W00D18, doi:10.1029/2008WR007043.

Allen, R., Morton, C., Kamble, B., Kilic, A., Huntington, J., Thau, D., et al. (2015). EEFlux: A Landsat-based Evapotranspiration mapping tool on the Google Earth Engine. In: *Proceedings of the Joint ASABE/IA Irrigation Symposium*. ASABE, St Joseph, MI, USA. doi:10.13031/irrig.20152143511

Archie, G. E. (1942). The electrical resistivity log as an aid in determining some reservoir characteristics. Transactions of the American Institute of Mining and Metallurgical Engineers, 146, 54–61.

Auerswald, K., Simon, S. and Stanjek, H. (2001). Influence of soil properties on electrical conductivity under humid water regimes. Soil Science, 166(6), 382–90.

Bartlett, A. C., Andales, A. A., Arabi, M. and Bauder, T. A. (2015). A smartphone app to extend use of a cloud-based irrigation scheduling tool. *Computers and Electronics in Agriculture*, 111, 127–30.

Boluwade, A., Madramootoo, C. and Yari, A. (2015). Application of unsupervised clustering techniques for management zone delineation: Case study of variable rate irrigation in Southern Alberta, Canada. Journal of Irrigation and Drainage Engineering, 142(1), 05015007.

Borghi, E., Avanzi, J. C., Bortolon, L., Junior, A. L. and Bortolon, E. S. (2016). Adoption and use of precision agriculture in Brazil: Perception of growers and service dealership. Journal of Agricultural Science, 8(11), 89.

Bouma, J. (1989). Using soil survey data for quantitative land evaluation. In *Advances in Soil Science* (pp. 177–213). New York, USA: Springer.

Brock, A., Brouder, S. M., Blumhoff, G. and Hofmann, B. S. (2005). Defining yield-based management zones for corn–soybean rotations. Agronomy Journal, 97(4), 1115–28.

Bronson, K. F., Booker, J. D., Bordovsky, J. P., Keeling, J. W., Wheeler, T. A., Boman, R. K., et al. (2006). Site-specific irrigation and nitrogen management for cotton production in the Southern High Plains. Agronomy Journal, 98(1), 212–19.

Camp, C. R., Sadler, E. J., Evans, D. E., Usrey, L. J. and Omary, M. (1998). Modified center pivot system for precision management of water and nutrients. Applied Engineering in Agriculture, 14(1), 23–31.

Chávez, J. L., Pierce, F. J., Elliott, T. V. and Evans, R. G. (2010a). A remote irrigation monitoring and control system for continuous move systems. Part A: Description and development. Precision Agriculture, 11(1), 1–10.

Chávez, J. L., Pierce, F. J., Elliott, T. V., Evans, R. G., Kim, Y. and Iversen, W. M. (2010b). A remote irrigation monitoring and control system (RIMCS) for continuous move systems. Part B: Field testing and results. Precision Agriculture, 11(1), 11–26.

Chávez, J. L., Pierce, F. J. and Evans, R. G. (2010c). Compensating inherent linear move water application errors using a variable rate irrigation system. Irrigation Science, 28(3), 203–10.

Cid-Garcia, N. M., Albornoz, V., Rios-Solis, Y. A. and Ortega, R. (2013). Rectangular shape management zone delineation using integer linear programming. Computers and Electronics in Agriculture, 93, 1–9.

Cohen, Jacob (1960). A coefficient of agreement for nominal scales. Educational and Psychological Measurement 20 (1): 37–46.

Córdoba, M., Bruno, C., Costa, J. and Balzarini, M. (2013). Subfield management class delineation using cluster analysis from spatial principal components of soil variables. Computers and Electronics in Agriculture, 97, 6–14.

Corwin, D. L. and Lesch, S. M. (2010). Delineating site-specific management units with proximal sensors. In *Geostatistical Applications for Precision Agriculture* (pp. 139–65). Dordrecht, Netherlands: Springer.

Corwin, D. L. and Lesch, S. M. (2003). Application of soil electrical conductivity to precision agriculture: Theory, principles, and guidelines. *Agronomy Journal*, 95, 455–71.

Debaene, G., Niedźwiecki, J., Pecio, A. and Żurek, A., (2014). Effect of the number of calibration samples on the prediction of several soil properties at the farm-scale. Geoderma, 214, 114–25.

Dong, X., Vuran, M. C. and Irmak, S. (2013). Autonomous precision agriculture through integration of wireless underground sensor networks with center pivot irrigation systems. Ad Hoc Networks, 11(7), 1975–87.

Drummond, S. T., Sudduth, K. A., Joshi, A., Birrell, S. J. and Kitchen, N. R. (2003). Statistical and neural methods for site-specific yield prediction. *Transactions of the ASAE*, 46(1), 5.

Dukes, M. D. and Perry, C. (2006). Uniformity testing of variable-rate center pivot irrigation control systems. *Precision Agriculture*, 7(3), 205.

Duncan, H. A., (2012). Locating the variability of soil water holding capacity and understanding its effects on deficit irrigation and cotton lint yield. Master's Thesis, University of Tennessee. http://trace.tennessee.edu/utk_gradthes/1286.

Earl, R., Taylor, J. C., Wood, G. A., Bradley, I., James, I. T., Waine, T., et al. (2003). Soil factors and their influence on within-field crop variability, part I: Field observation of soil variation. *Biosystems Engineering*, 84(4), 425–40.

Evans, R. G., LaRue, J., Stone, K. C. and King, B. A. (2013). Adoption of site-specific variable rate sprinkler irrigation systems. Irrigation Science, 31(4), 871–87.

Fortes, R., Millán, S., Prieto, M. H. and Campillo, C. (2015). A methodology based on apparent electrical conductivity and guided soil samples to improve irrigation zoning. Precision Agriculture, 16(4), 441–54.

Fraisse, C. W., Duke, H. R. and Heerman, D. F. (1995a). Laboratory evaluation of variable water application with pulse irrigation. Transactions of the ASAE, 38(5), 1363–9.

Fraisse, C. W., Heermann, D. F. and Duke, H. R. (1995b). Simulation of variable water application with linear-move irrigation systems. Transactions of the ASAE, 38(5), 1371–6.

Fraisse, C. W., Sudduth, K. A. and Kitchen, N. R. (2001). Delineation of site-specific management zones by unsupervised classification of topographic attributes and soil electrical conductivity. Transactions of the ASAE, 44(1), 155–66.

Haghverdi, A., Cornelis, W. M. and Ghahraman, B. (2012). A pseudo-continuous neural network approach for developing water retention pedotransfer functions with limited data. Journal of Hydrology, 442, 46–54.

Haghverdi, A., Leib, B. G., Washington-Allen, R. A., Ayers, P. D. and Buschermohle, M. J. (2015a). High-resolution prediction of soil available water content within the crop root zone. Journal of Hydrology, 530, 167–79.

Haghverdi, A., Leib, B. G., Washington-Allen, R. A., Ayers, P. D. and Buschermohle, M. J. (2015 b). Perspectives on delineating management zones for variable rate irrigation. *Computers and Electronics in Agriculture*, 117, 154–67.

Haghverdi, A., Leib, B. G., Washington-Allen, R. A., Buschermohle, M. J. and Ayers, P. D. (2016). Studying uniform and variable rate center pivot irrigation strategies with the aid of site-specific water production functions. *Computers and Electronics in Agriculture*, 123, 327–40.

Haghverdi, A., Öztürk, H. S. and Cornelis, W. M. (2014). Revisiting the pseudo continuous pedotransfer function concept: Impact of data quality and data mining method. *Geoderma*, 226, 31–8.

Han, Y. J., Khalilian, A., Owino, T. O., Farahani, H. J. and Moore, S. (2009). Development of Clemson variable-rate lateral irrigation system. *Computers and Electronics in Agriculture*, 68(1), 108–13.

Hedley, C. B. and Yule, I. J. (2009a). A method for spatial prediction of daily soil water status for precise irrigation scheduling. *Agricultural Water Management*, 96(12), 1737–45.

Hedley, C. B. and Yule, I. J. (2009b). Soil water status mapping and two variable-rate irrigation scenarios. *Precision Agriculture*, 10(4), 342–55.

Hedley, C. B., Yule, I. J., Tuohy, M. P. and Vogeler, I. (2009). Key performance indicators for simulated variable-rate irrigation of variable soils in humid regions. *Transactions of the ASABE*, 52(5), 1575–84.

Herbst, M., Diekkrüger, B. and Vereecken, H., (2006). Geostatistical co-regionalization of soil hydraulic properties in a micro-scale catchment using terrain attributes. *Geoderma*. 132(1), 206–21.

Higgins, C. W., Kelley, J., Barr, C. and Hillyer, C. (2016). Determining the minimum management scale of a commercial variable-rate irrigation system. *Transactions of the ASABE*, 59 (6): 1671–80.

Howell, T. A., Evett, S. R., O'Shaughnessy, S. A., Colaizzi, P. D. and Gowda, P. H. (2012). Advanced irrigation engineering: Precision and Precise. *Journal of Agricultural Science and Technology. A*, 2(1A), 1–9.

Iqbal, J., Thomasson, J. A., Jenkins, J. N., Owens, P. R. and Whisler, F. D. (2005). Spatial variability analysis of soil physical properties of alluvial soils. *Soil Science Society of America Journal*, 69(4), 1338–50.

Karmeli, D. and Peri, G. (1974). Basic principles of pulse irrigation. *Journal of the Irrigation and Drainage Division*, 100(3), 309–19.

Kerry, R., Oliver, M. A. and Frogbrook, Z. L. (2010). Sampling in precision agriculture. In *Geostatistical Applications for Precision Agriculture*, ed. M. A. Oliver (pp. 35–63). Dordrecht, Netherlands: Springer.

Kim, Y., Evans, R. G. and Iversen, W. M. (2009). Evaluation of closed-loop site-specific irrigation with wireless sensor network. *Journal of Irrigation and Drainage Engineering*, 135(1), 25–31.

King, B. A., Brady R. A., McCann I. R. and Stark J. C. (1995). Variable rate water application through sprinkler irrigation. In *Site-specific Management for Agricultural Systems*, eds. P. C. Robert, R. H. Rust and W. E. Larson (pp. 485–93). Madison, WI, American Society of Agronomy.

King, B. A., Stark, J. C. and Wall, R. W. (2006). Comparison of site-specific and conventional uniform irrigation management for potatoes. *Applied Engineering in Agriculture*, 22(5), 677–88.

King, B. A., Wall, R. W., Kincaid, D. C. and Westermann, D. T. (2005). Field testing of a variable rate sprinkler and control system for site-specific water and nutrient application. *Applied Engineering in Agriculture*, 21(5), 847–54.

King, B. and Kincaid, D. C. (2004). A variable flow rate sprinkler for site-specific irrigation management. *Applied Engineering in Agriculture*, 20(6), 765–70.

Kitchen, N. R., Sudduth, K. A., Myers, D. B., Drummond, S. T. and Hong, S. Y. (2005). Delineating productivity zones on claypan soil fields using apparent soil electrical conductivity. *Computers and Electronics in Agriculture*, 46(1), 285–308.

Kranz, W. L., Evans, R. G., Lamm, F. R., O'Shaughnessy, S. A. and Peters, R. T. (2012). A review of mechanical move sprinkler irrigation control and automation technologies. *Applied Engineering in Agriculture*, 28(3), 389–97.

Liang, X., Liakos, V., Wendroth, O. and Vellidis, G. (2016). Scheduling irrigation using an approach based on the van Genuchten model. *Agricultural Water Management*, 176, 170–9.

Monaghan, J. M., Daccache, A., Vickers, L. H., Hess, T. M., Weatherhead, E. K. and Grove, I. G., et al. (2013). More 'crop per drop': Constraints and opportunities for precision irrigation in European agriculture. *Journal of the Science of Food and Agriculture*, 93(5), 977–80.

Moral, F. J., Terrón, J. M. and Da Silva, J. M. (2010). Delineation of management zones using mobile measurements of soil apparent electrical conductivity and multivariate geostatistical techniques. *Soil and Tillage Research*, 106(2), 335–43.

Mouazen, A. M., Alhwaimel, S. A., Kuang, B. and Waine, T. (2014). Multiple on-line soil sensors and data fusion approach for delineation of water holding capacity zones for site specific irrigation. *Soil and Tillage Research*, 143, 95–105.

Nair, S., Wang, C., Segarra, E., Johnson, J. and Rejesus, R. (2012). Variable rate technology and cotton yield response in Texas. *Journal of Agricultural Science and Technology. B*, 2(9B), 1034.

O'Shaughnessy, S. A. and Evett, S. R. (2015). Zone edge effects with variable rate irrigation. Proceedings of the 27th Annual Central Plains Irrigation Conference, Colby, KS, USA. Available online at: https://www.ksre.k-state.edu/irrigate/oow/p15/OShaughnessy_15.pdf.

O'Shaughnessy, S. A., Evett, S. R. and Colaizzi, P. D. (2015). Dynamic prescription maps for site-specific variable rate irrigation of cotton. *Agricultural Water Management*, 159, 123–38.

O'Shaughnessy, S. A., Urrego, Y. F., Evett, S. R., Colaizzi, P. D. and Howell, T. A. (2013). Assessing application uniformity of a variable rate irrigation system in a windy location. *Applied Engineering in Agriculture*, 29(4), 497–510.

Omary, M., Camp, C. R. and Sadler, E. J. (1997). Center pivot irrigation system modification to provide variable water application depths. *Applied Engineering in Agriculture*, 13(2), 235–39.

Pan, L., Adamchuk, V. I., Martin, D. L., Schroeder, M. A. and Ferguson, R. B. (2013). Analysis of soil water availability by integrating spatial and temporal sensor-based data. *Precision Agriculture*, 14(4), 414–33.

Rivers, E.D. and Shipp R.F. (1972). Available water capacity of sandy and gravelly North Dakota soil. *Soil Science*, 113, 74–80.

Romano, N. and Santini, A. (2002). Field, In: *Methods of Soil Analysis*. Part 4, Physical Methods, Soil Sci. Soc. Am. Book Ser., vol. 5, edited by J. H. Dane and G. C. Topp (pp. 721–38). Madison, WI, USA: Soil Science Society of America.

Roth, R. L. and Gardner, B. R. (1989). Modified self-moving irrigation system for water-nitrogen crop production studies. *Applied Engineering in Agriculture* 5(2): 175–9.

Sadler, E. J., Evans, R., Stone, K. C. and Camp, C. R. (2005). Opportunities for conservation with precision irrigation. *Journal of Soil and Water Conservation*, 60(6), 371–8.

Saey, T., Van Meirvenne, M., Vermeersch, H., Ameloot, N. and Cockx, L. (2009). A pedotransfer function to evaluate the soil profile textural heterogeneity using proximally sensed apparent electrical conductivity. *Geoderma*, 150(3), 389–95.

Sudduth, K. A., Kitchen, N. R., Wiebold, W. J., Batchelor, W. D., Bollero, G. A., Bullock, D. G. and Thelen, K. D. (2005). Relating apparent electrical conductivity to soil properties across the north-central USA. *Computers and Electronics in Agriculture*, 46(1), 263–83.

Sui, R. and Fisher, D. K. (2015). Field test of a center pivot irrigation system. *Applied Engineering in Agriculture*, 31(1), 83–8.

Sui, R., Fisher, D. K. and Reddy, K. N. (2015). Yield response to variable rate irrigation in corn. *Journal of Agricultural Science*, 7(11), 11.

Van Genuchten, M.Th., (1980). A closed-form equation for predicting the hydraulic conductivity of unsaturated. *Soil Science Society of America Journal*, 43, 892–98.

Vellidis, G., Liakos, V., Andreis, J. H., Perry, C. D., Porter, W. M., Barnes, E. M., et al. (2016). Development and assessment of a smartphone application for irrigation scheduling in cotton. *Computers and Electronics in Agriculture*, 127, 249–59.

Vories, E., Rhine, M. and Straatmann, Z. (2016). Investigating irrigation scheduling for rice using variable rate irrigation. *Agricultural Water Management*. 179, 314–23.

Watcharaanantapong, P., Roberts, R. K., Lambert, D. M., Larson, J. A., Velandia, M., English, B. C., et al. (2014). Timing of precision agriculture technology adoption in US cotton production. *Precision Agriculture*, 15(4), 427–46.

Yari, A., Madramootoo, C. A., Woods, S. A. and Adamchuk, V. I. (2017). Performance evaluation of constant versus variable rate irrigation. *Irrigation and Drainage*. doi:10.1002/ird.2131

Zhang, X., Shi, L., Jia, X., Seielstad, G. and Helgason, C. (2010). Zone mapping application for precision-farming: A decision support tool for variable rate application. *Precision Agriculture*, 11(2), 103–14.

Chapter 15

Precision crop protection systems

E. C. Oerke, University of Bonn, Germany

1 Introduction

The production potential of agricultural and horticultural crops worldwide is reduced by abiotic and biotic factors. Various kinds of pests, that is, weeds, pathogens, viruses, arthropods and other animal pests reduce the quantity and quality of crop products. Oerke (2006) estimated that, without control, diseases, animal pests and weeds may reduce worldwide crop production by 16%, 18% and 34%, respectively; despite current crop protection practices, actual losses to diseases (caused by fungi, oomycetes and bacteria), animal pests and weeds account for 11%, 11% and 8% of the production potential. Pest management has to be effective, efficient and sustainable and can be realized by combining mechanical, biological and/or chemical tools and other supportive technologies in integrated pest management (IPM) systems (Oerke et al. 2010a).

Early detection of pests is essential to reduce their propagation and spread and their negative effects on yield formation – crop productivity, quantity and quality – by crop protection activities. Commonly the incidence of pests (as well as abiotic stress conditions, e.g. drought) in a field is heterogeneous in time and space. Pesticides applied uniformly can result in over- and under-dosing which in both cases is inefficient and uneconomical and may be an environmental burden. In site-specific management, information on spatial and temporal variability of pests in the field is used in a systematic approach to minimize the amount of pesticide applied by demand-oriented application without reducing the efficiency of crop protection and productivity (Oerke et al. 2010a). Provided the distribution of many weeds, diseases and animal pests is patchy, the advent of remote sensing techniques may help to control these pests only where and when needed. Site-specific application of pesticides has the potential to reduce their use and may cut the costs and ecological impacts in agricultural crop production systems (Gebbers and Adamchuck 2010).

http://dx.doi.org/10.19103/AS.2017.0032.16

This situation-specific, threshold-oriented and sustainable method of pest management requires large-scale (and geo-referenced) monitoring of pests in the crop for precise timing and application of control measures. Site-specific pest management uses information from electronic and other technologies in order to assess the crop's health status and the incidence of pests and – based on this information and decision support systems – to decide whether crop protection activities have to be implemented.

Heterogeneity in the incidence and spatial distribution of pests is often obvious for weed clusters, known for nematodes and other soil-borne pathogens, and largely unknown for other (airborne) diseases and insects and mites. For many insect-vectored virus diseases, airborne diseases like powdery mildews and cereal rusts, a rather random incidence in the field is likely. However, disease patterns may change with the stage of epidemic development (see below).

The detection and control of weeds, a technology in an advanced stage of transformation into practice is covered by Gerhards (Chapter 17). Crop protection systems which use information on within-field heterogeneity in pest occurrence need also the capability to react in a site-specific way in order to deliver the control activity where demanded. The use of variable-rate technology, an important method for site-specific application of pesticides, has been summarized in Chapter 8. Sensor technologies can be applied to limit pesticide applications to the target trees or grapevine in order to reduce the amount applied, for example, in arthropod control in fruit trees. Optical vegetation sensors are used to switch on and off according to the presence of target plants and their height

Table 1 Control of various pest groups using precision crop protection technologies – requirements and current status

		Diseases		Arthropod pests	
Trait	Weeds	Soil-borne	Air-borne	Soil-borne/ borer	Air-borne
Causal organism(s)	Plants	Nematodes, fungi, bacteria, viruses	Fungi, bacteria, viruses	Various orders of insects and mites	
Size [mm]	10–1000	0.0001–1	0.0001–0.1	0.1–100	0.1–100
Detection of	Plant(s)	Plant damage	Disease symptom(s)	Plant damage, pest activity	Plant injury, pest organism(s)
Size [mm]	10–1000	10–100	1–20	10–100	1–20
Mobility	Very low	Low	High – very high	Low – medium	High
Cycles per season	1(–2)	1–5	1–10	1–2(3)	1–10
Variability (s/t) in the field[1]	Moderate to high	Moderate to high	Low to high	Moderate to high	Moderate to high
Identification	XXX	–	X(–)	–(X)	?
Quantification	XXX	X(X)	XX	X(X)	
Prognosis/DSS[2]	X(X)	X(X)	XX	X	

[1] In space and time. [2] Decision support system.
XX: advanced stage; X: first steps/moderate knowledge; ?: not known/not feasible; –: knowledge low.

(Giles et al. 1989). Since commercial systems for automatic detection of plant diseases are not available yet, mechanical or optical devices for sensing plant biomass (Dammer 2010; Tackenberg et al. 2016) or crop growth stage (Scotford and Miller 2005) may be used to control spraying of fungicides (and growth regulators) adapted to the leaf area in order to apply equal amounts of active ingredient per biomass and to reduce the overall amount of pesticides. The sensor system actuates the dosing system of the sprayer, but ignores (spatial differences in) the presence of diseases.

The incidence of plant diseases and of arthropod pests is also often heterogeneous in space and time; however, the assessment and site-specific management of these pests is still a major challenge (Table 1). Pest groups differ significantly in size, mobility and number of generations (cycles) per growth period of crops. Microorganisms causing diseases as well as soil-borne pests cannot be detected as such, but by the response of the plant (disease symptom, crop damage) to pest attack. Soil-borne pests often are largely immobile, whereas the spores of fungal pathogens may be spread over thousands of kilometres by air currents. Basic information is still required on ground truth patterns of major airborne diseases and arthropods, epidemiology, spread and impact of latent infections, (spectral) signatures of infected crops for the identification of diseases and arthropod pests, and data handling for on-the-go applications (Oerke et al. 2010a). No commercial system for early detection of diseases and arthropod pests is currently available. This chapter summarizes the state-of-the-art of approaches for site-specific management of plant diseases caused by fungi, oomycetes, bacteria, viruses, nematodes and arthropod pests. Current knowledge on the spatial (and temporal) incidence of these pests and technologies suitable for their detection is given.

2 Variability of pest incidence and pest management strategies

2.1 Spatial and temporal heterogeneity

Heterogeneity in the incidence of pests in the field is a prerequisite for site-specific management. The incidence of diseases in the field often is rather heterogeneous in time and space – at least in early stages (Zadoks and van den Bosch 1994; Madden and Hughes 1995; Waggoner and Aylor 2000; Ojiambo and Scherm 2010). Heterogeneity results from the interaction between physical environment (e.g. edaphic factors, microclimatic conditions), crop development and the life cycles of the different pests. Heterogeneity may be limited to foci of primary inoculum or low levels of soil-borne pathogens as reported for the potato cyst nematode (Stafford et al. 2000). In later stages of epidemics in the vegetation period or at higher levels of overall infestation, the pattern of diseased plants may become more homogeneous.

Incidence of soil-borne diseases (caused by fungi, oomycetes and nematodes) and animal pests is often heterogeneous because they are adapted to or favoured by specific soil conditions, for example, soil texture, organic carbon (from crop debris) content and water status. Incidence and spatial patterns of airborne diseases (and mobile arthropod pests) depend on differences in crop canopy depending on plant status (influenced by soil conditions, availability of water and nutrients), microclimate of the crop, weather conditions and other factors.

In small fields spatial distribution of pests may be rather homogenous; however, conventional agricultural management practices have ignored the spatial variability also in larger fields. Seed-borne pathogens may be effectively controlled by seed dressing with non-systemic and systemic fungicides. This type of pesticide application reduces the amount of active ingredients applied per area and positions it to the site where needed, thereby reducing negative effects on the environment. Important fungal diseases like bunts and smuts of cereals can be controlled exclusively at this stage of the disease cycle. These pathogens may not be controlled when first disease symptoms occur (during anthesis and plants' maturation) as effective fungicides are lacking and, moreover, as plants have been already damaged and control is too late. Furthermore, seed dressing is active also against soil-borne pathogens by creating a protection zone around the treated seed via diffusion into the soil. Therefore, it is also used for targeted application of insecticides to control soil-borne insects, for example, in oilseed rape, sugar beet, potatoes and maize.

Airborne pathogens and pests often have a short generation time, produce a high number of spores/progenies and have several life cycles within the crop growth period. In order to limit the exponential spread, the detection and control of polycyclic pests have to be as early and efficient as possible, that is, at low levels of infestation of crop plants. In annual crops, polycyclic pests have to be identified and controlled in an online approach, for example, detection and control in one vehicle crossing, in order to limit the epidemic development. Monocyclic diseases may be managed by seeding cultivars differing in resistance to the pest and in yield potential (e.g. Bt crops, sugar beet resistant to cyst nematodes, cotton) and the use of seed dressing on appropriate sites of the field. In perennial crops, patches of soil-borne pests may be used for control activities in the off-season, for example, removal of diseased fruit trees to eradicate the pathogen and/or adaptation of the next crop to the spatial pattern by planting other cultivars (or even other species). Control options in the field tend to rely on pest control with synthetic pesticides that can be targeted easily (Swinton 2003). Biological control agents may be used under controlled conditions where mass release of predators and hyperparasites/parasitoids is common practice.

2.2 Sensing of pests

For a long period, the classification of plant diseases and pests has depended on inspection by human eye and the knowledge of the operator (e.g. Nutter 2001; Bock et al. 2010). The development of microscopic, serological and DNA-based technologies has significantly improved and facilitated the identification of plant pathogens and arthropod pests; however, these techniques are destructive and often time-consuming. Technical sensing of pests or pest-induced changes in the crop status may be used for different purposes, under different (environmental) conditions and by different users:

- Quarantine inspection (under controlled conditions); regulatory authorities
- Field crops; farm industry
- Speciality crops, especially perennial, high-value crops like orchards and vineyards; wine grower and gardener
- Post-harvest diseases and insect pests, in combination with control of plant products' quality; food industry
- Phenotyping (genotypes' reactions) for resistance to biotic (and abiotic) stress factors; plant breeders.

The detection of pests in the context of quarantine regulations is crucial in order to eradicate primary sources of inoculum from a plot, a field, a country or a continent. Protocols of the European and Mediterranean Plant Protection Organization for the detection of plant pathogens integrate phenotypic, serological and molecular techniques; these methods provide complementary information (Martinelli et al. 2014). Physical sensors provide the opportunity for using autonomous systems for quarantine inspections. Information from sensing of pests in the process of decision-making in pest management of field crops may be used I) in the same growth period (polycyclic pests), II) in the next growing season (soil-borne pests) and III) for estimating the yield (and production) losses expected to occur from pest-induced crop injury. In the food industry, post-harvest sensing of the quality and safety of (processed) plant products includes checking their suitability for storage, sorting/grading of quality, the detection of defects, bruises and diseases of fruits and vegetables as well as the assessment of mycotoxin contamination from fungi, for example, *Aspergillus flavus* in maize kernels (Yao et al. 2013).

Sensors are expected to be objective, accurate, precise, rapid and may be integrated into autonomous systems for the monitoring of crops for the incidence of plant diseases and arthropod pests, that is, checking and keeping continuous record of plant development and yield formation during critical stages. The systematic observation of crop performance by devices allows the operator to intervene when the process deviates from the performance or parameters fall below/exceed threshold levels. Sensors should be capable of I) detecting the deviation of the health status of the crop brought about by the pest, II) identifying the disease or causal pest organism and III) quantifying the severity of the pest attack. Detection of pests may be highly sensitive, but often lacks diagnostic potential. The identification of a disease requires the qualification to differentiate among various/all potential diseases according to disease-specific symptoms. The quantification of pests is expected to be basic for imaging systems and a challenge for non-imaging sensors.

2.2.1 Sensors

Sensors may be classified according to the range of the electromagnetic spectrum measured, that is, visible (VIS), near-infrared (NIR), short-wave infrared (SWIR), thermal infrared (TIR), radar; the scale (proximal, ground-based; unmanned aerial vehicles (UAV), air- and space-borne); and the type of data recording (non-imaging, imaging). Passive sensors measure the characteristic radiation of an object (thermography) or the reflectance of solar radiation; active sensors (radar, laser, fluorescence) emit their own measuring radiation and record its modifications due to interactions with the object of interest. They are less affected by environmental factors, but strongly depend on the geometry between sensor and object and the runtime of the electromagnetic radiation is twice that of a passive sensor. Other sensor types include mechanical and chemical sensors, electromagnetic induction measurements (for soil characterization) and acoustic sensors for recording the activity of insect pests.

Spectral information

Spectral information in the range of 350 to 2500 nm from (diseased) plants may be recorded by non-imaging and imaging spectral light detector systems. A non-imaging spectrometer may have a high spectral resolution (hundreds of narrow wavebands) in the full range; however, spectral information results from the average of the sensor's field of view (FOV). In contrast, imaging systems may record only one waveband or the sum over the spectrum from 400 to 700 nm (panchromatic), sensors may use the three basic colour components red, green and blue (typical bandwidth 60–80 nm), additional (NIR) bands (multispectral, discrete and somewhat narrow wavebands) or narrow spectral bands over a continuous spectral range (hyperspectral, spectral resolution <1 nm).

The mixed spectral information from different objects within the FOV complicates the interpretation of the signal, largely prevents to draw conclusions on the object(s) without additional information. Moreover, the incidence of tiny site(s) of primary infection is only a minor proportion of the overall signal which is dominated by healthy tissue (and background?). Therefore, reliable spectral differences only occur at high overall disease severities. Using a spectrometer, Oerke et al. (2016) could differentiate healthy and diseased grapevine leaves when 25–30% of leaf area was covered by downy mildew. With an imaging approach, individual spots (diameter 2–3 mm) of sporulating pathogen could be spotted.

For sensory detection of biotic and abiotic stress factors, the sensitivity and the specificity of the signal/techniques are crucial, in particular for the detection of plant diseases caused by microorganisms. Differentiation between diseases occurring on a crop is an essential feature for operational systems, and, moreover, the detection of host plant colonization by pathogens may become feasible even before the appearance of typical disease symptoms. The uniqueness of spectral signatures of plant diseases is not universally agreed in the literature (Nutter et al. 2010). Stress-causing agents, and likewise various pathogens often causes similar (disease) symptoms and changes in plant physiology (Nutter et al. 2010; Stafford 2000). Chlorosis (= reduction in chlorophyll content) and necrosis (= dead plant tissue) are caused not only by plant pests, but also by abiotic factors. However, in contrast to abiotic factors causing visible changes in plant appearance, that is, changes of the complete shoot, pathogen infection and infestation of plant tissue by arthropod pests often start at individual, discrete sites and these, often tiny, primary symptoms produce (characteristic?) spatial patterns on leaves or within the canopy which may be included as information for pest identification. Moreover, many important diseases are associated with characteristic symptoms produced by pathogen structures on the leaves, such as powdery mildew, downy mildew, rusts and smuts producing unique disease symptoms. Subtle spectral differences could be used to distinguish between necrotic tissue and grapevine tissue discoloured from resistance reactions to *Plasmopara viticola* (Oerke et al. 2016).

Thermography

Digital infrared thermography is used to assess and visualize differences in surface temperature of leaves, plants or crop canopies (Jones and Schofield 2008; Costa et al. 2013). Radiation emitted in the range of 8–12 (14) µm is detected by thermographic cameras and illustrated in false colour images. Thermal cameras are characterized by thermal sensitivity (thermal range, measurement precision), image resolution (pixel size, number of pixels, depth of focus) and scan speed (Oerke and Steiner 2010). Thermography can be applied on various scales ranging from microscopy, proximal ground-based equipment to air- and space-borne sensors. The major advantage of infrared thermal imaging is the non-invasive, non-contact and non-destructive nature of the technique to determine the temperature distribution of any object in a short period of time (Vadivambal and Jayas 2011).

Fluorescence measurements

Chlorophyll fluorescence is a non-invasive measurement of photosystem II activity which is highly sensitive to abiotic and biotic factors (Baker 2008; Murchie and Lawson 2013). The methodology may be used to monitor pathogenesis of relevant diseases also under field conditions. Rousseau et al. (2013) developed an image analysis procedure for quantifying the leaf area impacted by pathogens in order to use pulse-amplitude-modulation (PAM) chlorophyll fluorescence measurements of dark-adapted plants for high-throughput phenotyping.

The intensity and time course of fluorescence (spectra) of chlorophyll and other plant metabolites, for example, phenols, induced by various light sources (LED, laser, sunlight) may be recorded under controlled conditions as well as in the field. The instruments commonly are active sensor systems; passive detection of sun-induced chlorophyll fluorescence using remote spectroscopy is currently investigated to quantify changes in photochemical efficiency at canopy level under natural light conditions (Pinto et al. 2016). Pathogen attack affects the photosynthetic apparatus of crops – pigments, electron transport chain, enzymes of the Calvin cycle – directly by reducing the photosynthetic leaf area (necrosis), chlorophyll degradation (chlorosis), or indirectly through feedback regulation of electron transport chain.

Electronic nose

In plant science an electronic nose (e-nose) is a device to detect volatile organic compounds (VOCs) emitted by plants which have essential functions in their growth, communication, defence and survival (Baldwin et al. 2006). Plants have been shown to emit VOCs when affected by diseases and arthropod pests (Cardoza et al. 2002; Shulaev et al. 1997). As VOC emissions are also changed by abiotic stress – due to unsuitable temperature, water supply and air pollutants – a multitude of chemicals evaporated under normal atmospheric conditions of temperature and pressure are released and the composition of this cocktail is changed, sometimes with characteristic successions. A plant species may emit similar VOCs upon induction by different diseases and different plant species may emit similar VOCs. Although often unspecific, plant-emitted volatiles may be used for the detection and differentiation of diseases and arthropod infestation. The blend of VOCs may be specific for plant-parasite interactions as plants attacked by herbivorous arthropods release VOCs to attract specific parasitoids (Dicke et al. 1990; Dicke 2009) and predatory mites preferred the blend from leaves infested with their natural prey (De Boer et al. 2008).

The development and use of electronic noses (and tongues) in agriculture, horticulture and forestry have been reviewed by Li et al. (2010) and Wilson (2013). They are used in monitoring food quality and production processes, to detect crop diseases and to identify insect infestations (Li et al. 2010). Measurement techniques include the collection of VOCs and their assessment under laboratory conditions by gas chromatography and the use of e-noses (e-sensing), devices with an array of electronic sensors for gaseous chemicals and a mechanism of pattern recognition.

Acoustic sensors

Acoustic signals from the activity of pests may be detected by accelerometers, piezo-electric sensors, microphones and ultrasonic transducers (Liu et al. 2016).

2.2.2 Scales of non-destructive sensing

Remote sensing is the acquisition of information about an object or phenomenon by a recording device not in physical or intimate contact with the object or phenomenon under study, hence is non-invasive and non-destructive (Simonett 1983). Sensors of electromagnetic radiation may be used in pest management on various platforms – satellite, airplane, helicopter, blimp, unmanned aerial vehicle, machine-mounted, handheld and autonomous systems (Fig. 1). The distance between sensor and object determines the spatial resolution of the signal/information and its useability for pest management, damage assessment and basic research.

Proximal remote sensing may be defined as acquisition and classification of reflectance or transmittance data with an imaging sensor within a distance <1 m (often much less) from the object (Nansen 2016). Microscopy is essentially a type of proximal remote sensing, provided remote sensing is defined as above. Airborne or space-borne systems have lower spatial resolution than proximal sensor systems, but cover larger areas of landscape.

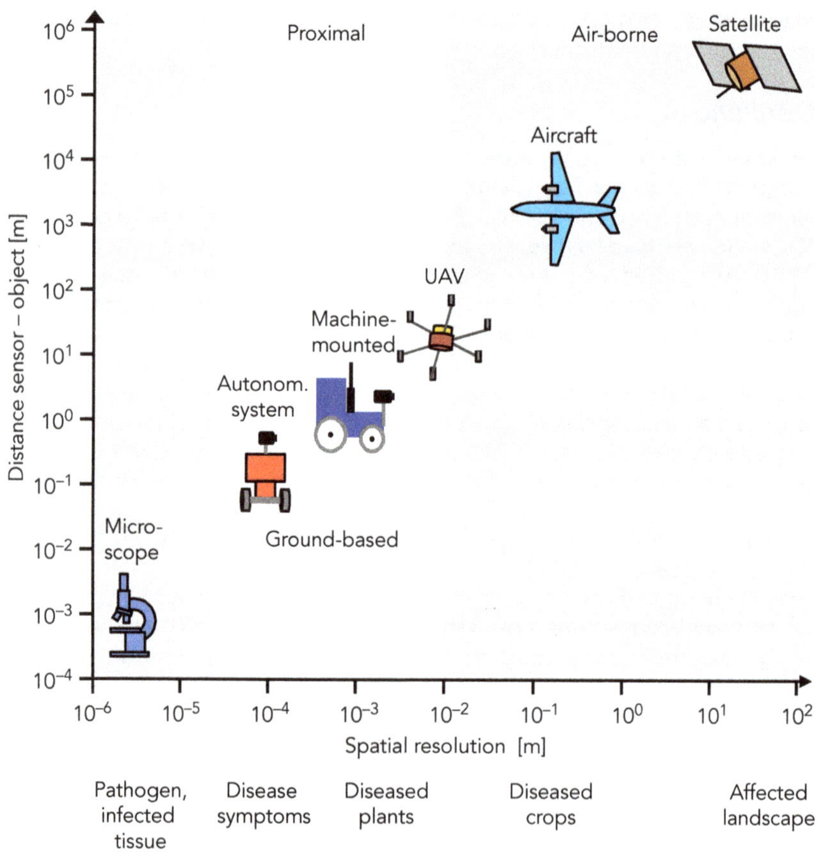

Figure 1 Scales of remote sensing platforms for pests and spatial resolution.

The maximal spatial resolution is defined by the minimum size of a pixel and hence the smallest identifiable symptom or structure. Pixels with a size larger than the target object size result in mixed information which is unsuitable for the definition of a spectral library, the prerequisite for a supervised classification.

The purpose of sensing defines the required spatial resolution of spectral data. As pathogens are microorganisms with typical size <0.01 mm which may be studied only microscopically, sensing of diseases on the plant and leaf level is based on disease-specific symptoms (or unspecific crop damage). The small size of primary site(s) of pest infestation of plants necessitates a high spatial resolution of sensors suitable for early detection of diseases. Whereas water shortage or nutrient deficiency affects the complete plant or patches of plant canopy, individual tiny spots on plant organs can be detected only when the pixel size is considerably lower than the size of the targeted symptom. Supposing a circular disease symptom, spatial resolution should be better than one-third of the symptom's diameter (Oerke et al. 2016). This requirement limits the suitability of air- and satellite-borne remote sensing and non-imaging techniques to the detection and quantification of crop damage resulting from diseases and pest infestations in later stages. Nutter et al. (2010) studied the incidence and spread of Asian soybean rust of soybean and *Cercospora* leaf blight of peanut from satellite imagery by measuring the increase in the area of the heavily affected crop biomass. Aerial and satellite-borne images, however, are hardly suitable for the detection of early disease symptoms with typical dimensions in the mm range on the leaf or plant scale. Additionally, reflectance data are scale-dependent and profiles developed at one spatial scale may not be transferred to larger or smaller scales (Voss et al. 2010; Nansen and Elliott 2016).

Abiotic stressors – for example, drought, temperature extremes, pollutants – affect all shoot parts in a similar way and may be detected (and quantified) at the plant or canopy level. In contrast, the infestation of plants by most pathogens (and arthropod pests) are highly restricted to specific plant parts and affect only a single site of primary infestation. Soil-borne pathogens – bacteria, oomycota, fungi, nematodes – attacking the plant's root system cause malfunction of roots which may affect the total shoot. In this respect, they are similar to abiotic stressors and often it is hard to distinguish between wilting induced by pathogens or by water shortage.

The spatial resolution for the detection of the primary infection sites of airborne diseases like powdery mildew, rust, downy mildew or leaf spot has to be significantly better than for root-infecting pathogens or drought. In multispectral imaging for the detection of citrus greening-infected citrus trees, Garcia-Ruiz et al. (2103) demonstrated that higher spatial resolution of an UAV-based sensor resulted in higher classification accuracy when compared with an airplane-mounted sensor.

3 Sensor use for disease management

Visual inspection of plants (in the field) and microscopic, microbiological, serological and molecular diagnostic techniques represent common practice for the detection, diagnosis and quantification of plant diseases (Bock et al. 2010; Nutter 2001). These methods are effective, but labour-intensive, time-consuming and may result in delayed diagnosis (Baron et al. 2016). Techniques based on imaging sensors are expected to detect early infections in the field (Behmann et al. 2015a; Martinelli et al. 2014).

3.1 Management of nematodes

Plant–parasitic nematodes belong to the phylum nematoda (roundworms) and are tiny soil-inhabiting organisms that parasitize plant roots (and shoots). As their mobility is low, crop protection problems from nematodes are confined to field sites suitable for propagation and survival depending on soil conditions. The distribution of cyst nematodes and root-knot nematodes (RKNs) in the field is heterogeneous in many crops, at least at low levels of nematode infestation: potato (Evans et al. 1999, 2003; Stafford et al. 2000), maize (Wyse-Pester et al. 2002), cotton (Wrather et al. 2010; Holguin et al. 2015a,b) and sugar beet (Hbirkou et al. 2011; Hillnhuetter et al. 2011).

The use of air- and satellite-borne remote sensing for the detection of damage caused by plant–parasitic nematodes and other soil-borne pathogens for optimization of pest management is favoured by several factors: I) impaired functionality of infested roots becomes visible in the foliage at different times in the growing season, II) infestations are clustered in the field, III) movement out of a cluster is slow due to low pathogen mobility, IV) introduction of new infection loci into a field is rare and V) precision detection used in one season can be applicable for future crops (Hillnhuetter et al. 2010).

3.1.1 Spatial patterns of nematode infestation

In South Carolina, USA, the only nematode species found to be associated with soybean was *Hoplolaimus columbus* but, in North Carolina, *H. stephanus* was the dominant species (Holguin et al. 2016). Variation in the response of tolerant soybean cultivars to nematode populations may be due to their genetic diversity. In cotton, aggregations of the Columbia lance nematode (CLN, *H. columbus*) were not affected by the crop species. As CLN and the reniform nematode (*Rotylenchulus reniformis*) have been associated with different soil textures, the sand content may be used to identify the key pest (Holguin et al. 2015b). In contrast, aggregation patterns of *R. reniformis* were favoured by suitable host plant species (cotton and soybean) (Holguin et al. 2015a). The relationships between RKN (*Meloidogyne incognita*) populations and soil terrain, for example, apparent electrical conductivity and edaphic parameters, suggest the potential for using soil properties to identify RKN risk zones (Ortiz et al. 2010, 2011).

Aerial hyperspectral sensing of sugar beet fields with patches infested with the beet cyst nematode (*Heterodera schachtii*) and *Rhizoctonia solani* causing *Rhizoctonia* crown and root rot demonstrated that remote sensing in combination with geographic information system technologies can be used effectively for the detection and mapping of crop damage resulting from the pathogens (Hillnhuetter et al. 2011). *Heterodera schachtii* prefers deep soil with medium to light textures and non-stagnant water conditions (Hbirkou et al. 2011). Management maps based on apparent electric conductivity and soil taxonomy may improve site-specific management strategies on nematode-infested fields. Hbirkou et al. (2012) found airborne hyperspectral imaging suitable for the characterization of spatial heterogeneity in soil organic carbon content at the field-scale, a parameter for the delineation of management zones.

3.1.2 Sensing for nematode management

Hillnhuetter et al. (2010) reviewed the state-of-the-art of remote sensing for the detection of soil-borne nematodes and fungal pathogens. Airborne infrared thermography and hyperspectral reflectance proved to have potential to detect patches of plant–parasitic nematodes and root rot fungi in sugar beet which may be used for pest management in the next growth seasons.

Mueller et al. (2010) and Overstreet et al. (2014) summarized the development and use of site-specific nematode management in cotton in the United States. Site-specific application of nematicides is a viable option for producers as a management tool against several nematode pathogens of cotton. Levels of *R. reniformis* are affected by modest differences in soil texture and elevation (Davis et al. 2013). In a cotton field with management zones, but levels of *R. reniformis* above the action threshold in all zones, a uniform rate of nematicide would have been recommended. Ortiz et al. (2012) demonstrated the value of site-specific application of nematicides based on management zones, although this approach might not be economically beneficial in fields with little variability in soil texture. Practical recommendations for the effective control of cotton nematodes by site-specific management were provided by Liu et al. (2015).

3.2 Spatial patterns of diseases caused by microorganisms

As the development of plant diseases depends on the presence of a susceptible host plant, an aggressive pathogen, and suitable environmental conditions, the occurrence of crop diseases may greatly vary in space and time. Especially polycyclic pathogens may reach epidemic levels under suitable conditions within a short period, whereas other sites may be less affected or the overall disease level is lower in other growth periods. Optical sensing technologies, therefore, may be useful in identifying the primary disease foci which are the sources of inoculum for further spread, and areas differing in disease intensity in order to estimate yield losses due to the diseases.

In early investigations on the spatial distribution of plant diseases incidence of powdery mildew caused by the anemochorous fungus *Blumeria graminis* was favoured by shelterbelts or neighbouring fields with winter cereals (Koch 1980; Bjerre 1999), whereas the soil-borne *Zymoseptoria tritici* (teleomorph *Mycosphaerella graminicola*) prevails in zones with high survival on plant debris and is favoured by crop density and use of growth regulators. Secher (1997) and Bjerre (1999) successfully used the heterogeneity of diseases for site-specific variation of the fungicide dose in disease management. Systematic investigations on mapping the diversity of diseases on the field and regional levels, for example, Lorestani et al. (2013) are rare. Distribution of yellow Sigatoka leaf spot of banana caused by *Mycosphaerella musicola* depends on elevation and concentration of sand and organic matter (Freitas et al. 2016).

Although *Fusarium* head blight of wheat is aggregated on a microscale and the within-field scale, the incidence depends on inoculum from the previous crop, soil cultivation, rain spells during wheat anthesis as well as airborne inoculum, factors likely to favour first field-to-field and regional variability (Del Ponte et al. 2003; Oerke et al. 2010b).

3.3 Spectral information for disease management

Plant diseases commonly affect leaf pigmentation, chemical composition and physical structure of plant tissue, characteristics which may be reflected in the spectral signature of diseased plant parts. Biotroph pathogens take up nutrients from living plant cells via haustoria and these cells may die only in late disease stages from nutrient depletion and exhaustion. Pathogen structures (mycelium, spores/sporangia) – hyalin for powdery mildews and downy mildews, pigmented for rusts – on the leaf surface produce characteristic visible disease symptoms. In contrast, necrotrophic pathogens first kill the plant cells they are feeding on subsequently. Necrotic plant tissue has collapsed, cellular integrity is lost and the size of necroses often increases with time. Root-infecting pathogens (bacteria, oomycetes, fungi, nematodes) impair the functionality of the root system and/or reduce the transport of water and nutrients in the xylem and often result in wilt diseases.

Early reviews of Jackson (1986), Hatfield and Pinter (1993), and Nilsson (1995) emphasized the potential of using the information of electromagnetic radiation after interaction with plants for plant pathology and crop protection purposes. Remotely sensed parameters may indicate deviations from the expected signal and may be even able to detect diseases pre-visually. At that time (imaging) sensors often were limited to air- and space-borne remote sensing; ground-based digital sensors have been introduced in the 1990s and later have considerably broadened the application of remote sensing techniques, first in crop science and subsequently in crop production.

Rapid progress in the sensor recording of crop pests has been reviewed putting emphasis on crop protection (West et al. 2003; Bock et al. 2010; Mahlein et al. 2012a; Martinelli et al. 2014), plant phenotyping (Li et al. 2014b; Fahlgren et al. 2015; Mutka and Bart 2015; Simko et al. 2017) and both (Sankaran et al. 2015; Baron et al. 2016; Mahlein 2016). Arredondo Valdes et al. (2016) summarized the efforts in citrus production to identify efficient methods for citrus greening detection. The potential of non-invasive techniques for post-harvest quality analysis and inspection of specialty crops, mainly fresh fruits and vegetables, has been reviewed by Hahn (2009) and Ruiz-Altisent et al. (2010). Liew et al. (2008) reported on the development of transgenic phytosensors ('reporter plants') to elucidate plant status in relation to environmental conditions. Most of these studies, however, were limited to laboratory or controlled environments at leaf level.

Remote sensing techniques differ in the potential to identify plant diseases; that is, chlorophyll fluorescence and thermography, although highly sensitive to changes in the plant metabolism caused by the disease, are not able to identify the cause – abiotic, various types of biotic pests – of these changes (Oerke et al. 2014). In contrast, spectral information – in combination with spatial information from images – and information on VOCs emitted from affected plants seem to be suitable for disease identification or categorization. Some papers emphasize the potential of sensors to detect plant diseases (and other stresses) prior to the appearance of visible symptoms (Mahlein 2016; Peteinatos et al. 2016). Many sensing systems have been developed for use in specialty crops, for example, citrus and blueberry (Lee and Ehsani 2015). The use of these ground and air-based systems is often not limited to disease detection, but also includes fruit recognition and yield mapping.

Recent improvements in the application of remote sensing for crop protection depend on the development of high-resolution optical sensors and data analysis methods able to cope with the resolution, size and complexity of the sensor signals (Plaza et al. 2009; Behmann et al. 2015b). Following Martinelli et al. (2014), the data analysis techniques adopted for plant disease detection may be grouped into (I) correlation and regression analysis of disease detection and severity; (II) use of general spectral vegetation indices or development of specific indices sensitive to disease symptoms; (III) data mining algorithms applied to spectral data processing and feature extraction/selection for data dimensionality reduction; and (IV) machine learning and classification techniques, parametric and non-parametric, supervised and unsupervised, for classifying presence/

absence of disease, various levels of disease severity, and differentiating among various diseases of a crop. Machine learning methods include support vector machines and neural networks for classification (supervised methods) and *k*-means and self-organizing maps for clustering (unsupervised). Singh et al. (2016) have summarized various approaches of machine learning for phenotyping in plants.

3.3.1 RGB and multispectral imaging

Aerial photography has been the primary remote sensing technique used in crop protection. Taubenhaus et al. (1929) used panchromatic images from cotton fields taken from an airplane to identify patches infested by the soil-borne fungus *Phymatotrichum omnivorum*. Aerial photographs were an effective technique to detect for example, stem canker in oilseed rape, leaf spot of sugar beet, mould and dwarf-mosaic virus in maize, virus yellows in sugar beet, barley yellow dwarf virus in winter wheat, barley yellow mosaic virus in barley, take-all and yellow rust in winter wheat (Ryerson et al. 1997). Coupled with spatial analysis, aerial colour-infrared photographs were successfully used to assess patterns of late blight in potato grown under pivot irrigation systems quantitatively (Johnson et al., 2003). Nicholas (2004) used (analog) multispectral and thermal remote imaging of wheat with *Septoria* leaf blotch (caused by *Z. tritici*) to determine the optimum time of fungicide application in a field plot experiment.

Airborne R, G and NIR digital imagery accurately identified root rot areas within cotton fields (Yang et al. 2005). Buffer zones around the root rot areas were generated to account for the spread of cotton root rot on the classification maps in site-specific disease management. Reflectance measurements in quinoa diseased with downy mildew (caused by *Peronospora farinosa* f.sp. *chenopodii*) gave a high correlation to yield due to its ability to measure pathogen-induced defoliation (Danielson and Munk 2004).

Camargo and Smith (2009a,b) developed an image-processing-based algorithm for automatic identification of visual symptoms of plant diseases. They used colour transformations and texture information in order to feed support vector machine-based classification methods. Linear discriminant analysis outperformed other approaches to classify pixels from RGB images to identify ozone damage of cloverleaves using colour and texture information (Opstad Kruse et al. 2014). Barbedo (2013, 2016) reviewed and discussed the main challenges in automatic plant disease identification based on visible range images.

Parameters derived from RGB images taken during grain filling were highly correlated with grain yield of wheat genotypes differing in susceptibility to *Puccinia striiformis* causing stripe rust (Zhou et al. 2015a). In experiments to detect *Fusarium* head blight-diseased wheat ears under field conditions, a multispectral (red, infrared) camera had a better calibration capacity than an RGB camera which resulted in a higher rate of false positives (Dammer et al. 2011).

The use of RGB images for the detection (and identification) of *Cercospora* leaf spot (CLS) of sugar beet has been investigated by Bauer et al. (2011) and de Coninck et al. (2012). Neumann et al. (2014) and Zhou et al. (2014, 2015b) proposed various feature combinations on a single leaf scale under field conditions to differentiate CLS from other leaf diseases of sugar beet and to discriminate CLS from the background and specular reflection, respectively.

A machine vision system for the detection of early powdery mildew on greenhouse tomatoes uses augmented lighting and reached detection rates of 85% (Wspanialy and

Moussa 2016). Processing of RGB images to detect and classify the severity of bacterial spot (due to *Xanthomonas* spp.) in field tomatoes included colour space transformation, filtering and use of a severity index (Borges et al. 2016). The results correlated strongly with experts' analysis and had smaller error. The detection sensitivity of proximal optical sensing for primary discrete foci of powdery mildew in grapevine was optimized by image recording at a view angle of 40–60° (Oberti et al. 2014).

A customized vision sensor measuring the polarization of starch accumulation in Huanglongbing-(HLB) diseased citrus leaves enhanced identification performance of citrus greening to >95.5% (Pourreza et al. 2015a,b). HLB detection based on the extraction of the texture and histograms of image colour space and C-support vector classification achieved 91% detection with low-cost and low-computation complexity (Deng et al. 2016).

3.3.2 Air- and satellite-borne remote sensing

Satellite-borne remote sensing technologies have been used successfully to detect, quantify and map disease stress (Nutter et al. 2010). GPS, GIS and remote sensing technologies, however, may also be used to extract pathogen-specific temporal and spatial signatures that have potential to identify the cause(s) of biotic and abiotic stress in crops. Nevertheless, in general satellite-borne remote sensing is used to monitor the health status of vegetation/forests and to detect (and quantify) crop damage resulting from high disease severity in annual and perennial crops. In this regard, remote sensing is a very cost-effective tool (Lee et al. 2010). In perennial crops the information may be used for pest management. In addition to spatial resolution, the (timely) availability of space and airborne data for decision-making in crop protection is limited because of the need for pre-processing, effects of unsuitable weather conditions. Mewes et al. (2011) demonstrated that even a few hyperspectral bands as well as bands with lower spectral resolution still allow an adequate detection of powdery mildew in wheat fields.

Satellite imagery has been used to detect orange rust (caused by *Puccinia kuehnii*) in sugarcane (Apan et al. 2004), take-all (caused by *Gaeumannomyces graminis*) in wheat (Chen et al. 2007), outbreaks and development of *Sclerotinia* rot (due to *Sclerotinia sclerotioum*) in a mustard growing region in Northwest India (Bhattacharya and Chattopadhyay 2013) and mapping of powdery mildew of wheat at a regional scale (Zhang et al. 2014b). Laudien et al. (2004) compared QuickBird satellite multispectral data and hyperspectral airborne and ground-based data suitable for detecting *Rhizoctonia* root and crown rot affected sugar beet on a regional, field and plant level.

The detection of powdery mildew (caused by *Blumeria graminis*) and leaf rust in a wheat field from multitemporal high-resolution satellite images demonstrated an increase in classification accuracy with disease severity during the growth period (Franke and Menz 2007). Zhang et al. (2011) used a spectral knowledge base of diseased winter wheat plants for linking disease severity from aerial measurements to satellite multispectral imagery suitable for scouting of larger regions. Scenes from two satellites demonstrated the importance of spatial resolution (IKONOS 1–4 m; Landsat 8 15–30 m). In particular, texture features extracted from high-resolution IKONOS imagery improved classification accuracy of health levels of *Robinia pseudoacacia* forests in the Yellow River delta of China exposed to dieback diseases (Wang et al. 2015a).

Airborne multispectral (CIR) imagery was applied to detect and map *Phytophthora* foot rot-infected citrus trees (Fletcher et al. 2001), citrus greasy spot (Du et al. 2004) and late blight in tomato (Zhang et al. 2003, 2005). Multi- and hyperspectral data were used to identify areas of trees with severe citrus greening (Li et al. 2012); to detect and map symptoms caused by beet cyst nematode and *Rhizoctonia* crown and root rot in sugar beet (Hillnhuetter et al. 2011; Reynolds et al. 2012); to quantify stripe rust severity in wheat (Huang et al. 2007, 2012); and to detect *Verticillium* wilt of olive (Calderon et al. 2013, 2015), Huanglongbing of citrus (Li et al. 2014a), red leaf blotch of almond (Lopez-Lopez et al. 2016) and leaf discolouration of grapevine from grapevine leaf roll-associated viruses in five vineyards and two seasons (MacDonald et al. 2016).

Multitemporal airborne multispectral imagery from cotton fields was used to examine the within-field consistency and variation of cotton root rot infections over 10-year intervals and to assess the feasibility to use historical imagery to create prescription maps for site-specific management of the soil-borne disease (Yang et al. 2016). Spatial patterns of the disease were similar between the two seasons. Buffer zones around the infected areas accounted for the potential expansion and temporal variation of the disease.

3.3.3 Proximal hyperspectral techniques for disease monitoring

Hyperspectral data include a high amount of redundant information as neighbouring narrow wavebands are highly correlated with each other (Bravo et al. 2003; Delalieux et al. 2007b; Mewes et al. 2011). Only a few wavebands of interest may be necessary for disease detection and identification. Thus the expensive, complex and accident-sensitive hyperspectral devices may be tailored to specific crops and their relevant diseases (Bravo et al. 2003; West et al. 2010). However, the system should be flexible enough to assess also new diseases, for example, to differentiate leaf rust of wheat (due to *P. triticina*) from stripe rust which has become a threat to European wheat production because of changes in the race spectrum (Hovmøller et al. 2016).

Spectroradiometer measurements

Using a multispectral radiometer, Steddom et al. (2003, 2005) and Laudien et al. (2003, 2004) indicated that spectral reflectance data may be useful for the detection of sugar beet diseases under field conditions. Other diseases of field crops could be successfully detected by using spectroradiometers; for example, panicle blast (caused by *Magnaporthe oryzae*) of rice (Kobayashi et al. 2001), late blight (due to *Phytophthora infestans*) of tomato (Zhang et al. 2003), and scab (due to *Venturia inaequalis*) on apple leaves (Delalieux et al. 2007a).

Spectral vegetation indices (SVIs) are widely used for monitoring, analysing and mapping temporal and spatial variation in vegetation. The calculation of ratios between relevant wavebands from different ranges of the spectrum results in a reduction of data dimension. The use of combinations of two or more SVIs improved the discrimination of three sugar beet diseases (Mahlein et al. 2010). These SVIs were used as features for an automatic discrimination by Rumpf et al. (2010). Developed usually for aerial remote sensing of vegetation, SVIs often are hardly suitable for proximal detection and identification of plant diseases. The development of disease-specific indices has been shown to give results superior to that from SVIs developed for remote sensing of vegetation (Mahlein et al. 2013).

Zhang et al. (2012b) used a spectroradiometer on detached wheat leaves for detecting powdery mildew. The photosynthetic reflection index was the most promising spectral vegetation index for the detection of yellow rust at the leaf level (Zhang et al. 2012a, 2014a). Cao et al. (2013, 2015) measured plots of wheat diseased with powdery mildew with a spectrometer 0.5 m above the canopy and calculated various SVIs for relating reflectance to powdery mildew severity and grain yield, respectively. Feng et al. (2016) constructed dual-green vegetation indices for powdery mildew quantification at the canopy level. Disease-specific indices for the detection of wheat leaf rust by ground-based spectroradiometer measurements outperformed SVIs from remote sensing (Ashourloo et al. 2014). SVIs optimized for the detection of wheat leaves infested with powdery mildew, yellow rust and aphids resulted in 85.2%, 91.6% and 93.5% classification accuracies, respectively (Huang et al. 2014).

Reflectance spectra of detached grapevine leaves from non-infected and grapevine leaf roll-associated virus-3 (GLRaV-3) infected plants were recorded by Naidu et al. (2009). The differences in leaf reflectance and their correlation with RT-PCR results for the presence of GLRaV-3 suggested spectral reflectance technique as a promising tool for diagnosis of grapevine leaf roll disease (GLD) in the field. Using a multispectral image of a grapevine growing area (Ningxia, China), the combination of spectrometer measurements and an artificial intelligence algorithm resulted in a classification accuracy of GLD severity from 75 to 94% (Hou et al. 2016). Calcante et al. (2012) compared two commercial sensors for green biomass with a Vis-NIR spectroradiometer for their potential to detect adaxial leaf symptoms of downy mildew of grapevine (caused by *Plasmopara viticola*). Using the linearity between percentage symptomatic leaf area (= pale yellow oil spots that become dry and necrotic and reduce photosynthesis) and the normalized difference vegetation index, the spectrometer gave better results than the commercial sensors which had even more problems to measure the severity of downy mildew on abaxial leaf sides.

Using a spectroradiometer on a cherry picker, Lelong et al. (2010) measured reflectance of oil palms infected by *Ganoderma* basal stem rot at 6–10 m height, corresponding to about 9 m² sampled on crown areas of about 60 m². Disease severity classes, graded on a 1–4 scale, and diseased and healthy palms could be discriminated with an accuracy >89%. Mid-infrared spectroscopy (2.5–25.0 μm) of leaves of palms with no, mild, moderate and severe symptoms of basal stem rot proved suitable to classify the samples into four disease levels (Liaghat et al. 2014).

Other applications of spectroradiometers for disease detection include bacterial leaf spot (caused by *Xanthomonas perforans*) of tomato (Jones et al. 2010), bacterial leaf blight of rice (Yang 2010), yellow mosaic disease (caused by mung bean yellow mosaic virus) of blackgram (Prabhakar et al. 2013a), and HLB-infected citrus trees (Sankaran et al. 2011). The spectral signature of citrus black spot (CBS) caused by *Phyllosticta citricarpa* did not change significantly during storage for two months and reflectance at 588 nm was identified for developing a monochrome vision-based sensor for CBS diagnosis (Pourreza et al. 2016).

Hyperspectral imaging

Due to technical progress, the number of reports on proximal hyperspectral imaging (HSI) for disease detection and identification has increased recently. Earlier reports include the detection of yellow rust in wheat (Bravo et al. 2003; Moshou et al. 2004) and the discrimination of sugar beet symptoms caused by the beet cyst nematode (*Heterodera schachtii*) and the basidiomycete *Rhizoctonia solani* (Hillnhuetter et al. 2012a) in the field. HSI has been used previously in the food industry for post-harvest quality control of fruits. Qin et al. (2009) used the technique for the detection of canker lesions on citrus fruits.

Panicle blast (caused by *Magnaporthe oryzae*) of rice was detected on various rice cultivars sampled from the field (Huang et al. 2015). A 'bag of spectra words' representation of hyperspectral data as input for a support vector machine classifier resulted in classification accuracies of up to 81% for six-class grading and 96% for two-class differentiation. Wahabzada et al. (2015, 2016) were able to differentiate among three barley leaf diseases.

Hyperspectral imaging under laboratory conditions could recognize *Fusarium* head blight of wheat from growth stage (GS, Lancashire et al. 1991) 71 to GS85 with an optimum at GS 75 (Bauriegel et al. 2011a,b). Disease severity was correlated to photosynthetic efficiency and infected ears were detected at >5% severity by chlorophyll fluorescence. SAM classification of hyperspectral data was successful at early stages, but higher disease severity. A developed head blight index may be suitable for field applications. An NIR (1000–1700 nm) hyperspectral camera enabled the differentiation and quantification of *Stagonospora nodorum* blotch on detached leaves of durum wheat on a conveyor belt 48 h after inoculation (Iori et al. 2015).

Foliar anthracnose of strawberry (caused by *Colletotrichum* spp.) was detected with an overall accuracy of 80% by using hyperspectral imaging and multispectral imaging using four wavebands selected previously (Yeh et al. 2016). The differentiation between healthy and symptomatic leaves was better than the detection of latent infections invisible to the naked eye.

Hyperspectral imaging was used to study the development of various *Alternaria* species on leaves of oilseed rape (Baranowski et al. 2105). The method was highly predictive and also detected *C. beticola* on sugar beet leaves with a classification accuracy of >98% (Arens et al. 2016). Spatial distributions of pigment contents in cucumber leaves with angular leaf spot (caused by *Pseudomonas syringae* pv. *lachrymans*) were successfully assessed (Zhao et al. 2016). HIS was suitable to identify early and late blight on tomato leaves caused by *Alternaria solani* and *Phytophthora infestans*, respectively (Xie et al. 2015). The combination of spatial and spectral information resulted in 97% classification accuracy. Similarly, Roscher et al. (2016) demonstrated a gain in accuracy when using geometry and spectral information. Data processing in sensor fusion, however, has to take into account different structures of, and demands on, the data.

In the food industry, HSI has been used for the detection of Fusarium damaged kernels (FDK) in wheat samples (Shahin and Symons 2011). The classification of kernels into sound and two FDK categories had >92% overall accuracy. Results based on six selected wavelengths were comparable to those based on the full spectrum. In fluorescence hyperspectral imaging, single maize kernels contaminated with aflatoxins had longer peak wavelength than non-contaminated kernels (Yao et al. 2013). Near-infrared (1000–2500 nm) hyperspectral imaging for the differentiation of kernels with aflatoxin levels >100 ppb or <10 ppb gave accuracies ranging from 80 to 96% (Wang et al. 2015b,c). Atas et al. (2012) proposed a compact machine vision system based on hyperspectral imaging and machine learning for the detection of aflatoxin contamination of chilli pepper.

Hyperspectral data of citrus fruits proved suitable for differentiating CBS, greasy spot, melanose, wind scar and a normal fruit (average accuracy 92%). Bulanon et al. (2013) derived a multispectral image algorithm to be adopted in real-time multispectral imaging for CBS detection. Similarly, Li et al. (2016) used hyperspectral imaging of early post-harvest decay symptoms (caused by *Penicillium digitatum*) in navel orange fruits in order to develop a fast classification method for an online multispectral imaging system. The use of four key wavelength images resulted in 98.6% correct classification with no false negatives.

The potential of hyperspectral imaging in phenotyping for disease resistance has been demonstrated for the systems including barley – powdery mildew (Wahabzada et al. 2016), grapevine – downy mildew (Oerke et al. 2016) and sugar beet – CLS (Leucker et al. 2016, 2017). The large amount of data recorded ('big data') stresses the importance of efficient methods for data mining and classification. Wahabzada et al. (2016) demonstrated the potential of advanced machine learning methods and the advantage of result visualization for intuitive interpretation of hyperspectral data.

3.4 Thermal imaging

Infrared thermography is highly suitable for the detection of plant diseases affecting the water status and transpiration of plants. The temperature of green plant parts results from the interactions between ambient temperature, light conditions, relative humidity and a plant's transpiration rate; plant surface temperature decreases due to the cooling effect of evapotranspiration (Jones and Schofield 2008). Pathogenesis may affect the status of stomata regulating the water loss of plants similar to abiotic stress factors, and may increase cuticular transpiration due to damage of the outer tissue layer. Thermal imaging is a passive sensing method to monitor spatial and temporal patterns of heterogeneities in transpiration due to plant diseases on different scales and may include pre-symptomatic detection of within-tissue differences on the leaf level. Oerke and Steiner (2010) have reviewed the potential and limits of thermography in sensing for plant diseases.

At the field (and regional) scale, thermography may be used in scouting for spatial heterogeneity in the incidence of soil-borne diseases characterized by patchiness in the occurrence of above-ground symptoms and low mobility in the field (Hillnhuetter et al. 2010). Aerial thermographic images demonstrated patches of sugar beets infested by *H. schachtii* with increased leaf temperature due to reduced water uptake of nematode-infected roots (Schmitz et al. 2004). Thermal imaging of leaves is especially suitable for the detection and quantification of wilt-causing fungi, for example, *Fusarium oxysporum* f.sp. *cucumerinum* in cucumber (Wang et al. 2012). Although chemical control is not possible, disease detection may be used to remove infected plants even pre-symptomatically in order to limit disease spread. A portable system was used to differentiate between pea accessions susceptible and resistant to *Fusarium oxysporum* f.sp. *pisi* under controlled conditions (Rispail and Rubiales 2016).

Thermography may be also applied for detecting primary disease foci caused by airborne pathogens as well as for monitoring of environmental conditions conducive to pathogen infections, for example, leaf wetness and leaf wetness duration, parameters useful for models quantifying disease risks (Oerke and Steiner 2010). Lenthe et al. (2007) assessed the microclimate of wheat canopies and the spatial and temporal dynamics of fungal diseases. The detection of *Fusarium* head blight symptoms on wheat ears by using infrared thermography in the field is based on the higher temperature of infected ears or kernels within ears compared to healthy ears (Oerke and Steiner 2010; Al Masri et al. 2017). This technique may contribute to a target-oriented sampling (e.g. for nucleic acid-based disease diagnosis) and to site-specific harvesting in order to reduce the contamination of grain with mycotoxins produced by *Fusarium* species.

On the tissue level, temperature changes due to pathogen attack or defence mechanisms have been reported for tobacco – tobacco mosaic virus, sugar beet – CLS (Chaerle et al. 2007b), cucumber – downy mildew caused by *Pseudoperonospora cubensis* (Lindenthal et al. 2005; Oerke et al. 2006), grapevine – downy mildew caused by *Plasmopara viticola* (Stoll et al. 2008a,b) and apple leaves – scab (Oerke et al. 2011). Depending on the stage of pathogenesis and the appearance of chloroses and necroses, infection resulted in typical spatial and temporal temperature patterns during disease development. For downy mildews and scab, deviations of temperature from healthy proved to be suitable to detect latent infection of leaf tissue. Thermal imaging was superior to fluorescence imaging in detecting and quantifying apple scab on leaves under controlled conditions (Belin et al. 2013).

The sensitivity to environmental conditions during measurements and the variability in the thermal response of plants during pathogenesis limit the practical use of thermography for disease detection. Furthermore, the thermal response (= changes in transpiration) of plants largely lacks diagnostic potential for the identification of plant diseases. The sensitivity of thermography, for example, for pre-symptomatic disease detection may be combined with other sensors or other diagnostic tools, for example, for quarantine stations or in the field. Classification models using features from multispectral cameras and a thermal camera in nadir position over citrus trees gave 87% accuracy in the detection of HLB-symptomatic trees (Sankaran et al. 2013).

As thermal imaging of plants is also influenced by measuring geometry (Costa et al. 2013), Raza et al. (2015) combined thermal and visible imaging with depth information to remotely detect tomato plants with powdery mildew due to *Oidium neolycopersici*. The combination of colour information, thermal and depth information improved the accuracy of detecting diseased plants.

The higher transpiration of dense cereal canopies may be used as a bio-indicator. Airborne thermography was suitable for the delineation of management zones within fields (due to differences in soil suitability) as well as for the assessment of plant vitality related to crop yield in fungicide trials (Stenzel et al. 2007; Oerke and Steiner 2010).

3.5 Recording of plant fluorescence

Modifications in photosynthetic activity due to the activity of plant pathogens on the cellular and tissue level lack the potential to identify the cause of this effect – physical and chemical factors may have effects similar to various diseases. Measurements of other wavelengths or fluorescence spectra may overcome this deficit. The high sensitivity of fluorescence techniques – reviewed by Chaerle et al. (2009), Scholes and Rolfe (2009) – is compromised additionally by the longer response time of the active sensor system, that is, the time between emission and detection of the measured signal.

Chlorophyll fluorescence

In laboratory experiments dark adaptation of leaves was obligatory for an accurate detection of powdery mildew and leaf rust of wheat by PAM fluorescence imaging (Kuckenberg et al. 2009b). Spot measurements of laser-induced chlorophyll fluorescence were not suitable for the differentiation of nitrogen deficiency, powdery mildew and leaf rust at wheat leaf and canopy level (Kuckenberg et al. 2009a).

Studies on the effect of *Monographella albescens* infection on the photosynthetic activity of rice leaf blades confirmed high sensitivity of photochemical quenching and yield of photochemistry to pathogenesis-related changes in chlorophyll fluorescence (Tatagiba et al. 2015). Avocado plant roots infected by *Rosellinia necatrix* showed altered chlorophyll fluorescence of leaves only in later stages of disease development (Granum et al. 2015).

Laser-induced fluorescence imaging showed a potential to monitor *Citrus sinensis* for the occurrence of citrus greening incited by Candidatus *Liberibacter asiaticus* (Pereira et al. 2011). Diseased and healthy plants had significant differences in colour descriptors suitable to identify citrus greening at a pre-symptomatic stage of the disease.

The spatio-temporal and methodological context of remotely sensed passive solar-induced chlorophyll fluorescence (SICF) is different from most of the methods described in literature so far (Porcar-Castell et al. 2014). Results of SICF images of cassava leaves recorded with a multispectral imaging system indicated that changes in the red-to-far-red fluorescence image ratio were correlated to the severity level of the viral cassava mosaic disease at an early stage (Raji et al. 2015). The technique has potential for monitoring the health status of crops from proximal sensing platforms.

Chaerle et al. (2007a) demonstrated that phenotyping of resistance differences in sugar beet to CLS relies on the use of attached leaves in chlorophyll fluorescence imaging (CFI). A CFI approach was successfully used to distinguish lettuce breeding lines resistant and susceptible to *Bremia lactucae* (causing downy mildew), respectively (Bauriegel et al. 2014). The use of CFI for studying plant–pathogen interactions has been reviewed by Rolfe and Scholes (2010).

Other fluorescence techniques

A few fluorescence features from a handheld fluorescence spectrometer and bagged decision tree classifier yielded 97% accuracy in the classification of healthy and HLB-symptomatic citrus trees (Sankaran and Ehsani 2013). Results from the United States and Brazil indicated that SVM could classify HLB-infected leaf fluorescence intensities with up to 90% classification accuracy (Wetterich et al. 2013). Multi-colour fluorescence imaging was successful in distinguishing between healthy and *Dickeya dadantii* (bacteria) infiltrated squash and melon leaves (Pérez-Bueno et al. 2016). The results indicated a potential to scale up this approach for plant phenotyping.

Fluorescence spectra of wheat infected by *Puccinia triticina* were suitable to detect leaf rust development in a pre-symptomatic stage as early as two days after inoculation (Roemer et al. 2011). A combination of piecewise curve fitting and support vector machines resulted in 93% classification accuracy. Fluorescence lifetime, image-resolved multispectral fluorescence and selected indices of a portable multi-parametric fluorescence device were tested for their suitability to detect powdery mildew and leaf rust in four spring barley genotypes (Buerling et al. 2011; Leufen et al. 2014). Fluorescence (410–560 nm) mean lifetimes revealed genotype-specific responses to infections 3 d ai. Blue and green fluorescence may help to differentiate between the diseases at later disease stages. Early detection of barley diseases was successful by using non-imaging hyperspectral and fluorescence measurements. A combined reflectance index gave the best discrimination between healthy and diseased plants for seven barley varieties (Yu et al. 2014). Leaf chlorophyll measurements indicated that mild disease symptoms may be detected prior to significant disease-induced chlorophyll losses.

Also Bellow et al. (2013) demonstrated that a pre-symptomatic and symptomatic sensing of downy mildew of grapevine was possible based on the measurement of violet–blue autofluorescence of grapevine. A handheld UV-LED-based field fluorimeter allowed non-destructive detection of stilbenoid phytoalexins associated with downy mildew infections in a vineyard (Latouche et al. 2015).

3.6 Sensing of VOCs

The effects of infectious diseases changing VOC emission of plants have been reviewed by Jansen et al. (2011) and Niederbacher et al. (2015). Most of these diseases affect the plant shoot; however, also root-infecting organisms may induce a systemic plant response. The oomycete *Phytophthora cinnamomi* produced VOCs during infection of lupine seedlings (Qiu et al. 2014).

E-noses were able to discriminate between healthy and infected leaves of cucumber, pepper and tomato (Laothawornkitkul et al. 2008) and to detect herbivore-induced VOCs from tomato (Miresmailli et al. 2010). Laothawornkitkul et al. (2010) identified three marker volatiles of potato leaves 3–4 days after infection by *P. infestans*. Potato leaves infected by *P. infestans* or infested by four different types of herbivores released different VOC blends characteristic for the attacking organism. Results of a principal component analysis suggested that the discrimination between different causes of plant injury is possible (Jansen et al. 2011).

Two commercial models demonstrated the feasibility of e-nose-based screening of kiwifruit for the early diagnosis of *Pseudomonas syringae* pv. *actinidiae* (bacterial canker) in asymptomatic plant material (Cellini et al. 2016b). After a preliminary validation *in vitro*, Cellini et al. (2016a) successfully developed a diagnostic protocol to scale up the analysis of VOCs for the discrimination of fire blight (*Erwinia amylovora*) and blossom blight (*Pseudomonas syringae* pv. *syringae*) on apple propagation material to real nursery conditions.

The PEN3 e-nose discriminated between healthy and inoculated root fragments of shade tree species and between various decay fungi in different soil types for most host-fungus combinations, but Baietto et al. (2013) conceded a relative ineffectiveness to detect low levels of root decay in some combinations of fungal colonization of tree species. An e-nose was useful in the detection of Douglas fir (*Pseudotsuga menziesii*) infected by the root-pathogenic fungus *Phellinus sulphurascens* for research, but provides limited benefits for identifying individual *P. sulphurascens* hazard trees, or for extensive ground surveys in the forest (Kelsey et al. 2016).

Maize ears infected by *Fusarium graminearum*, *F. verticillioides* and *F. subglutinans* released 22 volatiles differentially from healthy ears 4 to 8 days after infection. Becker et al. (2014) concluded that volatile biomarkers of *Fusarium* infection may be used for early detection of fungal infections even before disease symptoms become visible. The detection of VOC emission of injured plants is favoured by limited air volume and seems to be more feasible for greenhouses than for the open field. Jansen et al. (2010) demonstrated that GC-MS data can be automatically processed in order to accurately determine the concentrations of crop health-associated VOCs in a greenhouse.

E-sensing was suitable for the early diagnosis of storage diseases of strawberry fruits infected by *Botrytis* sp., *Penicillium* sp. and *Rhizopus* sp., and to classify the pathogenic fungal type in the fruits (Pan et al. 2016). The volatile profile of apples and pears infected with *P. expansum* (causing blue mould) and *R. stolonifer* indicated plant species-specific biomarkers which may be used for pre-symptomatic disease detection under storage conditions (López et al. 2015).

VOCs of potato tubers infected by *P. infestans* and *Fusarium coeruleum* differed from the profile of healthy tubers (De Lacy Costello et al. 2001). In potatoes infected by *Ralstonia solanacearum* (causing brown rot) or *Clavibacter michiganensis* subsp. *sepedonicus* (ring rot) 80 and 90% total samples were correctly classified under intermediate (100 tubers stored at 4°C in net bag) and real scale conditions (tubers stored at 4°C in 1.25 t bags). A strict relationship between disease severity and responses of the e-nose sensors could be verified (Biondi et al. 2014). Currently potato soft rot caused by *Pectobacterium* species cannot be detected reliably with non-destructive methods, neither in storage, nor in the open field. Lui et al. (2005) could differentiate between potatoes infected with *Pectobacterium* spp. and *F. sambucinum*. A portable field asymmetric ion mobility spectrometer could differentiate between healthy and soft rot-infected potato tubers (Rutolo et al. 2014). Similarly, a commercial array of 12 MOX sensors was suitable for pre-symptomatic detection of soft rot as well as when symptoms of infection first became evident (Rutolo et al. 2016).

3.7 Sensing of other information

Contreras-Medina et al. (2012) tested a field programmable gate array for the detection and differentiation of infectious diseases and nutrient deficiency. FPGAs have been used in computer vision for the classification of crop products (Pearson 2009).

Nuclear magnetic resonance and *X-ray imaging* techniques can be applied in plant science for detecting diseases of plants and plant products (Williamson et al. 1992; Sankaran et al. 2010). Goodman et al. (1992) used NMR microscopic imaging for identification of *Botrytis cinerea* in red raspberry. Magnetic resonance imaging was used to investigate internal bruise and spraing disease – caused by tobacco rattle virus and potato mop top virus – symptoms in potato tubers (Thybo et al. 2004), the quality of apples (Melado-Herreros et al. 2013) and to differentiate below-ground damage of sugar beets affected by *Heterodera schachtii* and *Rhizoctonia solani* (Hillnhuetter et al. 2012b). Navakar et al. (2009) used X-ray imaging to identify fungal infections in wheat. These laboratory techniques need expensive equipment, trained personnel and extensive sample preparation.

Optical coherence tomography (OCT) of melon seeds verified an additional subsurface layer of seeds infected with cucumber green mottle mosaic virus absent in normal seeds (Lee et al. 2011). The combination of OCT and advanced data analysis methods may be used in automatic screening of viral infections in seeds. Bio-photonic inspection using a 1310 nm swept source optical coherence tomograph was able to identify morphological differences between healthy leaves of *Diospyros kaki* (persimmon) and leaves with circular leaf spot caused by *Mycosphaerella nawae*. According to Wijesinghe et al. (2016) this method has potential to generate significant cost savings and good CLS control in persimmon fields.

Advanced mathematical models of weather data from wireless sensory and field-level surveillance data on groundnut leaf spot may support decision-making in disease management (Tripathy et al. 2014). A GIS approach for the interpolation of weather data recorded at a regional scale for use on the field level proved highly suitable in forecasting of late blight in potato (Racca et al. 2010). The integration of elevation, slope and aspect in the modelling of temperature and relative humidity results in disease risk maps.

4 Sensor use for the management of invertebrate pests

4.1 Variability of pest incidence in crops

Arthropods are often highly mobile and flying insects have considerably increased their three-dimensional ecological niche. Stern et al. (1959) were probably the first to recognize that the population density of an insect changes through time and space. The combined effects of variations in plant physiological stage and local climate, arthropod behaviour and population dynamics, and (tri-)trophic interactions often result in aggregated spatial distributions of pests, which can evolve over time due to pest-plant interactions (van Helden 2010). Stable and/or predictable patterns will certainly not occur for all pests. Outbreaks of insect pests in crops are frequently spatially aggregated (Schotzko and Quisenberry 1999; Winder et al. 1999; Korie et al. 2000; Nansen et al. 2005).

Early reports demonstrated patterns of aggregations on the field level for the Colorado beetle (*Leptinotarsa decemlineata*) in potato (Weisz et al. 1995), European corn borer (*Ostrinia nubilalis*) and western bean cutworm (*Richia albicosta*) (Walter and Peairs 1999). Spatial analysis of visual count data indicated that the adults and nymphs of *Eurygaster integriceps* (sunn pest) in cereals had aggregated distribution (Karimzadeh et al. 2011). Site-specific spraying reduced the insecticide input by 40–50% and increased the number of biocontrol agents. Infestation of *E. integriceps* was similar in site-specific and whole-field treatments and the per cent damaged grain was below the economic damage threshold.

Helicoverpa armigera was found to be concentrated at the borders of a 20 ha tomato field (Moral García 2006). The existence of field-level spatial variability in populations of key pests of soybean (bean leaf beetle *Ceratoma trifurcata*) and corn (western corn rootworm *Diabrotica virgifera virgifera*, northern corn rootworm *Diabrotica barberi*) suggested that a site-specific approach to IPM is possible (Park et al. 2008).

Variogram analysis confirmed moderate to high degrees of aggregation in dogwood borer (*Synanthedon scitula*) infestations on neighbouring apple trees with ranges of spatial autocorrelation from 2.8 to 17.0 m (Frank et al. 2011). Area-wide distribution of *Anastrepha ludens* (Mexican fruit fly) in citrus varied from year to year from low to moderate heterogeneity in geostatistical analysis (Vanoye-Eligio et al. 2016). De Alves et al. (2011) mapped the spatial variation of damage caused by the berry borer (*Hypothenemus hampei*) and leaf miner (*Leucoptera coffeella*) in a coffee agro-ecosystem.

As insect populations generally are spatially dynamic – changing density and location over time – the existing methods for mapping their distribution tend to be complicated, labour-intensive and uneconomic. Currently farmers have to inspect intensively crop plants in order to detect arthropods on plants or may use sweep nets, pheromones, sticky traps or other traps to monitor incidence and abundance of pests.

4.2 Sensing of arthropods for pest management

Monitoring insect pest populations is currently a key issue in agriculture and forestry protection. Several technologies have the potential to detect and identify invertebrate pests – arthropods, slugs and snails – on or in plant material and may be used for a targeted control of these pests. Acoustic sensing is suitable for detecting (and identifying) pests in soil, stored products and wood. Machine vision systems may be used to detect and identify arthropods on plants (Nansen and Elliott 2016). Despite intensive research, however, the suitability under practical conditions is relatively low as yet.

Nguy-Robertson et al. (2016) demonstrated the potential of an operational risk map for mite-vectored viruses due to pre-season hail events in wheat by using site-level hyperspectral reflectance from hailed wheat showing increased canopy albedo.

4.2.1 Insect sensors

In contrast to microorganisms arthropods can actively choose host plants as they sense their environment and can actively navigate in space. Insect behaviour may be used (for monitoring purposes) and manipulated in pest management. The potential of semiochemicals, that is, species-specific behaviour-modifying chemicals like pheromones, for the management of insect pests (e.g. members of Noctuidae, Curculionidae, tephritid fruit flies, etc.) in agriculture, horticulture, forestry, stored products and for insect vectors of diseases has been reviewed by Sarles et al. (2015), Guerrero et al. (2014), Witzgall et al. (2010) and El-Sayed et al. (2009).

Samplings verified the spatio-temporal distribution of *Lobesia botrana* (European grapevine moth) eggs, larvae and adult populations in an agro-ecosystem in central Italy including vineyards and olive groves (Sciarretta et al. 2011). In olive groves, hot spots of *L. botrana* males disappeared and the number of larval nests on vine inflorescences was decreased, when additional traps were deployed. Site-specific control, that is, treating only egg hot spots with *Bacillus thuringiensis* var. *kurstaki*, allowed for a decrease in the treated area of the vineyard and, consequently, the quantity of insecticide utilized.

Insect populations are controlled by air permeation and attract-and-kill techniques on at least one million hectares (Witzgall et al. 2010). The behaviour of arthropod pests can be modified by highly specific chemicals in order to lure the pest to the control agent, instead of spraying pesticides (or releasing predators or parasitoids) on the sites of infested plants.

Mate recognition and location in leafhoppers – important vectors of plant diseases – are mediated by species- and sex-specific vibrational signals exchanged in coordinated duets (Korinsek et al. 2016). An autonomous digital processing system proved to be capable of recognizing the male calls of the leafhopper *Aphrodes bicincta* and generating female replies. This mimicking of a duetting female attracted the males to the autonomous system which may be used as a vibrational trap.

4.2.2 Sensor use for arthropod management

E-nose and bionic nose

Plants emit phytochemicals as a defensive mechanism against herbivores. The role of plant volatiles in plant self-protection against pests by attracting parasitic beneficials has been described by Turlings et al. (1995). Das et al. (2013) have reviewed plant volatiles as a method of communication among plants and their potential in crop protection. A low-mass-fraction carbon composite sensor has been developed to detect phenol VOCs (Weerakoon and Chin 2012). Wehrenfennig et al. (2013) discussed the development and use of on-site biosensors and biomimetic approaches for airborne pheromone sensing, together with biomimetic VOC sensor systems.

Feeding of cotton bolls by different hemipteran species elicited a similar quantitative increase in emissions, but significant differences in the emissions of some individual compounds (Degenhardt et al. 2011). Under field conditions, feeding damage by *Nezara viridula* (stink bug) resulted in significantly greater volatile emissions compared with undamaged and mechanically damaged bolls indicating that physical damage alone did not account for the complete blend of volatiles released in response to biotic injury. Lampson et al. (2014a,b) developed an e-nose for the detection of cotton damaged by *N. viridula*, but improvements are required to address sensitivity to fluctuations in environmental conditions. The brown rice plant hopper (BRPH, *Nilaparvata lugens*) was successfully detected using bionic e-nose technology to sample BRPH volatiles – sulphur-containing organics, aromatics, sulphur- and chlorine-containing organics and nitrogen oxides – and advanced machine learning methods (Xu et al. 2014).

Specially trained dogs for scenting trees infested by the invasive Asian longhorn beetle *Anoplophora glabripennis* are used in Austria, Switzerland, and Germany to locate infested trees.

Acoustic detection

Accelerometers and piezoelectric sensors characterize vibration through measuring of acceleration of surfaces they are attached to (Mankin et al. 2011; Martin et al. 2013; Pearson et al. 2007). They are used not only for stored products, but also in perennial and annual crops. *Cephus cinctus* (wheat stem sawfly) has been detected in wheat under controlled conditions (Mankin et al. 2000) and *Rhynchophorus ferrugineus* (red palm weevil) in palm trees in their natural habitat (Gutierrez et al. 2010; Siriwardena et al. 2010; Martin et al. 2013). Laser Doppler vibrometers avoid the interference between sensor and specimens and were shown to be able to distinguish leaf vibration produced by resident caterpillars, intruding predators, wind and rain (Guedes et al. 2012). Using acoustic methods, Jalinas et al. (2015) could even assess the effect of the biological control agent *Beauveria bassiana* on internally feeding insects. Human and acoustic sensor-based detection of red palm weevil-infested palms gave average true positive rates of 75% and 80%, respectively (Hetzroni et al. 2016).

Impact acoustic measurement collects and analyses the sound of strikes between a (steel) plate and samples measured by microphones sensitive to the respective frequencies (Buerano et al. 2012). Acoustic emissions from the plate were used to differentiate between healthy and insect-infested wheat kernels for storage. Crawling and scraping activities of the Coleopteran stored-product pests *Sitophilus oryzae*, *Tribolium castaneum* and *Stegobium paniceum* could be detected by sound and vibration sensors (Fleurat-Lessard et al. 2006; Leblanc et al. 2009; Mankin et al. 2010). The storage insect pests *Prostephanus truncatus* and *Sitophilus zeamais* could be differentiated by acoustic signals from movement and feeding activities of last instars and adults in maize (Kiobia et al. 2015; Njoroge et al. 2016). Results suggest that automatic monitoring of infestations in bulk grain is feasible in small containers (Eliopoulos et al. 2015, 2016). Provided it can be transferred to larger storage establishments (e.g. silos), this kind of technique can assist in decision-making.

Wood-boring insect pests, such as the invasive Asian longhorned beetle (ALB) are difficult to detect because larvae mine inside deciduous trees, logs or wood packing material (Zorovic and Cokl 2015). Ultrasound transducers have been used for the detection of wood-penetrating insects. The parameter time-of-flight enabled Kim et al. (2008) to detect holes >13 mm. As entering holes of the insect are usually <5 mm, Oh and Lee (2014) developed an improved system able to detect holes 3 mm in diameter. Laser Doppler vibrometry was successfully used as a non-destructive diagnostic tool for detecting ALB infestations and proved to be superior to a piezoelectronic sensor for the detection of wood-boring larvae developed by Mankin et al. (2008).

Spectral measurements of infested plant material and arthropods

RGB images may be used for the identification of arthropod pests under the microscope or in the field by using various image-processing methods (e.g. Boissard et al. 2008; reviewed by Liu et al. 2016). Images from smartphones may be also used for counting the pest density on leaves (Wang et al. 2013). Multispectral remote sensing was suitable to detect plant damage from infestations with the less mobile mites in apples and cotton (Fitzgerald et al. 2000; Penuelas et al. 1995) and aphids in wheat (Michels et al. 2000; Riedell et al. 2000). However, typically it was not possible to distinguish arthropod-induced damage from changes of the crop status due to abiotic (and other biotic) stress factors.

Backoulou et al. (2011, 2016) differentiated among causes of wheat stress – infestation by and damage from Russian wheat aphid *Diuraphis noxia*, greenbug (*Schizaphis graminum*), drought, agronomic conditions – by using multispectral remote sensing in combination with topographic and edaphic information.

Extending the spectral resolution and the spectral range of hyperspectral imaging resulted in an accuracy >91% in the detection of kernels infested by the rice weevil *S. oryzae* and three other insect species (Singh et al. 2010). Food products studied for damage and internal infestation by insects include field peas (Nansen et al. 2014; Zhang et al. 2015), wheat kernels (Singh et al. 2009, 2010) and jujube (Wang et al. 2011a,b). Saranwong et al. (2010, 2011) and Haff et al. (2013) used HSI to detect eggs and larvae of *Drosophila melanogaster* (fruit fly) on mango. Spatial resolution was improved to 0.2 mm per pixel and achieved an overall error rate of 6.0%. Wang et al. (2011b) used a similar system on jujube and 94% of infested fruits could be correctly recognized.

Luo et al. (2013) successfully used continuous wavelet analysis for extracting relevant wavebands from spectroradiometer measurements of wheat plants damaged by various levels of the wheat aphid *Sitobion avenae.* Using similar equipment Prabhakar et al. (2013b) developed mealybug stress indices (MSI) for the detection of *Phenacoccus solenopsis*, a widespread mealybug infesting cotton. MSIs were superior to SVIs from remote sensing applications; however, the potential to distinguish various levels of pest infestation was low. Similar results were obtained for leafhopper damage in cotton when LH indices were superior to common SVIs (Prabhakar et al. 2011). Non-imaging hyperspectral measurements of blackgram (*Vigna mungo*) with yellow mosaic disease (due to mung bean yellow mosaic virus) also indicated to a potential for rapid and non-destructive estimation of leaf chlorophyll and yellow mosaic disease severity (Prabhakar et al. 2013a).

Hyperspectral reflectance profiles have been used successfully to classify different species of stored-grain insects (Singh et al. 2010), tobacco budworm (*Heliothis virescens*) and corn earworm (*Helicoverpa zea*) (Jia et al. 2007). HSI data were also used to classify three species of juvenile egg parasitoids *Trichogramma* on the basis of the host eggs (Nansen et al. 2014). Similarly the parasitization of pupae of nine fruit fly species by the ectoparasitic parasitoid *Pachycrepoideus vindemiae* could be detected (Nansen 2016).

Air- and space-borne remote sensing

Radar-based remote sensing has provided insight into the dynamics of migrating and dispersing insects across a wide range of orders (Beerwinkle et al. 1994, 1995; Chapman et al. 2002; Riley et al. 1991; Smith et al. 2000). Nansen and Elliott (2016), however, stated that most field-based applications of remote sensing cover the detection and quantification of plant response (= damage) to insect herbivory (Riedell and Blackmer 1999; Yang et al. 2005, 2009; Mirik et al. 2006, 2012; Nansen et al. 2009, 2013).

In air- and space-borne remote sensing insect herbivory has been shown to adversely affect photosynthesis of plants and result in increased leaf reflectance in the visible range (Carter and Knapp, 2001, Prabhakar et al. 2012). Greenbug infestations of wheat caused a decrease in the UV range, whereas Russian wheat aphid caused an increase of reflectance compared to non-infested control leaves (Riedell and Blackmer 1999). SVIs calculated from airborne hyperspectral imagery were used to detect differences in biomass and vigour of maize due to feeding and stem boring by *Ostrinia nubilalis* larvae (Carroll et al. 2008). Mirik et al. (2012) reported a high potential of remotely sensed data to identify and separate hot spots of wheat damaged by *D. noxia* from non-infested areas for site-specific control.

Large-scale spread of brown planthopper (BPH) *Nilaparvata lugens* was assessed by satellite-borne multi-colour images (Ghobadifar et al. 2016). Huang et al. (2015) confirmed that spectral indices have potential to detect BPH infestations in rice fields (irrespective of differences in N fertilization). In oilseed rape infested with *Myzus persicae* (green peach aphid) UAV-mounted multispectral imagery from 120 m height above ground (65 mm pixel resolution) gave higher classification accuracy than imagery from 15 m (8 mm spatial) resolution and hyperspectral imagery acquired from field plants under laboratory conditions (Severtson et al. 2016).

AVIRIS imagery was used in combination with spectral mixture analysis for detecting the damage caused by the strawberry spider mite (*Tetranychus turkestani*) in cotton fields (Fitzgerald et al. 2004). The procedure successfully distinguished between adjacent mite-free and mite-infested cotton fields and may be used to identify the position, spatial extent and severity of crop stresses. The modified soil-adjusted vegetation index was the best vegetation index tested for detecting armyworm damage in cotton satellite multispectral imagery (Zhang et al. 2016). The successful application of this method in an armyworm outbreak in the province of Hebei, PR China, demonstrated the feasibility of the method and its potential for implementation in practice.

The hemlock woolly adelgid (*Adelges tsugae*) is an invasive insect in the eastern United States and has spread to half of the eastern hemlock (*Tsuga canadensis*) natural range. Significant differences in reflectance between infested and non-infested foliage were observed in the weeks immediately following *A. tsugae* settlement on new growth (Williams et al. 2016). A Landsat time series was suitable to detect low-level infestations; agreement between classification of satellite data and test-survey locations was 79% (Williams et al. 2017).

Other sensors and sensor fusion

The olive fruit fly (*Bactrocera oleae* (Rossi)) causes crop losses worth billions of euros per year alone in Europe. Automatic monitoring traps may enhance efficient monitoring of flying pests by identifying and counting targeted pests by an embedded optoelectronic sensor as they enter the trap (Chen et al. 2014; Potamitis et al. 2014). The wing-beat recording is based on the interruption of the emitted light due to the partial occlusion from insect's wings as they enter the trap. In another approach an image sensor system may be easily fixed inside existing traps to wirelessly deliver images to a central station (Lopez et al. 2012).

A data logger for recording the internal temperature of date palms showed that red palm weevil-infested palms had temperatures 1.5–2.5°C higher than non-infested plants (El-Faki et al. 2016).

A combination of soil characteristics (measured by electromagnetic induction sensors), plant biomass via active optical sensors and topography via GPS was used successfully to predict the likelihood of the soil-borne, redheaded cockchafer (*Adoryphorus couloni*) in pasture in Australia (Cosby et al. 2016).

5 Perspectives

The need for efficient crop production due to limited resources and society's demand for reduced pesticide use are driving forces to develop crop protection systems which are more effective and use fewer pesticides. Present and future precision agriculture technologies will facilitate and support the development; however, there are challenges and limitations for the site-specific management of diseases and invertebrate pests. Technical constraints may be overcome by innovations, for example, miniaturization, whereas biological limitations are immanent to the system and have to be considered or bypassed in IPM systems.

Technical constraints and scopes

Satellite- and airborne sensors may be sufficient for regional and even within-field monitoring of soil-borne crop damage associated with biotic (and abiotic) stress; however, the spatial resolution of these systems is not good enough for information for within-season decision support systems. Proximal imaging sensors have been successfully used for the detection of diseases and arthropod pests under controlled conditions. Their application in the open field at an early time of epidemic development, that is, at low levels, however, is impeded by effects of environmental conditions – for example zenith angle of the sun as light source, rain and wind – and characteristics of the crop to be monitored. Whereas diseases from airborne pathogens often occur on the top of the crop canopy, soil-borne diseases spread from the bottom to the top of the plants (Fig. 2). Sensors have to have access to the primary disease foci producing the inoculum for an epidemic spread in order to initiate control measures in time to prevent economic losses. Even for disease symptoms on the canopy's top layers, the detection of symptoms is affected by the heterogeneity of the crop.

The spatial resolution of aerial sensing of crop canopies for diseases and arthropod pests is crucial for early detection of primary sites of infestation. Reducing the distance between sensor and target – thereby decreasing pixel size – however, increases the disturbing effect of wind-induced movement of plants and plant parts. Moreover, the canopy of many crops has a distinct three-dimensional structure at relevant growth stages which especially impedes not only the focusing of optical sensors but also the extraction of relevant information (Behmann et al. 2015a).

Recent progress in data processing has significantly contributed to improve the potential of sensor systems for crop protection. Sometimes the combination of several spectral vegetation indices gives sufficient progress in pest classification, for other tasks the development of pest-specific indices proved superior to SVIs for satellite data (Mahlein et al. 2013; Prabhakar et al. 2013b; Ashourloo et al. 2014; Huang et al. 2014; Oerke et al. 2016). Wahabzada et al. (2015) discussed also the necessity of data representations suitable for intuitively answering the primary task. Methods from machine learning and artificial intelligence currently used in other fields may be adopted for pest identification in future.

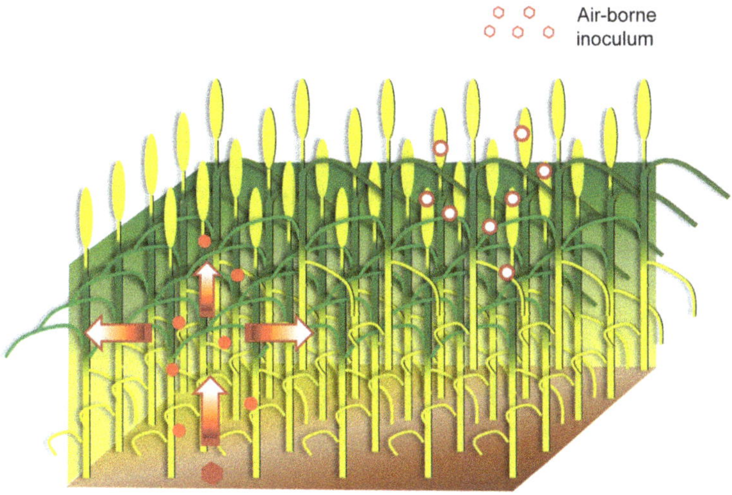

Figure 2 Spatial development of plant diseases in cereals depending on the inoculum source.

Many studies have focused on changes in crop biomass or degradation of leaf pigments for the detection of diseases and arthropod pests. Sensor systems relying on the detection (and quantification) of crop damage, incited by the pest of interest, are hardly suitable for within-season decision-making. In perennial crops, this information may be used for the next season, maps of soil-borne pests may be used to identify management zones also in crop rotations of annual crops. Nevertheless, sensors for pest management need not only detect low pest levels, but should also record data suitable for their identification. Currently the number of reports to cover the diversity in crops and pests is insufficient.

Sensor systems for use in practice should be robust and affordable; the reverse is true for laboratory instrumentation. Xie et al. (2015) used hyperspectral imaging to derive disease-specific indices based on a few sensitive wavebands for constructing low-cost, robust sensors able to detect and differentiate two tomato diseases. Sensor systems may be specific for diseases of one crop; however, they should be flexible enough to be adapted for the detection of symptoms of a new disease with increased importance.

Nowadays consultants and farmers use smartphones for communication and data mining. The device may be used also as a sensor, customized with filters and extended spectral range. Pethybridge and Nelson (2015) described 'Leaf Doctor', a new portable application for quantifying plant disease severity; Neumann et al. (2014) used smartphones for image recording and as an interface for the identification of CLS of sugar beet. More sophisticated sensor systems are being developed, especially for plant phenotyping purposes which may be suitable also for practical use. Artificial (insect) compound eyes with hemispherical FOV and high temporal resolution (up to 1500 fps) could be interesting miniature sensors not only for detecting fast moving insects (Floreano et al. 2013; Liu et al. 2016), but could also be used within the crop canopy.

Autonomous systems equipped with various sensors (and actuators) may become available for the scouting of fields for pests. E-nose even may be able to guide the system to sites of primary infestation. The German Federal Ministry of Food and Agriculture

(BMEL) support the development of an autonomous system for the detection of slugs and snails and the application of baits for rodents in field crops. A prototype 'SlugBot' developed at the Intelligent Autonomous Systems Laboratory, University of the West of England, not only detects slugs, but uses their rotting bodies to generate the electricity it needs to power itself.

As disease symptoms instead of the microorganisms themselves are detected by optical remote sensing, pathogen characteristics like virulence (to specific host genotypes) or fungicide resistance cannot be differentiated. However, aerial sampling of pathogen's spores by using UAVs and subsequent DNA-based identification seems to be promising not only to assess these additional characteristics, but also to detect (and identify) the pathogens before entering the crop. This approach of pathogen scouting has been described by West et al. (2010) for ground-based spore traps, but may be extended to UAV-assisted air sampling of pathogens used in research (Schmale III et al. 2008; Gonzalez et al. 2011; Aylor et al. 2011; Schmale III and Ross 2015).

Biological limitations

Sensors with potential for use in the management of plant diseases detect and identify the (visible) reaction of plants (symptoms) to the pathogen's attack. Plants, however, may be colonized by the pathogen several days before disease symptoms occur. Latent infections are hard to detect (and to identify) (Yeh et al. 2016), but have to be considered in decision-making for fungicide use. Otherwise, the pathogen developing in non-sprayed, but infected plants causes crop damage and, for polycyclic diseases proliferates to other sites in later stages. Buffer zones around the symptomatic patches have been shown to be appropriate in the management of soil-borne nematodes. For airborne pathogens, however, the problem is far more complicated (Fig. 3) and more basic research is needed to improve our knowledge on their spatial distributions within fields.

The potential for detecting latent infections by sensors is limited due to the marginal biomass of pathogens in plant tissue, especially in early stages of pathogenesis. Pre-symptomatic detection of infected plant tissue requires other spectral characteristics than visible, mature symptoms, as already implicated by the wording. The continuum from healthy plant tissue surrounding typical disease symptoms, to symptomatic tissue, and to heavily damaged tissue in space and time has been described for several diseases recorded at high spatial resolution (Mahlein et al. 2012b; Oerke et al. 2016). This phenomenon indicates that, depending on the symptom size, there is an optimum spatial resolution of the sensor for disease detection.

Insects are often highly mobile; however, their behaviour offers the opportunity for manipulations. Microorganisms, in contrast, are spread passively and have no usable behaviour. Strategies like attract and kill, push and pull, the use of pheromone traps for monitoring interfere with the communication system of insect pests which is often species-specific (Hassanali et al. 2008). The concept of bringing the insect to the control agent – in contrast to the application of insecticides onto pests – has also been tested for soil-inhabiting stages of insects, for example, *Diabrotica virgifera virgifera* (corn rootworm) which is attracted by CO_2 sources mimicking the respiration of host roots (Schumann et al. 2014). The adoption of these strategies into crop production practice sometimes suffers from higher labour requirements than for conventional insecticide sprays.

In integrated disease management fungicide applications often are timed according to weather conditions/forecast suitable for pathogen infection. This hazard-based approach

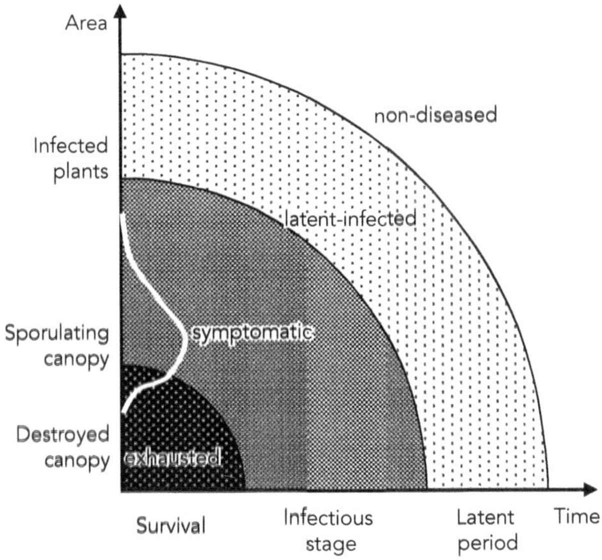

Figure 3 Role of latent infection of plant tissue by pathogens on the area which needs disease control; spatio-temporal sequence from latent infection/period, symptomatic area/infectious stage, destroyed canopy area/survival period.

is used for example, for the control of diseases with a low economic injury level and follows the idea to play it safe. Prognosis systems may have been further improved by assessing the presence and quality of pathogen inoculum, but do not account for the real threat of pathogen infection. A diversification of the control decision within the orchard or field, however, is barely conceivable.

Areas of application

Monitoring of the health status of plants and putative deviations from the optimum by sensor use can be realized under controlled conditions more easily than in the open field. Protected production systems often include high-value crops which make the use of expensive technology more profitable. Moreover, the protected environment allows for the successful application of biological control agents as well as sensor technologies like e-noses. Similarly technical solutions for the improvement of pest management have a better chance in perennial crops in orchards and vineyards where crop protection has a high proportion in overall production costs than in field-grown crops. Nevertheless, because of the large areas covered by arable crops, improvements are highly welcome also in this sector.

In annual crops, pests have to be controlled preferably in an online approach, for example, detection and control in one vehicle crossing. In perennial crops, knowledge on patches of soil-borne pests may be used for control activities in the off-season or for adaptation of the next crop to the spatial pattern, for example, by seeding cultivars differing in resistance to the pest and in yield potential (Bt crops, sugar beet resistant to cyst nematodes, cotton) on appropriate sites of the field. In fruit tree crops, diseased plants may be removed from the field or even cured. Genetically modified (GM) crops, that is, crops with introduced resistance to insect pests and viral diseases, compete with site-specific crop protection. In contrast, GM crops with tolerance to herbicides have favoured the uniform application of herbicides on a large scale.

Grower-friendly molecular and serological methods for pathogen monitoring must be cost-effective, simple to use, robust, rapid as well as have an appropriate specificity and sensitivity depending on the purpose of application (De Boer and Lopez 2012). In speciality crops as well as in post-harvest quality control, sensing techniques may be used in a complementary way for preliminary assessment of separating non-suspicious materials/areas.

Air- and space-borne imagery is used to detect and map the area and intensity of crop damage due to pests. However, the identification of the causal agent(s) requires additional ground-based information. Satellite data have been used for monitoring and forecasting two locust species, the desert and the Australian plague locust (Latchininsky 2013). The assessment of crop production losses due to pests and abiotic factors on a regional and global scale – also in the context of climate change detection – seems to be suitable to increase the availability, overall coverage and timeliness of relevant data.

In conclusion, sensor systems have a huge potential for the detection of diseases and animal pests and may be used in control strategies in order to avoid losses in crop productivity due to these organisms. In the recent past, the evolution of optical, mechanical and chemical sensor systems has improved their suitability and their use in research, but many approaches are not yet ready for practical applications. The large diversity of diseases and pests requires – and enables – the application of various methods for their detection and identification – no system is suitable for all challenges, and the possibility to translate a method from one crop-pest system to another is limited. Various research approaches are necessary to develop a couple of systems which are not only sensitive and specific, but also robust and reliable under production conditions. In addition to technical innovations for the development of suitable sensor systems, more knowledge on the variability of pests in space and time – their biology and epidemiology – is needed. The implementation of sensor technologies into practice – their acceptance by growers – finally depends on the benefit they provide the user with: cost savings, quality of the crop product and the production process.

6 References

Al Masri, A., Hau, B., Dehne, H. W., Mahlein, A. K. and Oerke, E. C. (2017) Impact of primary infection site of *Fusarium* species on head blight development in wheat ears evaluated by IR-thermography. *European Journal of Plant Pathology* 147, 855–68.

Apan, A., Held, A., Phinn, S. and Markley, J. (2004) Detecting sugarcane 'orange rust' disease using EO-1 Hyperion hyperspectral imagery. *International Journal of Remote Sensing* 25, 489–98.

Arens, N., Backhaus, A., Doell, S., Fischer, S., Seiffert, U. and Mock, H. P. (2016) Non-invasive presymptomatic detection of *Cercospora beticola* infection and identification of early metabolic responses in sugar beet. *Frontiers in Plant Science* 7, 1377, doi:10.3389/fpls.2016.01377.

Arredondo Valdes, R., Delgado Ortiz, J. C., Beltran Beache, M., Anguiano Cabello, J., Cerna Chavez, E., Rodriguez Pagaza, Y., et al. (2016) A review of techniques for detecting Huanglongbing (greening) in citrus. *Canadian Journal of Microbiology* 62, 803–11.

Ashourloo, D., Mobasheri, M. R. and Huete, A. (2014) Developing two spectral disease indices for detection of wheat leaf rust (*Puccinia triticina*). *Remote Sensing* 6, 4723–40.

Atas, M., Yardimci, Y. and Temizel, A. (2012) A new approach to aflatoxin detection in chili pepper by machine vision. *Computers and Electronics in Agriculture* 87, 129–41.

Aylor, D. E., Schmale, D. G., Shields, E. J., Newcomb, M. and Nappo, C. J. (2011) Tracking the potato late blight pathogen in the atmosphere using unmanned aerial vehicles and Lagrangian modeling. *Agricultural and Forest Meteorology* 151, 251–60.

Backoulou, G. F., Elliott, N. C., Giles, K., Phoofolo, M., Catana, V., Mirik, M., et al. (2011) Spatially discriminating Russian wheat aphid induced plant stress from other wheat stressing factors. *Computers and Electronics in Agriculture* 78, 123–9.

Backoulou, G. F., Elliott, N. C. and Giles, K. L. (2016) Using multispectral imagery to compare the spatial pattern of injury to wheat caused by Russian wheat aphid and greenbug. *Southwestern Entomologist* 41, 1–8.

Baietto, M., Pozzi, L., Wilson, A. D. and Bassi, D. (2013) Evaluation of a portable MOS electronic nose to detect root rots in shade tree species. *Computers and Electronics in Agriculture* 96, 117–25.

Baker, N. R. (2008) Chlorophyll fluorescence: a probe of photosynthesis *in vivo*. *Annual Review of Plant Biology* 59, 89–113.

Baldwin, I. T., Halitschke, R., Paschold, A., von Dahl, C. C. and Preston, C. A. (2006) Volatile signaling in plant-plant interactions: 'talking trees' in the genomics era. *Science* 311, 812–14.

Baranowski, P., Jedryczka, M., Mazurek, W., Babula-Skowronska, D., Siedliska, A. and Kaczmarek, J. (2015) Hyperspectral and thermal imaging of oilseed rape (*Brassica napus*) response to fungal species of the genus *Alternaria*. *PLoS ONE* 10, e0122913, doi: 10.1371/journal.pone.0122913.

Barbedo, J. G. A. (2013) Digital image processing techniques for detecting, quantifying and classifying plant diseases. *Springer Plus* 2, 660.

Barbedo, J. G. A. (2016) A review on the main challenges in automatic plant disease identification based on visible range images. *Biosystems Engineering* 144, 52–60.

Baron, M., Pineda, M. and Pérez-Bueno, M. L. (2016) Picturing pathogen infection in plants. *Zeitschrift fuer Naturforschung* 71, 355–68.

Bauer, S. D., Korc, F. and Foerstner, W. (2011) The potential of automatic methods of classification to identify leaf diseases from multispectral images. *Precision Agriculture* 12, 361–77.

Bauriegel, E., Brabandt, H., Garber, U. and Herppich, W. B. (2014) Chlorophyll fluorescence imaging to facilitate breeding of *Bremia lactucae*-resistant lettuce cultivars. *Computers and Electronics in Agriculture* 105, 74–82.

Bauriegel, E., Giebel, A., Geyer, M., Schmidt, U. and Herppich, W. B. (2011a) Early detection of Fusarium infection in wheat using hyper-spectral imaging. *Computers and Electronics in Agriculture* 75, 304–12.

Bauriegel, E., Giebel, A. and Herppich, W. B. (2011b) Hyperspectral and chlorophyll fluorescence imaging to analyse the impact of *Fusarium culmorum* on the photosynthetic integrity of infected wheat ears. *Sensors* 11, 3765–79.

Becker, E. M., Herrfurth, C., Irmisch, S., Koellner, T. G. Feussner, I., Karlovsky, P., et al. (2014) Infection of corn ears by *Fusarium* spp. induces the emission of volatile sesquiterpenes. *Journal of Agricultural and Food Chemistry* 62, 5226–36.

Beerwinkle, K. R., Lopez, J. D. J., Schleider, P. G. and Lingren, P. D. (1995) Annual patterns of aerial insect densities at altitudes from 500 to 2400 meters in east-central Texas indicated by continuously-operating vertically orientated radar. *Southwestern Entomologist. Supplement* 18, 63–80.

Beerwinkle, K. R., Lopez, J. D. J., Witz, J. A., Schleider, P. G., Eyster, R. S. and Lingren, P. D. (1994) Seasonal radar and meteorological observations associated with nocturnal insect flight at altitudes to 900 meters. *Environmental Entomology* 23, 676–83.

Behmann, J., Mahlein, A. K., Paulus, S., Kuhlmann, H., Oerke, E. C. and Pluemer, L. (2015a) Calibration of hyperspectral close-range pushbroom cameras for plant phenotyping. *Journal of Photogrammetry and Remote Sensing* 106, 172–82.

Behman, J., Mahlein, A. K., Rumpf, T., Roemer, C. and Pluemer, L. (2015b) A review of advanced machine learning methods for the detection of biotic stress in precision crop protection. *Precision Agriculture* 16, 239–60.

Belin, E., Rousseau, D., Boureau, T. and Caffier, V. (2013) Thermography versus chlorophyll fluorescence imaging for detection and quantification of apple scab. *Computers and Electronics in Agriculture* 90, 159–63.

Bellow, S., Latouche, G., Brown, S. C., Poutaraud, A. and Cerovic, Z. G. (2013) Optical detection of downy mildew in grapevine leaves: Daily kinetics of autofluorescence upon infection. *Journal of Experimental Botany* 64, 333–41.

Bhattacharya, B. K. and Chattopadhyay, C. (2013) A multi-stage tracking for mustard rot disease combining surface meteorology and satellite remote sensing. *Computers and Electronics in Agriculture* 90, 35–44.

Biondi, E., Blasioli, S., Galeone, A., Spinelli, F., Cellini, A., Lucchese, C., et al. (2014) Detection of potato brown rot and ring rot by electronic nose: From laboratory to real scale. *Talanta* 129, 422–30.

Bjerre, K. D. (1999) Disease maps and site-specific fungicide application in winter wheat, pp. 495–504. In J. V. Stafford (Ed.) *Precision Agriculture 99*. Proceedings of the 2nd European Conference on Precision Agriculture, Sheffield Academic Press, Sheffield.

Bock, C. H., Poole, G. H., Parker, P. E. and Gottwald, T. R. (2010) Plant disease severity estimated visually, by digital photography and image analysis, and by hyperspectral imaging. *Critical Reviews in Plant Sciences* 29, 59–107.

Boissard, P., Martin, V. and Moisan, S. (2008) A cognitive vision approach to early pest detection in greenhouse crops. *Computers and Electronics in Agriculture* 62, 81–93.

Borges, D. L., Guedes, S. T. C. D., Nascimento, A. R. and Melo-Pinto, P. (2016) Detecting and grading severity of bacterial spot caused by *Xanthomonas* spp. in tomato (*Solanum lycopersicon*) fields using visible spectrum images. *Computers and Electronics in Agriculture* 125, 149–59.

Bravo, C., Moshou, D., West, J., McCartney, A. and Ramon, H. (2003) Early disease detection in wheat fields using spectral reflectance. *Biosystems Engineering* 84, 137–45.

Buerano, J., Zalameda, J. and Ruiz, R. (2012) Microphone system optimization for free fall impact acoustic method in detection of rice kernel damage. *Computers and Electronics in Agriculture* 85, 140–8.

Bulanon, D. M., Burks, T. F., Kim, D. G. and Ritenour, M. A. (2013) Citrus black spot detection using hyperspectral image analysis. *Agricultural Engineering International: CIGR Journal* 15, 171–80.

Buerling, K., Hunsche, M. and Noga, G. (2011) Use of blue-green and chlorophyll fluorescence measurements for differentiation between nitrogen deficiency and pathogen infection in wheat. *Journal of Plant Physiology* 168, 1641–8.

Calcante, A., Mena, A. and Mazzetto, F. (2012) Evaluation of 'ground sensing' optical sensors for diagnosis of *Plasmopara viticola* on vines. *Spanish Journal of Agricultural Research* 10, 619–30.

Calderón, R., Navas-Cortés, J. A. and Zarco-Tejada, P. J. (2015) Early detection and quantification of Verticillium wilt in olive using hyperspectral and thermal imagery over large areas. *Remote Sensing* 7, 5584–610.

Calderón, R., Navas-Cortés, J. A., Lucena, C. and Zarco-Tejada, P. J. (2013) High-resolution airborne hyperspectral and thermal imagery for early detection of verticillium wilt of olive using fluorescence, temperature and narrow-band spectral indices. *Remote Sensing of Environment* 139, 231–45.

Camargo, A. and Smith, J. S. (2009a) An image-processing based algorithm to automatically identify plant disease visual symptoms. *Biosystems Engineering* 102, 9–21.

Camargo, A. and Smith, J. S. (2009b) Image pattern classification for the identification of disease causing agents in plants. *Computers and Electronics in Agriculture* 66, 121–5.

Cao, X., Luo, Y., Zhou, Y., Duan, X. and Cheng, D. (2013) Detection of powdery mildew in two winter wheat cultivars using canopy hyperspectral reflectance. *Crop Protection* 45, 124–31.

Cao, X., Luo, Y., Zhou, Y., Fan, J., Xu, X., West, J. S., et al. (2015) Detection of powdery mildew in two winter wheat plant densities and prediction of grain yield using canopy hyperspectral reflectance. *PLoS ONE* 10, e0121462, doi:10.1371/journal.pone.0121462.

Cardoza, J., Alborn, H. T. and Tumlinson, J. H. (2002) *In vivo* volatile emissions from peanut plants induced by simultaneous fungal infection and insect damage. *Journal of Chemical Ecology* 28, 161–74.

Carroll, M. W., Glaser, J. A., Hellmich, R. L., Hunt, T. E, Sappington, T. W., Copenhaver, K., et al. (2008) Use of spectral vegetation indices derived from airborne hyperspectral imagery for detection of European corn borer infestation in Iowa corn plots. *Journal of Economic Entomology* 101, 1614–23.

Carter, G. A. and Knapp, A. K. (2001) Leaf optical properties in higher plants: linking spectral characteristics to stress and chlorophyll concentration. *American Journal of Botany* 88, 677–84.

Cellini, A., Biondi, E., Blasioli, S., Rocchi, L., Farneti, B., Braschi, I., et al. (2016a) Early detection of bacterial diseases in apple plants by analysis of volatile organic compounds profiles and use of electronic nose. *Annals of Applied Biology* 168, 409–20.

Cellini, A., Biondi, E., Buriani, G., Farneti, B., Rodriguez-Estrada, M. T., Braschi, I., et al. (2016b) Characterization of volatile organic compounds emitted by kiwifruit plants infected with *Pseudomonas syringae* pv. *actinidiae* and their effects on host defences. *Trees – Structure and Function* 30, 795–806.

Chaerle, L., Hagenbeck, D., De Bruyne, E. and Van Der Straeten, D. (2007a) Chlorophyll fluorescence imaging for disease-resistance screening of sugar beet. *Plant Cell, Tissue and Organ Culture* 91, 97–106.

Chaerle, L., Leinonen, I., Jones, H. G. and van der Straeten, D. (2007b) Monitoring and screening plant populations with combined thermal and chlorophyll fluorescence imaging. *Journal of Experimental Botany* 58, 773–84.

Chaerle, L., Lenk, S., Leinonen, I., Jones, H. G., Van der Straeten, D. and Buschmann, C. (2009) Multi-sensor plant imaging: Towards the development of a stress-catalogue. *Biotechnology Journal* 4, 1152–67.

Chapman, J. W., Reynolds, D. R., Smith, A. D., Riley, J. R., Pedgley, D. E. and Woiwod, I. P. (2002) High-altitude migration of the diamondback moth *Plutella xylostella* to the U.K.: A study using radar, aerial netting, and ground trapping. *Ecological Entomology* 27, 641–50.

Chen, X., Ma, J., Qiao, H. and Zhao, Y. (2007) Detecting infestation of take-all disease in wheat using Landsat Thematic Mapper imagery. *International Journal of Remote Sensing* 28, 5183–9.

Chen, Y. P., Why, A., Batista, G., Mafra-Neto, A. and Keogh, E. (2014) Flying insect detection and classification with inexpensive sensors. *Journal of Visualized Experiments* 92, e52111, doi:10.3791/52111.

Contreras-Medina, L. M., Osornio-Rios, R. A., Torres-Pacheco, I., Romero-Troncoso, R. D. Guevara-Gonzalez, R. G. and Millan-Almaraz, J. R. (2012) Smart sensor for real-time quantification of common symptoms present in unhealthy plants. *Sensors* 12, 784–805.

Cosby, A. M., Falzon, G. A., Trotter, M. G., Stanley, J. N., Powell, K. S. and Lamb, D. W. (2016) Risk mapping of redheaded cockchafer (*Adoryphorus couloni*) (Burmeister) infestations using a combination of novel k-means clustering and on-the-go plant and soil sensing technologies. *Precision Agriculture* 17, 1–17.

Costa, J. M., Grant, O. M. and Chaves, M. M. (2013) Thermography to explore plant – environment interactions. *Journal of Experimental Botany* 64, 3937–49.

Dammer, K. H. (2010) Variable rate application of fungicides, pp. 349–65. In E. C. Oerke, R. Gerhards, G. Menz and R. A. Sikora (Eds) *Precision Crop Protection – The Challenge and Use of Heterogeneity*. Springer, Dordrecht, Netherlands.

Dammer, K. H., Moller, B., Rodemann, B. and Heppner, D. (2011) Detection of head blight (*Fusarium* ssp.) in winter wheat by color and multispectral image analyses. *Crop Protection* 30, 420–8.

Danielsen, S. and Munk, L. (2004) Evaluation of disease assessment methods in quinoa for their ability to predict yield loss caused by downy mildew. *Crop Protection* 23, 219–28.

Das, A., Lee, S. H., Hyun, T. K., Kim, S. W. and Kim, J. Y. (2013) Plant volatiles as method of communication. *Plant Biotechnology Reports* 7, 9–26.

Davis, R. F., Aryal, S. K., Perry, C. D., Sullivan, D. G., Timper, P., Ortiz, B. V., et al. (2013) Utilizing management zones for *Rotylenchulus reniformis* in cotton: Effects on nematode levels, crop damage, and *Pasteuria* sp. *Crop Protection* 50, 53–60.

De Alves, M. C., da Silva, F. M., Campos Moraes, J., Pozza, E. A., de Oliveira, M. S., Souza, J. C. S., et al. (2011) Geostatistical analysis of the spatial variation of the berry borer and leaf miner in a coffee agroecosystem. *Precision Agriculture* 12, 18–31.

De Boer, J. G., Hordijk, C. A., Posthumus, M. A. and Dicke, M. (2008) Prey and non-prey arthropods sharing a host plant: Effects on induced volatile emission and predator attraction. *Journal of Chemical Ecology* 34, 281–90.

De Boer, S. H. and López, M. M. (2012) New grower-friendly methods for plant pathogen monitoring. *Annual Review of Phytopathology* 50, 197–218.

De Coninck, B. M. A., Amand, O., Delauré, S. L., Lucas, S., Hias, N., Weyens, G., et al. (2012) The use of digital image analysis and real-time PCR fine-tunes bioassays for quantification of Cercospora leaf spot disease in sugar beet breeding. *Plant Pathology* 61, 76–84.

De Lacy Costello, B. P. J., Evans, P., Ewen, R. J., Gunson, H. E., Jones, P. R. H., Ratcliffe, N. M., et al. (2001) Gas chromatography mass spectrometry analyse of volatile compounds from potato tubers inoculated with *Phytophthora infestans* or *Fusarium coeruleum*. *Plant Pathology* 50, 489–96.

Degenhardt, D. C., Greene, J. K., Khalilian, A. and Reeves, R. B. (2011) Volatile emissions from developing cotton bolls in response to Hemipteran feeding damage. *Journal of Entomological Science* 46, 177–90.

Del Ponte, E. M., Shah, D. A. and Bergstrom, G. C. (2003) Spatial patterns of Fusarium head blight in New York wheat fields suggest role of airborne inoculum. *Plant Management Network*, published 18 April 2003, http://plantmanagementnetwork.org/pub/php/research/2003/fhb/bergstrom.pdf Last accessed 12 February 2017.

Delalieux, S., van Aardt, J., Keulemans, W., Schrevens, E. and Coppin, P. (2007a) Detection of biotic stress (*Venturia inaequalis*) in apple trees using hyperspectral data: Non-parametric statistical approaches and physiological implications. *European Journal of Agronomy* 27, 130–43.

Delalieux, S., van Aardt, J. A. N., Zarco-Tejada, P. J., Kempeneers, P., Verstraeten, W. W. and Coppin, P. (2007b). Development of robust hyperspectral indices for the detection of deviations of normal plant state. *European Association of Remote Sensing Laboratories (EARSeL) eProceedings* 6, 82–93.

Deng, X. L., Lan, Y. B., Hong, T. S. and Chen, J. X. (2016) Citrus greening detection using visible spectrum imaging and C-SVC. *Computers and Electronics in Agriculture* 130, 177–83.

Dicke, M. (2009) Behavioural and community ecology of plants that cry for help. *Plant Cell & Environment* 32, 654–65.

Dicke, M., Van der Maas, K. J., Takabayashi, J. and Vet, L. E. M. (1990) Learning affects response to volatile allelochemicals by predatory mites. *Proceedings of the Section Experimental and Applied Entomology of the Netherlands Entomological Society* 1, 31–6.

Du, Q., French, V., Skaria, M., Yang, C. H. and Everitt, J. H. (2004) Citrus pest stress monitoring using airborne hyperspectral imagery. In *Proceedings of 2004 International Geoscience and Remote Sensing Symposium*, Vol. VI, pp. 3981–4. IEEE, Piscataway, New Jersey.

El-Faki, M. S., El-Shafie, H. A. F. and Al-Hajhoj, M. B. R. (2016) Potentials for early detection of red palm weevil (Coleoptera: Curculionidae)-infested date palm (Arecaceae) using temperature differentials. *Canadian Entomologist* 148, 239–45.

Eliopoulos, P. A., Potamitis, I. and Kontodimas, D. C. (2016) Estimation of population density of stored grain pests via bioacoustic detection. *Crop Protection* 85, 71–8.

Eliopoulos, P. A., Potamitis, I., Kontodimas, D. C. and Givropoulou, E. G. (2015) Detection of adult beetles inside the stored wheat mass based on their acoustic emissions. *Journal of Economic Entomology* 108, 2808–14.

El-Sayed, A. M., Suckling, D. M., Byers, J. A., Jang, E. B. and Wearing, C. H. (2009) Potential of 'lure and kill' in long-term pest management and eradication of invasive species. *Journal of Economic Entomology* 102, 815–35.

Evans, K., Webster, R., Barker, A., Halford, P., Russell, M., Stafford, J., et al. (1999) Mapping infestations of potato cyst nematodes and the potential for patch treatment with nematicides, pp. 505–15. In J. V. Stafford (Ed.) *Precision Agriculture 99*. Proceedings of the 2nd European Conference on Precision Agriculture, Sheffield Academic Press, Sheffield.

Evans, K., Webster, R., Barker, A., Halford, P., Russell, M., Stafford, J., et al. (2003) Mapping infestations of potato cyst nematodes and the potential for spatially yarying application of nematicides. *Precision Agriculture* 4, 149–62.

Fahlgren, N., Gehan, M. A. and Baxter, I. (2015) Lights, camera, action: High-throughput plant phenotyping is ready for a close-up. *Current Opinion in Plant Biology* 24, 93–9.

Feng, W., Shen, W. Y., He, L., Duan, J. H., Guo, B. B., Li, Y. X., et al. (2016) Improved remote sensing detection of wheat powdery mildew using dual-green vegetation indices. *Precision Agriculture* 17, 608–27.

Fitzgerald, G. J., Maas, S. J. and deTar, W. R. (2000) Multispectral multitemporal remote sensing for spider mite detection in cotton. In P. C. Robert (Ed.) *5th International Conference on Precision Agriculture*. ASA-CSSA-SSSA, Madison, WI, USA(CD-ROM).

Fitzgerald, G. J., Maas, S. J. and Detar, W. R. (2004) Spider mite detection and canopy component mapping in cotton using hyperspectral imagery and spectral mixture analysis. *Precision Agriculture*, 5, 275–89.

Fletcher, R. S., Skaria, M., Escobar, D. E. and Everitt, J. H. (2001) Field spectra and airborne digital imagery for detecting *Phytophthora* foot rot infections in citrus trees. *HortScience* 36, 94–7.

Fleurat-Lessard, F., Tomasini, B., Kostine, L. and Fuzeau1, B. (2006) Acoustic detection and automatic identification of insect stages activity in grain bulks by noise spectra processing through classification algorithms. In *9th International Working Conference on Stored Product Protection*. Passo Fundo: Brazilian Post-Harvest Association (ABRAPOS). ISBN: 8560234004.

Floreano, D., Pericet, C., Viollet, S., Ruffier, F., Brueckner, A., Leitel, R., et al. (2013) Miniature curved artificial compound eyes. *Proceedings of the National Academy of Sciences of the United States of America* 110, 9267–72.

Frank, D., Brewster, C. C. and Leskey, T., Bergh, C. (2011) Factors influencing the temporal and spatial patterns of dogwood borer (Lepidoptera: Sesiidae) infestation in newly planted apple orchards. *Environmental Entomology* 40, 173–83.

Franke, J. and Menz, G. (2007). Multi-temporal wheat disease detection by multi-spectral remote sensing. *Precision Agriculture*, 8, 161–72.

Freitas, A. S., Pozza, E. A., Alves, M. C., Coelho, G., Rocha, H. S. and Pozza, A. A. A. (2016) Spatial distribution of Yellow Sigatoka Leaf Spot correlated with soil fertility and plant nutrition. *Precision Agriculture* 17, 93–107.

Garcia-Ruiz, F., Sankaran, S., Maja, J. M., Lee, W. S., Rasmussen, J. and Ehsani, R. (2013) Comparison of two aerial imaging platforms for identification of Huanglongbing-infected citrus trees. *Computers and Electronics in Agriculture* 91, 106–15.

Gebbers, R. and Adamchuk, V. I. (2010) Precision agriculture and food security. *Science* 327, 828–31.

Ghobadifar F., Aimrun, W. and Jebur, M. N. (2016) Development of an early warning system for brown planthopper (BPH) (*Nilaparvata lugens*) in rice farming using multispectral remote sensing. *Precision Agriculture* 17, 377–91.

Giles, D. K., Delwiche, M. J. and Dodd, R. B. (1989) Sprayer control by sensing orchard crop characteristics: Orchard architecture and spray liquid savings. *Journal of Agricultural Engineering Research* 43, 271–89.

Gonzalez, F., Castro, M. P., Narayan, P., Walker, R. and Zeller, L. (2011) Development of an autonomous unmanned aerial system to collect time-stamped samples from the atmosphere and localize potential pathogen sources. *Journal of Field Robotics* 28, 961–76.

Goodman, B. A., Williamson, B. and Chudek, J. A. (1992) Non-invasive observation of the development of fungal infection in fruit. *Protoplasma* 166, 107–9.

Granum, E., Pérez-Bueno, M. L., Calderón, C. E., Ramos, C., de Vicente, A., Cazorla, F. M., et al. (2015) Metabolic responses of avocado plants to stress induced by *Rosellinia necatrix* analysed by fluorescence and thermal imaging. *European Journal of Plant Pathology* 142, 625–32.

Guedes, R. N. C., Matheson, S. M., Frei, B., Smith, M. L. and Yack, J. E. (2012). Vibration detection and discrimination in the masked birch caterpillar (*Drepana arcuta*). *Journal of Comparative Physiology A* 198, 325–35.

Guerrero, A., Malo, E. A., Coll, J. and Quero, C. (2014) Semiochemical and natural product-based approaches to control *Spodoptera* spp. (Lepidoptera: Noctuidae). *Journal of Pest Science* 87, 231–47.

Gutierrez, A., Ruiz, V., Molto, E., Tapia, G. and Tellez, M. D. (2010) Development of a bioacoustic sensor for the early detection of red palm weevil (*Rhynchophorus ferrugineus* Olivier). *Crop Protection* 29, 671–6.

Haff, R., Saranwong, S., Thanapase, W., Janhiran, A., Kasemsumran, S. and Kawano, S. (2013) Automatic image analysis and spot classification for detection of fruit fly infestation in hyperspectral images of mangoes. *Postharvest Biology and Technology* 86, 23–8.

Hahn, F. (2009) Actual pathogen detection: Sensors and algorithms – a review. *Algorithms* 2, 301–38.

Hassanali, A., Herren, H., Khan, Z. R., Pickett, J. A. and Woodcock, C. M. (2008) Integrated pest management: The push – pull approach for controlling insect pests and weeds of cereals, and its potential for other agricultural systems including animal husbandry. *Philosophical Transactions of the Royal Society* B 363, 611–21.

Hatfield, J. L. and Pinter, P. J. (1993) Remote-sensing for crop protection. *Crop Protection* 12, 403–13.

Hbirkou, C., Paetzold, S., Mahlein, A. K. and Welp, G. (2012) Airborne hyperspectral imaging of spatial soil organic carbon heterogeneity at the field-scale. *Geoderma* 175, 21–8.

Hbirkou, C., Welp, G., Rehbein, K., Hillnhuetter, C., Daub, M., Oliver, M. A., et al. (2011) The effect of soil heterogeneity on the spatial distribution of *Heterodera schachtii* within sugar beet fields. *Applied Soil Science* 51, 25–34.

Hetzroni, A., Soroker, V. and Cohen, Y. (2016) Toward practical acoustic red palm weevil detection. *Computers and Electronics in Agriculture* 124, 100–6.

Hillnhuetter, C., Mahlein, A. K., Sikora, R. A. and Oerke, E. C. (2011) Remote sensing to detect plant stress induced by *Heterodera schachtii* and *Rhizoctonia solani* in sugar beet fields. *Field Crop Research* 122, 70–7.

Hillnhuetter, C., Mahlein, A.-K., Sikora, R. A. and Oerke, E. C. (2012a) Use of imaging spectroscopy to discriminate symptoms caused by *Heterodera schachtii* and *Rhizoctonia solani* on sugar beet. *Precision Agriculture* 13, 17–32.

Hillnhuetter, C., Schmitz, A., Kuehnhold, V. and Sikora, R. A. (2010) Remote sensing for the detection of soil-borne plant parasitic nematodes and fungal pathogens, pp. 151–65. In E. C. Oerke, R. Gerhards, G. Menz and R. A. Sikora (Eds) *Precision Crop Protection – the Challenge and Use of Heterogeneity*. Springer, Dordrecht, The Netherlands.

Hillnhuetter, C., Sikora, R. A., Oerke, E. C. and van Dusschoten, D. (2012b) Nuclear magnetic resonance: A tool for imaging belowground damage caused by *Heterodera schachtii* and *Rhizoctonia solani* on sugar beet. *Journal of Experimental Botany* 63, 319–27.

Holguin, C. M., Gerard, P., Mueller, J. D., Khalilian, A. and Agudelo, P. (2015a) Spatial distribution of reniform nematode in cotton as influenced by soil texture and crop rotations. *Phytopathology* 105, 674–83.

Holguin, C. M., Ma, X. Y., Mueller, J. D. and Agudelo, P. (2016) Distribution of *Hoplolaimus* species in soybean fields in South Carolina and North Carolina. *Plant Disease* 100, 149–53.

Holguin, C. M., Mueller, J. D., Khalilian, A. and Agudelo, P. (2015b) Population dynamics and spatial distribution of Columbia lance nematode in cotton. *Applied Soil Ecology* 95, 107–14.

Hou, J. W., Li, L. T. and He, J. (2016) Detection of grapevine leafroll disease based on 11-index imagery and colony clustering algorithm. *Precision Agriculture* 17, 488–50.

Hovmøller, M. S., Walter, S., Bayles, R. A., Hubbard A., Flath, K., Sommerfeldt, N., et al. (2016) Replacement of the European wheat yellow rust population by new races from the centre of diversity in the near-Himalayan region. *Plant Pathology* 65, 402–41.

Huang, J. R., Sun, J. Y., Liao, H. J. and Liu, X. D. (2015) Detection of brown planthopper infestation based on SPAD and spectral data from rice under different rates of nitrogen fertilizer. *Precision Agriculture* 16, 148–63.

Huang, L. S., Zhao, J. L., Zhang, D. Y., Yuan, L., Dong, Y. Y. and Zhang, J. C. (2012) Identifying and mapping stripe rust in winter wheat using multi-temporal airborne images. *International Journal of Agriculture and Biology* 14, 697–704.

Huang, S. P., Qi, L., Ma, X., Xue, K. N., Wang, W. J. and Zhu, X. Y. (2015) Hyperspectral image analysis based on BoSW model for rice panicle blast grading. *Computers and Electronics in Agriculture* 118, 167–78.

Huang, W. J., David, W. L., Niu, Z., Zhang, Y. J., Liu, L. Y. and Wang, J. H. (2007). Identification of yellow rust in wheat using in situ spectral reflectance measurements and airborne hyperspectral imaging. *Precision Agriculture* 8, 187–97.

Huang, W. J., Guan, Q. S., Luo, J. H., Zhang, J. C., Zhao, J. L., Liang, D., et al. (2014) New optimized spectral indices for identifying and monitoring winter wheat diseases. *IEEE Journal of Selected Topics in Applied Earth Observations and Remote Sensing* 7, 2516–24.

Iori, A., Scala, V., Cesare, D., Pinzari, F., D'Egidio, M. G., Fanelli, C., et al. (2015) Hyperspectral and molecular analysis of *Stagonospora nodorum* blotch disease in durum wheat. *European Journal of Plant Pathology* 141, 689–702.

Jackson, R. D. (1986) Remote sensing of biotic and abiotic plant stress. *Annual Review of Phytopathology* 24, 265–87.

Jalinas, J., Guerri-Agullo, B., Mankin, R. W., Lopez-Follana, R. and Lopez-Llorca, L. V. (2015) Acoustic assessment of *Beauveria bassiana* (Hypocreales: Clavicipitaceae) effects on *Rhynchophorus ferrugineus* (Coleoptera: Dryophthoridae) larval activity and mortality. *Journal of Economic Entomology* 108, 444–53.

Jansen, R., Hofstee, J. W., Bouwmeester, H. and van Henten, E. (2010) Automated signal processing applied to volatile-based inspection of greenhouse crops. *Sensors* 10, 7122–33.

Jansen, R. M. C., Wildt, J., Kappers, I. F., Bouwmeester, H. J., Hofstee, J. W. and van Henten, E. J. (2011) Detection of diseased plants by analysis of volatile organic compound emission. *Annual Review of Phytopathology* 49, 157–74.

Jia, F., Maghirang, E., Dowell, F., Abel, C. and Ramaswamy, S. (2007) Differentiating tobacco budworm and corn earworm using near-infrared spectroscopy. *Journal of Economic Entomology* 100, 759–64.

Johnson, D. A., Alldredge, J. R., Hamm, P. B. and Frazier, B. E. (2003) Aerial photography used for spatial pattern analysis of late blight infection in irrigated potato circles. *Phytopathology* 93, 805–12.

Jones, C. D., Jones, J. B. and Lee, W. S. (2010) Diagnosis of bacterial spot of tomato using spectral signatures. *Computers and Electronics in Agriculture* 74, 329–35.

Jones, H. G. and Schofield, P. (2008) Thermal and other remote sensing of plant stress. *General and Applied Plant Physiology* 34, 19–32.

Karimzadeh, R., Hejazi, M. J., Helali, H., Iranipour, S. and Mohammadi, S. A. (2011) Assessing the impact of site-specific spraying on control of *Eurygaster integriceps* (Hemiptera: Scutelleridae) damage and natural enemies. *Precision Agriculture* 12, 576–93.

Kelsey, R. G., Joseph, G., Westlind, D. and Thies, W. G. (2016) Ethanol and acetone from Douglas-fir roots stressed by *Phellinus sulphurascens* infection: Implications for detecting diseased trees and for beetle host selection. *Forest Ecology and Management* 360, 261–72.

Kim, K. M., Lee, J. J., Lee, S. J. and Yeo, H. (2008) Improvement of wood CT images by consideration of the skewing of ultrasound caused by growth ring angle. *Wood and Fiber Science* 40, 572–9.

Kiobia, D. O., Tumbo, S. D., Cantillo, J., Rohde, B. B., Mallikarjunan, P. K. and Mankin, R. W. (2015) Characterization of sounds in maize produced by internally feeding insects: Investigations to develop inexpensive devices for detection of *Prostephanus truncatus* (Coleoptera: Bostrichidae) and *Sitophilus zeamais* (Coleoptera: Curculionidae) in small-scale storage facilities in sub-Saharan Africa. *Florida Entomologist* 98, 405–9.

Kobayashi, T., Kanda, E., Kitada, K., Ishiguro, K. and Torigoe, Y. (2001). Detection of rice panicle blast with multi-spectral radiometer and the potential of using airborne multi-spectral scanner. *Phytopathology* 91, 316–23.

Koch, H. (1980) Raeumliche Befallsmuster beim Echten Mehltau der Gerste (Spatial patterns of powdery mildew in barley). *Zeitschrift fuer Pflanzenkrankheiten und Pflanzenschutz* 87, 731–7.

Korie, S., Perry, J. N., Mugglestone, M. A., Clark, S. J., Thomas, C. F. G. and Roff, M. N. (2000) Spatio-temporal associations in beetle and virus count data. *Journal of Agricultural Biological and Environmental Statistics* 5, 214–39.

Korinsek, G., Derlink, M., Virant-Doberlet, M. and Tuma, T. (2016) An autonomous system of detecting and attracting leafhopper males using species- and sex-specific substrate borne vibrational signals. *Computers and Electronics in Agriculture* 123, 29–39.

Kuckenberg, J., Tartachnyk, I. and Noga, G. (2009a) Detection and differentiation of nitrogen-deficiency, powdery mildew and leaf rust at wheat leaf and canopy level by laser-induced chlorophyll fluorescence. *Biosystems Engineering* 103, 121–8.

Kuckenberg, J., Tartachnyk, I. and Noga, G. (2009b) Temporal and spatial changes of chlorophyll fluorescence as a basis for early and precise detection of leaf rust and powdery mildew infections in wheat leaves. *Precision Agriculture* 10, 34–44.

Lampson, B. D., Han, Y. J., Khalilian, A., Greene, J. K., Degenhardt, D. C. and Hallstrom, J. O. (2014a) Development of a portable electronic nose for detection of pests and plant damage. *Computers and Electronics in Agriculture* 108, 87–94.

Lampson, B. D., Khalilian, A., Greene, J. K., Han, Y. J. and Degenhardt, D. C. (2014b) Development of a portable electronic nose for detection of cotton damaged by *Nezara viridula* (Hemiptera: Pentatomidae). *Journal of Insects* 2014, 8 pages, Article ID 297219.

Lancashire, P. D., Bleiholder, H., van den Boom, T., Langelueddeke, P., Stauss, R., Weber, E. and Witzenberger, A. (1991) A uniform decimal code for growth stages of crops and weeds. *Annals of Applied Biology* 119, 561–601.

Laothawornkitkul, J., Jansen, R. M. C., Smid, H. M., Bouwmeester, H. J., Muller, J. and van Bruggen, A. H. C. (2010) Volatile organic compounds as a diagnostic marker of late blight infected potato plants: A pilot study. *Crop Protection* 29, 872–8.

Laothawornkitkul, J., Moore, J. P., Taylor, J. E., Possell, M., Gibson, T. D., Hewitt, C. N. and Paul, N. D. (2008) Discrimination of plant volatile signatures by an electronic nose: A potential technology for plant pest and disease monitoring. *Environmental Science and Technology* 42, 8433–9.

Latchininsky, A. V. (2013) Locusts and remote sensing: A review. *Journal of Applied Remote Sensing* 7, 075099, doi:10.1117/1.JRS.7.075099.

Latouche, G., Debord, C., Ravnal, M., Milhade, C. and Cerovic, Z. G. (2015) First detection of the presence of naturally occurring grapevine downy mildew in the field by a fluorescence-based method. *Photochemical & Photobiological Sciences* 14, 1807–13.

Laudien, R., Bareth, G. and Doluschitz, R. (2003) Analysis of hyperspectral field data for detection of sugar beet diseases, pp. 375–81. In Z. Harnos, M. Herdon and T. B. Wiwczaroski (Eds) *Information Technology for a Better Agri-food Sector, Environment and Rural Living*. Proceedings 4th EFITA Conference, –Vol. I, Budapest, Hungary.

Laudien, R., Bareth, G. and Doluschitz, R. (2004) Comparison of remote sensing based analysis of crop diseases by using high resolution multispectral and hyperspectral data – case study *Rhizoctonia solani* in sugar beet, pp. 670–6. In S. A. Brandt (Ed.) *Geoinformatics 2004: Proceedings of the 12th International Conference on Geoinformatics: Geospatial Information Research: Bridging the Pacific and Atlantic*. University of Gavle, Sweden.

Leblanc, M. P., Gaunt, D. and Lessard, F. F. (2009). Experimental study of acoustic equipment for real-time insect detection in grain bins. *Grain Protector*. Chartres. http://www.grainprotector.com/index_htm_files/acoustic%20pest%20detection.pdf. Last retrieved 23 February 2017.

Lee, C., Lee, S. Y., Kim, J. Y, Jung, H. Y. and Kim, J. (2011) Optical sensing method for screening disease in melon seeds by using optical coherence tomography. *Sensors* 11, 9467–77.

Lee, W. S., Alchanatis, V., Yang, C., Hirafuji, M., Moshou, D. and Li, C. (2010). Sensing technologies for precision specialty crop production. *Computers and Electronics in Agriculture* 74, 2–33.

Lee, W. S. and Ehsani, R. (2015) Sensing systems for precision agriculture in Florida. *Computers and Electronics in Agriculture* 112, 2–9.

Lelong, C. C. D., Roger, J. M., Bregand, S., Dubertret, F., Lanore, M., Sitorus, N. A., et al. (2010) Evaluation of oil-palm fungal disease infestation with canopy hyperspectral reflectance data. *Sensors* 10, 734–47.

Lenthe, J. H., Oerke, E. C. and Dehne, H. W. (2007). Digital thermography for monitoring canopy health of wheat. *Precision Agriculture* 8, 15–26.

Leucker, M., Mahlein, A. K., Steiner, U. and Oerke, E. C. (2016) Improvement of lesion phenotyping in *Cercospora beticola* – sugar beet interaction by hyperspectral imaging. *Phytopathology* 106, 177–84.

Leucker, M., Wahabzada, M., Kersting, K., Peter, M., Beyer, W., Steiner, U., et al. (2017) Hyperspectral imaging reveals the effect of sugar beet QTLs on Cercospora leaf spot resistance. *Functional Plant Biology* 44, 1–9.

Leufen, G., Noga, G. and Hunsche, M. (2014) Proximal sensing of plant-pathogen interactions in spring barley with three fluorescence techniques. *Sensors* 14, 11135–52.

Li, H., Lee, W. S., Wang, K., Ehsani, R. and Yang, C. (2014a) Extended spectral angle mapping (ESAM) for citrus greening disease detection using airborne hyperspectral imaging. *Precision Agriculture* 15, 162–83.

Li, J. B., Huang, W. Q., Tian, X., Wang, C. P., Fan, S. X. and Zhao, C. J. (2016) Fast detection and visualization of early decay in citrus using Vis-NIR hyperspectral imaging. *Computers and Electronics in Agriculture* 127, 582–92.

Li, L., Zhang, Q. and Huang, D. F. (2014b) A review of imaging techniques for plant phenotyping. *Sensors* 14, 20078–111, doi:10.3390/s141120078.

Li, S., Simonian, A. and Chin, B. A. (2010) Sensors for agriculture and the food industry. *Electrochemical Society Interface* 19, 41–6.

Li, X. H., Lee, W. S., Li, M. Z., Ehsani, R., Mishra, A. R., Yang, C. H., et al. (2012) Spectral difference analysis and airborne imaging classification for citrus greening infected trees. *Computers and Electronics in Agriculture* 83, 32–46.

Liaghat, S., Mansor, S., Ehsani, R., Shafri, H. Z. M., Meon, S. and Sankaran, S. (2014) Mid-infrared spectroscopy for early detection of basal stem rot disease in oil palm. *Computers and Electronics in Agriculture* 101, 48–54.

Liew, O. W., Chong, P. C. J., Li, B. Q. and Asundi, A. K. (2008) Signature optical cues: emerging technologies for monitoring plant health. *Sensors* 8, 3205–39.

Lindenthal, M., Steiner, U., Dehne, H. W. and Oerke, E. C. (2005). Effect of downy mildew development on transpiration of cucumber leaves visualized by digital thermography. *Phytopathology* 95, 233–40.

Liu H. J., Lee, S. H. and Chahl, J. S. (2016) A review of recent sensing technologies to detect invertebrates on crops. *Precision Agriculture* 17, 1–32.

Liu, Z., Griffin, T. W., Kirkpatrick, T. L. and Monfort, W. S. (2015) Spatial econometric approaches to developing site-specific nematode management strategies in cotton production. *Precision Agriculture* 16, 587–600.

López, L., Echeverria, G., Usall, J. and Teixidó, N. (2015) The detection of fungal diseases in the 'Golden Smoothee' apple and 'Blanquilla' pear based on the volatile profile. *Postharvest Biology and Technology* 99, 120–30.

Lopez, O., Rach, M. M., Migallon, H., Malumbres, M. P., Bonastre, A. and Serrano, J. J. (2012) Monitoring pest insect traps by means of low-power image sensor technologies. *Sensors* 12, 15801–19.

López-López, M., Calderón, R., González-Dugo, V., Zarco-Tejada, P. J. and Fereres, E. (2016) Early detection and quantification of almond red leaf blotch using high-resolution hyperspectral and thermal imagery. *Remote Sensing* 8, 276, doi:10.3390/rs8040276.

Lorestani, E. Z., Kamkar, B., Razavi, S. E. and da Silva, J. A. T. (2013). Modeling and mapping diversity of pathogenic fungi of wheat fields using geographic information systems (GIS). *Crop Protection* 54, 74–83.

Lui, L. H., Vikram, A., Abu-Nada, A., Kushalappa, A. C., Raghavan, G. S. V. and Al-Mughrabi, K. (2005) Volatile metabolic profiling for discrimination of potato tubers inoculated with dry and soft rot pathogens. *American Journal of Potato Research* 82, 1–8.

Luo, J. H., Huang, W. J., Yuan, L., Zhao, C. J., Du, S., Zhang, J. C., et al. (2013) Evaluation of spectral indices and continuous wavelet analysis to quantify aphid infestation in wheat. *Precision Agriculture* 14, 151–61.

MacDonald, S. L., Staid, M., Staid, M. and Cooper, M. L. (2016) Remote hyperspectral imaging of grapevine leafroll-associated virus 3 in cabernet sauvignon vineyards. *Computers and Electronics in Agriculture* 130, 109–17.

Madden, L. V. and Hughes, G. (1995) Plant disease incidence: Distributions, heterogeneity, and temporal analysis. *Annual Review of Phytopathology* 33, 529–64.

Mahlein, A. K. (2016) Plant disease detection by imaging sensors – parallels and specific demands for precision agriculture and plant phenotyping. *Plant Disease* 100, 241–51.

Mahlein, A. K., Oerke, E. C., Steiner, U. and Dehne, H. W. (2012a) Recent advances in sensing plant diseases for precision crop protection. *European Journal of Plant Pathology* 133, 197–203.

Mahlein, A. K., Rumpf, T., Welke, P., Dehne, H. W., Pluemer, L., Steiner, U. and Oerke, E. C. (2013) Development of spectral indices for detecting and identifying plant diseases. *Remote Sensing of Environment* 128, 21–30.

Mahlein, A. K., Steiner, U., Dehne, H. W. and Oerke, E. C. (2010) Spectral signatures of sugar beet leaves for the detection and differentiation of diseases. *Precision Agriculture* 11, 413–31.

Mahlein, A. K., Steiner, U., Hillnhuetter, C., Dehne, H. W. and Oerke, E. C. (2012b) Hyperspectral imaging for small-scale analysis of symptoms caused by different sugar beet diseases. *Plant Methods* 8, 3.

Mankin, R. W., Brandhorst, H. J., Flanders, L. K., Zhang, M., Crocker, R. L., Lapointe, S. L., et al. (2000). Eavesdropping on insects hidden in soil and interior structures of plants. *Journal of Economic Entomology* 93, 1173–82.

Mankin, R. W., Hagstrum, D. W., Smith, M. T., Roda, A. L. and Kairo, M. T. K. (2011). Perspective and promise: A century of insect acoustic detection and monitoring. *American Entomologist* 57, 30–44.

Mankin, R. W., Hodges, R. D., Nagle, H. T., Schal, C., Pereira, R. M. and Koehler, P. G. (2010) Acoustic indicators for targeted detection of stored product and urban insect pests by inexpensive infrared, acoustic, and vibrational detection of movement. *Journal of Economic Entomology* 103, 1636–46.

Mankin, R. W., Smith, M. T., Tropp, J. M., Atkinson, E. B. and Jong, D. Y. (2008). Detection of *Anoplophora glabripennis* (Coleoptera: Cerambycidae) larvae in different host trees and tissues by automated analyses of sound-impulse frequency and temporal patterns. *Journal of Economic Entomology* 101, 838.

Martin, B., Juliet, V., Sankaranarayanan, P., Gopal, A. and Rajkumar, I. (2013) Wireless implementation of MEMS accelerometer to detect red palm weevil on palms, pp. 248–52. In *IEEE International Conference on Advanced Electronic Systems*, Pilani, India. IEEE, Piscataway, NJ.

Martinelli, F., Scalenghe, R., Davino, S., Panno, S., Scuderi, G., Ruisi, P., et al. (2014) Advanced methods of plant disease detection: A review. *Agronomy for Sustainable Development* 35, 1–25.

Melado-Herreros, A., Munoz-Garcia, M. A., Blanco, A., Val, J., Fernandez-Valle, M. E. and Barreiro, P. (2013) Assessment of watercore development in apples with MRI: Effect of fruit location in the canopy. *Postharvest Biology and Technology* 86, 125–33.

Mewes, T., Franke, J. and Menz, G. (2011) Spectral requirements on airborne hyperspectral remote sensing data for wheat disease detection. *Precision Agriculture* 12, 795–812.

Michels, G. J., Piccinni, G., Rush, C. M. and Fritts, D. A. (2000) Using infrared transducers to sense greenbug infestation in winter wheat. In P. C. Robert (Ed.) *5th International Conference on Precision Agriculture*. ASA-CSSA-SSSA, Madison, WI, USA(CD-ROM).

Miresmailli, S., Gries, R., Gries, G., Zamar, R. Z. and Isman, M. B. (2010) Herbivore-induced plant volatiles allow detection of *Trichoplusia ni* (Lepidoptera: Noctuidae) infestation on greenhouse tomato plants. *Pest Management Science* 66, 916–24.

Mirik, M., Ansley, R. J., Michels Jr., G. J. and Elliott, N. C. (2012) Spectral vegetation indices selected for quantifying Russian wheat aphid (*Diuraphis noxia*) feeding damage in wheat (*Triticum aestivum* L.). *Precision Agriculture* 13, 501–16.

Mirik, M., Michels, G. J., Kassymzhanova-Mirik, S., Elliott, N. C. and Bowling, R. (2006) Hyperspectral spectrometry as a means to differentiate uninfested and infested winter wheat by greenbug (Hemiptera: Aphididae). *Journal of Economic Entomology* 99, 1682–90.

Moral García, F. J. (2006) Analysis of the spatio – temporal distribution of Helicoverpa armigera Hb. in a tomato field using a stochastic approach. *Biosystems Engineering* 93, 253–9.

Moshou, D., Bravo, C., West, J., Wahlen, S., McCartney, A. and Ramon, H. (2004) Automatic detection of 'yellow rust' in wheat using reflectance measurements and neural networks. *Computers and Electronics in Agriculture* 44, 173–88.

Mueller, J. D., Khalilian, A., Monfort, W. S., Davis, R. F., Kirkpatrick, T. L., Ortiz, B. V., et al. (2010) Site-specific detection and management of nematodes, pp. 385–402. In Oerke, E. C., Gerhards, R., Menz, G. and Sikora, R. A. (Eds) *Precision Crop Protection – the Challenge and Use of Heterogeneity*. Springer, Dordrecht, the Netherlands.

Murchie, E. H. and Lawson, T. (2013) Chlorophyll fluorescence analysis: a guide to good practice and understanding some new applications. *Journal of Experimental Botany* 64, 3983–98.

Mutka, A. M. and Bart, R. S. (2015) Image-based phenotyping of plant disease symptoms. *Frontiers in Plant Science* 5, Article 734, doi:10.3389/fpls.2014.00734.

Naidu, R. A., Perry, E. M., Pierce, F. J. and Mekuria, T. (2009). The potential of spectral reflectance technique for the detection of grapevine leafroll-associated virus-3 in two red-berried wine grape cultivars. *Computers and Electronic in Agriculture* 66, 38–45.

Nansen, C. (2016) The potential and prospects of proximal remote sensing of arthropod pests. *Pest Management Science* 72, 653–9.

Nansen, C., Coelho Jr., A., Mendes, J. V. and Parra, J. R. (2014) Reflectance-based identification of parasitized host eggs and adult *Trichogramma* specimens. *Journal of Experimental Biology* 217, 1187–92.

Nansen, C. and Elliott, N. (2016) Remote sensing and reflectance profiling in entomology. *Annual Review of Entomology* 61, 139–58.

Nansen, C., Geremias, L. D., Xue, Y., Huang, F. and Parra, J. R. (2013) Agricultural case studies of classification accuracy, spectral resolution, and model over-fitting. *Applied Spectroscopy* 67, 1332–8.

Nansen, C., Macedo, T., Swanson, R. and Weaver, D. K. (2009) Use of spatial structure analysis of hyperspectral data cubes for detection of insect-induced stress in wheat plants. *International Journal of Remote Sensing* 30, 2447–64.

Nansen, C., Weaver, D. K., Sing, S. E., Runyon, J. B., Morrill, W. L., Grieshop, M. J., et al. (2005) Within-field spatial distribution of *Cephus cinctus* (Hymenoptera: Cephidae) larvae in Montana wheat fields. *Canadian Entomologist* 137, 202–14.

Nansen, C., Zhang, X., Aryamanesh, N. and Yan, G. (2014) Use of variogram analysis to classify field peas with and without internal defects caused by weevil infestation. *Journal of Food Engineering* 123, 17–22.

Navakar, D. S., Singh, C. B., Jayas, D. S. and White, N. D. G. (2009) Assessment of soft X-ray imaging for detection of fungal infection in wheat. *Biosystems Engineering* 103, 49–56.

Neumann, M., Hallau, L., Klatt, B., Kersting, K. and Bauckhage, C. (2014) Erosion band features for cell phone image based plant disease classification. In *22nd International Conference on Pattern Recognition*, Stockholm. IEEE, Piscataway, NJ.

Nguy-Robertson, A. L., Zygielbaum, A. I., McMechan, A. J., Hein, G. L., Wegulo, S. N., Stilwell, A. R., et al. (2016) Developing the framework for a risk map for mite vectored viruses in wheat resulting from pre-harvest hail damage. *Crop Protection* 89, 21–31.

Nicolas, H. (2004) Using remote sensing to determine of the date of a fungicide application on winter wheat. *Crop Protection* 23, 853–63.

Niederbacher, B., Winkler, J. B. and Schnitzler, J. P. (2015) Volatile organic compounds as non-invasive markers for plant phenotyping. *Journal of Experimental Botany* 66, 5403–16.

Nilsson, H. (1995) Remote sensing and image analysis in plant pathology. *Annual Review of Phytopathology* 33, 489–528.

Njoroge, A. W., Affognon, H., Mutungi, C., Rohde, B., Richter, U., Hensel, O., et al. (2016) Frequency and time pattern differences in acoustic signals produced by *Prostephanus truncates* (Horn) (Coleoptera: Bostrichidae) and *Sitophilus zeamais* (Motschulsky) (Coleoptera: Curculionidae) in stored maize. *Journal of Stored Products Research* 69, 31–40.

Nutter Jr., F. W. (2001) Disease assessment terms and concepts, pp. 312–23. In: O. C. Maloy and T. D. Murray (Eds) *Encyclopedia of Plant Pathology*. John Wiley and Sons, New York, NY.

Nutter, F., van Rij, N., Eggenberger, S. K. and Holah, N. (2010) Spatial and temporal dynamics of plant pathogens, pp. 27–50. In E. C. Oerke, R. Gerhards, G. Menz and R. A. Sikora (Eds) *Precision Crop Protection – the Challenge and Use of Heterogeneity*. Springer, Dordrecht, The Netherlands.

Oberti, R., Marchi, Tirelli, P., Calcante, A., Iriti, M. and Borghese, A. N. (2014) Automatic detection of powdery mildew on grapevine leaves by image analysis: Optimal view-angle range to increase the sensitivity. *Computers and Electronics in Agriculture* 104, 1–8.

Oerke, E. C. (2006) Crop losses to pests. *Journal of Agricultural Science* 144, 31–43.

Oerke, E. C., Froehling, P. and Steiner, U. (2011) Thermographic assessment of scab disease on apple leaves. *Precision Agriculture*, 12, 699–715.

Oerke, E. C., Gerhards, R., Menz, G. and Sikora, R. S. (2010a). Preface. In E. C. Oerke, R. Gerhards, G. Menz and R. A. Sikora (Eds) *Precision Crop Protection – the Challenge and Use of Heterogeneity*. Springer, Dordrecht, Netherlands.

Oerke, E. C., Herzog, K. and Toepfer, R. (2016) Hyperspectral phenotyping of the reaction of grapevine genotypes to *Plasmopara viticola*. *Journal of Experimental Botany* 67, 5529–43.

Oerke, E. C., Mahlein, A. K. and Steiner, U. (2014) Proximal sensing of plant diseases, pp. 55–68. In Gullino, M. L. and Bonantsm P. J. M. (Eds) *Detection and Diagnostics of Plant Pathogens. Plant Pathology in the 21st Century*, vol. 5. Springer Netherlands, Dordrecht, The Netherlands.

Oerke, E. C., Meier, A., Dehne, H. W., Sulyok, M., Krska, R. and Steiner U. (2010b) Spatial variability of Fusarium head blight pathogens and associated mycotoxins in wheat crops. *Plant Pathology* 59, 671–82.

Oerke, E. C. and Steiner, U. (2010). Potential of digital thermography for disease control, pp. 167–82. In E. C. Oerke, R. Gerhards, G. Menz and R. A. Sikora (Eds) *Precision Crop Protection – the Challenge and Use of Heterogeneity*. Springer, Dordrecht, The Netherlands.

Oerke, E. C., Steiner, U., Dehne, H.-W. and Lindenthal, M. (2006) Thermal imaging of cucumber leaves affected by downy mildew and environmental conditions. *Journal of Experimental Botany* 57, 2121–32.

Oh, J. and Lee, J. (2014) Feasibility of ultrasonic spectral analysis for detecting insect damage in wooden cultural heritage. *Journal of Wood Science* 60, 21–9.

Ojiambo, P. S. and Scherm, H. (2010) Efficiency of adaptive cluster sampling for estimating plant disease incidence. *Phytopathology* 100, 663–70.

Opstad Kruse, O. M., Prats-Montalban, J. M., Indahl, U. G., Kvaal, K., Ferrer Riquelme, A. J. and Futsaether C. M. (2014) Pixel classification methods for identifying and quantifying leaf surface injury from digital images. *Computers and Electronics in Agriculture* 108, 155–65.

Ortiz, B. V., Perry, C., Goovaerts, P., Vellidis, G. and Sullivan, D. (2010) Geostatistical modelling of the spatial variability and risk areas of southern root-knot nematodes in relation to soil properties. *Geoderma* 156, 243–52.

Ortiz, B. V., Perry, C., Sullivan, D., Lu, P., Kemerait, R., Davis, R. F., et al. (2012) Variable rate application of nematicides on cotton fields: a promising site-specific management strategy. *Journal of Nematology* 44, 31–9.

Ortiz, B. V., Sullivan, D. G., Perry, C. and Vellidis, G. (2011) Delineation of management zones for Southern root-knot nematode using fuzzy clustering of terrain and edaphic field characteristics. *Communications in Soil Science and Plant Analysis* 42, 1972–94.

Overstreet, C., McGawley, E. C., Khalilian, A., Kirkpatrick, T. L., Monfort, W. S., Henderson, W., et al. (2014) Site specific nematode management – development and success in cotton production in the United States. *Journal of Nematology* 46, 309–20.

Pan, L., Zhang, W., Zhu, N., Mao, S. B. and Tu, K. (2016) Early detection and classification of pathogenic fungal disease in post-harvest strawberry fruit by electronic nose and gas chromatography-mass spectrometry. *Food Research International* 62, 162–8.

Park, Y. L., Ellsbury, M. M., Krell R. K., Tollefson, J. J., Pedigo, L. P. and Wade French, B. (2008) Spatial variability in corn and soybean insect pests: Precision farming and insect pest management for the future. *Site-Specific Management Guidelines* (SSMG-27, updated August 2007), http://www.ipni.net/ssmg.

Pearson, T. (2009) Hardware-based image processing for high-speed inspection of grains. *Computers and Electronics in Agriculture* 69, 12–18.

Pearson, T. C., Cetin, A., Tewfik, A. and Haff, R. (2007) Feasibility of impact-acoustic emissions for detection of damaged wheat kernels. *Digital Signal Processing* 17, 617–33.

Penuelas, J., Filella, I., Lloret, P., Muñoz, F. and Vilajeliu, M. (1995) Reflectance assessment of mite effects on apple trees. *International Journal of Remote Sensing* 16, 2727–33.

Pereira, F. M. V., Milori, D. M. B. P., Pereira, E. R., Venancio, A. L., Russo, M. D. T., Cardinali, M. C. D., et al. (2011) Laser-induced fluorescence imaging method to monitor citrus greening disease. *Computers and Electronics in Agriculture* 79, 90–3.

Pérez-Bueno, M. L., Pineda, M., Cabeza, F. M. and Barón, M. (2016) Multicolor fluorescence imaging as a candidate for disease detection in plant phenotyping. *Frontiers in Plant Science* 7, 1790, doi:10.3389/fpls.2016.01790.

Peteinatos, G. G., Korsaeth, A., Berge, T. W. and Gerhards, R. (2016) Using optical sensors to identify water deprivation, nitrogen shortage, weed presence and fungal infection in wheat. *Agriculture* 6, 24.

Pethybridge, S. J. and Nelson, S. C. (2015) Leaf Doctor: A new portable application for quantifying plant disease severity. *Plant Disease* 99, 1310–16.

Pinto, F., Damm, A., Schickling, A., Panigada, C., Cogliati, S., Müller-Linow, M., et al. (2016) Sun-induced chlorophyll fluorescence from high-resolution imaging spectroscopy data to quantify spatio-temporal patterns of photosynthetic function in crop canopies. *Plant Cell & Environment* 39, 1500–12.

Plaza, A., Benediktsson, J. A., Boardman, J. W., Brazile, J., Bruzzone, L., Camps-Valls, G., et al. (2009) Recent advances in techniques for hyperspectral image processing. *Remote Sensing of Environment* 113, 110–22.

Porcar-Castell, A., Tyystjärvi, E., Atherton, J., van der Tol, C., Flexas, J., Pfuendel, E. E., et al. (2014) Linking chlorophyll a fluorescence to photosynthesis for remote sensing applications: Mechanisms and challenges. *Journal of Experimental Botany* 65, 4065–95.

Potamitis, I., Rigakis, I. and Fysarakis, K. (2014) The electronic McPhail trap. *Sensors* 14, 22285–99.

Pourreza, A., Lee, W. S., Ehsani, R., Schueller, J. K. and Raveh, E. (2015a) An optimum method for real-time in-field detection of Huanglongbing disease using a vision sensor. *Computers and Electronics in Agriculture* 110, 221–32.

Pourreza, A., Lee, W. S., Etxeberria, E. and Banerjee, A. (2015b) An evaluation of a vision-based sensor performance in Huanglongbing disease identification. *Biosystems Engineering* 130, 13–22.

Pourreza, A., Lee, W. S., Ritenour, M. A. and Roberts, P. (2016) Spectral characteristics of citrus black spot disease. *HortTechnology* 26, 254–60.

Prabhakar, M., Prasad, Y. G., Desai, S., Thirupathi, M., Gopika, K., Rao, G. R., et al. (2013a) Hyperspectral remote sensing of yellow mosaic severity and associated pigment losses in *Vigna mungo* using multinomial logistic regression models. *Crop Protection* 45, 132–40.

Prabhakar, M., Prasad, Y. G. and Rao, M. N. (2012) Remote sensing of biotic stress in crop plants and its applications for pest management, pp. 517–49. In B. Venkateswarlu, A. K. Shanker, C. Shanker and M. Maheswari (Eds) *Crop Stress and Its Management: Perspectives and Strategies*. Springer Netherlands, Dordrecht, the Netherlands.

Prabhakar, M., Prasad, Y. G., Thirupathi, M., Sreedevi, G., Dharajothi, B. and Venkateswarlu, B. (2011) Use of ground based hyperspectral remote sensing for detection of stress in cotton caused by leafhopper (Hemiptera: Cicadellidae). *Computers and Electronics in Agriculture* 79, 189–98.

Prabhakar, M., Prasad, Y. G., Vennila, S., Thirupathi, M., Sreedevi, G., Rao, G. R., et al. (2013b) Hyperspectral indices for assessing damage by the solenopsis mealybug (Hemiptera: Pseudococcidae) in cotton. *Computers and Electronics in Agriculture* 97, 61–70.

Qin, J., Burks, T. F., Ritenour, M. A. and Bonn, W. G. (2009) Detection of citrus canker using hyperspectral reflectance imaging with spectral information divergence. *Journal of Food Engineering* 93, 183–91.

Qiu, R., Qu, D., Trengove, R., Agarwal, M., Hardy, G. E. St. J. and Ren, Y. 2014. Headspace solid-phase microextraction and gas chromatography-mass spectrometry for analysis of VOCs produced by *Phytophthora cinnamomi*. *Plant Disease* 98, 1099–105.

Racca, P., Zeuner, T, Jung, J. and Kleinhenz, B. (2010) Model validation and use of geographic information systems in crop protection warning service, pp. 259–76. In E. C. Oerke, R. Gerhards, G. Menz and R. A. Sikora (Eds) *Precision Crop Protection – the Challenge and Use of Heterogeneity*. Springer, Doredrecht, The Netherlands.

Raji, S. N., Subhash, N., Ravi, V., Saravanan, R., Mohanan, C. N., Nita, S., et al. (2015) Detection of mosaic virus disease in cassava plants by sunlight-induced fluorescence imaging: a pilot study for proximal sensing. *International Journal of Remote Sensing* 36, 2880–97.

Raza, S.-e-A.., Prince, G., Clarkson, J. P. and Rajpoot, N. M. (2015) Automatic detection of diseased tomato plants using thermal and stereo visible light images. *PLoS ONE* 10, e0123262.

Reynolds, G. J., Windels, C. E., MacRae, I. V. and Laguette, S. (2012) Remote sensing for assessing Rhizoctonia crown and root rot severity in sugar beet. *Plant Disease* 96, 497–505.

Riedell, W. E. and Blackmer, T. M. (1999) Leaf reflectance spectra of cereal aphid-damaged wheat. *Crop Science* 39, 1835–40.

Riedell, W. E., Hesler, L. S., Osborne, S. T. and Blackmer, T. M. (2000) Remote sensing of insect damage in wheat. In P. C. Robert (Ed.) *5th International Conference on Precision Agriculture*. ASA-CSSA-SSSA, Madison, WI, USA (CD-ROM).

Riley, J. R., Cheng, X. N., Zhang, X. X, Reynolds, D. R., Xu, G. M., Smith, A. D., et al. (1991) The long-distance migration of *Nilaparvata lugens* (Stal) (Delphacidae) in China: radar observations of mass return flight in the autumn. *Ecological Entomology* 16, 471–89.

Rispail, N. and Rubiales, D. (2016) Rapid and efficient estimation of pea resistance to the soil-borne pathogen *Fusarium oxysporum* by infrared imaging. *Sensors* 16, 1282.

Roemer, C., Buerling, K., Hunsche, M., Rumpf, T., Noga, G. and Pluemer, L. (2011) Robust fitting of fluorescence spectra for pre-symptomatic wheat leaf rust detection with Support Vector Machines. *Computers and Electronics in Agriculture* 79, 180–8.

Rolfe, S. A. and Scholes, J. D. (2010) Chlorophyll fluorescence imaging of plant–pathogen interactions. *Protoplasma* 247, 163–75.

Roscher, R., Behmann, J., Mahlein, A. K., Dupuis, J., Kuhlmann, H. and Pluemer, L. (2016) Detection of disease symptoms on hyperspectral 3D plant models. *ISPRS Annals of the Photogrammetry, Remote Sensing and Spatial Information Sciences* III-7, pp. 89–96.

Rousseau, C., Belin, E., Bove, E., Rousseau, D., Fabre, F., Berruyer, R., et al. (2013) High throughput quantitative phenotyping of plant resistance using chlorophyll fluorescence image analysis. *Plant Methods* 9, 17.

Ruiz-Altisent, M., Ruiz-Garcia, L., Moreda, G. P., Lu, R. F., Hernandez-Sanchez, N., Correa, E. C., et al. (2010) Sensors for product characterization and quality of specialty crops - a review. *Computers and Electronics in Agriculture* 74, 176–94.

Rumpf. T., Mahlein, A. K., Steiner, U., Oerke, E. C., Dehne, H. W. and Pluemer, L. (2010) Early detection and classification of plant diseases with Support Vector Machines based on hyperspectral reflectance. *Computers and Electronics in Agriculture* 74, 91–9.

Rutolo, M., Covington, J. A., Clarkson, J. and Iliescu, D. (2014) Detection of potato storage disease via gas analysis: a pilot study using field asymmetric ion mobility spectrometry. *Sensors* 14, 15939–52.

Rutolo, M. F., Iliescu, D., Clarkson, J. P. and Covington, J. A. (2016) Early identification of potato storage disease using an array of metal-oxide based gas sensors. *Postharvest Biology and Technology* 116, 50–8.

Ryerson, R. A., Curran, P. J. and Stephens, P. R. (1997). Applications: Agriculture, pp. 365–97. In W. R. Philipson (Ed.), *Manual of Photographic Interpretation*. American Society for Photogrammetry and Remote Sensing, Bethesda, MD, USA.

Sankaran, S., Mishra, A., Maja, J. M. and Ehsani, R. (2011) Visible-near infrared spectroscopy for detection of Huanglongbing in citrus orchards. *Computers and Electronics in Agriculture* 77, 127–34.

Sankaran, S. and Ehsani, R. (2013) Detection of Huanglongbing-infected citrus leaves using statistical models with a fluorescence sensor. *Applied Spectroscopy* 67, 463–9.

Sankaran, S., Khot, L. R., Zuniga Espinoza, C., Jarolmasjed, S., Sathuvalli, V. R., Vandemark, G. J., et al. (2015) Low-altitude, high-resolution aerial imaging systems for row and field crop phenotyping: A review. *European Journal of Agronomy* 70, 112–23.

Sankaran, S., Maja, J. M., Buchanon, S. and Ehsani, R. (2013) Huanglongbing (citrus greening) detection using visible, near infrared and thermal imaging techniques. *Sensors* 13, 2117–30.

Sankaran, S., Mishra, A., Ehsani, R. and Davis, C. (2010). A review of advanced techniques for detecting plant diseases. *Computers and Electronics in Agriculture* 72, 1–13.

Saranwong, S., Haff, R., Thanapase, W., Janhiran, A., Kasemsumran, S. and Kawano, S. (2011) A feasibility study using simplified near infrared imaging to detect fruitfly larvae in intact fruit. *Near Infrared Spectroscopy* 19, 55–60.

Saranwong, S., Thanapase, W., Suttiwijitpukdee, N., Rittiron, R., Kasemsumran, S. and Kawano, S. (2010) Applying near infrared spectroscopy to the detection of fruitfly eggs and larvae in intact fruit. *Near Infrared Spectroscopy* 18, 271–80.

Sarles, L., Verhaeghe, A., Francis, F. and Verheggen, F. J. (2015) Semiochemicals of *Rhagoletis* fruit flies: Potential for integrated pest management. *Crop Protection* 78, 114–18.

Schmale III, D. G., Dingus, B. R. and Reinholtz, C. (2008) Development and application of an autonomous unmanned aerial vehicle for precise aerobiological sampling above agricultural fields. *Journal of Field Robotics* 25, 133–47.

Schmale III, D. G. and Ross, S. D. (2015) Highways in the sky: scales of atmospheric transport of plant pathogens. *Annual Review of Phytopathology* 53, 591–611.

Schmitz, A., Kiewnick, S., Schlang, J. and Sikora, R. A. (2004) Use of high resolutional digital thermography to detect *Heterodera schachtii* infestation in sugar beets. *Communications in Agriculture and Applied Biological Sciences* 69, 359–63.

Scholes, J. D. and Rolfe, S. A. (2009). Chlorophyll fluorescence imaging as tool for understanding the impact of fungal diseases on plant performance: A phenomics perspective. *Functional Plant Biology* 36, 880–92.

Schotzko, D. J. and Quisenberry, S. S. (1999) Pea leaf weevil (Coleoptera: Curculionidae) spatial distribution in peas. *Environmental Entomology* 28, 477–84.

Schumann, M., Patel, A. and Vidal, S. (2014) Soil application of an encapsulated CO_2 source and its potential for management of Western corn rootworm larvae. *Journal of Economic Entomology* 107, 230–9.

Sciarretta, A., Zinni, A. and Trematerra, P. (2011) Development of site-specific IPM against European grapevine moth *Lobesia botrana* (D. & S.) in vineyards. *Crop Protection* 30, 1469–77.

Scotford I. M. and Miller P. C. H. (2005) Applications of spectral reflectance techniques in northern European cereal production: A review. *Biosystems Engineering* 90, 235–50.

Secher, B. J. M. (1997) Site-specific control of diseases in winter wheat. *Aspects of Applied Biology* 48, 57–64.

Severtson, D., Callow, N., Flower, K., Neuhaus, A., Olejnik, M. and Nansen, C. (2016) Unmanned aerial vehicle canopy reflectance data detects potassium deficiency and green peach aphid susceptibility in canola. *Precision Agriculture* 17, 659–77.

Shahin, M. A. and Symons, S. J. (2011) Detection of Fusarium damaged kernels in Canada Western Red Spring wheat using visible/near-infrared hyperspectral imaging and principal component analysis. *Computers and Electronics in Agriculture* 75, 107–12.

Shulaev, V., Silverman, P. and Raskin, I. (1997) Airborne signalling by methyl salicylate in plant pathogen resistance. *Nature* 385, 718–21.

Simko, I., Jimenez-Berni, J. A. and Sirault, X. R. R. (2017) Phenomic approaches and tools for phytopathologists. *Phytopathology* 107, 6–17.

Simonett, D. S. (1983) *Manual of Remote Sensing*. American society of photogrammetry, Fall Church, VA, USA.

Singh, A., Ganapathysubramanian, B., Singh A. K. and Sarkar, S. (2016) Machine learning for high-throughput stress phenotyping in plants. *Trends in Plant Science* 21, 110–24.

Singh, C., Jayas, D., Paliwal, J. and White, N. (2010) Identification of insect-damaged wheat kernels using short-wave near-infrared hyperspectral and digital colour imaging. *Computers and Electronics in Agriculture* 73, 118–25.

Singh, C. B., Jayas, D. S., Paliwal, J. and White N. D. G. (2009) Detection of insect-damaged wheat kernels using near-infrared hyperspectral imaging. *Journal of Stored Products Research* 45, 151–8.

Siriwardena, K. A. P., Fernando, L. C. P., Nanayakkara, N., Perera, K. F. G., Kumara, A. D. N. T. and Nanayakkara, T. (2010) Portable acoustic device for detection of coconut palms infested by *Rynchophorus ferrugineus* (Coleoptera: Curculionidae). *Crop Protection* 29, 25–9.

Smith, A. D., Reynolds, D. R. and Riley, J. R. (2000) The use of vertical-looking radar to continuously monitor the insect fauna flying at altitude over southern England. *Bulletin of Entomological Research* 90, 265–77.

Stafford, J. V. (2000) Implementing precision agriculture in the 21st century. *Journal of Agricultural Engineering Research* 76, 267–75.

Stafford, J. V., Evans, K., Barker, A., Halford, P. D. and Russell, M. D (2000) Changes in the within-field spatial distribution of potato cyst nematodes before and after cropping with potatoes and the consequences for modulating nematicide application. *Aspects of Applied Biology* 59, 1–7.

Steddom, K., Bredehoeft, M. W., Khan, M. and Rush, C. M. (2005) Comparison of visual and multispectral radiometric disease evaluations of Cercospora leaf spot of sugar beet. *Plant Disease* 89, 153–8.

Steddom, K., Heidel, G., Jones, D. and Rush, C. M. (2003) Remote detection of rhizomania in sugar beets. *Phytopathology* 93, 720–6.

Stenzel, I., Steiner, U., Dehne, H. W. and Oerke, E. C. (2007) Occurrence of fungal leaf pathogens in sugar beet fields monitored with digital infrared thermography, pp. 529–35. In Stafford, J. V. (Ed.) *Precision agriculture'07*. Proceedings of the 6th European Conference on Precision Agriculture. Wageningen Academic Publishers, The Netherlands.

Stern, V. M., Smith, R. F., van den Bosch, R. and Hagen, K. S. (1959) The integrated control concept. *Hilgardia* 29: 81–101.

Stoll, M., Schultz, H. R., Baecker, G. and Berkelmann-Loehnertz, B. (2008a) Early pathogen detection under different water status and the assessment of spray application in vineyards through the use of thermal imagery. *Precision Agriculture* 9, 407–17.

Stoll, M., Schultz, H. R. and Berkelmann-Loehnertz, B. (2008b) Exploring the sensitivity of thermal imaging for Plasmopara viticola pathogen detection under different water status. *Functional Plant Biology* 35, 281–8.

Swinton, S. M. (2003) Site-specific pest management, pp. 155–68. In F. den Hond, P. Groenewegen and N. M. van Straalen (Eds) *Pesticides: Problems, Improvements, Alternatives*. Wiley-Blackwell, Oxford, UK.

Tackenberg, M., Volkmar, C. and Dammer, K. H. (2016) Sensor-based variable-rate fungicide application in winter wheat. *Pest Management Science* 72, 1888–96.

Tatagiba, S. D., DaMatta, F. M. and Rodrigues, F. A. (2015) Leaf gas exchange and chlorophyll a fluorescence imaging of rice leaves infected with *Monographella albescens*. *Phytopathology* 105, 180–8.

Taubenhaus, J. J., Ezekiel, W. N. and Neblette, C. B. (1929) Airplane photography in the study of cotton root rot. *Phytopathology* 19, 1025–9.

Thybo, A. K., Jespersen, S. N., Laerke, P. E. and Stodkilde-Jorgensen, H. J. (2004) Nondestructive detection of internal bruise and spraing disease symptoms in potatoes using magnetic resonance imaging. *Magnetic Resonance Imaging* 22, 1311–17.

Tripathy, A. K., Adinarayana, J., Vijayalakshmi, K., Merchant, S. N., Desai, U. B., Ninomiya, S., et al. (2014) Knowledge discovery and leaf spot dynamics of groundnut crop through wireless sensor network and data mining techniques. *Computers and Electronics in Agriculture* 107, 104–14.

Turlings, T. C. J., Loughrin, J. H., McCall, P. J., Rose, U. S. R., Lewis, W. J. and Tumlinson, J. H. (1995) How caterpillar-damaged plants protect themselves by attracting parasitic wasps. *Proceedings of the National Academy of Sciences of the United States of America* 92, 4169–74.

Vadivambal, R. and Jayas, D. S. (2011) Applications of thermal imaging in agriculture and food industry – a review. *Food and Bioprocess Technology* 4, 186–99.

van Helden, M. (2010) Spatial and temporal dynamics of arthropods in arable fields, pp. 51–64. In Oerke, E. C., Gerhards, R., Menz, G. and Sikora, R. A. (Eds) *Precision Crop Protection – the Challenge and Use of Heterogeneity*. Springer, Dordrecht, The Netherlands.

Vanoye-Eligio, V., Barrientos-Lozano, L., Mora-Olivo, A., Sanchez-Ramos, G. and Chacon-Hernandez, J. C. (2016) Spatial heterogeneity of *Anastrepha ludens* populations over a large citrus region including a sterile insect release area in northeastern Mexico. *Precision Agriculture* 17, doi 10.1007/s11119-016-9493-2.

Voss, K., Franke, J., Mewes, T., Menz, G. and Kuehbauch, W. (2010) Remote sensing for precision crop protrection – a matter of scale, pp. 101–18. In E. C. Oerke, R. Gerhards, G. Menz and R. A. Sikora (Eds) *Precision Crop Protection – The Challenge and Use of Heterogeneity*. Springer, Dordrecht, the Netherlands.

Waggoner P. E. and Aylor D. E. (2000) Epidemiology, a science of patterns. *Annual Review of Phytopathology* 38, 1–24.

Wahabzada, M., Mahlein, A. K., Bauckhage, C., Steiner, U., Oerke, E. C. and Kersting, K. (2015) Metro maps of plant disease dynamics – automated mining of differences using hyperspectral images. *PLoS ONE* 10, e0116902, doi:10.1371/journal.pone.0116902.

Wahabzada, M., Mahlein, A. K., Bauckhage, C., Steiner, U., Oerke, E. C. and Kersting, K. (2016) Plant phenotyping using probabilistic topic models: Uncovering the hyperspectral language of plants. *Scientific Reports* 6, 22482, doi:10.1038/srep22482.

Walter, S. M. and Peairs, F. B. (1999) Spatial variability of trap catches of three insect pests in sprinkler irrigated maize in eastern Colorado, pp. 229–37. In J. V. Stafford (Ed.) *Precision Agriculture 99*. Proceedings of the 2nd European Conference on Precision Agriculture, Sheffield Academic Press, Sheffield, UK.

Wang, H., Pu, R. L., Zhu, Q., Ren, L. L. and Zhang, Z. Z. (2015a) Mapping health levels of *Robinia pseudoacacia* forests in the Yellow River delta, China, using IKONOS and Landsat 8 OLI imagery. *International Journal of Remote Sensing* 36, 1114–35.

Wang, J., Nakano, K. and Ohashi, S. (2011a). Non-destructive detection of internal insect infestation in jujubes using visible and near-infrared spectroscopy. *Postharvest Biology and Technology* 59, 272–9.

Wang, J., Nakano, K., Ohashi, S., Kubota, Y., Takizawa, K. and Sasaki, Y. (2011b) Detection of external insect infestations in jujube fruit using hyperspectral reflectance imaging. *Biosystems Engineering* 108, 345–51.

Wang, K., Zhang, S., Wang, Z., Liu, Z. and Yang, F. (2013). Mobile smart device-based vegetable disease and insect pest recognition method. *Intelligent Automation & Soft Computing* 19, 263–73.

Wang, M., Ling, N., Dong, X., Zhu, Y., Shen, Q. and Guo, S. (2012) Thermographic visualization of leaf response in cucumber plants infected with the soil-borne pathogen *Fusarium oxysporum* f.sp. *cucumerinum*. *Plant Physiology and Biochemistry* 61, 153–61.

Wang, W., Lawrence, K. C., Ni, X., Yoon, S. C., Heitschmidt, G. W. and Feldner, P. (2015b) Near-infrared hyperspectral imaging for detecting Aflatoxin B1 of maize kernels. *Food Control* 51, 347–55.

Wang, W., Ni, X., Lawrence, K. C., Yoon, S. C., Heitschmidt, G. W. and Feldner, P. (2015c) Feasibility of detecting aflatoxin B1 in single maize kernels using hyperspectral imaging. *Journal of Food Engineering* 166, 182–92.

Weerakoon, K. A. and Chin, B. A. (2012) A chemical switch for detecting insect infestation. *Pest Management Science* 68, 906–13.

Wehrenfennig, C., Schott, M., Gasch, T., Duering, R., Vilcinskas, A. and Kohl, C. D. (2013) On-site airborne pheromone sensing. *Analytical and Bioanalytical Chemistry* 405, 6389–403.

Weisz, R., Fleischer, S. and Smilowitz, Z. (1995) Site-specific integrated pest management for high-value crops: Sample units for map generation using the Colorado potato beetle (Coleoptera: Chrysomelidae) as a model system. *Journal of Economic Entomology* 88, 1069–80.

West, J. S., Bravo, C., Oberti, R., Lemaire, D., Moushou, D. and McCartney, H. A. (2003). The potential of optical canopy measurement for targeted control of field crop diseases. *Annual Review of Phytopathology* 41, 593–614.

West, J. S., Bravo, C., Oberti, R., Moshou, D., Ramon, H. and McCartney, H. A. (2010) Detection of fungal diseases optically and pathogen inoculum by air sampling, pp. 135–49. In E. C. Oerke, R. Gerhards, G. Menz and R. A. Sikora (Eds) *Precision Crop Protection – The Challenge and Use of Heterogeneity*. Springer, Dordrecht, The Netherlands.

Wetterich, C. B., Kumar, R., Sankaran, S., Belasque Junior, J., Ehsani, R. and Marcassa, L. G. (2013) A comparative study on application of computer vision and fluorescence imaging spectroscopy for detection of Huanglongbing citrus disease in the USA and Brazil. *Journal of Spectroscopy*, 2013, 6 pages, article ID 841738, doi.org/10.1155/2013/841738.

Wijesinghe, R. E., Lee, S. Y., Kim, P., Jung, H. Y., Jeon, M. and Kim, J. (2016) Optical inspection and morphological analysis of *Diospyros kaki* plant leaves for the detection of circular leaf spot disease. *Sensors* 16, 1282.

Williams, J. P., Hanavan, R. P., Rock, B. N., Minocha, S. C. and Linder, E. (2016) Influence of hemlock woolly adelgid infestation on the physiological and reflectance characteristics of eastern hemlock. *Canadian Journal of Forest Research* 46, 410–26.

Williams, J. P., Hanavan, R. P., Rock, B. N., Minocha, S. C. and Linder, E. (2017) Low-level *Adelges tsugae* infestation detection in New England through partition modeling of Landsat data. *Remote Sensing of Environment* 190, 13–25.

Williamson, B., Goodman, B. A. and Chudek, J. A. (1992) Nuclear magnetic resonance (NMR) micro-imaging of ripening red raspberry fruits. *New Phytologist* 120, 21–8.

Wilson, A. D. (2013) Diverse applications of electronic-nose technologies in agriculture and forestry. *Sensors* 13, 2295–348.

Winder, L., Perry, J. N. and Holland, J. M. (1999) The spatial and temporal distribution of the grain aphid *Sitobion avenae* in winter wheat. *Entomologia Experimentalis et Applicata* 93, 277–90.

Witzgall, P., Kirsch, P. and Cork, A. (2010) Sex pheromones and their impact on pest management. *Journal of Chemical Ecology* 36, 80–100.

Wrather, J. A., Stevens, W. E., Vories, E. D., Kirkpatrick, T. L., Mueller, J. D. and Mauromoustakos, A. (2010) The use of root gall ratings to determine high risk zones in cotton fields infested by *Meloidogyne incognita*. *Crop Science* 50, 2575–9.

Wspanialy, P. and Moussa, M. (2016) Early powdery mildew detection system for application in greenhouse automation. *Computers and Electronics in Agriculture* 127, 487–94.

Wyse-Pester, D. Y, Wiles, L. J. and Westra, P. (2002) The potential for mapping nematode distributions for site-specific management. *Journal of Nematology* 34, 80–7.

Xie, C., Shao, Y., Li, X. and He, Y. (2015) Detection of early blight and late blight diseases on tomato leaves using hyperspectral imaging. *Scientific Reports* 5, 16564, doi 10.1038/srep16564.

Xu, S., Zhou, Z., Lu, H. Z. Luo, X. W., Lan, Y. B., Zhang, Y., et al. (2014) Estimation of the age and amount of brown rice plant hoppers based on bionic electronic nose use. *Sensors* 14, 18114–30.

Yang, C., Fernandez, C. J. and Everitt, J. H. (2005) Mapping *Phymatotrichum* root rot of cotton using airborne three-band digital imagery. *Transactions of the ASAE* 48, 1619–26.

Yang, C., Odvody, G. N., Thomasson, J. A., Isakeit, T. and Nichols, R. L. (2016) Change detection of cotton root rot infection over 10-year intervals using airborne multispectral imagery. *Computers and Electronics in Agriculture* 123, 154–62.

Yang, C. M. (2010) Assessment of the severity of bacterial leaf blight in rice using canopy hyperspectral reflectance. *Precision Agriculture* 11, 61–81.

Yang, Z., Rao, M. N., Elliott, N. C., Kindler, S. D. and Popham, T. W. (2005) Using ground-based multispectral radiometry to detect stress in wheat caused by greenbug (Homoptera: Aphididae) infestation. *Computers and Electronics in Agriculture* 47, 121–35.

Yang, Z., Rao, M. N., Elliott, N. C., Kindler, S. D. and Popham, T. W. (2009) Differentiating stress induced by greenbugs and Russian wheat aphids in wheat using remote sensing. *Computers and Electronics in Agriculture* 67, 64–70.

Yao, H., Hruska, Z., Kincaid, R., Brown, R. L., Bhatnagar, D. and Cleveland, T. E. (2013) Detecting maize inoculated with toxigenic and atoxigenic fungal strains with fluorescence hyperspectral imagery. *Biosystems Engineering* 115, 125–35.

Yeh, Y. H., Chung, W. C., Liao, J. Y., Chung, C. L., Kuo, Y. F. and Lin, T. T. (2016) Strawberry foliar anthracnose assessment by hyperspectral imaging. *Computers and Electronics in Agriculture* 122, 1–9.

Yu, K., Leufen, G., Hunsche, M., Noga, G., Chen, X. P. and Bareth, G. (2014) Investigation of leaf diseases and estimation of chlorophyll concentration in seven barley varieties using fluorescence and hyperspectral indices. *Remote Sensing* 6, 64–86.

Zadoks, J. C. and van den Bosch, F. (1994) The spread of plant disease: A theory on foci. *Annual Review of Phytopathology* 32, 503–21.

Zhang, J., Pu, R., Huang, W., Lin, Y., Luo, J. and Wang, J. (2012a). Using in situ hyperspectral data for detecting and discriminating yellow rust disease from nutrient stresses. *Field Crops Research* 134, 165–74.

Zhang, J., Pu, R., Wang, J., Huang, W., Yuan, L. and Luo, J. (2012b) Detecting powdery mildew of winter wheat using leaf level hyperspectral measurements. *Computers and Electronics in Agriculture* 85, 13–23.

Zhang, J., Yuan, L., Pu, R., Loraamm, R., Yang, G. and Wang, J. (2014a) Comparison between wavelet spectral features and conventional spectral features in detecting yellow rust for winter wheat. *Computers and Electronics in Agriculture* 100, 79–87.

Zhang, J. C., Huang, W. J., Li, J. Y., Yang, G. J., Luo, J. H., Gu, X. H., et al. (2011) Development, evaluation and application of a spectral knowledge base to detect yellow rust in winter wheat. *Precision Agriculture* 12, 716–31.

Zhang, J. C., Huang, Y. B., Yuan, L., Yang, G. J., Chen, L. P. and Zhao, C. J. (2016) Using satellite multispectral imagery for damage mapping of armyworm (*Spodoptera frugiperda*) in maize at a regional scale. *Pest Management Science* 72, 335–48.

Zhang, J. C., Pu, R. L., Yuan, L., Wang, J. H., Huang, W. J. and Yang, G. J. (2014b) Monitoring powdery mildew of winter wheat by using moderate resolution multi-temporal satellite imagery. *PLoS ONE* 9, e93107.

Zhang, M. H., Qin, Z. H. and Liu, X. (2005) Remote sensed spectral imagery to detect late blight in field tomatoes. *Precision Agriculture* 6, 489–508.

Zhang, M. H., Qin, Z. H., Liu, X. and Ustin, S. L. (2003) Detection of stress in tomatoes induced by late blight disease in California, USA, using hyperspectral remote sensing. *International Journal of Applied Earth Observation and Geoinformation* 4, 295–310.

Zhang, X., Nansen, C., Aryamanesh, N., Yan, G. and Boussaid, F. (2015) Importance of spatial and spectral data reduction in detection of internal defects in food products. *Applied Spectroscopy* 69, 473–80.

Zhao, Y. R, Li, X., Yu, K. Q., Cheng, F. and He, Y. (2016) Hyperspectral imaging for determining pigment contents in cucumber leaves in response to angular leaf spot disease. *Scientific Reports* 6, 27790, doi 10.1038/srep27790.

Zhou, R., Kaneko, S., Tanaka, F., Kayamori, M. and Shimizu, M. (2014) Disease detection of Cercospora leaf spot by robust template matching. *Computers and Electronics in Agriculture* 108, 58–70.

Zhou, B., Elazab, A., Bort, J., Vergara, O., Serret, M. D. and Araus, J. L. (2015a) Low-cost assessment of wheat resistance to yellow rust through conventional RGB images. *Computers and Electronics in Agriculture* 116, 20–9.

Zhou, R., Kaneko, S., Tanaka, F., Kayamori, M. and Shimizu, M. (2015b) Image-based field monitoring of Cercospora leaf spot in sugar beet by robust template matching and pattern recognition. *Computers and Electronics in Agriculture* 116, 65–79.

Zorovic, M. and Cokl, A. (2015) Laser vibrometry as a diagnostic tool for detecting wood-boring beetle larvae. *Journal of Pest Science* 88, 107–12.

Chapter 16

Precision weed management systems

Roland Gerhards, University of Hohenheim, Germany

1 Introduction

Weed populations in arable fields have often been found to be distributed heterogeneously in time and space (Marshall 1988; Thornton et al. 1990; Wiles et al. 1992; Johnson et al. 1996; Gerhards et al. 1997; Christensen and Heisel 1998; Nordmeyer et al. 2003; Gerhards and Christensen 2003). They often occur in aggregated patches of varying size or in stripes along the direction of cultivation. Herbicides and mechanical weed control methods, however, are usually applied uniformly across the whole field. Precision farming technologies provide a great potential for herbicide saving and make guidance of mechanical hoes more accurate. There are two approaches for precision weed management in arable crops. In the offline approach, weed distribution is mapped and georeferenced. Data are interpolated to estimate weed infestation between the sampling points and to create continuous weed distribution maps. A weed control threshold is set to determine sections in the field where weed control methods are warranted. Those application maps are then used to direct a patch sprayer or vehicles for mechanical weed control. In many studies, this map-based (offline) approach for site-specific weed management has successfully been applied, resulting in 20–90% untreated area (Timmermann et al. 2003; Gerhards and Oebel 2006). In the real-time (online) approach, weed detection and patch spraying or hoeing are performed simultaneously with a sensor mounted on the spray boom or hoe.

2 Weed detection

Weed seedling distribution in the field was often assessed using discrete weed mapping or continuous-area sampling (Rew and Cousens 2001). In most studies, discrete weed

http://dx.doi.org/10.19103/AS.2017.0032.17

mapping has been applied in a regular sampling grid established in the field. The grid size has varied from a few metres up to approximately 50 m and has been dependent on the width of the spray boom used for site-specific herbicide application. Density and/or coverage of emerged weed seedlings were counted and measured prior to and after post-emergence herbicide application in a sampling frame of 0.5 m^{-2} placed at all grid intersection points. Continuous sampling of weed distributions included assessing weed patches during harvest operations (Barroso et al. 2005) and walking around the borders of aggregated patches of weed species such as *Avena fatua* using a data logger connected to a differential global positioning system (DGPS) (Colliver et al. 1996).

A major improvement of weed mapping for site-specific weed management was the development of precise and powerful sensor techniques to automatically and continuously determine in-field variation of weed seedling populations (Lamb and Brown, 2001; Vrindts and de Baerde-maeker 1997; Biller 1998; Chapron et al. 1999; Sökefeld et al. 2000; Sogaard and Heisel 2002; Gerhards and Oebel, 2006). Airborne remote sensing has been applied to identify *Avena fatua* L. and *Avena sterilis* ssp. *ludoviciana* (Durieu) Nyman populations in wheat but could not detect densities of less than 19 plants m^{-2} (Lamb and Brown 2001). A finer resolution of the sensor, however, is required to detect low-density weed seedling populations. Lately, unmanned aerial vehicles (UAVs) have been used for weed mapping in agricultural fields with a higher geometric pixel resolution (Rasmussen et al. 2013) (Fig. 1). RGB images taken from these UAVs reach a geometric resolution of less than 5 mm per pixel on the ground. This is sufficient to assess coverage, biomass and leaf area in arable fields. Based on spectral information, size, height, position and leaf geometry, crop and weeds can be identified using UAV images (Fig. 2).

In most studies using automatic weed detection systems, sensor data have been measured from a distance of a few metres to achieve a higher resolution in the spectral data (Fig. 3). Felton and McCloy (1992), Vrindts and de Baerdemaeker (1997) and Biller (1998) used spectrometers to measure the reflectance in the green, red and near-infrared (NIR) wavebands. Green plants were characterized by a high reflectance in the green and NIR bands and a low reflectance in the red spectrum. Soil, dead mulch and stone, however, reflected more light in the red waveband and less in the infrared spectrum.

Figure 1 A hexacopter (XR6 drone) carrying a high-resolution RGB camera was used to map weed distributions in maize and sugar beet fields at the University of Hohenheim in 2016; photo: Robin Mink (unpublished).

Figure 2 RGB image of a Sony α 6000 camera with 27 Mpix showing green weed patches of mostly *Alopecurus myosuroides* in a 5.3 ha maize field at 2-leaf stage at the University of Hohenheim in 2016; processed image: Martin Weis (unpublished).

However, the most promising approach for weed detection is a continuous ground-based and image analysis method (Weis et al. 2008). For automatic weed detection, digital bi-spectral cameras (VIS/NIR) were developed and mounted on the vehicle. With each bi-spectral camera, two images were taken simultaneously in the NIR spectrum (770–1150 nm) and in the red spectrum (610–670 nm). The images of both cameras were normalized and subtracted (NIR-VIS) in real time (Fig. 4). With this camera, a high contrast

Figure 3 Field tractor at the University of Hohenheim with two booms on each side carrying multispectral cameras, spectrometers, fluorescence sensors, RTK-GPS and a robotic operation system (ROS) to trigger each sensor at the same spot. Sensor data were then fused on the tractor; feature extraction, data analysis and classification were done later in the laboratory; photo: Roland Gerhards (unpublished).

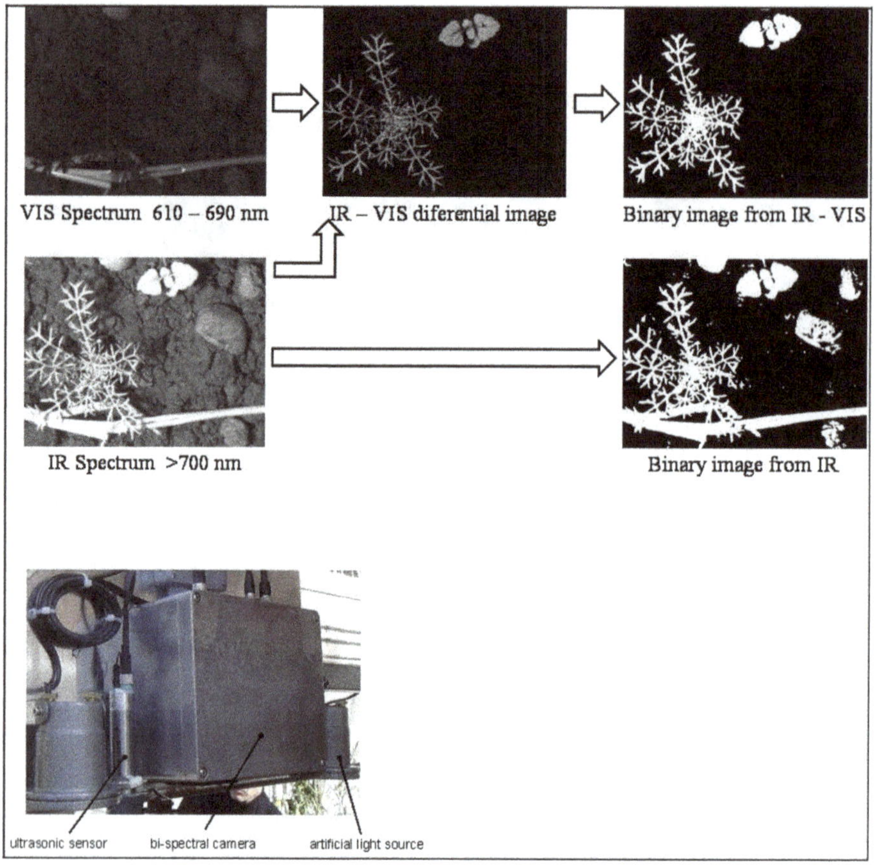

Figure 4 Principle of a bi-spectral camera for weed identification (after Gerhards and Oebel 2006).

between green plants and soil, dead mulch and stones has been achieved even under variable illumination and soil moisture conditions. Artificial lightening was not necessary.

A grey-level threshold was set automatically and the contours of all plants have been extracted from the picture. Objects that were smaller or bigger than plants have been automatically removed from the image. Simple geometric features, such as compactness, circumference, area, shape features and structural features of the contours, were calculated and computed from the most dominant weed species in maize, winter wheat, winter barley and sugar beet fields in Germany. This database has been used to classify images of weed species in field trials with maize, sugar beet and cereals. Cameras were mounted on a boom at a distance of 3 m and images were taken every 1 s at a speed of 8 km h^{-1}, resulting in sampling of approximately 10% area compared to mostly less than 1% for regular grid sampling. With the best classifier, on average 89% of the weed species and crops were correctly classified into the groups 'crop', annual grass 'weed', 'annual broadleaf weed' and 'perennial weed'. Weed distribution maps using manual sampling and automatic camera weed identification have been created (Fig. 5).

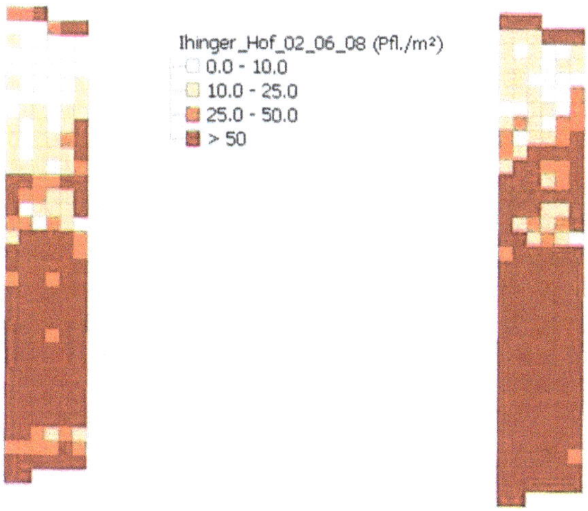

Figure 5 Weed distribution maps in a sugar beet field of 5.3 ha at Ihinger Hof Research Station, 2008, using visual weed sampling (left) and automatic imaging classification (right); untreated area was 37%.

Interpolated weed maps were reclassified based on weed infestation levels (Gerhards et al. 1997). An application map was created to provide a decision rule for the patch sprayer (Fig. 6).

Gerhards and Oebel (2006) grouped the weed species into three classes including grass weeds, annual broadleaves and perennial weed species, and used simple weed density thresholds for all three classes of weed species to vary the herbicide rate between 0 and 100% of the recommended dose. Perennials such as *Convolvulus arvensis* L. and *Cirsium arvense* L. have been found to be highly aggregated in small annual grains, maize

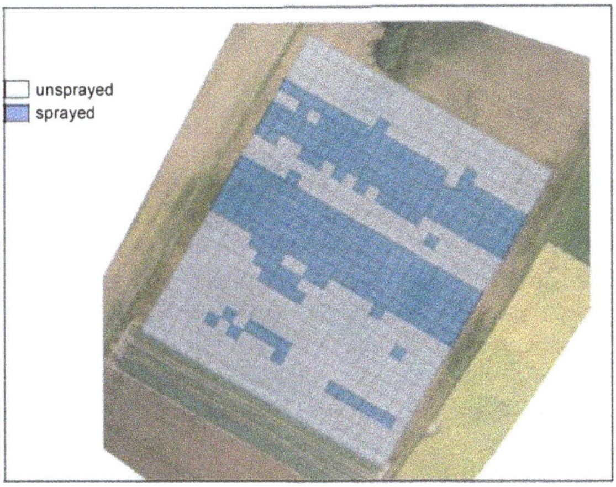

Figure 6 Herbicide application map against grass weeds in a 6.3 ha winter-wheat field at the University of Hohenheim; areas with more than five weeds m^{-2} have been sprayed.

and sugar beets with less than 20% of the field being infested. Grass weeds covered on average 30–40% of the fields at infestation levels higher than the weed control thresholds and annual broadleaf weeds have been found on 20–90% of the fields (Timmermann et al. 2003, Gerhards and Oebel 2006).

Studies at Bonn University (Timmermann et al. 2003) proved that weed patches were relatively stable in location and size over a period of five years. Weed species distributions were found to be heterogeneous in all crops studied with high-density patches requiring application of full herbicide doses and areas where herbicide application was not warranted. The average weed density in the experiments varied from 19 to 58 weeds m^{-2}. Herbicide savings ranged from 6% in maize up to 77% in winter wheat for herbicides against broadleaf weeds and from 22% in winter rape to 65% in winter barley for herbicides against grass weeds. In 2005, herbicide savings varied from 19% in winter rape up to 81% in winter wheat for herbicides against broadleaf weeds and from 20% in winter rape to 79% in winter wheat for herbicides against grass weeds. Highest savings were realized in cereals followed by sugar beet, maize and winter rape. Efficacy of site-specific weed control in this study was measured by counting the weed density before and approximately 3 weeks after weed control in the sprayed and unsprayed areas. Efficacy of site-specific weed control is the relative reduction in weed density between dates one and two. In all experiments of this study, site-specific post-emergent herbicide application resulted in an efficacy of 85% to 98%. Despite additional costs for weed sensing and application technology, site-specific weed control resulted in higher economic return compared to conventional uniform applications (Oerke et al. 2010). The results of these experiments show that site-specific weed management reduced costs for weed control and resulted in a less impact on the environment. However, for a broader acceptance of site-specific weed management in practical agriculture, an online system would be needed, which combines weed species detection and herbicide application in one operation.

Christensen et al. (2003) determined the economic optimal herbicide rate with respect to the spatial heterogeneous weed distribution, weed competition and population dynamics. This strategy was tested in a five-year experiment and resulted in highest crop yields, lowest soil seed banks and equal weed control costs compared to conventional decision models.

3 Dynamics of weed populations

The dynamics of weed populations are influenced by the biological characteristics of weed species, farming practices such as tillage and crop rotation, weed management and soil parameters (Mortensen et al. 1998; Nordmeyer and Niemann 1992; Timmermann et al. 2002). The major weed species have developed specific adaptation and survival strategies to persist in cropping systems (Radosevich et al. 1997). Those strategies include the production of a high number of seeds over a long period of time and seed dormancy (e.g. *Chenopodium album*). In addition, successful weed species have the capacity to survive under variable environments based on high phenotypic and genetic plasticity to invade new sites (e.g. *Abutilon theophrasti*). Many weeds are able to strongly compete for space, light, water and nutrients with the crops by high growth rates and efficiency in using water and nutrients. Several weeds produce mature seeds in a much shorter time than crops so that the seeds are spread long before a dense crop stand has been established (e.g. *Galinsoga parviflora*). Other weed species, such as *Cirsium arvense* and *Agropyron repens,* have the ability to persist and spread via seeds and vegetative reproduction tissues. Those perennial weeds can emerge much faster than annual plants. These are only a few reasons for spatial and temporal dynamics of weed populations.

Nordmeyer and Niemann (1992) found that black grass (*Alopecurus myosuroides*) populations mostly occurred at locations in the field where the clay content was relatively high. Timmermann et al. (2002) reported that the crop rotation had a long-term effect on weed density and weed species composition. In fields that had been planted with 50% maize in the rotation more than 20 years ago, the density of *Chenopodium album* was still much higher than in fields with a high percentage of winter annual grains in the rotation. The crop rotation had a very strong effect on the organic matter content. Fields that had been planted with potatoes were lower in organic matter content than fields where mostly grains were planted. The difference in organic matter content again had a strong influence on weed species composition. *Galium aparine* predominantly occurred in fields with high organic matter content (Timmermann et al. 2002).

4 Spatial and temporal stability of weed distributions

Krohmann et al. (2002) studied the dynamics of weed seedling distribution over 5 years in a rotation of maize, sugar beet, winter wheat and winter barley and in continuous maize cultivation. They found that weed distribution maps obtained in maize and sugar beet were suitable for site-specific weed control in winter wheat and winter barley. Ritter and Gerhards (2008) reported that populations of *Alopecurus myosuroides* did not significantly change in density, location and size when site-specific weed control methods were applied over a period of 8 years in a rotation of winter annual cereals, maize and sugar beet. In all of three fields studied, weed seedling distribution was heterogeneous. Density was higher in maize and sugar beet than in winter cereals. High-density patches with densities higher than 25 plants per m^2 consistently recur over the years in the same areas in fields. Weed density reduction due to herbicides and other weed control methods were satisfactory in each year, indicating that site-specific weed control methods are sustainable for long-term weed suppression. Herbicide savings against *Alopecurus myosuroides* ranged from 50% in sugar beet to 75% in winter barley.

Ritter and Gerhards (2008) also studied weed population dynamics of *Galium aparine* L. and *Alopecurus myosuroides* under the influence of site-specific weed management. It was found that most of the tested population parameters were dependent on weed density. It was presumed that individual weeds without competition evolve better and produce more seeds, but this study showed opposite results. With increasing weed density, weed biomass and fecundity increased in this study. The authors assumed that individual weeds benefit from growing in patches, which could cause reduced competition by the crop or lower herbicide efficacy.

An understanding of fundamental weed population biology would improve our ability to develop site-specific management decisions. Weed population models have been applied to quantify the effects of site-specific weed management practices (Paice et al. 1997). However, the mechanism of weed patch stability is rather under-exploited. A few results have been reported showing that efficacy of weed control methods was lower in weed patches in low-density locations (Mortensen et al. 1998). Krohmann et al. (2002) found that the persistence of weed populations was also attributed to weed seedlings that emerged after weed control methods had been applied. Those individuals were able to produce viable seeds in maize and sugar beet but not in winter wheat and winter barley. The authors assumed that competition of the crop was higher in winter annual grains and therefore late-emerging weed seedlings were suppressed.

Few studies have attempted to quantify spatial stability of weed patches in agricultural fields. If weed patches were consistent in density and location over years, maps from one year could be used to direct sampling plans and to regulate weed control methods in subsequent years. Wilson and Brain (1991) found that the pattern of black grass (*A. myosuroides*) patches persisted during a 10-year study. Persistence of patches was attributed to the poor ability of black grass to colonize new locations when effective herbicides were applied. The pattern of patches was most stable in fields planted with cereals. Pester et al. (1995) observed a significant stability for velvetleaf populations using Pearson, Spearman rank, and chi-square correlation analyses to quantify year-by-year relationships between weed density at individual X, Y-coordinates of the sampling grid in four fields. Walter (1996) also used the chi-square correlation method and found that field violet (*Viola arvensis*), common lamb's quarters (*Chenopodium album* L.) and prostrate knotweed (*Polygonum aviculare* L.) distributions were stable in cereal grain fields over three years.

Gerhards et al. (1996) studied the spatial stability of velvetleaf, (*A. theophrasti* Medik.), hemp dogbane (*Apocynum cannabinum* L.), common sunflower (*Helianthus annuus* L.), yellow foxtail (*Setaria glauca* L.) and green foxtail (*Setaria viridis* L.) over four years (1992–5) in two fields in eastern Nebraska. The first field was planted to soybean in 1992 and corn in 1993, 1994 and 1995. The second field was planted to corn in 1992 and 1994 and soybean in 1993 and 1995. Weed density was sampled prior to post-emergence herbicide application at approximately 800 locations per year in each field on a regular 7-m grid. The same locations were sampled every year. Weed density at locations between the sample sites was determined by linear triangulation interpolation. Weed seedling distribution was significantly aggregated with large areas being weed-free in both fields. Common sunflower, velvetleaf and hemp dogbane patches were very persistent in the first field over the four years. Foxtail distribution and density continuously increased in each of the four years in the first field and decreased in the second field. A geographic information system (GIS) was used to overlay maps from each year for a species. This showed that 36% of the sampled area was free of common sunflower, 62.5% was free of hemp dogbane and 11.5% was free of velvetleaf in the first field, but only 1% was free of velvetleaf in the second field. The persistence of broadleaf weed patches observed in this study suggests that weed seedling distributions mapped in one year are good predictors of future seedling distributions.

Heijting et al. (2007) found strong spatial correlations for *Echinochloa crus-galli*, *Chenopodium album*, *C. polyspermum* and *Solanum nigrum* in three years of continuous maize cultivation. They attributed spatial and temporal stabilities of weed populations to their high recruitment capacity.

5 Spraying technologies for precision weed control

Sprayers and boom sections can be turned on and off according to the information in the spray map (Stafford and Miller 1993; Paice et al. 1997). Herbicide rate can be varied according to the size and density of weed species distribution in the field using pressure regulation systems, direct injection systems, automatic variation of nozzles with different sizes and pulse-modulated nozzles (Kifferle and Stahli 2001). These variable rate technologies allow real-time variation of herbicide rate in a range of +/−100% from a predefined rate (Kifferle and Stahli 2001; Escolà et al. 2013). For site-specific application of herbicides, map-based approaches (offline) and real-time technologies (online) have been investigated. Application maps for offline systems are created based on interpolated maps of weed distribution and weed control thresholds. With the technology developed at the University of Bonn, three different application maps could be realized at the same time using a multiple sprayer with three separated hydraulic circuits (Fig. 7). The sprayer allows varying the herbicide mixture on the go. Each of the sprayer circuits has a boom width of 21 m, divided into 7 sections of 3 m. Each sprayer circuit and each section are separately turned on and off by a control unit via solenoid valves. The main hydraulic circuit of each of the three sprayer circuits is similar to that on a conventional sprayer with an output from the main pump fed to a pressure control valve, which regulated the concentration that was set by the spray computer. During the herbicide application, the spray control system is linked to an on-board computer loaded with the weed treatment maps. A differential mode GPS is used for determining the real-time location of the patch sprayer. The on-board computer compares the actual position of the sprayer with the information in the weed treatment maps, and signals are transmitted to the control unit via a data bus to open each individual solenoid valve when herbicide application is warranted. In the same way, the herbicide dose is adjusted to the recommended rate in the treatment map. In a two-year study, herbicide use with this map-based approach was reduced in winter cereals by 6–81% for herbicides against broadleaf weeds and 20–79% for grass weed herbicides. Highest savings were achieved in cereals followed by sugar beet, maize and winter rape. Efficacy of weed control varied from 85% to 98%, indicating that site-specific weed management will not result in higher infestation levels in the following crops (Gerhards and Oebel 2006).

Figure 7 GPS-controlled patch sprayer with three separated hydraulic circuits.

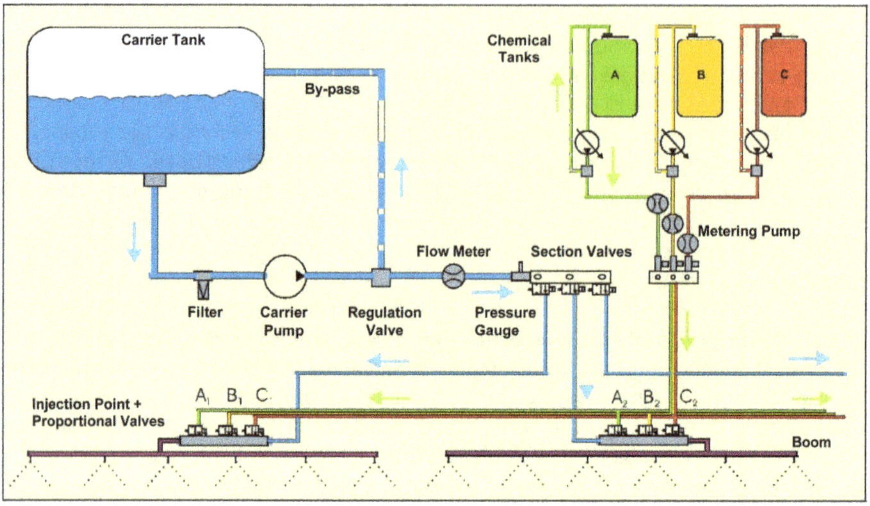

Figure 8 Schematic view of the hydraulic system for the direct injection of herbicides (after Sökefeld 2009).

A different approach is to employ sprayers with an integrated direct injection system (Fig. 8). In injection sprayers, herbicides and carrier (water) are kept separate. According to the information of the weed treatment map (offline application) or directly to the images of the camera system (online application), the herbicides are metered into the carrier and mixed immediately before entering the nozzles (Sökefeld 2009).

In current direct injection systems, there are three crucial factors that influence the accuracy of real-time patch application. The first one is lag time, i.e., the time it takes for the mixed solution to flow from the injection point to the spray nozzles. Commercial direct injection sprayers operate with one central injection point close to the main pump; thus lag times are rather large. Depending on the working width of the sprayer and the position of the injection point, a lag time up to 40 s was measured. The second factor is the occurrence of non-uniform mixtures caused by inadequate mixing of herbicides and carrier in the boom due to varying physical properties of different herbicides such as formulation, viscosity or solubility. The third problem is the cleaning of the pipes containing concentrated herbicide solutions. One very promising development of a direct-injection-sprayer prototype was introduced by Dammann GmbH (Germany) in 2015. This sprayer has three injection units for premixed herbicides close to the nozzles (Krebs et al. 2015).

Figure 8 shows an experimental direct injection system with one injection point for each boom section (3 m each). With this configuration, a lag time between 4 s and 7 s was obtained. Even a shorter lag time between 1.4 s and 0.8 s can be achieved by the injection of the herbicide at each nozzle (Sökefeld 2009).

Another way to vary herbicide dose over a wide range is by logarithmic spraying. For dose–response studies, approximately 512-fold dose variation is needed to assess the impact of herbicides and natural compounds on weeds and crops (Fig. 9). Products are tested at very low doses and at about the 10-fold recommended field rate. EU pesticide legislation requires dose–response data as part of the registration procedure. Based

Figure 9 Logarithmic plot sprayer with stepwise dose reduction and 2-m boom for dose–response studies with herbicides and natural products in the field; photo: R. Gerhards.

on these data, the baseline-sensitivity and recommended field rate of pesticides are determined. Dose–response studies are also conducted in standard biotests to determine if weeds have developed resistance to herbicides. None of the commercial spraying technologies can realize this large variation in pesticide rates. A typical logarithmic sprayer consists of a mixture tank and a pressurized water input (Tind et al. 2009; Poulsen and Jensen 2010). Water is continuously added to the mixture tank, steadily diluting the concentration applied. One of the major drawbacks of the typical logarithmic sprayer is that the pesticide dose decreases continuously. This causes difficulty in bioassays since multiple measurements within an equal dose are not possible. Peteinatos (2016) developed and tested a logarithmic sprayer with a stepwise dose reduction. A predefined volume of an arbitrary dose would be sprayed. Part of this mixture is transferred to a second spray bottle and refilled with water. Afterwards a second dose of the same volume would be applied, but with a predefined dilution ratio. This procedure is repeated *n* times. After 10 dose reduction steps, the measured dose varied only 1.06% from the expected dose. The new logarithmic spraying technology can significantly improve dose–response studies. Thus, pest and crop data on pesticide response can be attributed to a certain rate.

6 Precision weed hoeing and harrowing

Abundance of several weed species including grass weeds (*Alopecurus myosuroides*, *Apera spica-venti*, *Lolium multiflorum*) and perennial weed species (*Cirsium arvense*, *Elymus repens*) have increased in European cereal production systems during the last decade (Gerhards 2010). This has been mainly due to higher percentages of cereal crops in crop rotations and less-intensive soil tillage operations (Massa et al. 2013). Although chemical weed control still plays a dominant role in weed management strategies, there is a strong need for alternative measures. Negative impacts on the environment and the risk of herbicide residues in the food chain as well as strong increase of herbicide-resistant weed populations support the call for alternative weed control strategies.

6.1 Precision harrowing

Among mechanical weed control measures, weed harrowing using a spring tine harrow is most common because of its high labour efficiency. However, field studies carried out in Norway in spring wheat showed that weed control efficacy was fairly low, 26% after pre-emergence and 47% after post-emergence harrowing (Brandsaeter et al. 2012) and the success of harrowing depends on the balance between weed control and crop soil cover, which is reflected in the selectivity (Rasmussen et al. 2008). Efficacy of weed harrowing is usually lower for grass weeds such as *Apera spica-venti* and *Alopecurus myosuroides* (Tillett 2005) and tap-rooted weeds such as *Papaver rhoeas*, *Tripleurospermum inodorum* and volunteer oilseed rape (Welsh et al. 1996; Wilson et al. 1993) than for other annual broadleaved weed species. More intensive tillage operations are needed to effectively control perennial weed species such as *Cirsium arvense* and *Elymus repens*. Thus, farmers need effective alternative weed control methods to suppress weed competition.

Field studies have been conducted at the University of Hohenheim (Germany) in collaboration with J. Rasmussen, University of Copenhagen (Denmark), to study the short-term effects of harrowing on crop cover and weed density and the response of harrowing on grain yield. Studies were mostly set up in spring cereals. However, a few studies were also performed in winter wheat (Figs. 10 and 11). The crop was harrowed at different growth stages starting at 2- to 3-leaf stage until the end of tillering. Intensity of harrowing was varied by increasing the working depth (e.g. deeper penetration of the tines into the soil), the forward speed and the number of passes (Rasmussen et al. 2008). Data were analysed using models created by Rasmussen (1995) and Rasmussen et al. (2008). With the use of sensor technology and automatic plant species discrimination, crop and weed response can be precisely quantified (Weis et al. 2008; Rueda Ayala et al. 2013). The curve for leaf cover decline of the crop relative to increasing intensity with the harrow (Fig. 10a) allows calculation of the percentage of crop soil cover, and the curve for weed density decline (Fig. 10b) allows calculation of the percentage of weed control. The selectivity (Fig. 10c) is calculated as the relationship between crop soil cover and weed control.

A significant yield gain was achieved for two growth stages at high levels of weed control and relatively high levels of crop soil cover compared to the untreated control (Fig. 11). This can be explained by lower weed competition and possibly by stimulation of crop growth due to harrowing. Yield decreased at the highest intensities of weed harrowing which was probably due to crop damage (Fig. 11).

Due to the high variability of weed populations and soil type within arable fields, a prototype for site-specific and automatic adjustment of weed harrowing was developed (Fig. 12). A digital sensor (a) measures the resistance force of the soil to the forward movement of the harrow; the data are transmitted to a control unit (b) that generates an algorithm for the motors (c) to change the tine angle and perform more aggressive treatments according to the soil characteristics. Positioning of the whole system is detected with a real-time kinematic differential global positioning system (RTK-DGPS) (d). After the data acquisition with the sensors, real-time adjustment of the aggressiveness of cultivation

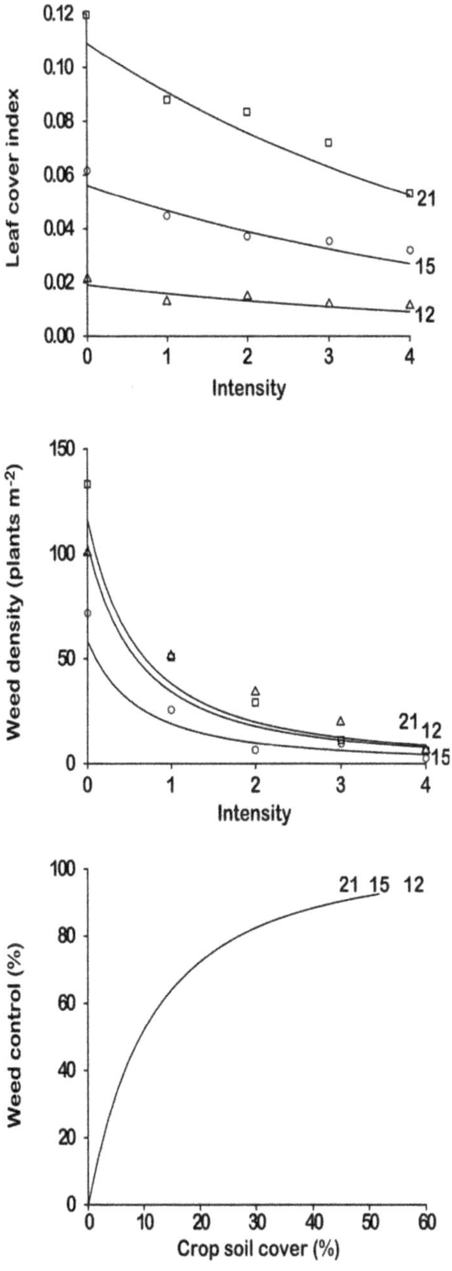

Figure 10 Crop leaf cover (top), weed density (middle) and selectivity (bottom) in response to increasing weed harrowing intensity in terms of increasing number of passes with a weed harrow at three wheat growth stages (BBCH 12, 15 and 21) in the winter wheat experiment at the University of Hohenheim in 2009 (Rueda Ayala et al. 2013).

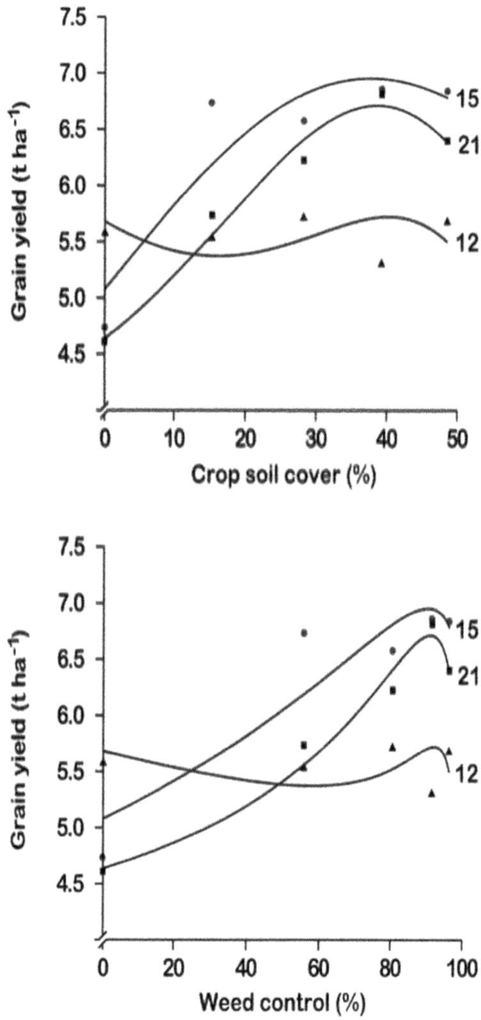

Figure 11 Crop yield relative to percentage of crop soil cover and weed control assessed immediately after harrowing in the winter wheat experiment at the University of Hohenheim, 2009. The numbers in the figure represent the crop growth stage at the time of harrowing (Rueda Ayala et al. 2013).

(intensity) of the harrow may be possible, according to the site-specific demand for weed control (Weis et al. 2008). To make the adjustments, a predictive model is required to estimate the positive and negative crop yield contributions from harrowing relative to the intensity of cultivation. This model requires site-specific knowledge about 1) the competitive ability of the weeds, 2) selectivity and 3) recovery as outlined in Rasmussen (1995). These input variables, however, still need further investigation.

Figure 12 Prototype of an automatically controlled flexible tine harrow containing a soil sensor (a), a computing control unit (b), a motor to change the tine angle (c) and an RTK-DGPS for positioning (d) (Rueda Ayala et al. 2013).

6.2 Precision hoeing

A promising alternative to chemical weed control and weed harrowing in cereals is weed hoeing which can be done between (inter-row hoeing) and within crop rows (intra-row hoeing). Hoeing can control larger weeds and grass weeds and the risk of crop damage is lower than that for harrowing (Lötjönen and Mikkola 2000). Hence weed hoeing is less sensitive to timing of the treatment because even larger weeds will be uprooted (Melander et al. 2003). Weed control efficacy of hoeing is less affected by soil moisture and soil type than harrowing (Rasmussen and Ascard 1995) because weeds are cut by the hoe blades, whereas the harrow covers weeds with soil or uproots smaller weeds. This requires dry and loose soil conditions. Efficacy of weed hoeing is rather high. Dierauer and Stöppler-Zimmer (1994) found that 90% of the weeds between the crop rows were uprooted and 75% of the weeds within the crop row were covered by soil after two passes of hoeing in maize and peas. Melander et al. (2003) suggested that a higher speed could increase the soil coverage of intra-row weeds and thus may lead to higher weed suppression.

Current cereal crop seeding is carried out at 12–15 cm row width. The seed spacing within the rows follows a random distribution because of bulk metering (Mülle and Heege 1981). It is hypothesized that, by using new innovative seeding technologies, cereal plants will get more even spacing and thus crop plants will be better anchored in the soil and hence, crop losses due to hoeing will be decreased. Furthermore, arranging cereal crop plants with more even intra-row spacing will result in higher yields (Kristensen et al. 2008) and better ability to suppress weeds (Weiner et al. 2001). Weiner et al. (2001) tested weed suppression ability of spring wheat varieties (*Triticum aestivum* L.) at different sowing densities and sowing patterns. They found that weed biomass decreased with increasing crop density and uniformity of sowing pattern. Weed biomass was 30–60% lower at crop densities of 600 seeds m^{-2} compared to conventional seeding. Equal patterns of crop plants can also be realized by precise sowing technologies for cereals. Horsch was one of the first companies offering precision drilling technologies.

Current hoeing systems in cereals require wider crop row distances. The recommended row distance varies between 18 and 25 cm (Lötjönen and Mikkola 2000; Rasmussen 2004; Melander 2006). Even row spacings of 30 and 50 cm are common in organic cereal production when using the so-called wide-row-spacing (*Weite Reihe*) practice. Grain yield is usually decreased in cropping systems with wider row distances. Lötjönen and Mikkola (2000), Mülle and Heege (1981) and Håkansson (1984) determined a yield loss of 12–15% in spring barley when row spacing was increased from 12.5 to 25 cm. Besides yield reduction, higher weed biomass was measured in cropping practices with wider row distances (Blair et al. 1997). Weed dry matter was found to be higher in cereal crops sown at 20 or 25 cm row distance compared with 12 or 12.5-cm rows (Rasmussen 1995; Hammarström et al. 1993). This is probably a consequence of lower competitive ability of the crop at wide row spacing.

For several crops with wide row spacing, including maize, sugar beet and several vegetable crops, vision-based and global navigation satellite system (GNSS)-based row guidance systems for inter-row hoeing have been described (Tillett et al. 2002; Søgaard and Olsen 2003; Griepentrog et al. 2007). Sensors identify the position of crop rows and a hydraulic system steers inter-row hoes close to crop rows. Selective intra-row weeding requires the use of sensor technologies to differentiate between crops and weeds and automatic steering systems to guide the hoe exactly to areas where weeds occur. It is difficult to remove weeds close to the crop without damaging the crop.

Intra-row weeding can be carried out in several different ways. Gobor et al. (2005) have designed a prototype of the rotary hoe for intra-row weed control driven by an electric motor at different speeds depending on the plant species identified in the crop row. Several authors including Sun et al. (2010) have created crop seed/plant maps using RTK-GNSS technology that can later be used for intra-row mechanical weed control and spot spraying (Søgaard and Lund 2007; Sun et al. 2010; Perez-Ruiz et al. 2012). Few complete autonomous systems have been developed for intra-row weeding. Ruckelshausen et al. (2006) integrated a multi-sensor approach for weed identification in maize with a robotic mechatronic system for selective weed control with maize rows.

Instead of manual steering, as is common practice, automatic machine and tool guidance systems are required to improve the accuracy of hoeing, resulting in less crop damage and higher weed control efficacy. Automatic steering systems are based on highly accurate GNSS technology combined with GIS and optical and vision-based sensor systems for crop row detection (Tillett et al. 2002; Søgaard and Lund 2007). They are capable of controlling hoes automatically along and within a crop row (Wiltshire et al. 2003; Gerhards and Oebel 2006). Current automatic steering systems have an accuracy of ±4 cm deviation from a centre line at a driving speed of up to 10 km/h (Melander 2006; Griepentrog et al. 2007).

Weed hoeing using automatic steering technologies (Fig. 13) reduced weed densities in soybean experiments at the University of Hohenheim by 89% and in sugar beet field trials by 87% compared to 85% weed control efficacy in soybean and sugar beet with

Figure 13 Kress-Robocrop® hoe with an automatic camera-based steering system in sugar beet at 8-leaf stage at the University of Hohenheim.

conventional weeding systems. Speed of weed hoeing could be increased from 4 km h^{-1} with conventional hoes to 7 and 10 km h^{-1}, when automatic steering systems were used. Precision hoeing technologies increased soybean yield by 23% and sugar beet yield by 37%. After conventional hoeing and harrowing, soybean yields were increased by 28% and sugar beet yield by 26% (Kunz et al. 2015).

An autonomous system for online weed hoeing has been developed for mechanical weed control in soybean, sugar beet and maize. The system consists of two completely autonomous camera-guided hoeing technologies, one for inter-row weeding and one for intra-row weeding. Inter-row weeding is controlled by a camera system identifying crop rows and steering the inter-row hoe close to the crop row. Intra-row weeding is controlled by a multispectral camera identifying weeds and crops based on shape features. Image analysis software coupled with a microcontroller classifies plants in the image as 'weed' or 'crop' and determines the exact position of the crop in relation to the hoe. The intra-row hoe is automatically controlled by a rotary pulse generator. The system has been trained for weed identification and weed control in soybean, sugar beet and maize. A specific database of plant shapes had to be developed for each crop. A safety area around the crop had to be defined.

Weed control in the inter-row area is performed by duckfoot blades. In the intra-row area, a modified finger weeder is used for selective weed control (Fig. 14).

Figure 14 An autonomous system for weed hoeing in soybean, sugar beet and maize; the system consists of two completely autonomous camera-guided hoeing technologies, one for inter-row weeding with goose-foot blades and one for intra-row weeding with a motor-driven finger weeder; photo: R. Gerhards.

7 Future trends and conclusion

Modern Information technologies offer great potential for precise weed control in arable crops with respect to the spatial and temporal variabilities of weed populations within agricultural fields. Those technologies include optical sensors for weed detection, GPS-controlled aerial vehicles, sensor-guided hoes and harrows and patch/spot-sprayers. Most sensors and UAVs have become more efficient and robust in the past few years. Prices for these technologies have decreased significantly. Therefore, it is expected to become profitable for farmers to invest in precision farming technologies for weed control.

Precision weed management could help to control herbicide-resistant weed populations because it reduces the selection pressure of herbicides. In mechanical weeding, new precision farming technologies could be the basis for robotic weeders between and within crop rows.

8 References

Barroso. J., Ruiz. D., Fernandez-Quintanilla, C., Leguizamon, E. S., Hernaz, P. Ribeiro, A., et al. (2005). Comparison of different sampling methodologies for site-specific management of Avena sterilis. *Weed Research* 45, 165–74.

Biller, R. H. (1998). Pflanzenunterscheidung und gezielter Einsatz von Herbiziden (*Plant identification and spot spraying*). Forschungs-Report 1, 34.

Blair, A. M., Jones, P. A. and Orson, J. H., Caseley JC (1997). Integration of row widths, chemical and mechanical weed control and the effect on winter wheat yield. In *Aspects of Applied Biology* 50, Optimising Cereal Inputs: Its Scientific Basis. Part 2, Crop Protection and Systems, 385–92.

Brandsaeter, L. O., Mangerud, K. and Rasmussen, J. (2012). Interactions between pre- and post-emergence weed harrowing in spring cereals. *Weed Research* 52, 338–47.

Chapron, M., Requena-Esteso, M., Boissard, P. and Assemat, L. (1999). A method for recognizing vegetal species from multispectral images. In: *Precision Agriculture1999*, 2nd European Conference on Precision Agriculture, Part 1. (ed. J. V. Stafford). Sheffield Academic Press: Sheffield, UK, pp. 239–48.

Christensen, S. and Heisel, T. (1998). Patch spraying using historical, manual and real-time monitoring of weeds in cereals in cereals. *Zeitschrift für Pflanzenkrankheiten und Pflanzenschutz, Sonderheft* XVI, 257–63.

Christensen, S., Heisel, T., Walter, A. M. and Graglia, E. (2003). A decision algorithm for patch spraying. *Weed Research* 43, 276–84.

Colliver, C. T., Maxwell, B. D., Tyler, D. A., Roberts, D. W. and Long, D. S. (1996). Georeferencing wild oat infestations in small grains: Accuracy and efficiency of three survey techniques. In: *Proceedings of the 3rd International Conference on Precision Agriculture*, Madison, WI, USA: ASA-CSSA-SSSA, pp. 453–63.

Dierauer, H. U. and Stöppler-Zimmer, H. (1994). *Unkrautregulierung ohne Chemie*. (Controlling weeds without herbicides). Stuttgart, Germany: Ulmer

Escolà, A., Rosell-Polo, J. R., Planas, S., Gil, E., Pomar, J. and Camp, F., et al. (2013). Variable rate sprayer. Part 1 – Orchard prototype: Design, implementation and validation. *Computers and Electronics in Agriculture* 95, 122–35.

Felton W. L. and McCloy, K. R. (1992). Spot spraying. *Agricultural Engineering* 11, 9–12.

Gerhards, R. (2010) Unkrautkontrolle in Weizen (Weed Control in wheat.) In O. Christen. *Handbuch zu Weizen*. Frankfurt, Germany: DLG-Verlag.

Gerhards, R., Pester, D. Y.. and Mortensen, D. A. (1996). Characterizing spatial stability of weed populations using interpolated maps. *Weed Science* 45, 108–19.

Gerhards, R., Sökefeld, M., Schulze-Lohne, K., Mortensen, D. A. and Kühbauch, W. (1997) Site specific weed control in winter wheat. *Journal of Agronomy and Crop Science* 178, 219–25.

Gerhards, R. and Christensen, S. (2003). Real-time weed detection, decision making and patch spraying in maize, sugar-beet winter wheat and winter barley. *Weed Research* 43, 1–8.

Gerhards, R. and Oebel, H. (2006). Practical experiences with a system for site-specific weed control in arable crops using real-time image analysis and GPS-controlled patch spraying. *Weed Research* 46, 185–93.

Gobor, Z., Schmittman, O. and Schulze Lammers, P. (2005) Mechanical weeding – concept of inter-row and intra-row hoeing. 5th European Conference on Precision Agriculture, Book of abstracts ISBN 91-7072-141-6, Uppsala (Sweden) 9–11 June 2005, pp. 99–101.

Griepentrog, H. W., Noerremark, M., Nielsen, J. and Soriano Ibarra, J. (2007). Autonomous inter-row hoeing using GPS based side-shift control. *Agricultural Engineering International – The CIGR E-Journal IX* (Manuscript ATOE 07, 005).

Håkansson, S (1984) Row spacing, seed distribution in the row, amount of weeds – influence on production in stand of cereals. Weeds and Weed Control. 25th Swedish Weed Conference, Vol. 1, pp. 17–34. Uppsala, Sweden: Swedish University of Agricultural Sciences.

Hammarström, L., Gillberg, B. and Pettersson, H. (1993). Inter-row cultivation in cereals. Communications of the Fourth Conference IFOAM – Non-Chemical Weed Control, Dijon. pp. 315–17.

Heijting, S., Van Der Werf, W., Stein, A. and Kropff, M. J. (2007). Are weed patches stable in locations? Application of an explicitly two-dimensional methodology. *Weed Research* 47, 381–95.

Johnson, G. A., Mortensen, D. A. and Martin, A. R. (1996). A simulation of herbicide use based on weed spatial distribution. *Weed Research* 35, 197–205.

Kifferle, G. and Stahli, W. (2001). Spritz- und Sprühverfahren in Pflanzenschutz und Flüssigdüngung bei Flächenkulturen. (Spray application technologies in arable crops.) Books on demand, Norderstedt, Germany ISBN 3-8311-2538-4, 229 p.

Krebs, M., Rautmann, D., Nordmeyer, H. and Wegener, J. K. (2015). Entwicklung eines Direkteinspeisungssystems ohne Verzögerungszeiten zur Pflanzenschutzmittelapplikation. (Development of a fast direct injection system of pesticides.) *Landtechnik* 70(6), 238–53.

Kristensen, L., Olsen, J. M. and Weiner, J. (2008). Crop density, sowing pattern, and Nitrogen fertilization effects on weed suppression and yield in spring wheat. *Weed Science* 56(1) 97-102.

Krohmann, P., Timmermann, C., Gerhards, R. and Kühbauch, W. (2002). Ursachen für die Persistenz von Unkrautpopulationen. (Reasons for persistence of weed populations.) *Journal of Plant Diseases and Protection*, Sonderheft XVIII, 261–8.

Kunz, C,, Weber, J. F. and Gerhards, R. (2015). Benefits of precision farming technologies in mechanical weed control in soybean and sugar beet – comparison of precision hoeing with conventional mechanical weed control. *Agronomy* 5, 130–42.

Lamb, D. W. and Brown, R. B. (2001). Remote sensing and mapping of weeds in crops. *Journal of Agricultural Engineering Research* 78, 117–25.

Lötjönen, T. and Mikkola, H. J. (2000). Three mechanical weed control techniques in spring cereals. *Agricultural and Food Science in Finland* 9, (4), 269–78.

Marshall, E. J. P. (1988). Field-scale estimates of grass populations in arable land. *Weed Research* 28, 191–8.

Massa, D., Kaiser, Y. I., Andújar-Sánchez, D., Carmona-Alférez, R., Mehrtens, J. and Gerhards, R. (2013). Development of a geo-Referenced database for weed mapping and analysis of agronomic factors affecting Herbicide resistance in Apera spica-venti L Beauv. (silky windgrass).. *Agronomy* 3, 13–27.

Melander B. (2006) Current achievements and future directions of physical weed control in Europe. Association Française de Protection des Plantes.3rd International conference on non-chemical crop protection methods, Lille, France, pp. 49–58.

Melander, B., Cirujeda, A. and Jørgensen, M. H. (2003). Effects of inter-row hoeing and fertilizer placement on weed growth and yield of winter wheat. *Weed Research* 43, 428–38.

Mortensen, D. A., Dieleman, J. A. and Johnson, G. A. (1998). Weed spatial variation and weed management. In *Integrated Weed and Soil Management*, edited by J. L. Hatfield, D. D. Buhler and B. A. Stewart, Ann Arbor Press, Chelsea, MI, USA, p. 293.

Mülle, G. and Heege, H. J. (1981). Kornverteilung über die Fläche und Ertrag bei Getreide. (Distribution of seed density and yield within arable fields.) Z. Acker-und Pflanzenbau, *Verlag Paul Parey* 150, 97–112.

Nordmeyer, H. and Niemann, P. (1992). Möglichkeiten der gezielten Teilflächenbehandlung mit Herbiziden auf Grundlage von Unkrautverteilung und Bodenvariabilität. (Potentials for site-specific weed control based on weed distributions and soil variability). Zeitschrift für Pflanzenkrankheiten und Pflanzenschutz Sonderheft XIII, 539–47.

Nordmeyer, H., Zuck, A. and Häusler, A. (2003). Experiences of site-specific weed control in winter weeds. In J. Stafford and A. Werner (eds), *Proceedings of 3rd European Conference on Precision Agriculture*. Wageningen Academic Publishers, The Netherlands.

Oerke, E.-C., Gerhards, R., Menz, G. and Sikora, R. A. (eds) (2010). *Precision Crop Protection – the Challenge and Use of Heterogeneity*, 1st edn, 345p., Hardcover, ISBN: <DR0-S-I01_isbn>978-90-481-9276-2</DR0-S-I01_isbn>, Springer: Dordrecht, Heidelberg, London and New York.

Paice, M. E. R., Day, W., Rew, L. J. and Howard, A. (1997). A simulation model for evaluating the concept of patch spraying. *Weed Research*, 373–88.

Perez-Ruiz, M., Slaughter, D. C., Gliever, C. and Upadhyaya, S. (2012). Tractor-based Realtime Kinematic-Global Positioning System (RTK-GPS) guidance system for geospatial mapping of row crop transplants. *Biosystems Engineering* 111(1), 64–71.

Pester, D. Y., Mortensen, D. A. and Gotway, C. A. (1995). Statistical methods to quantify spatial stability of weed populations, *North Central Weed Science Society* 50, 52.

Peteinatos, G. G. (2016). Sensor based identification of stress in crops and weeds and utilisation of sensors for precision pest management. Dissertation, University of Hohenheim, Germany.

Poulsen, R. T. and Jensen, J. E. (2010). Using logarithmic spraying to visualise pesticide efficiency. In: ENDURE IPM Training Guide, ENDURE Network of Excellence, S. 245-252. http://www.endure-network.eu/endure_publications/endure_ipm_training_guide/tools, 04 February 2015.

Radosevich, S., Holt, J. and Ghersa, C. (1997). Weed *Ecology – Implication for Weed Management.* 2nd edn, John Wiley & Sons, New York, 589 p.

Rasmussen, I. A. (2004). The effect of sowing date, stale seedbed, row width and mechanical weed control on weeds and yields of organic winter wheat. *Weed Research*, 44: 12–20.

Rasmussen, J. (1995). Selective weed harrowing in cereals. *Biological Agriculture and Horticulture* 12, 29–46.

Rasmussen, J. and Ascard, J. (1995). Weed control in organic farming systems. In *Ecology and Integrated Farming Systems* (eds), D. M. Glen, M. P. Greaves, H. M. Anderson). John Wiley & Sons, Chichester, UK, 49–67.

Rasmussen, J., Bibby, B. M. and Schou, A. P. (2008). Investigating the selectivity of weed harrowing with new methods. *Weed Research* 48, 523–32.

Rasmussen, J., Nielsen, J., Garcia-Ruiz, F., Christensen, S. and Streibig, J. C. (2013). Use of small unmanned aircraft systems (UAS) in weed research. *Weed Research* 53, 242–8.

Rew, L. J. and Cousens, R. D. (2001). Spatial distribution of weeds in arable crops: a current sampling and analytical methods appropriate. *Weed Research* 41, 1–18.

Ritter, C. and Gerhards, R. (2008). Population dynamics of Galium aparine L. and A. myosuroides (Huds.) under the influence of site-specific weed management. *Journal of Plant Diseases and Protection* Special Issue XXI, 209–14.

Ruckelshausen, A., Klose, R., Linz, A., Marquering, J., Thiel, M. and Tölke, S. (2006). Autonome Roboter zur Unkrautbekämpfung. (Robotic weeding). *Zeitschrift für Pflanzenkrankheiten und Pflanzenschutz*, Sonderheft XX, 173–80.

Rueda Ayala, V., Weis, M., Keller, M., Andujar, D. and Gerhards, R. (2013). Development and testing of a decision making based method to adjust automatically the harrowing intensity. *Sensors* 13, 6254–71.

Søgaard, H. T. and Heisel, T. (2002). Maschine vision identification of weed species based on active shape models. In *European Weed Research Symposium* (ed. H. H. van Laar), Wageningen University, The Netherlands, 402–3.

Søgaard, H. T. and Olsen, H. J. (2003). Determination of crop rows by image analysis without segmentation. *Computers and Electronics in Agriculture* 38 (2), 141–58.

Søgaard, H. and Lund, I. (2007). Application accuracy of a machine vision-controlled robotic micro-dosing system. *Biosystems Engineering* 96:315–22.

Sökefeld, M., Gerhards, R. and Kühbauch, W. (2000). Teilschlagspezifische Unkrautkontrolle – von der Unkrauterfassung bis zur Herbizidapplikation. (Site-specific weed control – from weed detection to patch spraying). *Journal of Plant Diseases and Protection*, Special Issue XVII, 227–32.

Sökefeld, M. (2009). Variable rate technology for herbicide application. In E. Oerke, G. Menz, R. Gerhards, R. Sikora (eds), *Precision Crop Protection*. Springer, Verlag: Dordrecht, Heidelberg, London and New York, 335–48.

Stafford, J. V. and Miller, P. C. H. (1993). Spatially selective application of herbicide to cereal crops. *Computers and Electronics in Agriculture* 9, 217–29.

Sun, H., Slaughter, D. C., Perez-Ruiz, M., Gliever, C., Upadhyaya, S. K. and Smith, R. F. (2010). RTK GPS mapping of transplanted row crops. *Computers & Electronics in Agriculture* 71(1): 32–7.

Thornton, P. K., Fawcett, R. H., Dent, J. B. and Perkins, T. J. (1990). Spatial weed distribution and economic thresholds for weed control. *Crop Protection* 9, 337–42.

Tillett, N. D. (2005). *Cost-effective Weed Control in Cereals using Vision Guided Inter-row Hoeing and Band Spraying Systems*. London, UK: HGCA Project Report No.370.

Tillett, N. D., Hague, T. and Miles, S. J. (2002). Inter-row vision guidance for mechanical weed control in sugar beet. *Computers and Electronics in Agriculture* 33, 163–77.

Timmermann, C., Gerhards, R. and Kühbauch, W. (2002). Ursachen für Ertragsunterschiede in Ackerschlägen. (Reasons for yield variability in arable fields.) *Journal of Agronomy & Crop Science* 187, 1–9.

Timmermann, C., Gerhards, R. and Kühbauch, W. (2003). The economic impact of the site specific weed control. *Precision Agriculture* 4, 249–60.

Tind, T., Matthiesen, T. J., Jensen, J. E., Ritz, C. and Streibig, J. C. (2009). Using a selectivity index to evaluate logarithmic spraying in grass seed crops. *Pest Management Science* 65, 1257–62.

Vrindts, E. and de Baerdemaeker, J. (1997). Optical discrimination of crop, weed and soil for on-line weed detection. In *Precision Agriculture 1997, Proceedings of the 1st European Conference on Precision Agriculture*, edited by J. Stafford, BIOS Scientific Publishers, Oxford UK, Volume 2: Technology, IT and Management, pp. 537–44.

Walter, W. (1996). Temporal and spatial stability of weeds, In *Proceedings of 2nd International Weed Congress*, edited by H. Brown, Flakkebjerg, Denmark, pp. 125–30

Weiner, J., Griepentrog, H. W. and Kristensen, L. (2001). Suppression of weeds by spring wheatTriticum aestivum increases with crop density and spatial uniformity. *Journal of Applied Ecology* 38, 784–90.

Weis, M., Gutjahr, C., Rueda-Ayala, V., Gerhards, R., Ritter, C. and Schölderle, F. (2008). Precision farming for weed management: techniques. *Gesunde Pflanzen* 60(4), 171–80.

Welsh, J. P., Bulson, H. A. J., Stopes, C. E., Froud-Williams, R. J. and Murdoch, A. J. (1996). Weed control in organic winter wheat using a spring-tine weeder. In *Proceedings of 2nd International Weed Congress*, edited by H. Brown, Flakkebjerg, Denmark, pp. 1127–32.

Wiles, L. J., Oliver, G. W., York, A. C., Gold, H. J. and Wilkerson, G. G. (1992). Spatial distribution of broadleaf weeds in North Carolina soybean (Glycine max) fields. *Weed Science* 40, 554–7.

Wilson, B. J. and Brain, P. (1991). Long-term stability of distribution of Alopecurus myosuroides Huds within cereal fields. *Weed Research* 31, 367–73.

Wilson, B. J., Wright, K. J. and Butler, R. C. (1993). The effect of different frequencies of harrowing in the autumn or spring on winter wheat, and on the control of Stellaria media (L.) Vill., Galium aparine L. and Brassica napus L. *Weed Research* 33, 501–6.

Wiltshire, J. J. J., Tillett, N. D. and Hague, T. (2003). Agronomic evaluation of precise mechanical hoeing and chemical weed control in sugar beet. *Weed Research* 43, 236–44.

Chapter 17

Precision livestock farming and pasture management systems

Mark Trotter, Central Queensland University Institute for Future Farming Systems, Australia

1 Introduction

1.1 What is precision livestock farming?

Much of this textbook has focused on the application of precision agriculture (PA) in the cropping and horticultural industries. However, at the same time that the concepts and technologies associated with managing spatial and temporal variability in the plant industries have been evolving, there has also been a growing effort in the development of precision livestock and pasture management.

Within this topic, there are two broad conceptual areas. These are (1) precision livestock farming (PLF) which focuses on individual animal management and the opportunities to be gained over herd- or flock-based management and (2) precision pasture/rangeland management, which is primarily concerned with the management of landscape spatial and temporal variability whilst considering the influence that animals bring to these systems. These concepts can be mapped across the various livestock industries from highly intensive pork and poultry to extensive rangeland grazing systems (Fig. 1). There is a significant body of literature already developed around the intensive animal industries with several leading publications summarising, and journal special issues reporting on PLF in this context (Wathes et al., 2008; Halachmi and Guarino, 2016; Berckmans, 2008, 2014; Banhazi et al., 2012). In comparison, there is far less literature concerning the

http://dx.doi.org/10.19103/AS.2017.0032.18

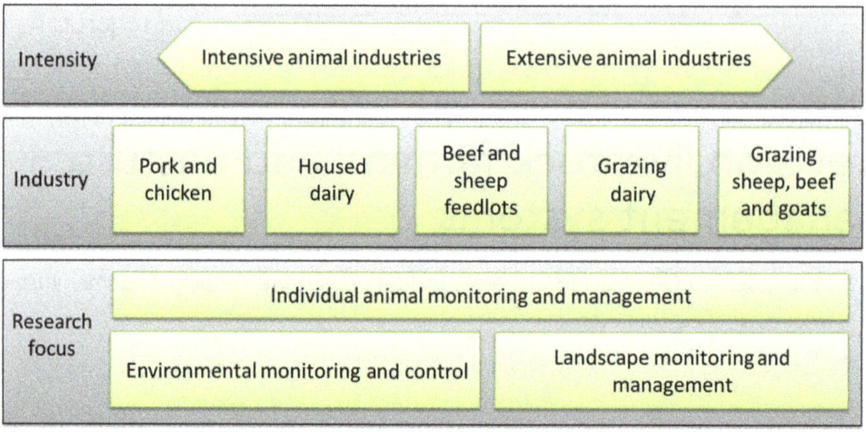

Figure 1 The research focus areas of different livestock industries across their relative intensities.

application of PA to extensively grazed livestock (Laca, 2009; Bocquier et al., 2014) and less again considering grasslands (Schellberg and Verbruggen, 2014). The conceptual thinking, research and commercial technologies around PA as applied to grazing animal and pasture management are generally less well developed than intensive animal PLF and will be the focus of this chapter. There are however numerous technologies evolving in the intensive industries which have application into the extensive grazing systems, these will be considered along the way. This is particularly the case when considering individual animal monitoring and management.

1.2 Opportunities and challenges in PLF and individual animal management

The value in individual animal management comes about because of the variation that is found in herds and flocks which have traditionally formed the management unit (with all animals receiving the same treatment). This traditional cohort-based management is analogous to the field-scale management which is the target of precision management in cropping and horticultural systems. Much like PA seeks to assess and manage the variation in soils and vegetation within a paddock of corn or wheat, individual animal management seeks to first measure the variability of animal production, disease or welfare and then prescribe targeted management strategies and treatments. This chapter will review a range of technologies which are currently applied or in development, which facilitate this individual animal management across the spectrum of extensive grazing production systems.

1.3 Opportunities and challenges in precision pasture and rangeland management

Pastures, grassland and rangelands are labels provided to managed vegetation production systems which involve the grazing of domestic livestock. These systems range from extremely low stocking rates (animals per unit area) in arid desert environments through to intensively managed, rotationally grazed, high input pasture systems in high rainfall areas with much higher stocking rates. All have one thing in common: they are production units which integrate three key sub-systems: the soil sub-system, the plant sub-system and the animal sub-system. In crop production, traditional PA focuses on managing just two of the above (soil and plant), the inclusion of the animal brings an entirely new complex of problems and challenges when seeking to manage spatial and temporal variability.

When considering PA management for grasslands, the fact that these areas differ fundamentally in their spatial heterogeneity cannot be ignored. Most land used for grazing livestock is subject to this as it cannot be easily or economically used for cropping. This can be the result of several factors including terrain, soil type, soil depth, vegetation and the density of rocks all of which limit cultivation. As a consequence, grazing lands are often more spatially variable in key landscape characteristics than is observed in cropping or horticultural production systems and this has important implications when considering the application of precision management concepts.

In addition to the natural variation, the inclusion of animals into the grazing system further increases spatial variability. Unless intensively managed, animals do not use the landscape evenly. They have preferential locations for key activities such as grazing and resting and as a consequence they can have a significant influence on the vegetation and soil sub-systems.

An excellent example of how animals influence and interact with the grazing landscape can be seen in Fig. 2. This field has been subject to grazing by sheep for over 150 years. Over time, the grazing sheep have harvested nutrients from the landscape in preferred locations. These can be seen as the day-time grazing areas, in some ways this is analogous to a cropping yield map where nutrient removal can be quantified. However, as well as removing nutrients, livestock are also responsible for redistributing nutrients within the field and across the farm. This can be most obviously seen in the correlation between camping areas (locations where animals rest and sleep) and key nutrients levels in this field. Extremely high nutrient levels can be seen on the top of hills where sheep congregate to sleep, they spend a disproportionate amount time per unit area in these zones and, as a consequence of defecation and urination, cause a nutrient build up.

Whilst complex, there are a number of current and emerging technologies being developed to refine decision making in grazing systems. This chapter will review a number of innovations being applied or developed to refine the spatial and temporal management of pastures and rangelands.

1.4 Precision livestock and pasture management for sustainability

One of the primary drivers of the development of precision livestock and pasture management is the need to increase food production to meet the growing world population. This is a similar motivation for the plant-based industries; however, there is one key difference. Changes in food preference mean that livestock products will be preferentially sought after. Whilst overall demand for food is anticipated to increase with the expected population growth of 33% to 2050 (7.2–9.6 billion), the demand for milk is expected to grow by 58% and meat by 73% (Gerber et al., 2013). PLF and pasture management offer a solution to the need to sustainably intensify production to meet this demand.

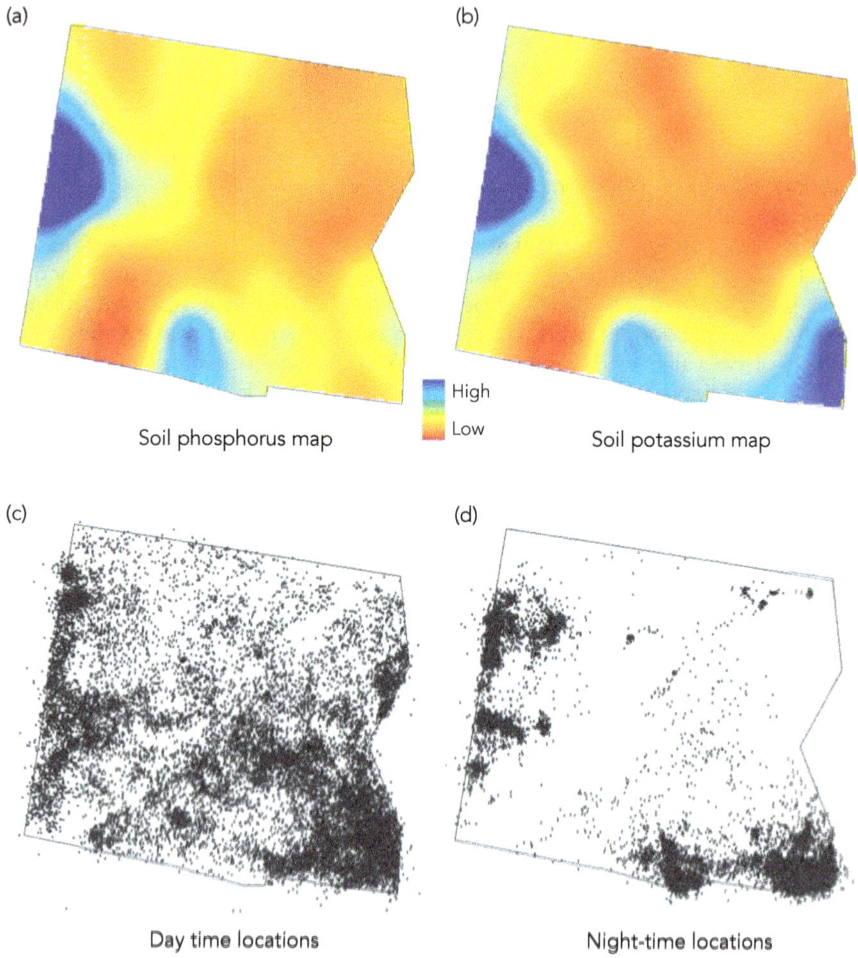

Figure 2 Grazing systems are inherently complex. This complexity is brought about by natural variation but also by the animals themselves. In this 40 ha paddock, gridded soil sampling reveals the high and low areas of key soil nutrients (P and K). GPS tracking of sheep over a 2 month period shows that the high nutrient areas are highly used by the sheep not only during the day time but also particularly at night. This concentrated utilisation of preferred areas results in nutrient redistribution within a field.

2 Individual animal management in extensive grazing systems

The goal of individual animal management strategies in extensive systems is to identify the variability in key attributes of livestock and exploit these through genetic selection or management to optimise productivity. Some of the key technologies being explored include individual animal identification through a radio frequency identification (RFID) which is then leveraged in other technologies such as monitoring of milk production or live-weight through walk-over-weigh, parental matching and individual management through automated sorting and supplement feeding. More novel technologies such as individual animal thermal and behavioural monitoring with on-animal sensing are increasingly being adopted in the high value animal industries.

2.1 Animal identification technologies and individual animal data recording

Individual animal identification in the sheep and beef industries has traditionally been limited to stud or seed stock breeders. In this context, it has predominantly been used to determine pedigree and record performance for genetic selection. The dairy industry was one of the first to apply individual animal performance recording in a commercial herd context with milk production data used to allocate rations and cull non-performing animals. Historically, the identification of each cow was achieved using visual ear tags, collars, leg bands or freeze brands (Hooven, 1968) with data initially recorded on paper and then in electronic databases. These visual identification systems have now largely been replaced or augmented with electronic identification (Eradus and Jansen, 1999) and the same technologies are now being applied in the more extensive commercial beef and sheep industries (Morris et al., 2012).

2.1.1 Radio frequency identification

The primary system now used across the globe for electronic animal data recording is based on radio frequency identification (RFID). RFID-based systems are now widely used in many areas of agriculture and post-farm gate food processing (Ruiz-Garcia and Lunadei, 2011) and provide the backbone of several sensing technologies which require individual animal identification. In its simplest form, a device which emits a unique radio frequency code is attached to an animal or product. The device can then be scanned using a radio antenna and the identity of that animal or product at different points in the agricultural production chain can be determined. There are several variations of the RFID system; however, they largely fall into two categories: active tags and passive tags. Active RFID tags contain an energy source (i.e. a battery) and broadcast a radio signal whilst passive RFID tags contain an energy harvesting coil which requires stimulation (by a radio signal) to broadcast a radio signal (Roberts, 2006). There are three main form factors of RFID used in commercial livestock, the most common is the ear tag, with rumen bolus and injectable implant used largely in high value animals.

In some countries, RFID systems have been imposed on livestock production systems as a legislative requirement. Australia's National Livestock Identification System is based on passive RFID technology and was originally developed to provide lifetime animal traceability to maintain access to high value markets. Although all cattle producers are required to use the system when trading their livestock, only a small proportion have taken advantage of this technology to record the individual animal performance data on-farm and refine management decisions.

There are many commercially available animal database systems which rely on the RFID technology (in part or whole) to identify and manage the information collected from feedlot and grazing livestock (e.g. www.practicalsystems.com.au/farm-software/stockbook/). Data entry is either manually undertaken (in association with a visual identification system) or fully automated, as is the case with many stockyard/crush/race-based animal scanning and weighing systems. Scanning can be used to record treatments (e.g. vaccination and application of pesticides) or can be integrated with weigh scales to provide information on live weight when animals periodically pass through the stockyards.

2.1.2 New animal identification technologies

RFID systems based on ear tags and rumen bolus are not without their limitations. Retention rates of RFID ear tags have been reported to vary with manufacturer but losses and malfunctions as high as 25% over 3 years have been reported (Roberts et al., 2012). Rumen bolus RFID largely overcome these accidental loss issues, however there are concerns amongst meat processors around managing them through abattoirs. Injectable RFID have yet to overcome concerns around recovery in the meat processing chain and potential for contamination of human food. These challenges have spawned research into several new technologies aiming to provide a more robust and reliable individual animal identification system (Awad, 2016). Facial recognition (Corkery et al., 2007; Kumar et al., 2017), retinal eye scanning (Gonzales Barron et al., 2008) and muzzle scanning (Barry et al., 2007; Kumar et al., 2018) have all been investigated with further research required before they could be considered for widespread commercial application.

2.2 Applications of individual animal identification and RFID in extensive production systems

2.2.1 Monitoring individual animal production

In a similar way that Global Navigation Satellite Systems (GNSSs) have enabled the monitoring of productivity of small units of land area, RFID has also allowed producers to monitor the productivity of individual animals. This 'yield' data takes many forms, from milk to red-meat and fibre (wool) quantification. In some situations such as dairy, RFID is linked with milk flow and quality measurements to provide daily summaries of the individual cow's production. At the other end of the spectrum, extensive rangeland beef production systems might simply involve the recording of pregnancy status of a cow to determine if she will produce a calf. There is significant interest in developing technologies that can provide a similar level of information as is possible in the dairy industry for the more extensive red-meat and animal fibre sectors.

2.2.2 Remote autonomous live-weight monitoring

Walk-over-weigh (WOW) technologies were first commercialised in the dairy industry where they were used to monitor live-weight variation in cows moving through the milking parlour (Peiper et al., 1993). These systems are similar to traditional weigh scales used in central handling facilities (stockyards). However, the animal is allowed to move freely along the platform as opposed to being held in a static position between gates in a race. The WOW system is designed for deployment in the field with a set of yards usually set up to direct animals onto the scales. The yards are commonly set up around an attractant, usually water or supplement feed and allow livestock to pass in and out freely. The challenge in WOW systems is recording a valid weight of a dynamically moving animal with contact limited by the irregular tread pattern of hooves across the platform. WOW systems can be used without individual animal identification. However, most integrate RFID tags and readers to provide in-depth information on the individual variation in body weights of livestock. Where individual animal identification is not used, the target information being sought is usually herd or flock average weights and weight change (Brown et al., 2012). These non-RFID systems are likely to be superseded as the reliability of individual animal identification and measurement increases, and costs decrease.

There are several WOW systems currently available for integration with dairy systems which can be adapted for the red-meat industry (Menzies et al., 2017; González et al., 2014). However, there are only two commercially available WOW package offered which include the weigh scales, telemetry, data management and visualisation as an integrated system suitable for extensive grazing (https://livestock.tru-test.com/en-au/product/remote-wow and https://www.datamuster.net.au/). These same systems offer auto-drafting (automated sorting) capabilities which allow producers to automatically remove animals from the main herd and hold them in the yards or in smaller paddocks nearby. Auto-drafting enables individual management as cattle can be selected based on RFID or weight criteria if not fitted with a tag. A typical application of this technology would be when a grazier wants to sell cattle, those meeting the market specifications (e.g. calves at 300–400 kg) would be drafted into a holding yard whilst the breeding cows (e.g. identified either by RFID or weight range >450 kg) would be allowed back into the main paddock. The grazier can determine when sufficient animals have been diverted into the holding yard and then arrange for transport of these animals to market. This process has significant labour savings over traditional mustering of the entire paddock and then manual sorting of cows and calves.

More advanced applications of WOW and auto-drafting have also been evaluated in a research context where animals have been allocated to different supplement treatment groups by drafting into separate yards. In this regime, all animals (controls and treatments) use the same paddocks which is a major advantage when conducting trials in large heterogeneous rangelands (Quigley et al., 2015). WOW has also been used to identify the calving date of extensively grazed beef breeding cows (Aldridge et al., 2017).

2.2.3 Parent association with offspring

One key application of RFID in extensive breeding systems has been to identify the association between dam and offspring. This information is important as it allows producers to identify the highest producing offspring and then target selection of the dam's genetics. In more intensive grazing situations and particularly stud or seed stock production systems, the dam and offspring (either cow and calf or ewe and lamb) are often manually determined either at birth or by yard sorting. Both of these are labour intensive and because they rely on human interpretation (particularly yard sorting) are prone to error.

One concept which has been developed to address this issue uses RFID and the tendency for dam and offspring to pass closely together when entering attractive areas such as water and supplement. This concept has been applied to both sheep (Morris et al., 2012) and cattle (Menzies et al., 2017). RFID reading panels are set up in a similar way to a WOW system, the time and order of passing the RFID is recorded and then analysed to provide an association between individual animals; accuracies of over 90% have been reported (Menzies et al., 2017). This provides the required pedigree information to enable critical genetic selection and culling of poor performing individuals.

2.2.4 Body composition monitoring of grazing livestock

A key determinant of a livestock grazing system productivity is the body condition of the animals at critical times. For animals about to be used for red-meat production, the composition of the carcass in terms of muscle and fat can have a significant impact on the price received as markets have unique specifications for these parameters. Body composition is also critical for breeding cows and ewes with adequate levels of muscle and fat required at certain times of the reproductive cycle to optimise conception and birth. Three-dimensional imaging systems using stereoscopic visible and thermal along with Light Detection and Ranging (LiDAR) systems have been investigated and found feasible in the dairy and pig industry (Halachmi et al., 2013; Spoliansky et al., 2016; Guo et al., 2017) with good correlations reported between sensor and manual body condition score (BCS) estimates. Developing sensors to measure this elusive trait has proven challenging for the extensive grazing industries, primarily due to the need to have the animal pass through a platform enabling image capture. Some proof-of-concept studies have demonstrated the potential for red-meat industries (Menesatti et al., 2014); however, the practical deployment and ability to deal with coat variation (particularly fleece in sheep) remains a challenge. The integration of BCS sensors into a WOW platform has been proposed and could potentially provide not only composition but also live-weight estimates without the need for load bars.

2.3 Emerging technologies for sensing animal location, behaviour and state

2.3.1 Developments in the dairy and feedlot beef industry

The increasing development and application of wearable technologies for humans is being followed with interest amongst animal producers and researchers. The dairy industry has long been a leader in the adoption of technologies; this is particularly the case when it comes to using on-animal sensors, with numerous systems currently being used. These began with leg-borne pedometer-type sensors (Kiddy, 1977) which were initially focused on detecting oestrus behaviours to optimise artificial insemination programmes. There are now several neck collar (Kamphuis et al., 2012) and ear tag (Bikker et al., 2014) form factor sensors which provide real-time or near-real-time sensing of a number of key biological states. Oestrus detection remains the primary focus for most. However, several have now developed disease detection algorithms as part of their system (Rutten et al., 2013).

Most collars and ear tags use an accelerometer sensor to detect and interpret movement (Fig. 3) and a short-range radio to send this data to a local base station before it is transmitted to a server via the local cellular network. Some devices use radio signals to provide a location through triangulation via radio signal time and angle of arrival (Shane et al., 2016; Wolfger et al., 2017).

2.3.2 Behavioural monitoring in extensive grazing systems

The use of on-animal sensors in the extensive grazing sector has largely been limited to research applications, although several attempts at commercial developments have been

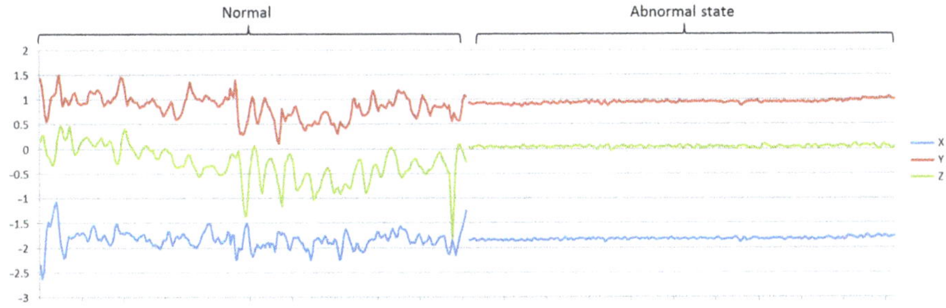

Figure 3 The accelerometer data collected from a 3-axis sensor deployed on a cow's collar. The three traces represent 10 s of data collected at the same time 3 days apart. This sensor was programmed to collect at 25 Hz. The first section shows a period when the cow was behaving normally (grazing/walking). The second section shows the behaviour when it was in an abnormal state (lying down and ill with viral infection). This is a small sample of a larger data set that demonstrates the clear differences in sensor signals at a point in time. More subtle behavioural detection relies on analysing the data over longer periods of time.

made (Trotter, 2013). One of the key challenges posed by extensive grazing systems is that the radio telemetry required to deliver the data must be capable of transmitting over long distances with low power utilisation. Whilst wireless sensor network systems (where data is hopped back to a central node via linked nodes) have been proposed in the research (Wark et al., 2007; Guo et al., 2006; González et al., 2014), there are none currently available as commercial systems. Some livestock monitoring systems have been adapted from more intensive applications to provide short-range connectivity, taking advantage of the natural behaviour of livestock to revisit known sites, particularly water points.

The HerdDogg ear tag-based system (Fig. 4) is one such platform which collects accelerometer data, stores on the device and then transfers this data using Bluetooth radio when the animal approaches a small base station located at a water trough. Whilst not providing real-time data, it does provide critical information to graziers which cannot be collected currently through other means.

One of the key differences between sensor deployment in dairy situations and more extensive grazing enterprises is the importance of location data. GNSSs have been used to track the location of grazing livestock in a research context for well over a decade (Swain et al., 2011). More recent attempts have been made at using terrestrial radio-location systems (Menzies et al., 2016) to locate animals in real-time (Fig. 5). However, despite several attempts using a variety of technologies, this concept is yet to see widespread commercial application (Trotter, 2013). The potential applications and benefits of location data provided either as historical or in real-time have been variously explored and include simple applications such as improved labour use efficiency through targeted mustering; identification of key reproductive behaviours of interest such as oestrus (Fogarty et al., 2015), parturition (Dobos et al., 2014), and detection of critical problems such as disease (Falzon et al., 2013) and predation (Manning et al., 2014). More complex uses include understanding landscape utilisation patterns (Trotter et al., 2010b) and, ultimately, the integration with other data layers (González et al., 2014) to improve decision-making production efficiency.

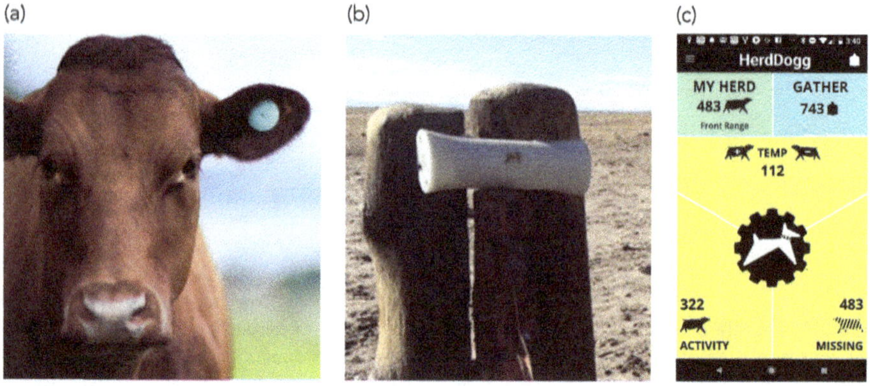

Figure 4 An early version of the HerdDogg ear tag (a) (the blue circle in ear) which contains accelerometer and temperature sensors. It downloads data using Bluetooth radio to a small base station (b) located near a water trough when the animal visits. The producer then drives past the water point and the data is transferred to a smart phone (c) where it is viewable in a format enabling informed decision making.

Figure 5 The Agtrix/Taggle animal location tracking system data interface showing the current location of several animals (cow icon) and the track of locations for one animal over the previous hours (blue line). Originally published in Trotter (2013).

2.3.3 Internal animal sensing

Three technologies that have been used with some success in the research community are internal thermal, motion and biochemical sensors. These sensors can be divided into two main types based on the deployment mode: those inserted orally which are then held in the rumen (Nogami et al., 2017; Earley et al., 2015) and those inserted as subcutaneous devices (Lee et al., 2016). Rumen sensors are commonly used to detect motion, temperature and biochemical states. The most common biochemical sensor is pH (Mottram et al., 2008) which is associated with metabolic disease. Subcutaneous microchips are used to measure body temperature, although they are frequently reported to be subject to error due to their proximity to the skin and influenced by ambient temperature (Lee et al., 2016). Both sensor types have significant potential but there are concerns amongst red-meat processors around the capture of these devices in the abattoir and prevention of them entering the human food chain.

2.3.4 Autonomous animal location management

The remote management of livestock location has been pursued in a research context for several decades. Autonomous animal location management (AALM) or virtual fencing as it is more commonly known involves the use of a positioning device to detect the location of an animal and then provide stimulus to move it away from a location if its presence there is undesirable. The most advanced AALM systems have used GNSS positioning to determine the animal's location in relation to the permitted/unpermitted zone (Bishop-Hurley et al., 2007). The most common stimulus applied has been an audio warning cue followed by a mild electrical shock, other stimulus combinations have been tested but largely failed (Umstatter, 2011). The goal of most audio/electrical stimulus systems is to have the animal learn to respond to the audio cue with only occasional application of the electric shock. In most cases, this seems to have worked (Anderson, 2007).

The first research into AALM began in the 1970s. Since this time, there have been numerous attempts to commercialise the technology but most have met with limited success. At the time of writing this chapter, there are at least three commercial entities proposing the development of AALM systems with a web presence (e.g. www.agersens.com), although it is unclear to what stage each has developed their hardware.

The potential applications for AALM systems are significant. One of the key uses is the control of grazing pressure. In some pasture and rangeland systems, the control of spatial grazing activity can significantly improve productivity. In the most intensive pastures such as grazed dairy systems. AALMs may enable producers to manage their fields without the need to use electric fences. At the other end of the spectrum, rangeland producers are interested in applying AALM to increase the spatial utilisation of areas that are not favoured by livestock (Anderson, 2007). Other environmental applications have also been proposed including the use of AALM to reduce access of grazing livestock to sensitive areas such as stream banks (Wark et al., 2009) or areas susceptible to overgrazing and erosion.

3 Precision pasture and range management

So far, developments in precision livestock management which seek to measure and manage the animal have been discussed. Precision pasture management focuses on monitoring and managing the spatial and temporal variability found in the soils and plant sub-systems of a grazing enterprise. Many of the concepts being developed and applied in this area are similar to those undertaken in PA in a cropping situation, although, as discussed previously, the inclusion of a grazing animal brings about new levels of complexity.

3.1 GIS for basic farm planning and management

The most basic of PA innovations in grazing systems is the simple geographical information system (GIS). In a similar way to its evolution in the cropping sector, the first recording of spatial data in pastures was undertaken using survey equipment before progression to early computer-based GIS. This application began in a research context before being developed for use by graziers on commercial properties. As has occurred in cropping, there are now several GIS commercially available. Some are part of larger farm data management systems whilst others are dedicated stand-alone farm mapping systems. In recent years, there has been a significant growth in the development and application of cloud-based GIS with mobile device applications designed for in-field use to record spatial data (e.g. weed locations) and more general farm activities (e.g. livestock movements to new fields). Cloud-based systems offer significant benefits but are limited by poor connectivity in rural areas.

In more extensive rangeland operations, one of the key applications of GIS is to understand total grazing pressure as limited by distance to water. Sheep and cattle grazing in rangelands are generally limited in their utilisation of the landscape by water availability (Holechek et al., 1989). In many rangelands, artificial water points are provided (often accessed from groundwater) and animals can be found to graze up to limited distances from these water points. Because of the high cost of implementation and maintenance of water points, only a limited number can usually be deployed across the landscape. Using a GIS and basic information around the likely distances travelled by animals, appropriate stocking rates can be estimated for areas in which water limits grazing distribution.

4 Case study 1: using a basic farm GIS to determine sustainable long-term stocking rates

The FarmMap4D Spatial Hub is a cloud-based farm GIS that provides graziers with the ability to digitise basic farm infrastructure such as fence lines and water points (Fig. 6). Government spatial data such as land-type classification can be accessed and mapped along with the property boundaries and internal infrastructure (Fig. 7). This land-type classification is critical as its variation can be associated with safe stocking rates. This is the maximum number of animals that a grazier can run on his property sustainably before they are likely to cause issues such as overgrazing and subsequent soil loss. Land-type zones are similar to the zonal productivity maps developed in PA for cropping from yield mapping technologies. They provide an indication of the likely stocking rate (number of animals per unit area) of a given paddock. Some land types might be able to grow enough biomass to feed 10 animals for 12 months whilst others can only support 5 animals. However, land type alone does not determine stocking rates as the grazing distribution of livestock can be limited by access to water. This restriction is taken into account by fitting

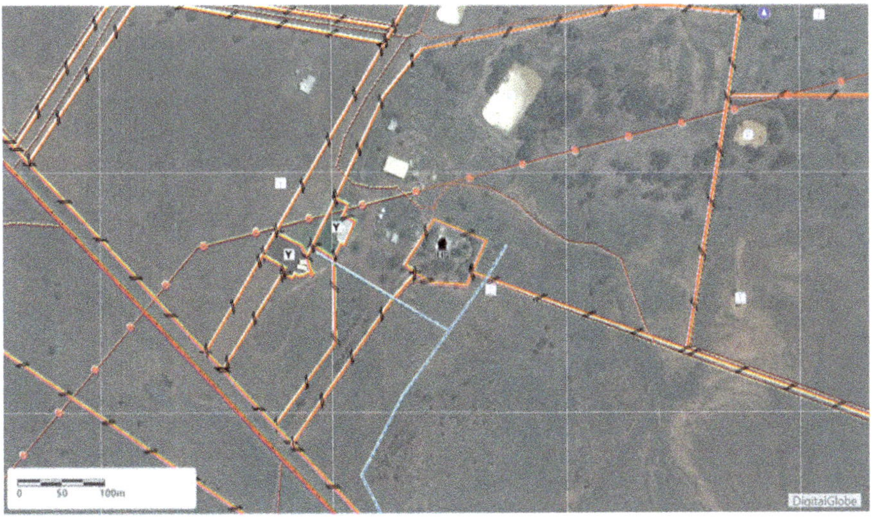

Figure 6 An example of a cloud-based farm GIS designed for grazing systems. The FarmMap4D system provides producers with the ability to digitally map property infrastructure such as fences and livestock watering points.

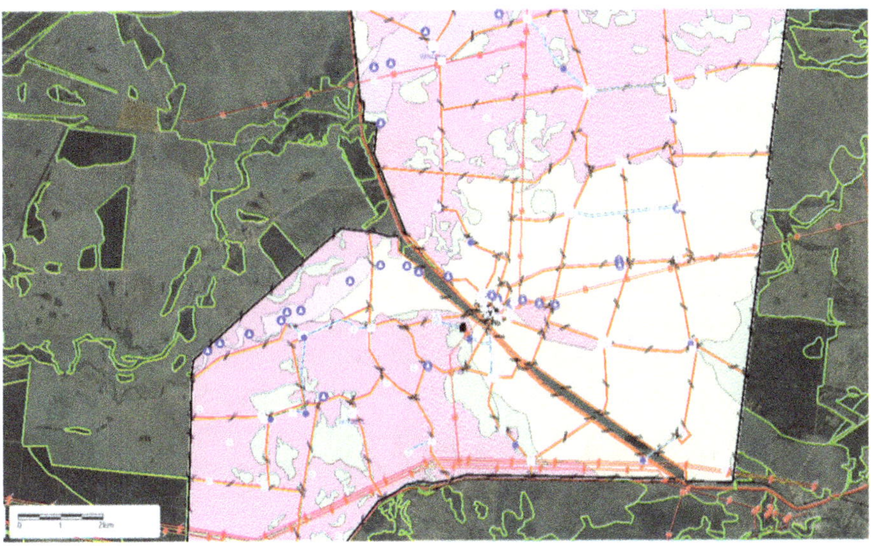

Figure 7 Cloud-based farm GIS such as the FarmMap4D enables easy access to government data sets such as this land-type classification for integration into the farm map. The different colours on the property map represent different land types. Like a zonal map in cropping, each has a different productivity potential with each having a varying maximum recommended stocking rate.

grazing distribution buffers around each water source (Fig. 8). These are concentric rings which represent the likely utilisation of an area as it gets further away from the water point. Spatial analysis is then used to calculate the maximum safe stocking rate as limited by the different land types in a field and the grazing distribution. This information is critical to enable producers to better manage their landscapes and avoid over-stocking which leads to overgrazing, reduced ground cover and then soil loss through wind and water erosion. It also ensures the long-term productivity and profitability of the grazing operation.

4.1 Measuring pasture/range vegetation

One of the most critical pieces of information required in a sustainably managed grazing system is the amount of biomass available for animals to graze at any given time. In case study 1, the long-term stocking rates that are used to guide management decisions in extensive rangeland grazing systems were explored. In more intensively managed rangelands and pastures, these stocking rate decisions are frequently made by measuring the actual feed available at any given time and moving stock between paddocks to avoid overgrazing and optimise animal production. The most common measurement unit used is the dry weight per unit area (e.g. kg/ha), although leaf number and plant height remain critical measures in specific pasture or rangeland types. Individual pasture types, environments and management strategies each have a unique set of biomass thresholds within which graziers attempt to keep their pasture. For example a high rainfall, temperate perennial Fescue-based pasture managed as a rotational grazing system will likely target

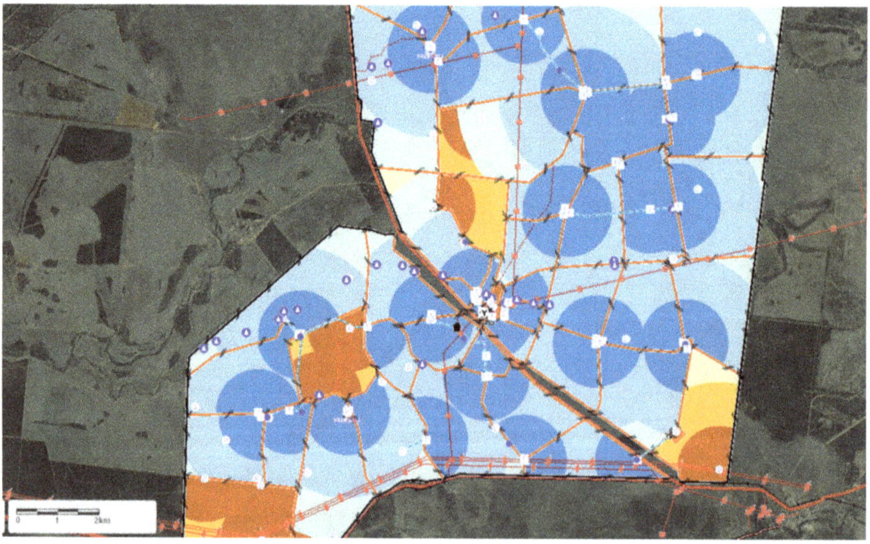

Figure 8 The spatial analysis functionality within this farm GIS allows producers to map the likely grazing areas around each livestock water point. The dark blue areas represent the zones most likely to be completely used (100%) by grazing cattle or sheep. The medium blue areas are likely to be used to a lesser extent (75%) and the light blue are areas likely to be only infrequently used (25%). This restriction on landscape utilisation means that the maximum safe stocking rate (animals per unit area) is actually less than if it were calculated using the whole field.

biomass levels below 2500 kg/ha and above 1000 kg/ha. The lower limit (1000 kg/ha) is set based on maintaining adequate ground cover to avoid overgrazing and soil loss as well as providing adequate leaf area to capture sunlight and optimise regrowth. The upper limit is set so as to avoid the development of low-quality plant material (older leaves) as well as reduction in growth rate due to leaf shading and competition. A grazing manager will rotate animals around several fields attempting to graze each paddock within these limits. The critical challenge for many graziers is obtaining accurate measurements of the amount of biomass in each paddock. Traditionally this has been done using a visual estimate based on experience and previous destructive sampling. However, visual estimates are notoriously inaccurate depending on the individual's experience and destructive sampling (to improve accuracy) is frequently considered too laborious. Consequently there have been several systems developed that aim to remotely detect this information for graziers. Early sensor systems were based on physical contact with the pasture sward (e.g. rising and falling plate meters). Whilst these have been an important development and continue to be used today, this section will focus on newer technologies which apply remote and proximal sensing principles to collect the desired information.

4.1.1 Remote sensing of pasture biomass

Remote sensing systems for measuring pasture biomass consist of two key components: the first is the sensor system which measures the pasture; and the second is the platform which carries the sensor and positions it to collect the data. Remote sensing in this context is defined as a non-destructive, non-contact system which collects data from a distance greater than that can be achieved from a ground vehicle. Ground vehicle and handheld instruments will be considered as proximal sensors later. However it should be noted that there is a blurring of the lines when considering how some sensors are being used on unmanned aerial vehicles (UAVs).

4.1.2 Platforms used to collect data

There are several remote sensing platforms that have been developed or adapted to quantify pasture biomass. The most commonly used are satellite- or space-borne instruments, although airborne platforms remain important in some contexts. Some of the earliest applications of satellite imaging systems were the objective assessment of grassland vegetation (Deering and Rouse, 1975). In the commercial context of delivering information for use by graziers, satellite systems remain the most commonly applied. This is likely due to the relatively low cost and the way in which this data is constantly collected without the need for a grazier to undertake a dedicated sampling process. Piloted airborne systems are mainly used in a research context. Although UAVs hold significant promise, they are yet to be used extensively for sensing biomass in pasture outside a research and demonstration context. Whilst the platforms are a critical factor in the spatial and temporal resolution of remote sensing systems, it is the sensing systems themselves that provide the critical data and limit the spectral resolution.

4.1.3 Remote sensing systems

There are a number of sensing systems applied to measure pasture biomass, the majority rely on measuring parts of the electromagnetic spectrum to differentiate between high and low biomass levels. For a full explanation of how these systems work, see Part 1 of this book.

These systems can be classified as either passive, where they measure the reflection from an ambient source (e.g. optical systems using sunlight) or active where they measure the reflection of energy emitted by the sensing system itself (e.g. microwave or radar systems).

4.1.4 Optical imaging systems

Optical imaging systems can be broadly classified as either multi-spectral (MS) or hyperspectral (HS). Most of the research and commercial activity related to measuring pasture biomass has been undertaken using MS systems whilst HS systems have only been intermittently applied for this purpose. MS imaging systems have been used in grassland research and commercial application for decades. These systems measure the reflectance of light in specific regions of the EM spectrum. The most commonly used regions are the red (~600–700 nm) and near infrared (~750–900 nm) with different sensors collecting reflectance data in discrete bands within these ranges. One of the longest-running and most commonly used, the moderate-resolution imaging spectro-radiometer (MODIS) system collects reflectance data between 620–670 nm (red) and 841–876 nm (near infrared). This MODIS data is used to calculate the Normalised Difference Vegetation Index (NDVI) which is applied in the Pastures from Space™ (PFS) programme (Mata et al., 2004). The PFS programme uses the NDVI data to quantify biomass, enabling graziers to improve livestock management (see case study 2 for details). Other MS systems commonly used include the Landsat 8 and the newly released Sentinel systems. One of the recent developments in MS sensing is based on the nano-satellite approach, this involves the deployment of numerous small form factor and low-cost satellites. Early applications have been tested in a PA cropping context (Houborg and McCabe, 2016) and, whilst the spectral resolution is limited, there appears to be significant potential in this kind of platform.

The deployment of MS imaging systems on UAVs in a pasture context has been explored (Von Bueren et al., 2015). However, specific application to quantifying biomass is limited (Von Bueren and Yule, 2013). There are clear benefits in terms of the higher spatial resolution and essentially on-demand temporal resolution. However, the complex data processing requirements (Laliberte et al., 2011) mean that there is little application outside of the research context. Another optical sensing system that has been evaluated for UAV deployment is structure-from-motion photogrammetry technique. This image processing system uses high-density digital photographs to build a three-dimensional model of vegetation and has proven successful in a research context (Cunliffe et al., 2016).

HS imaging systems measure reflected light like MS systems, however they do so at a much finer spectral resolution. HS systems collect data across a large number (>200) of small bandwidth regions of the EM spectrum. The Airborne Visible/Infrared Imaging Spectrometer (AVARIS) is a HS sensor that has been used as part of a modelling system to quantify rangeland biomass (Phillips et al., 2013). Pullanagari et al. (2017b) reported the use of an airborne HS system to successfully quantify the proportion of dead material in a mixed (green and dead) pasture sward. Advances in sensing technology will mean that this class of sensor will inevitably become more important in the future.

4.1.5 Microwave or radar systems

Microwave or radar systems have been only sparingly applied to measure pasture biomass and generally only in a proof-of-concept research context. These systems are active sensors in that they emit energy in distinct wavelengths (and polarisations) and then measure the reflectance (often termed backscatter) of these radio signals. The first applications of microwave systems in pastures were based on airborne systems (Hill et al., 1999). Satellite-based systems are now more commonly applied, with several studies exploring the use of microwave sensors for biomass estimation (Ali et al., 2016).

One of the key benefits of microwave sensing is the capacity to provide measurements through cloud cover that precludes the use of standard optical sensors. This is of particular benefit to pastures and rangelands which experience heavy cloud cover during growing seasons often resulting in high rainfall and emphasising the need to more accurately measure biomass. One example of the application of radar in grasslands was undertaken by McNeill et al. (2010) in New Zealand dairy pastures who reported prediction errors of between 8% and 18%. Ali et al. (2016) provided a detailed account of the specific types of microwave sensing that have been applied to grasslands and managed pastures along with some of the limitations and challenges. Whilst it does hold significant potential for application in pastures and rangelands, there are still no commercial products available which incorporate this technology to provide information to producers.

4.1.6 Trade-offs in spatial and temporal resolution and potential for sensor integration within remote sensing platforms

One of the key issues which continues to challenge the use of remote sensing imagery in pastures is the trade-off that usually exists between spatial, spectral and temporal resolution (Ali et al., 2016). Ideally, a remote sensing platform would provide data at a spatial resolution which at least enabled detection of the vegetation within the paddock (i.e. pixel size less than the field size). Better still would be a system which enabled some measure of sub-field variability (see case study 2 for a practical exploration of this). In terms of temporal variability, the ideal platform would provide data at least at an interval equivalent to the management decisions being made. However, those sensors and platforms designed to collect high spatial or spectral resolution generally have lower temporal resolution. The Landsat platform and sensing system has 30 m × 30 m spatial resolution but generally has a repeat cycle of ~16 days whilst the MODIS platforms and sensing systems have a spatial resolution of 250 m × 250 m but collect images on a daily basis. There are some studies which have explored the integration of low- and high-resolution optical reflectance imagery; however, the fusion of data remains in the research domain. Further to this, there have been some studies which have explored the integration of optical reflectance and microwave imagery (Schmidt et al., 2016). This concept holds significant potential for the measurement of biomass in both pastures and rangelands.

4.1.7 Proximal sensing systems

Traditionally, proximal sensing systems have been defined as those being within the 'proximity' of the target. The current use of the definition is somewhat looser but generally captures those instruments which are ground based and either deployed from a vehicle or by hand. Proximal sensors can be broadly categorised as either optical reflectance sensors or sward structure sensors in terms of how they measure the pasture. A further categorisation occurs around their footprint as either imaging or point sensors.

4.1.8 Proximal optical reflectance sensors

Like remote sensing systems, proximal sensors can be divided into either active (generate their own illumination) or passive (use sunlight as an illumination source). There are very few examples of passive instruments being deployed for monitoring pasture biomass and none outside of a research context (Handcock et al., 2016). The challenge with the use of passive MS systems is the need to calibrate for daily variations in illumination (Gobbett et al., 2013). This complication along with the relatively high cost means there is limited commercial application of this type of sensor.

There has been significantly more interest in the development of active optical sensors (AOS) for measuring pasture biomass. Active sensors have emerged from commercial deployment in the cropping industries; details of how they operate can be found in Part 1 of this book. One of the first to explore this technology for measuring pasture biomass was Flynn et al. (2008). They reported a significant relationship with a correlation coefficient of 0.68 between Greenseeker® RT500 Sensor (Trimble, CA, USA)-derived NDVI and total dry biomass (both green and senescent material). One of the known limitations of NDVI is that it measures photosynthetically active material and as such is only directly measuring the green fraction of the sward. The use of AOS to estimate the green fraction of the sward has been investigated (Trotter et al., 2010a) with higher correlation coefficients achieved (0.80), albeit with more refined reflection indices (Soil Adjusted Vegetation Index – SAVI). An exploration of the relationships between green and dead sward fractions and the limitations of AOS were discussed by Trotter et al. (2012). However, a critical deficiency of AOS is the inability to quantify the senescent component of the sward. The advent of relatively affordable AOS means that graziers can now explore the use of this class of sensors for day-to-day management (Trotter, 2013).

Numerous researchers have explored the potential for AOS to be linked to GPS and provide a spatial measure of pasture biomass (Serrano et al., 2016; Trotter et al., 2010a). This has obvious benefits in terms of providing whole field estimates of pasture biomass and understanding spatial variability, although this can only be achieved by driving a vehicle (or walking with a handheld instrument). The deployment of AOS on low-flying aircraft in a cropping context (Lamb et al., 2011) has proven the potential for this class of sensor to be deployed using an aerial platform. Should this innovation be further developed for deployment from a UAV platform, there would be obvious applications and benefits for the grazing industries.

4.1.9 Proximal sward structure sensors

An alternative approach to optical sensing systems are sward structure sensors. Rather than sensing the reflectance characteristics of the pasture, this class of sensor measures the biophysical structure of the sward. This might include plant height, leaf density or canopy volume. Some of these sensing systems still use the EM spectrum (but not in terms of proportional reflectance), whilst others rely on ultrasound or digital image processing.

One of the first proximal pasture sensors to be commercialised and used by the dairy industry was the C-Dax Pasture Meter (Jago et al., 2013), originally developed by Massey University in New Zealand (Yule et al., 2005). This sensor uses interrupted light beams to sense the height of a pasture at an extremely high sample rate (200 Hz). The device is towed by a vehicle across the field being assessed to collect a high-density data set which can then be converted using a standard linear calibration equation, for example kg dry matter (DM)/ha = Pasture height × 18.6 + 750 (C-DAX, 2016). There are a number of tested calibrations available depending on region, species and season which are suggested to improve the accuracy of estimation in each circumstance (C-DAX, 2016). One of the key benefits of the C-Dax system is that it maintains contact with the ground providing more accurate estimates for sward height than sensors mounted on vehicles. This is however also one of its limitations, in that the presence of surface rocks and other obstacles is problematic. An alternative system providing similar pasture height measurements has been developed which utilises an ultrasound sensor (http://pasturereader.com.au/). This device also records plant height at high frequency with linear calibrations applied to determine pasture biomass. Whilst the non-contact nature of this sensor enables deployment in more difficult terrain, the independent movement of the vehicle on which it is mounted compared to the slides of the C-Dax sensor are likely to make it less accurate.

LiDAR sensors measure distance from a sensor to target using the time of flight of a concentrated light source. These systems have been evaluated for quantification of biomass and have so far proven highly promising (Radtke et al., 2010; Schaefer and Lamb, 2016). The critical limitation of these sensors has so far been cost. However, the recent development of consumer-grade sensors at a fraction of the cost (Trotter et al., 2016) has meant that these systems may become commercially available for graziers.

4.1.10 Integrating proximal sensors

One key development that is likely to improve the accuracy of sensor-based biomass estimation is the fusion of multiple instruments. Moeckel et al. (2017) reported that the combination of an HS reflectance sensor and ultrasonic height sensor provided better calibration accuracy than either instrument alone. Similarly, Schaefer and Lamb (2016) found a combination of proximal LiDAR and active optical sensor to be better than either instrument alone. One potentially powerful sensor fusion approach includes the integration of remote and proximal sensing systems. Moeckel et al. (2017) simulated the integration of ultrasound and satellite-based NDVI using a proximal sensor. However, there are no known working examples of this in the current literature.

4.1.11 Why can a mobile phone not be used to measure biomass?

A common complaint amongst graziers is the inability to use the camera system in their mobile phone as a biomass sensor. Their assumption that the advanced optics and processing power of modern mobile devices should be able to provide this service is not unfounded. They are applied in other industries (e.g. citrus industry, Gong et al., 2013). However, there remain several key problems with their development for pastures. Using the optical reflectance bands is limited due to the lack of an NIR band and, although the processing capabilities of modern phones is good, it remains insufficient to process three-dimensional imagery using techniques such as structure from motion (Cunliffe et al., 2016) at the moment. However, it is likely that these limitations will be overcome in the future.

4.2 Measuring pasture quality

The amount of biomass available to grazing livestock is important. However, the quality or nutritive value of this biomass determines the intake and, ultimately, productivity of the grazing animals. In contrast to the degree of investigation into biomass quantity, the reported research activity into quality is minimal. 'Quality' is an umbrella term used to describe a number of specific biochemical characteristics of the biomass; these include digestibility, carbohydrate, protein and mineral content.

To a certain extent, the proportion of green and senescent material will influence the overall quality of the pasture and can be measured using some of the sensors previously discussed for biomass quantification (Pullanagari et al., 2017a). The species composition of a pasture plays a significant role in determining overall quality and there have been several attempts to develop image-based species composition sensors to achieve this (Himstedt et al., 2009).

A more direct approach to sensing quality parameters has been investigated with both proximal and remote optical sensing systems. Both MS (Pullanagari et al., 2012a) and HS (Pullanagari et al., 2012b; Zhao et al., 2007) proximal sensing systems have provided reasonable results when correlated to a number of pasture quality parameters. The reported accuracy of sensor-based quality measures is variable. For crude protein, Zhao et al. (2007) reported a coefficient of determination (CD) of 0.61, Pullanagari et al. (2012a) a CD of 0.72 and Pullanagari et al. (2012b) a CD of 0.78. Sensor calibration with metabolisable energy (ME) has provided better results, with Pullanagari et al. (2012a) reporting a CD of 0.65 and Pullanagari et al. (2012b) a CD of 0.83. However, the high cost of these optical proximal sensing systems makes them prohibitive to use in a commercial context. HS imaging systems have been explored for pasture quality parameters. Pullanagari et al. (2015) reported CD of 0.77 for CP and 0.79 for ME. One key advantage of these imaging systems is the ability to collect large amounts of data across the entire landscape.

4.3 Estimating pasture growth rates

In addition to the amount and quality of biomass available, livestock managers are also interested in the change in quantity over time due to growth. When plants are actively growing, this is commonly referred to as pasture growth rate (PGR) and represents the amount of biomass generated by the sward as it captures energy and builds lignin, carbohydrate and other cell structures. PGR can be directly inferred using sensors to measure the beginning and end of a growth period (Trotter et al., 2010a). However this direct approach is limited to times when livestock are not grazing the field as animal intake occurs simultaneously and masks the true PGR.

Climate and soil water balance modelling are currently used to provide broad-scale estimates of PGR. Examples of these simple systems include New Zealand's 'Pasture Growth Forecaster' (http://pasture-growth-forecaster.dairynz.co.nz/, last accessed 31 January 2018 – Fig. 9) and the Australian 'Rainfall to Pasture Outlook Tool' (http://rainfall.mla.com.au/Station/AllLocations, last accessed 31 January 2018). As an alternative approach, the PastureBase Ireland system (https://pasturebase.teagasc.ie/V2/, last accessed 31 January 2018) uses growth estimate data from many farms to calculate regional average growth rates (Hanrahan et al., 2017 – Fig. 10). The coarse spatial resolution of these systems is a limitation with data only provided at a regional level.

More complex systems integrate both plant growth modelling and remote sensing. In Australian rangelands, the Australian Grassland and Rangeland Assessment by Spatial Simulation system (Aussie GRASS) has been developed to provide estimates of PGR (Carter et al., 2000) based on the GRASP plant growth model (Rickert et al., 2000). This

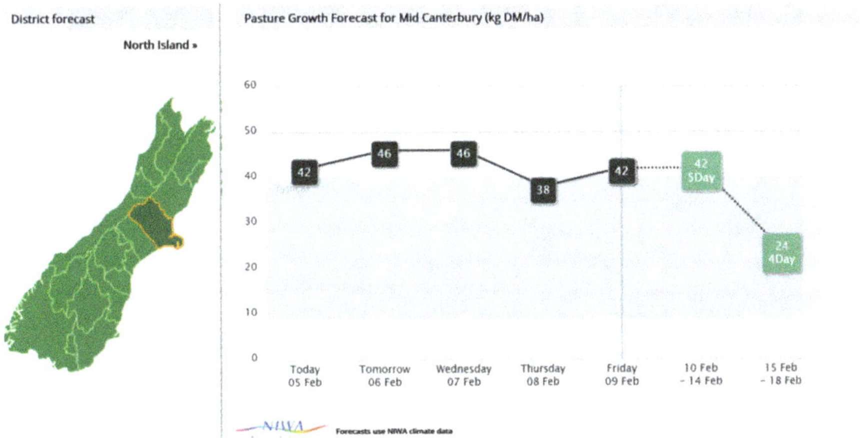

Figure 9 The 'Pasture Growth Forecaster' provides estimate and forecasts of pasture growth rates using rainfall and soil water balance modelling. This simple system provides important data to producers managing high input pastures for grazing dairy operations. One key limitation is the spatial resolution with data provided at only a regional average.

modelling system integrates interpolated meteorological data with satellite MS imagery and is made available to producers through a simplified web form, the FORAGE system, which provides local estimates (https://www.longpaddock.qld.gov.au/forage/, last accessed 31 January 2018).

In southern Australian temperate pastures, a more complex system, the PFS programme, integrates both satellite imagery and plant growth modelling to provide estimates of PGR at a relatively high spatial resolution. A full description of the process used to integrate meteorological data with MS satellite imaging was described by Hill et al. (2004). The basic protocol involves using NDVI data from a given land area (currently a MODIS or Landsat pixel) to estimate the potential fraction of absorbed photosynthetically active radiation (fAPAR). In this situation, NDVI provides an indication of the amount of photosynthetically active biomass on the ground (or green leaf area) which will limit this area's ability to absorb energy and grow. The amount of solar radiation actually delivered to the area is estimated from interpolated meteorological data and combined with the fAPAR to provide an estimate of the maximum likely energy captured for a given unit of time. This is then adjusted by a soil moisture index and temperature index derived from interpolated meteorological data which limit the potential growth response. The final estimate of utilisable APAR is then converted to kg of DM per ha per day using a constant. The constant has been established using a calibration process established through a field sampling process (Hill et al., 1989).

One of the key limitations of the current PGR estimation systems is that they largely rely on meteorological data collected by government agencies. This typically has a coarse spatial resolution. The soil moisture estimates which drive these models also rely on indirect measurements which further limit the accuracy of the system. The integration of local climate data including information collected at a farm or even field level could greatly improve the estimates of PGR as reported by Rahman et al. (2014).

Figure 10 The regional pasture growth rate (PGR) map provided by 'PastureBase Ireland'. Growth rates have been derived from actual differences between two pasture biomass estimates taken over time on farms across the country.

5 Case study 2: integrating satellite MS imaging with plant growth modelling to manage livestock rotations

The Pastures from Space™ (PFS) platform is one of the few commercial systems providing regular near-real-time satellite information to graziers (https://pfs.landgate.wa.gov.au/, last accessed 31 January 2018). The system provides two types of data: the first is a measure of green dry biomass, referred to as 'food on offer' (FOO) and is measured in kg/ha. The second is an estimate of the change in pasture biomass due to growth, this is referred to as PGR measured in kg/ha/day. Graziers need both these pieces of information if they are to manage the feed demands of their livestock whilst optimising the biomass production and maintain the sustainability of their pastures.

The FOO data is simply generated by applying a calibration curve developed from field sampling pastures at the same time that satellite data has been collected to generate NDVI values. The PGR is more complex and uses both the satellite NDVI data and meteorological data to model the estimated plant growth rates. A full explanation is provided in Hill et al. (2004). Basically, the NDVI is used to measure the potential energy absorption which is then limited by a solar radiation, temperature and soil moisture index derived from the meteorological information which is interpolated from regional weather stations.

The current PFS system uses data from two different imaging systems. The coarse resolution data is derived from the MODIS MS imaging system on board the Terra and Aqua satellites. When processed, this provides a spatial (pixel) resolution of 250 m × 250 m (Fig. 11). This can be a problem when field sizes are small as the pixels cross over multiple fields resulting in error in the data provided to producers seeking to understand the feed base in each paddock. Whilst this low spatial resolution is a limitation of the system, the relatively high temporal resolution (images collected daily) is a key benefit as the proprietary processing selects the most suitable image to provide to graziers on a weekly basis. The higher resolution imagery is derived from the Landsat MS imaging system and has a spatial resolution of approximately 30 m × 30 m when processed (Fig. 12). Whilst this imagery has obvious benefits in terms of understanding sub-paddock-scale variability, it is limited in that its temporal resolution is low with an image taken on an approximate 14 day return cycle (depending on location). These less frequent images can be affected by cloud as opposed to MODIS for which a cloud-free image can generally be collected at least once during the 7 day period.

The data from the MODIS imaging system is used to develop paddock average FOO (Fig. 13) and PGR (Fig. 14) maps.

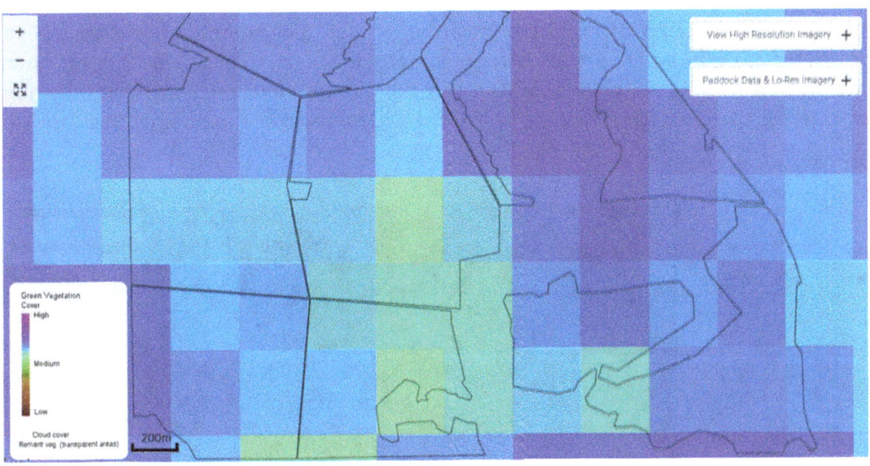

Figure 11 NDVI derived from the MODIS imaging system. Spatial resolution is 250 m × 250 m. Each pixel represents 6.25 ha and the problems with cross paddock pixels are apparent. There is only limited ability to recognise the spatial variability in the landscape. However, the high temporal resolution (MODIS captures images daily) makes this a useful imaging system in terms of collecting regular data. Source: https://pfs.landgate.wa.gov.au/.

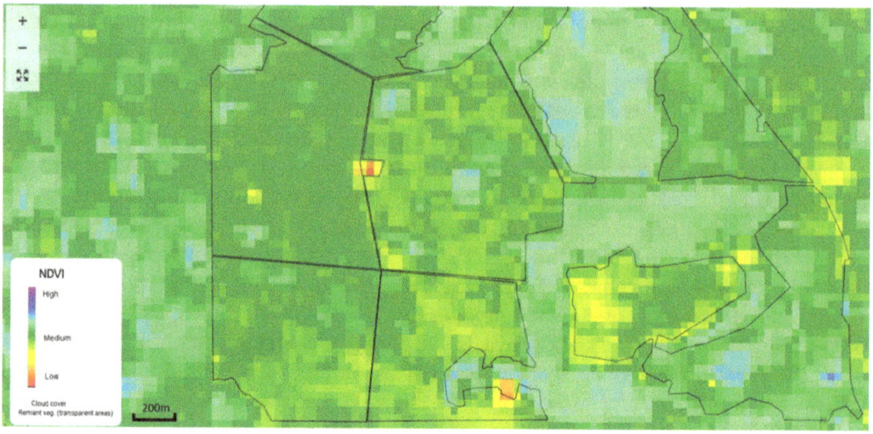

Figure 12 NDVI derived from the Landsat imaging system where spatial resolution is 30 m × 30 m. The higher spatial resolution that the Landsat imagery provides compared to that of MODIS shows clear benefits in terms of understanding the sub-paddock scale variability present in the field. However, the lower temporal resolution of the data (Landsat captures an image only once every ~14 days) means that it cannot be relied upon for timely decision making. Source: https://pfs.landgate.wa.gov.au/.

As an example of how producers apply this system, the data from paddock 5 will be used to calculate the number of days a flock might be left in this field. Paddock 1 has a FOO of 1768 kg/ha and a PGR of 85 kg/ha/day. If a grazier has a flock of 1500 ewes and wants to estimate how many days that they can be grazed in the field, a simple calculator within the PFS system can be used to determine this. In this situation, assuming the ewes consume 2 kg of FOO per day, the grazier can expect to hold them in this paddock for 28 days until the residual biomass target is reached (Fig. 15). This information allows graziers to optimise their short-term stocking rates whilst ensuring that they do not overgraze the paddock. Overgrazing beyond the ideal residual biomass (in this case set to

Figure 13 Food on offer (FOO) paddock average map derived from a calibration of NDVI to green dry biomass. Source: https://pfs.landgate.wa.gov.au/.

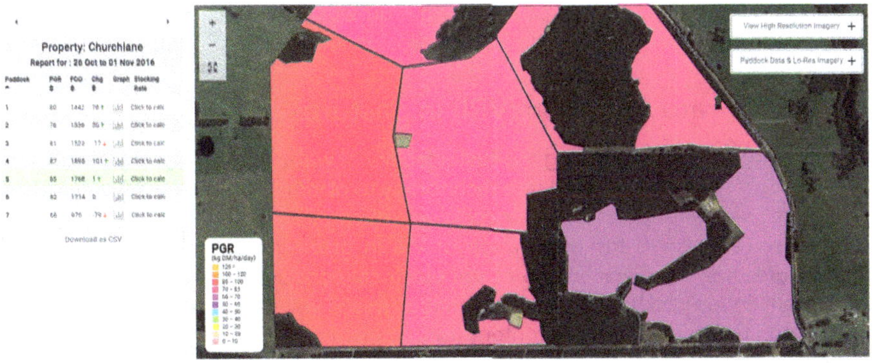

Figure 14 Pasture growth rate (PGR) paddock average map derived from NDVI and meteorological data. Source: https://pfs.landgate.wa.gov.au/.

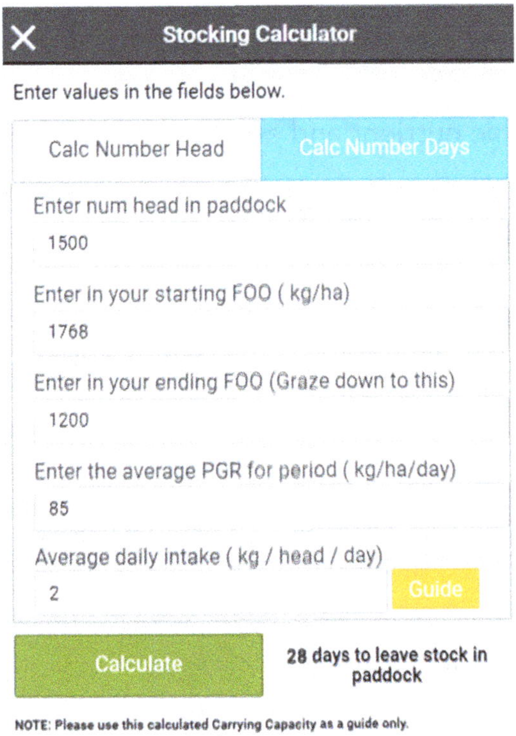

Figure 15 Pastures from Space incorporates a simple stocking rate calculator which allows graziers to input details such as the number of animals in a herd/flock, their estimated feed intake and the desired residual biomass (ending FOO) so that the number of days that the paddock can be grazed can be estimated. Source: https://pfs.landgate.wa.gov.au/.

1200 kg/ha) will result in reduced post-grazing regrowth rates and in extreme cases, when residuals become extremely low, can lead to erosion and soil loss.

5.1 Variable rate irrigation (VRI) in pastures

With increasing concern regarding the sustainability of water resources, irrigated cropping has increasingly turned to variable rate technology to optimise water use efficiency. Similar drivers exist for the adoption of this technology in irrigated pastures; however, there has been far less research in the application of VRI in grazing systems. The application of VRI in pastures has involved the use of soil electrical conductivity mapping to determine management zones which reflect spatial variation in soil water-holding capacity along with real-time monitoring of soil water content using moisture probes (Hedley and Yule, 2012). Variable rates of water are then applied to each zone based on requirements. Savings in water from the application of VRI to pastures are thought to range between 21% and 27% (Hedley and Yule, 2012). These savings come about due to reduced water application on areas that require less water as well as avoiding the application of water to non-productive areas such as cow tracks, drinking troughs and natural watercourses. Despite the little published research available concerning variable rate irrigation in pastures, there appears to be increasing industry engagement with this technology with several commercial suppliers of VRI systems promoting its use on grazed dairy farms in New Zealand (e.g. www.precisionirrigation.co.nz/, last accessed 31 January 2018).

5.1.1 Site-specific nutrient and ameliorant application

As outlined in the introduction to this chapter, the spatial variability of key attributes of grazing systems is generally higher than those observed in cropping and horticulture systems. This is no less the case for soil nutrient status. This variability poses both a challenge and an opportunity. The challenge is that spatial variation can result in lost productivity from areas low in key nutrients. On the flip side of this, some areas can be so high in some nutrients that they pose a significant environmental problem. The opportunity in this situation is the potential to develop site-specific nutrient management systems that can detect nutrient deficiencies and apply the required ameliorant to increase production of limited areas. At the same time, a site-specific nutrient management system can be used to reduce applications in areas which are in excess. This could potentially reduce fertiliser costs and alleviate environmental problems. One of the key challenges faced in the development and determination of the value of site-specific management (SSM) strategies in grazing systems is the absence of the equivalent of a yield map which has driven much of the thinking in the cropping and horticultural industries. To overcome this, research has focused on direct soil nutrient mapping and, to a lesser extent, indirect sensing of key elements.

5.2 Direct soil nutrient mapping

The current best management practice undertaken by graziers managing productive temperate pastures involves field-level soil nutrient testing. This means transect sampling of the field, bulking of the soil cores before mixing and sub-sampling for laboratory assessment. This technique has been used to improve the profitability of many grazing operations (Laurence, 2013). While useful for providing a general indication of the nutrient status of the field, this technique ignores the significant degree of spatial variability that has been observed in grazing systems at a sub-field scale.

Several studies have demonstrated the extent of variation in soil nutrients through GIS analysis of soil samples (Fu et al., 2010, 2013; Serrano et al., 2011; McCormick et al., 2009). In two typical fields used for grazing livestock in Australia, Trotter et al. (2014b) reported variation in phosphorus (P) and sulphur (S) through spatial soil sampling. In the native pasture, the key element was P which ranged from a low of 13 to a high of 121 mg/kg (mean = 31 mg/kg, CV = 59%). The critical P level for this field was 30 mg/kg meaning that the field average suggested that the P level was adequate. In the improved pasture, S was the key nutrient and ranged from 4 to 17 mg/kg (mean = 8 mg/kg CV = 35%). The critical S level was 8 mg/kg indicating that S levels were adequate. In both cases, the mean nutrient levels were at or above the critical limit (P = 30 mg/kg and S = 8 mg/kg) suggesting that fertiliser was not required. However, the native field had 56% of the sample sites below the critical P limit and the improved field had 50% of sites below the critical S level. This provides an insight into the potential fertiliser savings, with approximately half of each field not requiring nutrient addition and productivity gains with the remaining half potentially capable of improving production if nutrient levels could be increased to critical limits. Bernardi et al. (2016) undertook a similar process across multiple fields in a Lucerne pasture in Brazil but further developed the concept by calculating the variability in fertiliser required to ameliorate key deficiencies. P fertiliser requirements were found to vary from 0 to 1167 kg/ha (mean = 408) of single super phosphate with a CV of 102% (Fig. 16). Potassium (K) fertiliser requirements varied even more, from 0 to 525 kg/ha (mean = 126) of potassium chloride with a CV of 133% (Fig. 17). It is clear that considerable variation in fertiliser requirement exists and would translate

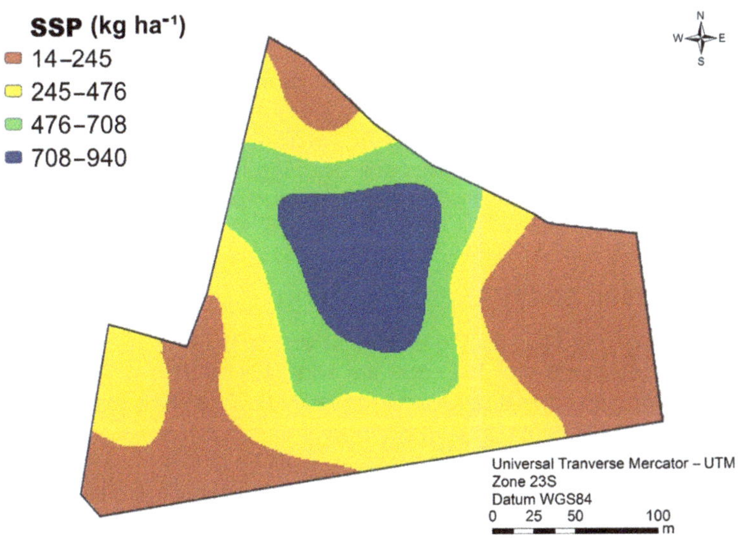

Figure 16 A site-specific management (SSM) prescription map for single super phosphate (SSP) application rate across a series of alfalfa (lucerne) fields developed in response to a gridded soil survey, originally published in Bernardi et al. (2016). The scale of soil nutrient variability and subsequent fertiliser recommendation is of particular interest with some areas requiring considerably more SSP than others. This demonstrates the potential value that SSM strategies could bring to nutrient management in grazing systems as considerable cost savings and production benefits could be realised.

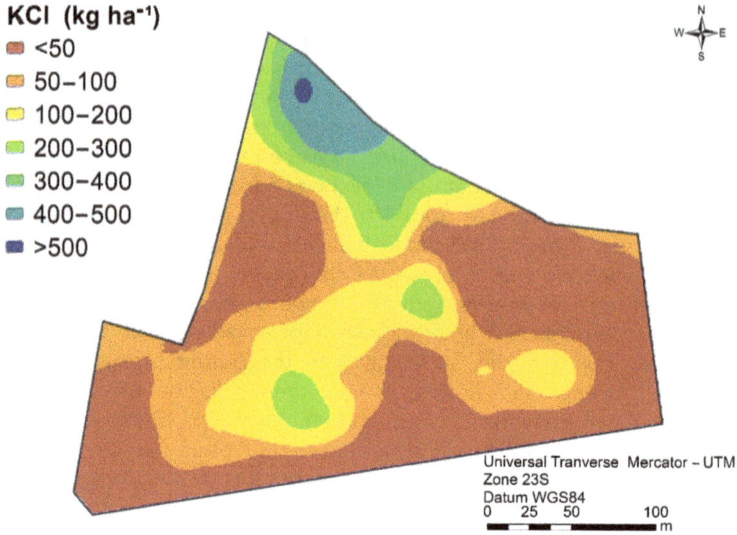

Figure 17 A site-specific management prescription map for potassium fertiliser (KCl) application rate across a series of alfalfa (lucerne) fields developed in response to a gridded soil survey, originally published in Bernardi et al. (2016). Potassium is a key nutrient in alfalfa production systems with large amounts exported in the forage product. Comparison of this map to the SSP (phosphorus) map in Fig. 16 reveals different spatial patterns. This poses a significant challenge in the development of SSM fertiliser strategies which will need to deliver varying rates across multiple limiting nutrients.

into a large potential savings and potential production increases if an SSM strategy was to be developed for this pasture.

These studies have demonstrated the value of SSM of nutrients in grazing systems. However, the widespread use of gridded soil sampling in a commercial context remains limited. The cost of soil sampling and analysis in some countries is thought to be a critical limitation.

5.3 Sensor-based soil nutrient mapping

The alternative to gridded soil testing is remote sensing to develop management zones from which a limited number of soil cores are taken for validation. This practice is well developed in a cropping context but less so for pastures. One of the main tools employed in the delineation for zones in cropping is the yield map but this has also been achieved using remote sensing and soil survey techniques. Outside of forage harvesting situations, direct yield maps (spatial variability in product removal) remain a challenge in grazing systems. However, the use of electrical conductivity (EC) mapping has been explored to better direct soil sampling (Serrano et al., 2017). The correlation between EC and key nutrients (e.g. P, K and S) has not been favourable. There have been some significant relationships with pH, although the correlation coefficients have been weak, for example Serrano et al. (2010) reported at CD of 0.30 and Serrano et al. (2017) reported CD of 0.50.

There is increasing interest in the application of remote sensing for both the development of management zones and the direct sensing of nutrient status. There is an extensive history of published work regarding the classification of grassland types using remote sensing imagery (Ali et al., 2016). However, there has been little published research involving the application of remote sensing imagery to improve nutrient management. The concept was first suggested as far back as the 1980s when the application of satellite imagery to agriculture was in its infancy (Vickery et al., 1980). Since then, there have been several papers promoting the concept (Simpson et al., 2011; Trotter, 2013) but few actively seeking to implement it (Cicore et al., 2016).

A more direct approach to detecting nutrient status has been proposed by Pullanagari et al. (2015) who applied airborne HS imaging to estimate foliar nutrient levels across a New Zealand hill-country pasture. They found good correlation between the analysed spectral data and folia N ($r^2 = 0.80$), P ($r^2 = 0.76$), K ($r^2 = 0.83$) and S ($r^2 = 0.86$). There is a clear opportunity for this sensing system to be applied to either develop zone management strategies or directly provide an indication of nutrient deficiencies that require amelioration. This system is now being rolled out as part of a commercial product (https://www.ravensdown.co.nz/services/spreading/intellispread – last accessed 23 March 2018).

5.4 Understanding spatial variability in fertiliser response

Understanding the spatial variation in soil nutrients provides an indication of the potential growth response likely from the application of fertiliser. However, in some cases, soil nutrients may not be the key limiting factor to increasing pasture growth. Soil water-holding capacity, along with solar radiation load (a function of aspect) and pasture species all contribute to the responsiveness of a given area to fertiliser (Murray et al., 2007). Murray and Yule (2007b) undertook a comprehensive simulation of a well-characterised grazing property in New Zealand. They tested several fertiliser management strategies including traditional blanket rate approach (all areas receive the same rate) and a number of SSM techniques using different levels of technical implementation. The blanket rate approach was simulated to produce 7918 kg DM/ha/year from total fertiliser application of 671 t over the entire farm. An SSM strategy using the same total amount of fertiliser (672 t) resulted in a simulated production of 8436 kg DM/ha/year equating to an increase of 7% in total pasture production with the same fertiliser input costs. A fully optimised SSM strategy involved increasing the total fertiliser applied to 738 t (10% increase) which resulted in a production gain of 24%. The economics of these results are discussed in detail by Murray and Yule (2007a). It is apparent that considering grassland responsiveness to fertiliser addition holds significant promise for increasing production efficiency.

5.5 Precision management of nutrients for environmental outcomes

One critical issue with fertiliser use in some pastures is the loss of excess nutrients into waterways. In some countries and regions, legislation now limits the use of certain fertilisers, particularly nitrogen (Hoogendoorn et al., 2011). In these areas, there is significant interest in developing systems that can improve nitrogen use efficiency in a similar way to those in cropping systems which use AOS and reference strips. One of the key sources of N loss in grazing systems is when excess N leaches from urine patches. Urine patches have extremely high soil N levels and do not require additional N fertiliser. A New Zealand-based entity (www.agrioptics.co.nz) has adapted an existing active optical sensor system to identify urine patches through the increased NDVI calculated as a result of higher growth. The system then avoids allocation of the liquid N fertiliser to the urine patches reducing unnecessary N application and reducing the potential additional leaching of excess N (http://www.agrioptics.co.nz/portfolio/smart-n/ – last accessed 23 March 2018). This simple system will likely provide fertiliser savings in high input pastures. However, there are likely to be further gains from SSM systems which take into account the underlying response potential of variable soils and the seasonal conditions.

5.5.1 Integrating sensor data across soils, plants and animal systems

One of the key opportunities which has yet to be fully explored is the integrated quantification of spatial variability across the various sub-systems (soils, plants and animals). This feature has been explored to a limited extent in a research context (Handcock et al., 2009; Trotter and Lamb, 2008) but has not yet been developed to the point where graziers have access.

One of the key pieces of information that has transformed PA in the cropping industry is the yield map. As previously discussed, the equivalent for grazing systems is not currently available. However, some attempts have been made to provide analogous data. Trotter et al. (2014a) reported the integration of live-weight gain, GNSS and behavioural modelling data to create an annualised red-meat yield map (Fig. 18). Similar conceptual

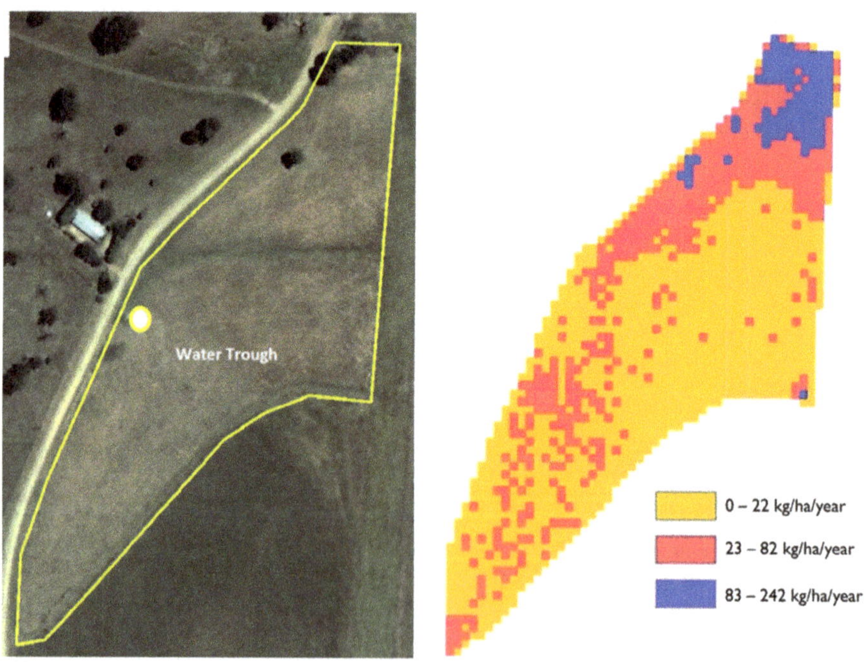

Figure 18 A red-meat yield map derived from integrating live-weight gain data with GNSS tracking data. This provides graziers with a conceptual understanding of the spatial variability in the productivity of their landscape, the equivalent of a yield map in cropping (Trotter et al., 2014a).

processes have been reported for dairy farming systems where fields were ranked based on milk productivity (Haultain, 2014).

The integration of sensor data measuring the spatial and temporal variability of key characteristics of soils, plants and animals will likely be a focus of some of the most revolutionary research to be undertaken for the grazing industry.

6 Future trends and conclusion

The scope for future research in precision livestock and pasture management is enormous. To some extent, this statement is based on the need for grazing systems to catch up with the technologies and innovation already present in the cropping and horticultural industries. However, the very complexity and spatial heterogeneity of grazing systems which has yet to be fully exploited provide significant opportunities for future research and production gains.

To take advantage of this variability, a range of new sensing platforms need to be developed in a way which provides affordable solutions to graziers. As well as the further development of the numerous soil, plant and animal sensors already available, a range of new platforms warrant research investment. The full extent currently being explored is too large to detail, but some of the most interesting include spatial soil moisture monitoring technologies (Bertoldi et al., 2014); pasture weed and pest mapping and control technologies (Cosby et al., 2016; Gebhardt et al., 2006); audio sensing of animal vocalisations (Meen et al., 2015); systems for measuring social interactions in herds and flocks (Swain et al., 2015); quantifying the intake of grazing animals for selection of higher feed efficiency use (Greenwood et al., 2014); interpreting animal behaviour data as an indicator of resource status (Manning et al., 2017; Roberts et al., 2015); and the integration of new genetic selection traits based on optimising spatial landscape utilisation (Bailey et al., 2015). The deployment mode of sensors and management tools will also dramatically change in the next few years with significant advances in autonomous vehicles (Nyamuryekung'e et al., 2016) and telecommunications across remote areas being two primary drivers of innovation.

Most importantly, the emerging and current sensors will continue to evolve and be extended in the way they are applied. One production system that remains largely unexplored is the integrated cropping and livestock production system (also known as mixed farming). The complexity of rotating crops and pasture phases, or grazing of crops, will mean that an integration of precision cropping and precision livestock concepts will be required (McEntee et al., 2013). There is an increasing interest in the application of sensors to animal welfare (Jukan et al., 2017; Morris et al., 2012) and sustainability indicators to influence and maintain social licence. There is also a growing demand for not just traceability but also validation of state across the entire food production system; this will see an increase in sensor integration to measure animal and animal product traits across the entire value chain.

One of the most critical issues facing the livestock industries is the adoption of technologies (Jago et al., 2013). Many of the previously mentioned innovations that could have a significant impact on production efficiency and sustainability will amount to nothing if they are not packaged and delivered to graziers in a manner that enables rapid intuitive learning and operation. There needs to be a significant investment in the development of digital literacy skills amongst producers. At the same time, a better understanding amongst those involved in research and development of the needs and capabilities of the great diversity of graziers is required (Eastwood et al., 2013).

The modern extensive grazing livestock producer manages some of the most complex agro-ecosystems ever developed. The continued development of appropriately delivered precision livestock and pasture technologies holds significant promise for these land managers in increasing production efficiency and sustainability.

7 Where to look for further information

Most of the leading journals relating to PA regularly feature articles concerning precision livestock and pasture management tools and systems. Two of the most valuable are *Precision Agriculture* and *Computers and Electronics in Agriculture*. Other important journals include *Animal Production Science, Crop and Pasture Science* and *Grass and Forage Science.*

The International Conference on Precision Agriculture held in the United States regularly hosts a small session on precision livestock and pasture management. Perhaps the best source of information on the current state of the art is *the European Precision Livestock Conference* held every 2 years. Reviewing the proceedings of these conferences will provide the reader with an excellent understanding of the latest technologies, although much is focused on the intensive animal industries.

Several smaller regional conferences and symposia now provide updates on developments in the more extensive grazing industries. *The Australian and New Zealand Spatially Enabled Livestock Management Symposium* is held annually with abstracts available on the internet.

8 References

Aldridge, M. N., Lee, S. J., Taylor, J. D., Popplewell, G. I., Job, F. R. and Pitchford, W. S. 2017. The use of walk over weigh to predict calving date in extensively managed beef herds. *Animal Production Science*, 57, 583–91.

Ali, I., Cawkwell, F., Dwyer, E., Barrett, B. and Green, S. 2016. Satellite remote sensing of grasslands: From observation to management—a review. *Journal of Plant Ecology*, 9, 649–71.

Anderson, D. M. 2007. Virtual fencing – past, present and future. *The Rangeland Journal*, 29, 65–78.

Awad, A. I. 2016. From classical methods to animal biometrics: A review on cattle identification and tracking. *Computers and Electronics in Agriculture*, 123, 423–35.

Bailey, D. W., Lunt, S., Lipka, A., Thomas, M. G., Medrano, J. F., Cánovas, A., et al. 2015. Genetic influences on cattle grazing distribution: Association of genetic markers with terrain use in cattle. *Rangeland Ecology and Management*, 68, 142–9.

Banhazi, T. M., Lehr, H., Black, J. L., Crabtree, H., Schofield, P., Tscharke, M., et al. 2012. Precision Livestock Farming: An international review of scientific and commercial aspects. *International Journal of Agricultural and Biological Engineering*, 5(3), 1–9.

Barry, B., Gonzales-Barron, U. A., McDonnell, K., Butler, F. and Ward, S. 2007. Using muzzle pattern recognition as a biometric approach for cattle identification. *Transactions of the ASABE*, 50, 1073–80.

Berckmans, D. 2008. Precision livestock farming (PLF). *Computers and Electronics in Agriculture*, 62(1), 1–80.

Berckmans, D. 2014. Precision livestock farming technologies for welfare management in intensive livestock systems. *OIE Revue Scientifique et Technique*, 33, 189–96.
Bernardi, A. C. C., Bettiol, G. M., Ferreira, R. P., Santos, K. E. L., Rabello, L. M. and Inamasu, R. Y. 2016. Spatial variability of soil properties and yield of a grazed alfalfa pasture in Brazil. *Precision Agriculture*, 17, 737–52.
Bertoldi, G., Della Chiesa, S., Notarnicola, C., Pasolli, L., Niedrist, G. and Tappeiner, U. 2014. Estimation of soil moisture patterns in mountain grasslands by means of SAR RADARSAT2 images and hydrological modeling. *Journal of Hydrology*, 516, 245–57.
Bikker, J., van Laar, H., Rump, P., Doorenbos, J., Van Meurs, K., Griffioen, G., et al. 2014. Technical note: Evaluation of an ear-attached movement sensor to record cow feeding behavior and activity. *Journal of Dairy Science*, 97, 2974–9.
Bishop-Hurley, G., Swain, D. L., Anderson, D., Sikka, P., Crossman, C. and Corke, P. 2007. Virtual fencing applications: Implementing and testing an automated cattle control system. *Computers and Electronics in Agriculture*, 56, 14–22.
Bocquier, F., Debus, N., Lurette, A., Maton, C., Viudes, G., Moulin, C. H., et al. 2014. Precision farming in extensive livestock systems. *Productions Animales*, 27, 101–12.
Brown, D. J., Savage, D. B., Hinch, G. N. and Semple, S. J. 2012. Mob-based walk-over weights: Similar to the average of individual static weights? *Animal Production Science*, 52, 613–18.
Carter, J., Hall, W., Brook, K., McKeon, G., Day, K. and Paull, C. 2000. Aussie GRASS: Australian grassland and rangeland assessment by spatial simulation. In: Hammer, G., Nicholls, N. and Mitchell, C. (Eds), *Applications of Seasonal Climate Forecasting in Agricultural and Natural Ecosystems – the Australian Experience*. the Netherlands: Kluwer Academic Press, pp. 329–49.
C-DAX. 2016. *User Manual*. New Zealand: C-DAX.
Cicore, P., Serrano, J., Shahidian, S., Sousa, A., Costa, J. L. and da Silva, J. R. M. 2016. Assessment of the spatial variability in tall wheatgrass forage using LANDSAT 8 satellite imagery to delineate potential management zones. *Environmental Monitoring and Assessment*, 188, 513.
Corkery, G. P., Gonzales-Barron, U. A., Butler, F., McDonnell, K. and Ward, S. 2007. A preliminary investigation on face recognition as a biometric identifier of sheep. *Transactions of the ASABE*, 50, 313–20.
Cosby, A. M., Falzon, G. A., Trotter, M. G., Stanley, J. N., Powell, K. S. and Lamb, D. W. 2016. Risk mapping of redheaded cockchafer (Adoryphorus couloni) (Burmeister) infestations using a combination of novel k-means clustering and on-the-go plant and soil sensing technologies. *Precision Agriculture*, 17, 1–17.
Cunliffe, A. M., Brazier, R. E. and Anderson, K. 2016. Ultra-fine grain landscape-scale quantification of dryland vegetation structure with drone-acquired structure-from-motion photogrammetry. *Remote Sensing of Environment*, 183, 129–43.
Deering, D. and Rouse, J. 1975. Measuring 'forage production' of grazing units from Landsat MSS data. In: *10th International Symposium on Remote Sensing of Environment*. Ann Arbor, MI: Environmental Research Institute of Michigan, pp. 1169–78.
Dobos, R., Dickson, S., Bailey, D. W. and Trotter, M. 2014. The use of GNSS technology to identify lambing behaviour in pregnant grazing Merino ewes. *Animal Production Science*, 54, 1722–7.
Earley, B., Prendiville, D. J., Lowe, J., Spahr, C. and Kettlewell, P. 2015. Radiotelemetry systems for measuring the body temperature of cattle. In: Guarino, M. and Berckmans, D. (Eds), *Precision Livestock Farming 2015 – Papers Presented at the 7th European Conference on Precision Livestock Farming*, ECPLF 2015. Milan, Italy: EC-PLF, pp. 758–67.
Eastwood, C., Trotter, M. and Scott, N. 2013. Understanding the user: Learning from on-farm application of precision farming technologies in the Australian livestock sector. *Australian Journal of Multidisciplinary Engineering*, 10, 41–50.
Eradus, W. J. and Jansen, M. B. 1999. Animal identification and monitoring. *Computers and Electronics in Agriculture*, 24, 91–8.
Falzon, G., Schneider, D., Trotter, M. and Lamb, D. W. 2013. Correlating movement patterns of merino sheep to faecal egg counts using global positioning system tracking collars and functional data analysis. *Small Ruminant Research*, 111, 171–4.

Flynn, E. S., Dougherty, C. T. and Wendroth, O. 2008. Assessment of pasture biomass with normalised difference vegetation index from active ground-based sensors. *Agronomy Journal*, 100, 114–21.

Fogarty, E. S., Manning, J. K., Trotter, M. G., Schneider, D. A., Thomson, P. C., Bush, R. D., et al. 2015. GNSS technology and its application for improved reproductive management in extensive sheep systems. *Animal Production Science*, 55, 1272–80.

Fu, W., Tunney, H. and Zhang, C. 2010. Spatial variation of soil test phosphorus in a long-term grazed experimental grassland field. *Journal of Plant Nutrition and Soil Science*, 173, 323–31.

Fu, W., Zhao, K., Jiang, P., Ye, Z., Tunney, H. and Zhang, C. 2013. Field-scale variability of soil test phosphorus and other nutrients in grasslands under long-term agricultural managements. *Soil Research*, 51, 503–12.

Gebhardt, S., Schellberg, J., Lock, R. and Kühbauch, W. 2006. Identification of broad-leaved dock (*Rumex obtusifolius* L.) on grassland by means of digital image processing. *Precision Agriculture*, 7, 165–78.

Gerber, P., Steinfeld, H., Henderson, B., Mottet, B., Opio, C., Dijkman, J., et al. 2013. *Tackling climate Change Through Livestock – A Global Assessment of Emissions and Mitigation Opportunities*. Rome, Italy: Food and Agriculture Organization of the United Nations.

Gobbett, D. L., Handcock, R. N., Zerger, A., Crossman, C., Valencia, P., Wark, T., et al. 2013. Prototyping an operational system with multiple sensors for pasture monitoring. *Journal of Sensor and Actuator Networks*, 2, 388–408.

Gong, A., Yu, J., He, Y. and Qiu, Z. 2013. Citrus yield estimation based on images processed by an Android mobile phone. *Biosystems Engineering*, 115, 162–70.

Gonzales Barron, U., Corkery, G., Barry, B., Butler, F., McDonnell, K. and Ward, S. 2008. Assessment of retinal recognition technology as a biometric method for sheep identification. *Computers and Electronics in Agriculture*, 60, 156–66.

González, L. A., Bishop-Hurley, G., Henry, D. and Charmley, E. 2014. Wireless sensor networks to study, monitor and manage cattle in grazing systems. *Animal Production Science*, 54, 1687–93.

Greenwood, P. L., Valencia, P., Overs, L., Paull, D. R. and Purvis, I. W. 2014. New ways of measuring intake, efficiency and behaviour of grazing livestock. *Animal Production Science*, 54, 1796–804.

Guo, Y., Corke, P., Poulton, G., Wark, T., Bishop-Hurley, G. and Swain, D. 2006. Animal behaviour understanding using wireless sensor networks. In: *Proceeding of 31st IEEE Conference on local Computer Networks*. Tampa, FL: IEEE, pp. 607–14.

Guo, H., Ma, X., Ma, Q., Wang, K., Su, W. and Zhu, D. 2017. LSSA_CAU: An interactive 3D point clouds analysis software for body measurement of livestock with similar forms of cows or pigs. *Computers and Electronics in Agriculture*, 138, 60–8.

Halachmi, I. and Guarino, M. 2016. Editorial: Precision livestock farming: A 'per animal' approach using advanced monitoring technologies. *Animal*, 10, 1482–3.

Halachmi, I., Klopčič, M., Polak, P., Roberts, D. J. and Bewley, J. M. 2013. Automatic assessment of dairy cattle body condition score using thermal imaging. *Computers and Electronics in Agriculture*, 99, 35–40.

Handcock, R., Swain, D., Bishop-Hurley, G., Patison, K., Wark, T., Valencia, P., et al. 2009. Monitoring animal behaviour and environmental interactions using wireless sensor networks, GPS collars and satellite remote sensing. *Sensors*, 9, 3586–603.

Handcock, R. N., Gobbett, D., González, L. A., Bishop-Hurley, G. J. and McGavin, S. L. 2016. A pilot project combining multispectral proximal sensors and digital cameras for monitoring tropical pastures. *Biogeosciences*, 13, 4673.

Hanrahan, L., Geoghegan, A., O'Donovan, M., Griffith, V., Ruelle, E., Wallace, M., et al. 2017. PastureBase Ireland: A grassland decision support system and national database. *Computers and Electronics in Agriculture*, 136, 193–201.

Haultain, J. 2014. Ranking paddock performance using data automatically collected in a New Zealand dairy farm milking system. A thesis presented in partial fulfilment of the requirements for the degree of Masters in Agriculture at Massey University, Manawatu, New Zealand.

Hedley, C. and Yule, I. 2012. Farmer uptake of variable rate irrigation technologies in New Zealand. *10th International Conference on Precision Agriculture*. https://www.ispag.org/proceedings/?action=year_abstracts (Last accessed February 2018).

Hill, N. S., Stuedemann, J. A., Ware, G. O. and Petersen, J. C. 1989. Pasture sampling requirements for near infrared reflectance spectroscopy estimates of botanical composition. *Crop Science*, 29, 774–7.

Hill, M. J., Donald, G. E. and Vickery, P. J. 1999. Relating radar backscatter to biophysical properties of temperate perennial grassland. *Remote Sensing of Environment*, 67, 15–31.

Hill, M. J., Donald, G. E., Hyder, M. W. and Smith, R. C. G. 2004. Estimation of pasture growth rate in the south west of Western Australia from AVHRR NDVI and climate data. *Remote Sensing of the Environment*, 93, 528–45.

Himstedt, M., Fricke, T. and Wachendorf, M. 2009. Determining the contribution of legumes in legume–grass mixtures using digital image analysis. *Crop Science*, 49, 1910–16.

Holechek, J. L., Pieper, R. D. and Herbel, C. H. 1989. *Range Management. Principles and Practices*. Englewood Cliffs, NJ: Prentice-Hall.

Hoogendoorn, C. J., Betteridge, K., Ledgard, S. F., Costall, D. A., Park, Z. A. and Theobald, P. W. 2011. Nitrogen leaching from sheep-, cattle- and deer-grazed pastures in the Lake Taupo catchment in New Zealand. *Animal Production Science*, 51, 416–25.

Hooven, N. 1968. Freeze branding for animal identification. *Journal of Dairy Science*, 51, 146–52.

Houborg, R. and McCabe, M. 2016. High-resolution NDVI from planet's constellation of earth observing nano-satellites: A new data source for precision agriculture. *Remote Sensing*, 8, (768) 1–19.

Jago, J., Eastwood, C., Kerrisk, K. and Yule, I. 2013. Precision dairy farming in Australasia: Adoption, risks and opportunities. *Animal Production Science*, 53, 907–16.

Jukan, A., Masip-Bruin, X. and Amla, N. 2017. Smart computing and sensing technologies for animal welfare: A systematic review. *ACM Computing Surveys*, 50(1), Article No. 10.

Kamphuis, C., DelaRue, B., Burke, C. and Jago, J. 2012. Field evaluation of 2 collar-mounted activity meters for detecting cows in estrus on a large pasture-grazed dairy farm. *Journal of Dairy Science*, 95, 3045–56.

Kiddy, C. A. 1977. Variation in physical activity as an indication of estrus in dairy cows. *Journal of Dairy Science*, 60, 235–43.

Kumar, S., Singh, S. K., Singh, R. and Singh, A. K. 2017. Recognition of cattle using face images. In: Kumar, S., Singh, S. K., Singh, R. and Singh, A. K. (Eds), *Animal Biometrics: Techniques and Applications*. Singapore: Springer, pp. 79–110.

Kumar, S., Pandey, A., Satwik, K. S. R., Kumar, S., Singh, S. K., Singh, A. K., et al. 2018. Deep learning framework for recognition of cattle using muzzle point image pattern. *Measurement*, 116, 1–17.

Laca, E. A. 2009. Precision livestock production: Tools and concepts. *Revista Brasileira de Zootecnia*, 38, 123–32.

Laliberte, A. S., Goforth, M. A., Steele, C. M. and Rango, A. 2011. Multispectral remote sensing from unmanned aircraft: Image processing workflows and applications for rangeland environments. *Remote Sensing*, 3, 2529–51.

Lamb, D. W., Schneider, D. A., Trotter, M. G., Schaefer, M. T. and Yule, I. J. 2011. Extended-altitude, aerial mapping of crop NDVI using an active optical sensor: A case study using a Raptor™ sensor over wheat. *Computers and Electronics in Agriculture*, 77, 69–73.

Laurence, H. 2013. A precision fertiliser plan: Real measurements, real costs, real results. In: Currie, L. and Christensen, C. (Eds), *26th Annual Fertiliser and Lime Research Centre Workshop – Accurate and Efficient Use of Nutrients on Farms*. Massey University Palmerston North.

Lee, Y., Bok, J. D., Lee, H. J., Lee, H. G., Kim, D., Lee, I., et al. 2016. Body temperature monitoring using subcutaneously implanted thermo-loggers from Holstein steers. *Asian-Australasian Journal of Animal Sciences*, 29, 299–306.

Manning, J. K., Fogarty, E. S., Trotter, M. G., Schneider, D. A., Thomson, P. C., Bush, R. D., et al. 2014. A pilot study into the use of global navigation satellite system technology to quantify the behavioural responses of sheep during simulated dog predation events. *Animal Production Science*, 54, 1676–81.

Manning, J., Cronin, G., González, L., Hall, E., Merchant, A. and Ingram, L. 2017. The behavioural responses of beef cattle (*Bos taurus*) to declining pasture availability and the use of GNSS technology to determine grazing preference. *Agriculture*, 7, 45.

Mata, G., Henry, D., Gherardi, S. and Smith, R. 2004. Pastures from space – introduction to the technology on farm. *Animal Production in Australia*, 25, 281.

McCormick, S., Jordan, C. and Bailey, J. S. 2009. Within and between-field spatial variation in soil phosphorus in permanent grassland. *Precision Agriculture*, 10, 262–76.

McEntee, P., Belford, R., Mandel, R., Harper, J. and Trotter, M. 2013. Sub-paddock scale spatial variability between the pasture and cropping phases of mixed farming systems in Australia. In: Stafford, J. V. (Ed.), *Precision Agriculture'13 Proceedings of the 9th European Conference on Precision Agriculture*. Wageningen, the Netherlands: Wageningen Academic Publishers, pp. 389–94.

McNeill, S., Pairman, D., Belliss, S., Dalley, D. and Dynes, R. 2010. Robust estimation of pasture biomass using dual-polarisation TerrASAR-X imagery. Geoscience and Remote Sensing Symposium (IGARSS), IEEE, pp. 3094–7.

Meen, G. H., Schellekens, M. A., Slegers, M. H. M., Leenders, N. L. G., van Erp-van der Kooij, E. and Noldus, L. P. J. J. 2015. Sound analysis in dairy cattle vocalisation as a potential welfare monitor. *Computers and Electronics in Agriculture*, 118, 111–15.

Menesatti, P., Costa, C., Antonucci, F., Steri, R., Pallottino, F. and Catillo, G. 2014. A low-cost stereovision system to estimate size and weight of live sheep. *Computers and Electronics in Agriculture*, 103, 33–8.

Menzies, D., Patison, K. P., Fox, D. R. and Swain, D. L. 2016. A scoping study to assess the precision of an automated radiolocation animal tracking system. *Computers and Electronics in Agriculture*, 124, 175–83.

Menzies, D., Patison, K., Corbet, N. and Swain, D. 2017. Using temporal associations to determine maternal parentage in extensive beef herds. *Animal Production Science*, 58(5), 943–9.

Moeckel, T., Safari, H., Reddersen, B., Fricke, T. and Wachendorf, M. 2017. Fusion of ultrasonic and spectral sensor data for improving the estimation of biomass in grasslands with heterogeneous sward structure. *Remote Sensing*, 9(1), 98.

Morris, J. E., Cronin, G. M. and Bush, R. D. 2012. Improving sheep production and welfare in extensive systems through precision sheep management. *Animal Production Science*, 52, 665–70.

Mottram, T., Lowe, J., McGowan, M. and Phillips, N. 2008. Technical note: A wireless telemetric method of monitoring clinical acidosis in dairy cows. *Computers and Electronics in Agriculture*, 64, 45–8.

Murray, R. I. and Yule, I. J. 2007a. Developing variable rate application technology: Economic impact for farm owners and topdressing operators. *New Zealand Journal of Agricultural Research*, 50, 65–72.

Murray, R. I. and Yule, I. J. 2007b. Developing variable rate application technology: Scenario development and agronomic evaluation. *New Zealand Journal of Agricultural Research*, 50, 53–63.

Murray, R. I., Yule, I. J. and Gillingham, A. G. 2007. Developing variable rate application technology: Modelling annual pasture production on hill country. *New Zealand Journal of Agricultural Research*, 50, 41–52.

Nogami, H., Arai, S., Okada, H., Zhan, L. and Itoh, T. 2017. Minimized bolus-type wireless sensor node with a built-in three-axis acceleration meter for monitoring a cow's rumen conditions. *Sensors (Switzerland)*, 17(4), 687.

Nyamuryekung'e, S., Cibils, A. F., Estell, R. E. and Gonzalez, A. L. 2016. Use of an unmanned aerial vehicle-mounted video camera to assess feeding behavior of raramuri criollo cows. *Rangeland Ecology and Management*, 69, 386–9.

Peiper, U. M., Edan, Y., Devir, S., Barak, M. and Maltz, E. 1993. Automatic weighing of dairy cows. *Journal of Agricultural Engineering Research*, 56, 13–24.

Phillips, R., West, M., Saliendra, N., Rundquist, B. and Pool, D. 2013. Prediction of senescent rangeland canopy structural attributes with airborne hyperspectral imagery. *GIScience and Remote Sensing*, 50, 133–53.

Pullanagari, R., Yule, I., Hedley, M., Tuohy, M., Dynes, R. and King, W. 2012a. Multi-spectral radiometry to estimate pasture quality components. *Precision Agriculture*, 13(4), 442–56.

Pullanagari, R. R., Yule, I. J., Tuohy, M. P., Hedley, M. J., Dynes, R. A. and King, W. M. 2012b. In-field hyperspectral proximal sensing for estimating quality parameters of mixed pasture. *Precision Agriculture*, 13, 351–69.

Pullanagari, R., Kereszturi, G., Yule, I. J. and Irwin, M. 2015. Determination of pasture quality using airborne hyperspectral imaging. *SPIE Remote Sensing*. International Society for Optics and Photonics, 963706-963706-5.

Pullanagari, R., Kereszturi, G. and Yule, I. 2017a. Quantification of dead vegetation fraction in mixed pastures using AisaFENIX imaging spectroscopy data. *International Journal of Applied Earth Observation and Geoinformation*, 58, 26–35.

Pullanagari, R. R., Kereszturi, G. and Yule, I. J. 2017b. Quantification of dead vegetation fraction in mixed pastures using AisaFENIX imaging spectroscopy data. *International Journal of Applied Earth Observation and Geoinformation*, 58, 26–35.

Quigley, S. P., Poppi, D. P. and Schatz, T. J. 2015. Validation and demonstration of a diagnostic tool for phosphorus status of beef cattle. *Meat and Livestock Australia*. https://www.mla.com.au/research-and-development/search-rd-reports/final-report-details/Productivity-On-Farm/Validation-and-demonstration-of-a-diagnostic-tool-for-phosphorus-status-of-beef-cattle/1061 (Last accessed February 2018).

Radtke, P. J., Boland, H. T. and Scaglia, G. 2010. An evaluation of overhead laser scanning to estimate herbage removals in pasture quadrats. *Agricultural and Forest Meteorology*, 150, 1523–8.

Rahman, M. M., Lamb, D. W., Stanley, J. N. and Trotter, M. G. 2014. Use of proximal sensors to evaluate at the sub-paddock scale a pasture growth-rate model based on light-use efficiency. *Crop and Pasture Science*, 65, 400–9.

Rickert, K., Stuth, J. and McKeon, G. 2000. Field and laboratory methods for grassland and animal production research. In: Mannetje, L. and Jones, R. M. (Eds), *Modelling Pasture and Animal Production*. Wallingford, UK: CABI Publishing, pp. 29–66.

Roberts, C. M. 2006. Radio frequency identification (RFID). *Computers and Security*, 25, 18–26.

Roberts, A. J., Wallace, L. E., Harbac, M. and Paterson, J. A. 2012. Retention and readability of radio frequency identification transponders in beef cows over a 5-year period 1. *Professional Animal Scientist*, 28, 221–6.

Roberts, J., Trotter, M., Schneider, D., Lamb, D., Hinch, G. and Dobos, R. 2015. Daily grazing time of free-ranging cattle as an indicator of available feed. *Precision Livestock Farming 2015 – Papers Presented at the 7th European Conference on Precision Livestock Farming*. EC-PLF, Milan, Italy, pp. 491–500.

Ruiz-Garcia, L. and Lunadei, L. 2011. The role of RFID in agriculture: Applications, limitations and challenges. *Computers and Electronics in Agriculture*, 79, 42–50.

Rutten, C., Velthuis, A., Steeneveld, W. and Hogeveen, H. 2013. Invited review: Sensors to support health management on dairy farms. *Journal of Dairy Science*, 96, 1928–52.

Schaefer, M. T. and Lamb, D. W. 2016. A combination of plant NDVI and LiDAR measurements improve the estimation of pasture biomass in tall fescue (*Festuca arundinacea* var. *Fletcher*). *Remote Sensing*, 8(2), 109.

Schellberg, J. and Verbruggen, E. 2014. Frontiers and perspectives on research strategies in grassland technology. *Crop and Pasture Science*, 65, 508–23.

Schmidt, M., Carter, J., Stone, G. and O'Reagain, P. 2016. Integration of optical and X-band radar data for pasture biomass estimation in an open Savannah woodland. *Remote Sensing*, 8, 989.

Serrano, J., Peça, J., Marques da Silva, J. and Shaidian, S. 2010. Mapping soil and pasture variability with an electromagnetic induction sensor. *Computers and Electronics in Agriculture*, 73, 7–16.

Serrano, J., Peça, J., Marques da Silva, J. and Shahidian, S. 2011. Spatial and temporal stability of soil phosphate concentration and pasture dry matter yield. *Precision Agriculture* 12(2), 214–32.

Serrano, J. M., Shahidian, S. and Marques da Silva, J. R. 2016. Monitoring pasture variability: Optical OptRx® crop sensor versus Grassmaster II capacitance probe. *Environmental Monitoring and Assessment*, 188, 117.

Serrano, J. M., Shahidian, S. and Marques da Silva, J. 2017. Spatial variability and temporal stability of apparent soil electrical conductivity in a Mediterranean pasture. *Precision Agriculture*, 18, 245–63.

Shane, D. D., White, B. J., Larson, R. L., Amrine, D. E. and Kramer, J. L. 2016. Probabilities of cattle participating in eating and drinking behavior when located at feeding and watering locations by a real time location system. *Computers and Electronics in Agriculture*, 127, 460–6.

Simpson, R. J., Oberson, A., Culvenor, R. A., Ryan, M. H., Veneklaas, E. J., Lambers, H., et al. 2011. Strategies and agronomic interventions to improve the phosphorus-use efficiency of farming systems. *Plant and Soil*, 349(1–2) 89–120.

Spoliansky, R., Edan, Y., Parmet, Y. and Halachmi, I. 2016. Development of automatic body condition scoring using a low-cost 3-dimensional Kinect camera. *Journal of Dairy Science*, 99, 7714–25.

Swain, D., Friend, M., Bishop-Hurley, G., Handcock, R. and Wark, T. 2011. Tracking livestock using global positioning systems–are we still lost? *Animal Production Science*, 51, 167–75.

Swain, D. L., Patison, K., Heath, B., Bishop-Hurley, G. and Finger, A. 2015. Pregnant cattle associations and links to maternal reciprocity. *Applied Animal Behaviour Science*, 168, 10–17.

Trotter, M. 2013. PA innovations in livestock, grazing systems and rangeland management to improve landscape productivity and sustainability. *Agricultural Science*, 25, 27–31.

Trotter, M. G. and Lamb, D. W. 2008. GPS tracking for monitoring animal, plant and soil interactions in livestock systems. In: Kholsa, R. (Ed.), *9th International Conference on Precision Agriculture*. Denver, Colorado.

Trotter, M. G., Lamb, D. W., Donald, G. E. and Schneider, D. A. 2010a. Evaluating an active optical sensor for quantifying and mapping green herbage mass and growth in a perennial grass pasture. *Crop and Pasture Science*, 61, 389–98.

Trotter, M. G., Lamb, D. W., Hinch, G. N. and Guppy, C. N. 2010b. Global Navigation Satellite Systems (GNSS) livestock tracking: System development and data interpretation. *Animal Production Science*, 50, 616–23.

Trotter, M. G., Schneider, D., Lamb, D., Edwards, C. and McPhee, M. 2012. Examining the potential for active optical sensors to provide biomass estimation in improved and native pastures. In: *Australian Agronomy Conference*. Armidale, Australia: The Regional Institute.

Trotter, M., Economou, Z., Barwick, J., Dobos, R. and Lamb, D. 2014a. A 'six inch front harvester' yield map: Developing measures of spatial variability in productivity that are meaningful to graziers. In: Roberts, J., Charters, S. and Trotter, M. (Eds), *2014 Australian and New Zealand Spatially Enabled Livestock Management Symposium*. Hamilton, New Zealand: Faculty of Environment Society and Design, Lincoln University.

Trotter, M., Guppy, C., Haling, R., Edwards, C., Trotter, T. and Lamb, D. 2014b. Spatial variability in pH and key soil nutrients: Is this an opportunity to increase fertiliser and lime use efficiency in grazing systems? *Crop and Pasture Science*, 65, 817–27.

Trotter, M., Andersson, K., Welche, M., Chau, M., Frizzel, L. and Schneider, D. 2016. Evaluating low-cost Lidar and Active Optical Sensors for pasture and forage biomass assessment. *13th International Conference on Precision Agriculture*. International Society for Precision Agriculture. https://www.ispag.org/proceedings/?action=year_abstracts (Last accessed February 2018).

Umstatter, C. 2011. The evolution of virtual fences: A review. *Computers and Electronics in Agriculture*, 75, 10–22.

Vickery, P. J., Hedges, D. A. and Duggin, M. J. 1980. Assessment of the fertilizer requirement of improved pasture from remote sensing information. *Remote Sensing of Environment*, 9, 131–48.

Von Bueren, S. and Yule, I. 2013. Multispectral aerial imaging of pasture quality and biomass using unmanned aerial vehicles (UAV). *Accurate and Efficient Use of Nutrients on Farms, Occasional Report*, 26.

Von Bueren, S. K., Burkart, A., Hueni, A., Rascher, U., Tuohy, M. P. and Yule, I. J. 2015. Deploying four optical UAV-based sensors over grassland: Challenges and limitations. *Biogeosciences*, 12, 163–75.

Wark, T., Corke, P., Sikka, P., Klingbeil, L., Gou, Y., Crossman, C., et al. 2007. Transforming agriculture through pervasive wireless sensor networks. *Pervasive Computing*, 6, 50–7.

Wark, T., Swain, D., Crossman, C., Valencia, P., Bishop-Hurley, G. and Handcock, R. 2009. Sensor and actuator networks: Protecting environmentally sensitive areas. *IEEE Pervasive Computing*, 8.

Wathes, C. M., Kristensen, H. H., Aerts, J. M. and Berckmans, D. 2008. Is precision livestock farming an engineer's daydream or nightmare, an animal's friend or foe, and a farmer's panacea or pitfall? *Computers and Electronics in Agriculture*, 64, 2–10.

Wolfger, B., Jones, B. W., Orsel, K. and Bewley, J. M. 2017. Technical note: Evaluation of an ear-attached real-time location monitoring system. *Journal of Dairy Science*, 100, 2219–24.

Yule, I., Lawrence, H. and Murray, R. 2005. Yield mapping to improve dairy farm production. *Paper No.* 051074. St Joseph, MI: ASABE.

Zhao, D., Starks, P. J., Brown, M. A., Phillips, W. A. and Coleman, S. W. 2007. Assessment of forage biomass and quality parameters of bermuda grass using proximal sensing of pasture canopy reflectance. *Grassland Science*, 53, 39–49.

Chapter 18

The economics of precision agriculture

James Lowenberg-DeBoer, Harper Adams University, UK

1 Introduction

While precision agriculture (PA) has enormous potential for reducing the environmental footprint of farming, awareness is growing with respect to the social implications of the technology and the widespread adoption of PA depends on economics. Farming is a business and technology is adopted if it provides benefits for the farmer and farm household. Sometimes those benefits are qualitative (e.g. more time for recreation, reduced fatigue, less stress), but often those benefits are in monetary terms (i.e. more stable cash flow, higher and less variable profits). Farming is subject to what has been called 'economic physics' in the sense that someone must pay the bills regardless of the ideological framework within which the farmer is functioning just as everyone is subject to gravity whether or not they approve or think it is fair. Labour and capital must be compensated. The opportunity costs of land and other natural resources must be covered. That compensation might be in cash or in kind. Consequently, the objective of this chapter is to summarize the state of the art in economics of PA. Assuming that adoption is one of the best indicators of technology profitability and benefits, the topic will be approached by (1) reviewing the status of PA adoption worldwide, (2) assessing the degree to which PA adoption was predicted by economic studies and (3) describing future adoption as predicted by recent PA economics research. The chapter will be of interest to PA researchers, research administrators seeking to identify fruitful research areas, educators, farmers and agribusinesses.

Many definitions have been provided for 'precision agriculture' (e.g. Robert et al., 1994; National Research Council, 1997; Lowenberg-DeBoer and Swinton, 1997; Whelan and McBratney, 2000; Thorp, 2014), but most centre on use of electronic information technology to better manage spatial and temporal variability in crops, livestock, forestry and other biological systems. PA is not one technology, but a toolbox of technologies. The 'classic' PA technologies

http://dx.doi.org/10.19103/AS.2017.0032.19

originally proposed in the early 1990s were yield monitoring, remote sensing with satellites or aircraft, and variable rate technology (VRT) for fertilizer, pesticide and seed, supported by use of the Global Positioning System (GPS) for location finding and Geographic Information Systems (GIS). In the late 1990s, Global Navigation Satellite System (GNSS) guidance was introduced first in the form of lightbars and later as autosteer. GNSS guidance was soon joined by related technologies like sprayer boom control and planter row shutoffs. At the time of writing, improved GNSS (the United States, European, Russian and Chinese systems), artificial intelligence (AI), analysis of 'big data' and robotics are being added to the PA toolbox.

This analysis will focus on PA for crops because that has received the most attention by economists, but the principles outlined can be applied to livestock, forestry and other biological production systems. The analysis will concentrate on the use of PA for mechanized agriculture because that is where most PA technology has been deployed thus far, but observations will be made about adapting PA technology for small- and medium-sized farms in developing countries. PA has implications for the whole food value chain, from input supply to processing and marketing, but this chapter will concentrate on the primary production stages. The main focus of this chapter will be on 'action' technologies like GNSS guidance and VRT that make a physical change in crop management (e.g. GNSS guidance reduces pass to pass overlap, VRT applied inputs where and when they are needed). Many of the technologies in the PA toolkit are information-gathering tools (e.g. yield monitoring, satellite images, drones). The profitability of information gathering depends on the economics of the action technologies that implement decisions made with the information gathered.

The beginnings of PA are often dated to the release in the United States of GPS for civilian use in the late 1980s. The first published PA economic studies are from the mid-1990s. In the early days of PA, at least two quasi-exhaustive reviews of the economics research were conducted. Lambert and Lowenberg-DeBoer completed a review in 2000. Griffin et al. (2004) updated that review. Those two reviews found that about two-thirds of research articles reported benefits from PA technology, but only 20% of the articles included management, learning and human capital costs in their profitability estimates. The two reviews identified better returns to PA for crops that are more intensively managed and have a higher gross revenue per unit area. For example, the reviews show that returns to PA for cotton, sugar beets and corn were relatively high, but low for extensive wheat production in the US Great Plains. The reviews also showed mixed results for VRT fertilizer. At the time of the 2000 and 2004 PA economics reviews, GNSS guidance for agricultural equipment had only recently been commercialized in the United States. The few studies of GNSS guidance profitability had identified it as a highly profitable PA technology. Both reviews also highlighted the variability of profit estimates depending on the combination of PA technologies and the economic methodology used.

After 2004, the number of economic studies exploded and overall reviews of PA economics became more difficult. Partial reviews of the recent PA economics literature can be found in Pierpaoli et al. (2013), Griffin et al. (2017a) and Balafoutis et al. (2017). There were some reviews of specific PA technologies and technology packages. For example, Tenkorang and Lowenberg-DeBoer (2008) reviewed hundreds of agricultural remote sensing articles and reports. They found only a few that reported economic benefits and an even smaller number that explained how that economic benefit was estimated. For twelve studies they found enough information to apply standard budgeting methods and found that remote sensing increased profit estimates by about US$31.74/ha. They argued for more detailed reporting of budget assumptions and replicating studies over several years and at multiple geographic locations.

As governments became aware of precision agriculture, there were numerous policy-oriented studies which attempted to identify the economic benefits of PA. For example, in 2014 the EU Directorate for Internal Policies reviewed a scattering of studies from Europe and elsewhere in the world (Zarco-Tejada et al., 2014). They found that there seemed to be modest, but consistent benefit, to GNSS guidance almost everywhere and that controlled traffic using GNSS guidance may be the single biggest PA benefit for EU agriculture, but the profitability of VRT depended on the crop, soil conditions and many other factors. The OECD (2016) and the UK Agriculture and Horticulture Development Board (Awan, 2016) reported similar results. The US Department of Agriculture (USDA) reported that PA benefits for corn producers were positive, but small (Schimmelpfennig, 2016). This was based on an in-depth analysis of data from the USDA Agricultural Resource Management Survey (ARMS) data.

In spite of the massive literature on the economics of precision agriculture, questions about the profitability of PA continue. Complaints about the slow adoption are a regular part of almost every analysis of the potential for PA and that slow adoption is often linked to uncertainty about benefits and profitability. For example, Lowenberg-DeBoer (1998) focused on the 'uneven' adoption path of some PA technologies because they were coming on the market in an immature form and before supporting services were available. Cook et al. (2000) noted that slow adoption of PA by Australian farmers was in part due to perceived lack of benefit. Zhang et al. (2002) pointed out that in spite of substantial research investment in PA in the United States, Canada, Australia and the European Union, only a small portion of farmers in those countries use the technology. Kitchen (2008) noted that precision agriculture adoption has been hindered because the agronomy and engineering have not come together in ways that compensate the farm investment of time and money. Tey and Brindal (2013) review some 50 papers focused on the adoption of PA technologies and find that perceived profitability of PA significantly influences adoption decisions. The economic analysis of PA adoption in Nebraska by Castle et al. (2017) was motivated by lack of clarity related to the profitability of the technology.

Given the continuing questions about the profitability and other economic benefits of PA, the overall goal of this chapter and the extant literature on the economics of PA, the specific objectives of this chapter are to:

1 review the adoption of PA technology worldwide,
2 identify the degree to which economics research has correctly identified technologies that will be widely adopted and
3 overview the current (last decade) PA economics research and what it suggests as the likely PA adoption trends of the next decade.

2 Adoption of PA technology

Adoption of PA technology must be inferred from sparse data. PA adoption is not tracked regularly by any national agricultural statistics service. For example the USDA collects information on technology use as part of the ARMS which focuses on one crop at a time, with irregular intervals between surveys. Consequently since 2000, there is PA technology use data for corn in 2000, 2005 and 2010; rice: 2006 and 2013; cotton: 2000, 2003 and 2007; peanuts: 2004 and 2013; soybeans: 2000, 2002, 2006 and 2012; and spring wheat: 2000, 2004 and 2009. Some countries do occasional surveys, like the DEFRA farming practices survey in the United Kingdom in 2012. In many countries, PA adoption data depends on one-of-a-kind surveys by researchers or information from manufacturers and agribusiness. Understanding trends in the PA adoption requires piecing together information from disparate sources and irregular intervals.

The best historical time series of PA adoption data is from the CropLife/Purdue dealership survey which has been done regularly since the mid-1990s (Erickson et al., 2018). This survey has been supported by Meister Publication/CropLife magazine and depends on their almost exhaustive list of US agricultural input suppliers. There have typically been 200–300 responses, mostly from agricultural input supply firms. Responses have been concentrated in the Midwest. From 1996 to 2015, it was purely a paper survey sent through the post. In 2017, a paper copy was still mailed, but a web-based survey was also made available to respondents. From 1996 to 2009, the survey was done annually. After 2009, it was done biennially. The survey is not a panel; the set of respondents differs from year to year so some sampling variability is evident. The survey has typically asked only if a technology was used, not the intensity of use. If two dealership both have ten fertilizer applicators, one dealership has GNSS autosteer on all ten and the other has autosteer only on one of the ten, both are still counted as using GNSS autosteer. Clearly, the CropLife/Purdue survey is not a perfect source of PA adoption data, but it is relatively consistent over time with some of the same question asked over many years. A survey of agricultural input suppliers is particularly relevant for PA adoption in the United States, because many American farmers access PA technology through hiring services rather than by owning and operating the equipment directly.

For example, Fig. 1 shows the use of GPS guidance and offering of sensor services by US agricultural input dealerships. The CropLife/Purdue data shows the adoption of GNSS guidance clearly. GNSS light bars were introduced for ground-based agricultural equipment in the United States in the late 1990s and by 2004 over 60% of the dealerships

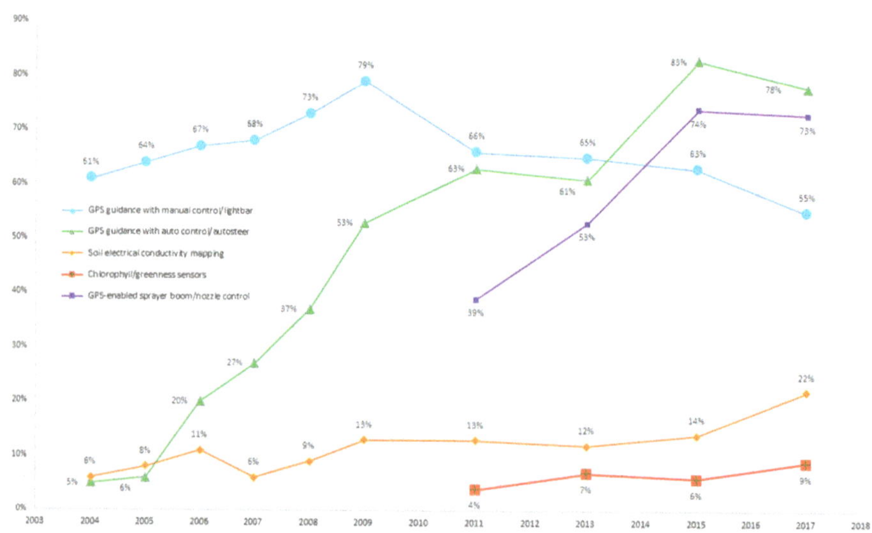

Figure 1 GPS guidance and sensor services at US agricultural dealerships, 2004–17. Source: Erickson et al. (2018).

were using them. By 2009, almost 80% of dealerships were using light bars. GNSS autosteer was introduced into the US Midwest about 2003. The CropLife/Purdue survey tracks the growth in autosteer use from only 5% in 2004 to around 80% in 2015 and 2017. The decline in light bar use after 2009 seems to show a case of technology replacement, in which dealerships gained experience with guidance through lightbars, but after 2009 have replaced lightbars with autosteer. The decline in autosteer use from 83% in 2015 to 78% in 2017 is probably more due to differences in the sample, than it is to any dis-adoption of this technology. The growth in sprayer section control is also documented in the rise from 39% in 2011 to the mid-70% in 2015 and 2017. Clearly, GNSS guidance is a widely adopted PA technology and it was adopted relatively quickly. In the case of both lightbars and autosteer, within about 10 years of introduction, GNSS was used by around 80% of the dealers. The CropLife/Purdue data indicate that complaints about slow adoption of PA technology should not apply to GNSS guidance.

The sensor data in Fig. 1 show different adoption patterns. The number of dealerships offering electrical conductivity mapping rose very gradually from 6% in 2004 to 22% in 2017. Chlorophyll/greenness sensors used in nitrogen management rose from 4% in 2011 to 9% in 2017. The CropLife/Purdue data indicate that the slow adoption complaint could apply to PA sensors.

Figure 2 shows the VRT services offered by the US agricultural input supply dealerships. Farmer use of these services in the United States is described below. A few agricultural input dealers started offering VRT fertilizer in the early 1990s. By 1999, the percentage offering that service was over 30%. That percentage has risen gradually to almost 80%. In the United States agricultural lime application is often done by a different set of contractors with specialized equipment that can spread crushed limestone and so it is tracked separately from granulated dry fertilizer in this survey. Controller VRT lime has followed the same pattern as VRT fertilizer, but 10–15% lower. The use of VRT lime is

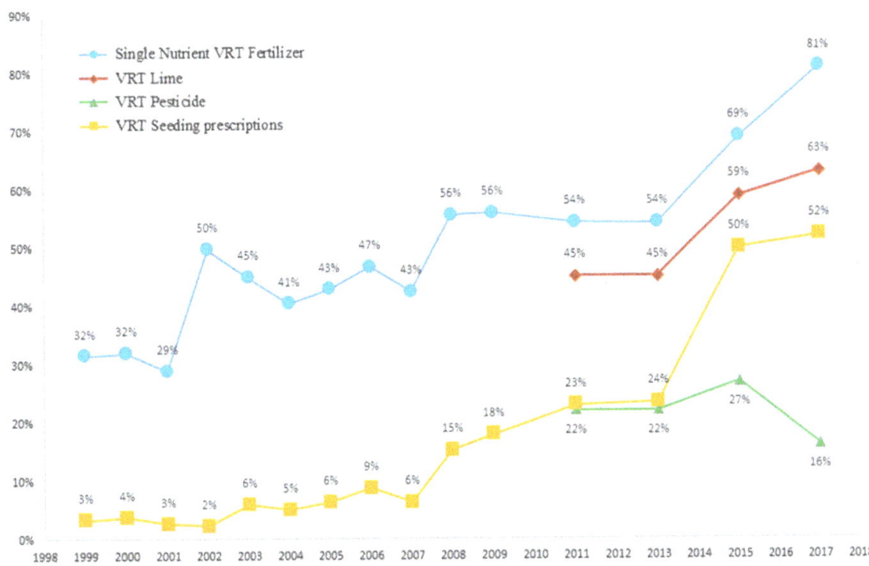

Figure 2 VRT services at US agricultural dealerships, 1999–2017. Source: Erickson et al. (2018).

probably underestimated by the CropLife Purdue survey because it does not separately track application of pelleted lime or 'manual' site-specific lime application. Pelleted lime can be applied by VRT fertilizer equipment along with other granulated products. Also in many US soils, pH differences often are observed in relatively large areas corresponding to soil type or topographical position. On those soils, farmers sometimes manually mark the area to be limed (e.g. by disking or ploughing around the designated area).

In the United States agricultural input suppliers often sell seed, but they rarely provide seeding services. Most farmers have their own planters and many of those planters are capable of varying seeding rates by management zone. The VRT seeding line in Fig. 2 shows the growing interest in 'prescription seeding'. In this case, the agricultural input dealer works with the farm to develop a map specifying seeding rates and sometimes varieties by management zone.

In spite of the irregular survey schedule and the crop focus, rather than a farm focus, the USDA ARMS data is probably the best information available anywhere in the world on farmer adoption of PA technology (USDA, 2018). Compared to the CropLife/Purdue results, the USDA data tells the same story about GNSS guidance and a different story about VRT. Figure 3 shows a scatter plot of the USDA ARMS estimates of GNSS guidance use by farmers producing six different crops: corn, soybeans, winter wheat, rice, cotton and peanuts. The USDA data does not distinguish between lightbar manual guidance and autosteer. The adoption estimates for 2001 and 2002 predate general commercial availability of autosteer in the United States and are mostly manual lightbar use. The more recent estimates are a mix of lightbar and autosteer adoption. The USDA guidance adoption estimates show a very clear trend that is very similar to the dealership adoption pattern in Fig. 1, especially the GNSS autosteer line. Dealerships probably started

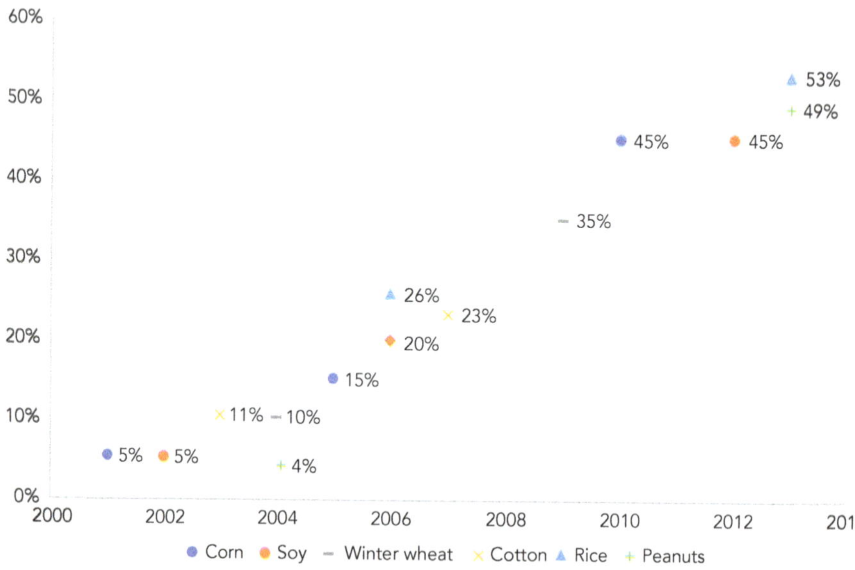

Figure 3 Percentage of planted area by crop on which GPS guidance or autosteer was used, 2000–13. Source: Based on data from USDA ARMS – https://data.ers.usda.gov/reports.aspx?ID=17883

adopting lightbars earlier than farmers and so by 2004, when the question was added to the CropLife/Purdue survey, the adoption rate was already over 60%.

Figure 4 is a scatter plot of VRT use for any purpose by farmers producing six crops. It shows a broad scatter of points mostly below 20% of planted area with only slight evidence of a trend towards greater adoption. In those years when data on the purpose of VRT is available, it shows that it is mostly for fertilizer application. This slight trend towards greater adoption of VRT is in spite of heavy media coverage of VRT technology, strong farmer interest in the technology and government subsidies for VRT in some areas. The slight upward trend in VRT occurred in the 2008–13 period of relatively high farm crop prices. Some evidence suggests that when crop prices declined sharply after 2014, farmers cut back on VRT and the intensive soil sampling on which it is usually based (Erickson et al., 2018).

The sparse data from many different parts of the world confirm the rapid and widespread adoption of GNSS guidance for mechanized agriculture and the relatively slow and irregular pace of VRT adoption. For example:

- In a one-time survey of agricultural input supply dealerships in Ontario, Canada, Mitchell et al. (2017) found that 74% of the dealerships used autosteer, 66% used section boom control and 88% offered VRT services, but VRT was used on only 28% of the crop area where they provided custom fertilizer spreading. The same survey found that lightbars were used by 30% of respondents.
- Miller et al. (2017) showed a similar pattern of adoption among farmers in the US state of Kansas using data from the Kansas Farm Management Association (KFMA). Use of GPS autosteer grew quickly from commercial introduction in the early 2000s

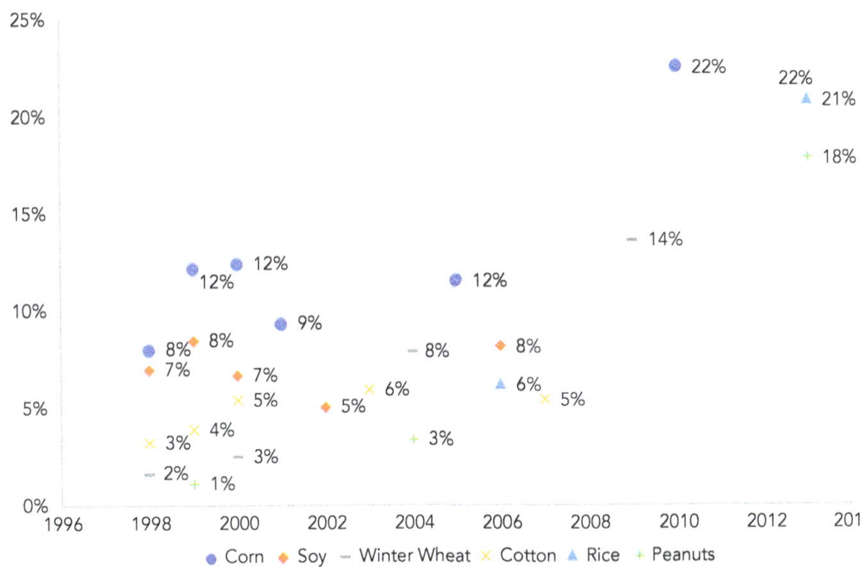

Figure 4 VRT for any purpose 1998 to 2013, Corn, Soybean, Winter Wheat, Cotton, Rice and Peanuts. Source: Based on data from USDA ARMS – https://data.ers.usda.gov/reports.aspx?ID=17883

to 66% by 2015, while VRT fertilizer grew slowly from the mid-1990s to about 25% in 2015. In this dataset, use of GNSS lightbars plateaus at about 40% in 2010. It should be noted that farm management association members are usually more interested in technology and improving farm management than the average US farmer and, consequently, may show somewhat higher adoption levels in the early stages of the adoption process.

- In the US state of Nebraska, Castle et al. (2017) studied technology adoption among farmer clients of Nebraska Farm Business, Inc. (NFBI). Over about a decade, use of GNSS autosteer among this group grew to over 70%, while use of VRT N, P, K and lime grew to about 50% in the two decades from the mid-1990s to 2014. Lightbar use peaked in this data at about 35% in 2010 and has declined slightly since then. The 50% VRT fertilizer adoption is high by worldwide standards and may be related to willingness to experiment among NFBI clients. Like the KFMA, NFBI members tend to be more interested in technology and improved management than the average US farmer.
- In Brazil, a phone survey of 922 large-scale grain producers showed that the majority used GNSS guidance and many used VRT fertilizer based on relatively large grid sizes (Molin, 2016). About 23% of them reported using autosteer and 14% lightbars. Thirteen percent reported using a VRT controller for fertilizer or pesticide.
- The National Agricultural Research Institute (INTA) of Argentina reports PA adoption in terms of number of machines instead of percent of farmers or percent of cropland, but the pattern is similar to that observed in the United States and Canada (Méndez et al., 2014; Scaramuzza et al., 2017). The number of GNSS guidance units is much greater than the number of VRT machines. The number of GNSS lightbars used for sprayers in Argentina has increased from 10 in 1998 to 18 342 in 2015. The number of GNSS autosteer units increased from 3 in 2004 to 5767 in 2015. The number of self-propelled variable rate urea and ammonium nitrate solution applicators increased from 50 in 2005 to only 854 in 2013. The number of seeders capable of variable rate starter fertilizer rose from 3 in 1997 to 3263 in 2015.
- In Australia, Llewellyn and Ousman (2014) found that 77% of Australian grain growers were using autosteer. The same survey found that while almost half of the Australian growers were varying fertilizer rates within fields, only about 15% were using computer controlled variable rate technology. This survey was done in 2012 in the main grain growing regions of Australia.
- In the United Kingdom in 2012 a nationwide survey of farming practices found that 22% of farmers were using GNSS guidance and 16% were using VRT (DEFRA, 2013), with higher adoption of both GNSS guidance and VRT on larger and more grain-focused farms. An internet survey in 2016 found similar adoption levels though the sample was dominated by younger respondents (Pickthall and Trivett, 2017).
- Elsewhere in Europe the PA adoption data is not recent, but shows the same pattern. A farmer survey documented that in Germany in 2010, 32% of respondents used GNSS guidance and 8% used VRT (Lawson et al., 2011). The same survey showed that in Denmark at that time 9% of respondents used GNSS guidance and 4% use VRT. In Finland, 4% use GNSS guidance and 3% used VRT.

While GNSS guidance is being adopted quickly almost wherever agriculture is mechanized, VRT is more likely to be found in 'hot spots' where the profit potential and soil variability combine to motivate adoption. The classic example of a VRT adoption hot spot is sugar beet producers in the Red River Valley of the North in the US states of Minnesota and North Dakota. Franzen (2017) documents that 53% of the sugar beet area in the Valley was managed with VRT nitrogen and the grid soil sampled area was even larger. Because soils in the Valley are originally an ancient lakebed, they are relatively uniform and some grid sampled fields were not judged variable enough to justify VRT application. Llewellyn and Ousman (2014) note that while overall adoption of computer controlled VRT in Australia is only 15% of grain farmers, it is 37% in the Victoria Mallee area where within-field variability is particularly noticeable.

Like GNSS guidance, adoption of other PA technology that depends only on location information is quite high. For example, the 2017 CropLife/Purdue survey found that 73% of responding dealerships used GNSS sprayer boom control. In the 2017 Ontario dealership survey, 69% use GNSS sprayer boom control and 25% use sprayer turn compensation. In Argentina in 2016, there were 9035 autosteer units and 3375 sprayers equipped with boom control. Since sprayer boom control, turn compensation and seed row shutoffs were introduced commercially quite recently, the USDA ARMS data and most other PA surveys do not yet include questions regarding these technologies.

PA information-gathering technology is often used to guide VRT and like VRT it has a mixed adoption picture.

- With VRT fertilizer, combine harvester yield monitors were part of the original PA technology introduced in the early 1990s. In industrialized countries, yield monitors have become standard equipment on many new combines, but many of those yield monitors are not used at all or used without GNSS so that the data cannot be mapped. For example the USDA ARMS data shows that for corn in 2010, 61% of respondents used a combine equipped with a yield monitor, but only 34% made yield maps. In 2006, 45% of US soybean farmer respondents used a combine equipped with a yield monitor, but 20% made yield maps. For rice in 2006, 29% of respondents used a yield monitor, but only 10% made maps. For winter wheat in 2009, 36% of respondents used a yield monitor, but only 6% made maps. Farmers worldwide complain that they often do not know how to use the yield map data and consequently lack the motivation to collect good quality yield data for maps.
- Efforts to use satellite remote sensing data for agricultural management has been made since the 1970s and some agricultural service providers have adopted the technology, but regular use of that data by farmers is still not widespread. The 2017 CropLife/Purdue survey showed that 59% of dealerships offered satellite or aerial imagery as a service to farmers. The 2017 Ontario survey revealed that 42% of the dealerships used satellite or aerial imagery for internal dealership decision-making. The most recent USDA ARMS data on the subject shows that in 2005 only 6% of corn farmers used an aerial or satellite image, in 2002 only 1% of soybean farmers, in 2003 7% of cotton farmers and in 2004, 3% of wheat farmers. Lawson et al. (2011) indicated that aerial images were used in 2010 by 10% of the German respondents, 1% of the Danish respondents and none of the Finnish respondents.
- Interest in drones and unmanned aerial vehicles has grown substantially. PA conferences and scientific journals are packed with drone presentations and articles. Some dealerships have started offering drone imaging services. The 2017 CropLife/Purdue survey showed that 32% offer drone image services. The 2017 Ontario dealership survey indicated that 21% offered drone imagery for fertilizer prescription map development.

- In the United States and Canada most agricultural input suppliers offer some kind of intensive soil sampling, usually either grid or zone based. The 2017 CropLife/ Purdue survey showed 82% of dealerships offering this service. The 2017 Ontario survey showed that 81% of the dealerships offered this service. The USDA ARMS data suggests that farmer's use of intensive soil sampling data for maps is substantially less than the service offerings: corn farmers, 2010, 22%; soybean producers, 2012, 19%; rice farmers, 2013, 12%; cotton farmers, 2007, 5%; peanut growers, 2013, 24%; and winter wheat producers, 2009, 6%. In the United Kingdom, the 2012 DEFRA survey showed 14% of farmers using some type of precision agriculture-based soil map. Lawson et al. (2011) indicated that German respondents reported 14% using grid sampling and 5% zone sampling, Danish respondents reported 5% using grid sampling and 1% zone sampling and Finnish respondent reported no grid sampling and 1% using zone sampling. Molin (2016) reports that 79% of the Brazilian large-scale farmers reported doing PA soil mapping, but the coarse scale of the sampling (often over 5 ha) raises questions about the accuracy of those soil maps.
- Soil electrical conductivity (EC) measurement for nitrogen management is a technique that predates the PA era, but dealership EC services offerings are modest and overall farmers have not been enthusiastic users of those services. The 2017 CropLife/ Purdue survey found that 34% of US dealerships offered EC soil mapping. The 2017 Ontario survey found that 11% of the dealerships offered EC mapping. USDA ARMS surveys from 2002 for soybean, 2003 for cotton and 2010 for corn showed about 1% of respondents using EC-based soil maps. Farmers and agronomists complain about difficulty of interpretation of soil EC.
- Griffin et al. (2017b) used data from the KFMA to show that VRT fertilizer application is more likely to be adopted in combination with yield monitoring and intensive soil sampling.

Overall, adoption by farmers and agribusiness of GNSS guidance and other technologies that depend primarily on location information is strong worldwide. The story on adoption of technologies that depend on a broader array of information is mixed. VRT fertilizer technologies have been adopted in some niches where the economic motivation is compelling and the practical constraints have not been overwhelming. The use of VRT N by sugar beet producers in the Red River Valley of Minnesota and North Dakota, USA, is a good example of strong VRT niche adoption. More generally for grain and oilseed producers, it is difficult to find adoption of information-intensive VRT fertilizer technology that is more than 20% of farmers or of cropland. Adoption of VRT for pesticides is even more limited. VRT seeding is growing fast, but is not yet widespread. The next section considers the degree to which economic studies have predicted the observed PA adoption pattern.

3 PA adoption and economics

In general, the current PA pattern was accurately predicted by economic studies done in the 1990s and in the early years of the twenty-first century. Those studies that looked beyond partial budgeting of variable cost and revenue changes, to include changes in management costs, training expense and personnel-related issues were particularly useful. Since GNSS guidance was introduced for ground-based agricultural equipment in the late

1990s, almost all economic studies have shown positive economic benefits which could be quantified and substantial qualitative benefits which were more difficult to measure. Consequently, the rapid adoption of this and related technologies should not come as a surprise. Since the early 1990s, VRT fertilizer economics studies have shown mixed results and the adoption record reflects that mixed picture with strong adoption in some niches, but with widespread adoption awaiting technical breakthroughs that reduce costs and enhance effectiveness. This section will review the track record of those early PA economic studies.

The first published study of GNSS guidance economics was done at Purdue University with data supplied by the companies introducing the initial products (Lowenberg-DeBoer, 1999). The study focused on the benefits of GNSS lightbar systems because autosteer was not yet on the North American market. It summarized the qualitative benefits of GNSS lightbar guidance and quantified the benefits of reduced skip and overlap minus operating costs compared to the foam and disk markers used up to that time. It showed that for custom operators, reduction in skip and overlap would pay for GNSS guidance and show a modest benefit. The situation was similar for the many farmers who had already invested in GNSS for yield monitoring. The study predicted the growth of GNSS autosteer technology in agriculture based on technology that was already being used in construction and mining at that time.

Subsequent studies added detail to the benefit estimates and the potential of related technologies. For example, Watson and Lowenberg-DeBoer (2004) added autosteer technology to the economic assessment, and considered the potential for farm expansion and controlled traffic. They concluded that autosteer would be profitable for a substantial group of US farmers. Griffin et al. (2005) put the GNSS guidance decision in a whole farm context and evaluated timeliness benefits for guidance. Batte and Ehsani (2006) added sprayer boom section control to the analysis and considered the impact of irregular field shape and field obstructions like waterways, trees and rocks. Bergtold et al. (2009) estimated benefits of autosteer in cotton production. Ortiz et al. (2013) estimated benefit of autosteer for peanut production.

The mixed results of VRT fertilizer economics studies are in contrast to the consistently positive results found in GNSS guidance studies. One of the earliest reviews of VRT fertilizer economics was done by Swinton and Lowenberg-DeBoer (1998). They reviewed nine VRT fertilizer studies with economic analysis done in the 1987–96 period. Because budgeting assumptions in the studies varied widely across studies, they attempted to standardize the methodology before making comparisons. Differences in assumptions included density of soil sampling, useful life of soil sampling information and cost of soil sampling, analysis, VRT application and recordkeeping. They found that the studies available at that time showed that VRT fertilizer was not profitable for low margin, extensive crops like wheat and barley, sometimes profitable for corn and quite profitable for higher value crops like sugar beet. They emphasized that the cost of soil nutrient information was a key factor in the results and that lowering the cost and increasing resolution of nutrient data with soil and optical sensors was a key requirement to widespread adoption of VRT fertilizer technology.

Subsequent, PA economic reviews by Lambert and Lowenberg-DeBoer (2000), Griffin et al. (2004) and Griffin and Lowenberg-DeBoer (2005) reinforced the conclusions of the Swinton and Lowenberg-DeBoer review. Lambert and Lowenberg-DeBoer reviewed 108 studies reporting economic analysis of PA technologies. Seventy-three percent of the studies reported on VRT technology mostly using partial budgets methods based on field trial data or simulation. Fewer than half (i.e. 44%) of the studies included information costs and only 30% included training and other human capacity expenses. Sixty-nine percent of the studies reported positive VRT fertilizer benefits. Four out of five sugar beet studies reported a benefit, 72% of the corn studies and only 20% of wheat studies.

Griffin et al. (2004) and Griffin and Lowenberg-DeBoer (2005) emphasized the economic advantage of 'embodied knowledge' technologies, compared with 'information-intensive' technologies like VRT. Most of the successful agricultural technology of the twentieth century was of the embodied knowledge type. For example, hybrid seed, fertilizer and most pesticides embodied substantial scientific knowledge, but did not require much expertise to use. Among PA technologies, GNSS guidance is an embodied knowledge technology and VRT fertilizer is a knowledge-intensive technology. They argued that VRT would need to become an embodied knowledge technology to achieve widespread adoption. One example of embodied knowledge VRT would be fertilizer application using a sensor mounted on the front of the machine, a computer using an algorithm on the applicator processing that sensor data, driving variable rate spreading out the back.

The economic studies in the first two decades of PA (starting from the first civilian use of GPS in the late 1980s) successfully identified the key economic opportunities for the technology, the long-term adoption patterns and the important constraints to widespread adoption. Those studies identified the potential of GNSS guidance and other locational technologies and questioned whether VRT as a knowledge-intensive technology could achieve general adoption. The economic studies were much less successful in identifying the short-term (i.e. year-to-year) patterns that speeded up or slowed down the adoption process in a specific geographic area, crop or farm type. The studies with a short term focus (e.g. Pierpaoli et al., 2013; Tey and Brindal, 2013) have been more successful in identifying the factors that could ex-post describe PA adoption trajectories. No one has been particularly successful predicting ex-ante short-run PA adoption patterns.

4 Predicting future trends based on recent studies of PA profitability

Understanding and predicting long-term PA adoption trends is important for farmers, agribusiness and research. Farmers need to make long-term plans when purchasing or leasing land, developing farm infrastructure and buying equipment. Understanding technology trends will help them make long-term decisions that facilitate the technology transition and avoid making major investments in technology that is about to become obsolete. Agribusiness profits often peak when supplying the technology and services demanded when large numbers of producers adopt a technology or innovation. Being ready to supply the innovation, and related inputs and services, is key to profitability. Research funding is often to be found for science that addresses key constraints limiting adoption of promising technologies. The PA economic studies of the last decade provide some insights on future PA adoption and are discussed in the following sections.

4.1 Sensor-based fertilizer VRT

Farmers intuitively like the idea of VRT fertilizer. It makes sense to manage different areas within fields according to the productive capacity and the constraints of that area.

However, both economic analysis and the attempts to market VRT fertilizer services and equipment show that current technology does not realize the VRT vision for most producers of agronomic crops. The information costs needed to develop VRT fertilizer prescription maps is too high and the crop response may be limited by factors beyond soil nutrients. On-the-go sensor-based systems would make VRT fertilizer more like the embodied knowledge agricultural technology that has a track record of success. On-the-go sensor-based systems could augment accuracy by increasing the information resolution and reduce costs by turning the fertilizer prescription over to an algorithm.

Unfortunately, economic analysis of currently available on-the-go sensor systems suggests that improvement is still needed. For example, Biermacher et al. (2009) analysed on-farm winter wheat experiments in Oklahoma done during the period from 1998 to 2006. Those experiments tested an optical sensor and a nitrogen fertilizer optimization algorithm developed at Oklahoma State University. The study found that sensor-based nitrogen management has substantial economic potential, but the current system is about a breakeven with traditional nitrogen management. Boyer et al. (2011) analysed winter wheat trials using optical sensors to determine nitrogen fertilizer rates done in Oklahoma from 2005 to 2009. They found that yields with the optical sensors were often slightly lower than those in uniform rate treatments, suggesting that the sensor algorithm underestimated the nitrogen needed. Scharf et al. (2011) report that in fifty-five on-farm trials in the US state of Missouri, sensor-driven nitrogen application resulted in higher maize yields and lower nitrogen use than farmer's typical practices. Scharf et al. (2011) reported that on average, sensor-driven nitrogen application resulted in a US$42/ha increase in the value of the crop minus fertilizer cost, but did not deduct the cost of the sensor system and increased management complexity. An analysis of optical sensing-based nitrogen VRT for cotton production in three states in the United States showed that while the VRT reduced fertilizer use in some cases, it did not increase profits (Stefanini et al., 2015). Kuang et al. (2014) reported that an online visible and near-infrared spectroscopy of soil pH provided negligible economic benefits for barley production in Denmark.

One of the weak points of sensor-only systems for VRT fertilizer is that they do not take into account the productive capacity of the soil in adjusting fertilizer rates. For instance, crop colour may signal lack of nitrogen to the sensor, but in fact the problem is soil depth or another constraint. Diacono et al. (2013) reviewed 17 studies of PA for nitrogen management on wheat and recommended combining sensor data with soil maps, remote sensing images, yield maps and other data to tailor the nitrogen application to the site-specific constraints. They reported that, in the studies reviewed, on-the-go N sensing had a US$5–US$60/ha profit advantage over uniform application. McFadden et al. (2018) provide a Bayesian approach to including in-season information on yield potential in the VRT fertilizer algorithm. Some crop nitrogen sensor companies are starting to introduce systems that combine sensor data with other field information.

4.2 Agricultural robotics

Economic logic indicates that robotics should be adopted quickly when the technology is adequate for repetitive agricultural tasks like seeding, weeding and harvesting. Everywhere in the world (even in developing countries) farm labour is increasingly expensive and hard to find. Many developed countries (e.g. the United States, the United Kingdom) have depended on immigrant farm labour for many years, but governments face increasing political pressure to reduce both legal and undocumented immigration.

Farms in developed countries are unlikely to be able to increase wages enough to attract local workers. Robots are the most likely alternative.

A key difference between the automation of conventional mechanized equipment which occurred with GNSS guidance, and agricultural robotics, is that a robotic agriculture is likely to use many relatively small machines instead of a few large ones (Auernhammer et al., 1995; Blackmore et al., 2004; Lowenberg-DeBoer, 2002; Gardener, 2011). Once machines are autonomous and the human beings are removed, the economic motivation for large machines almost disappears. Implementing cropping tasks with swarms of small robots will change agronomic practices and the geography of agriculture. For example with robotic pesticide application, it might be possible to spray each pest individually instead of broadcast application. This could reduce the amount of pesticide applied by over 90% and reduce the negative effects on beneficial species. With relatively small robots the economic advantage of large rectangular fields will be diminished. In the United States, the irregularly shaped fields with good soils, reliable rainfall and proximity to markets in New York and Pennsylvania could be competitive with the large rectangular fields of the Midwest and Great Plains that have more weather risk and greater distance to market. Small robots are less likely to trigger the safety and liability concerns that would keep humans on larger equipment to 'hit the off switch in case of a problem'.

Unfortunately, other than robotic milking, there is little on-farm data on robots in production agriculture. Milking robots were commercialized over 20 years ago. Numerous reports and studies have shown that they are practical and profitable for some dairy farms depending on labour cost, labour availability and other factors (e.g. Salfer et al., 2017; Hansen, 2015; Bijl et al. 2007; Armstrong and Daughety, 1997). For robots in crop production most economic studies have been based on economic engineering budgets, developed from technical parameters, not from field experience. For example, Pedersen et al. (2006) estimated benefits for weeding sugar beets, crop scouting in cereals and grass cutting on golf courses. They found that the robotic options were potentially economically competitive, with the cost of RTK-GNSS and the limited capacity of the small robots as the main constraints. With a similar methodology, Pedersen et al. (2017) found that early seeding with a small robot that could be in the field before soil conditions were dry enough for larger machinery and re-seeding with a similar robot would be profitable.

To put the robotics studies in context, from the earliest PA economic studies some observers have questioned profitability estimates based on simulation and economic engineering (e.g. Swinton and Lowenberg-DeBoer, 1998). The critics pointed out that simulation and economic engineering was a simplification of reality that would not reflect all of the constraints and risks faced on farms. However, Lambert and Lowenberg-DeBoer (2000) found that PA economic estimates based on simulation had roughly the same distribution of profitable/mixed/unprofitable results as those based on farm data.

4.3 Artificial intelligence and 'Big Data' in agriculture

AI and Big Data have been hot topics at agricultural research and industry conferences for several years. There are many academic papers on the subject. Coble et al. (2017) argued that analysis of Big Data will affect every stage in the agricultural value chain. Wolfert et al. (2017) found 114 refereed articles in English or Chinese relevant to the subject. An online search shows many blogs on the subject. In the United States agricultural input suppliers see data management as a growth area. In the 2017 CropLife survey, 58% of dealerships archived yield, soil and other data for farmers, 41% worked with farmers to analyse the data from their own farms, 17% pooled data across farmer customers and 10% participated in analysis of farm data beyond the dealership customer base. Only 31% said yield monitor data analysis was profitable for the dealership and only 20% indicated that profit/cost mapping was profitable. While it is logically appealing to use AI to make sense of the masses of data collected by sensors and other sources in agriculture, assessing the profitability and other economic benefits of this approach is more difficult. Problems in this economic assessment include:

- Defining exactly what constitutes Big Data – In most definitions, the key characteristic of Big Data is the massive amount of sensor data, but many mechanized farms have been collecting yield monitor data and accessing satellite images for years. Coble et al. (2017) argue that beyond size, Big Data in agriculture will be defined by volume, velocity, variety and veracity.
- What is AI in agriculture? – Computer algorithms have been used for many years in agriculture. For example since the 1960s, some farmers in the United States have used linear programming to plan crop rotations and make equipment choices (McCarl et al., 1977). Most animal feed in the industrialized world is formulated by computer algorithms.
- As with remote sensing, drones, yield monitoring and other information-gathering technologies in agriculture, the profitability of AI and Big Data analysis depend on the decisions made and actions taken. AI and Big Data analysis do not by themselves add to the bottom line.
- Most of the available economic analyses of AI and Big Data are based on simulation and economic engineering, rather than on field data because there are very few good examples of practical use of these concepts in commercial agriculture. For example, Li and Yost (2000) simulated the use of AI to optimize nitrogen and irrigation management, and showed a substantial economic gain. As in the case of agricultural robotics, the fact that much of the economic analysis is based on simulation rather than on-farm data is not enough of a reason to dismiss the idea, but it is a reason to press for more on-farm data.

4.4 Low-cost PA tools for medium and small farmers

So far, PA has been applied mostly on larger scale farms. One of the key challenges and opportunities is to adapt the technology for the millions of medium and small-scale farmers throughout the world (World Bank, 2017; Lowenberg-DeBoer, 2018). Boosting medium- and small-scale farm productivity through PA could help the world improve food security in the face of a population increase projected to more than nine billion. The science needed to adapt those PA tools for smaller, often unmechanized farms, is an opportunity for applied researchers. Providing the PA tools for those millions of farmers is a major opportunity for agribusiness.

The harbinger of the coming wave of PA tools for medium and small farmers is the adaptation of handheld optical sensors for use in managing nitrogen fertilizer in the developing world. Ortiz-Monasterio and Raun (2007) tested use of a sensor with wheat farmers in the Yaqui Valley in Mexico. The use of the sensor when combined with a nitrogen-rich strip for calibration saved farmers an average of US$56/ha over two crop seasons. Teboh et al. (2012) reviewed some attempts to use the N sensor technology in Africa. As with many innovations for agriculture in the developing world, one of the key challenges for PA is developing the supply chain that will provide technology, inputs and technical support at a cost that is within the purchasing power of medium and small farmers. The typical approach of many multinational agribusinesses is to sell a 'stripped down' version of their first-world technology in the developing world. Experience with other technologies suggests that going back to the original science and re-engineering a solution that fits developing countries conditions is often more successful than trying to sell first-world technology that was designed for very different conditions.

5 Future trends and conclusion

This chapter was built on the observation that widespread adoption of a technology is a good indicator of profitability or other economic benefits. With that point in mind, this chapter reviewed the publicly available data on use of precision agriculture in production of agronomic crops, identified the degree to which economic research has correctly identified PA technologies that would eventually be adopted by farmers and highlighted several PA technologies that recent economic research has identified as having profit, and consequently adoption, potential. The chapter focused on the economics of agronomic crops (e.g. cereals, oils seeds, cotton) because most PA economics research has focused on these crops, but the insights apply to PA for livestock, horticulture, forestry and other biological production systems. The chapter concentrated on PA economics for relatively large-scale mechanized farms because the economics of PA has been studied mostly for those types of farms, but has some observations about how the economics of PA might work out in medium and small farms.

PA adoption data show that GNSS guidance and related technologies like sprayer boom control and seeder row shutoffs are being adopted rapidly by mechanized farmers worldwide. Documented adoption rates are reaching the 70–80% range where a technology has become standard practice. Those same data show that use of VRT for fertilizer and other inputs is growing slowly. VRT is used by many producers of some specialized crops, like sugar beets in the US states of Minnesota and North Dakota, but for cereals, oilseeds and cotton it is difficult to find a region where more than 20% of farmers use VRT or 20% of cropland is managed with VRT.

A review of the economics research in the first two decades of PA suggests that by the late 1990s economists understood that GNSS guidance and related technologies were likely to be widely adopted in the longer run. By the late 1990s, economists had also started raising questions about the profitability of VRT for fertilizer and other inputs. They had identified the potential for reducing skip and overlap in input application to pay the GNSS guidance bills, and adoption trends to be reinforced by the reduced operator fatigue and other harder to quantify benefits. They had also raised questions about the cost of intensive soil sampling and other spatial information needed to make VRT decisions, the expense of human intervention in making VRT recommendation maps and the unpredictability of crop response to any of the inputs managed with VRT. PA economic research has been much less successful in understanding short-run adoption behaviour. A social science perspective has been able to ex-post identify some of the factors affecting PA adoption short run, but no one has been successful in predicting short-run PA adoption.

In the last decade economists have continued to study the economics of new PA technology as it emerged. The PA economics literature is now so voluminous that the kind of almost exhaustive study done by Lambert and Lowenberg-DeBoer (2000) and Griffin et al. (2004) is no longer possible. PA technology that shows economic promise and the potential for widespread adoption over the next decade includes sensor-driven VRT N, agricultural robotics, AI and Big Data, and PA for medium and small farms.

6 Where to look for further information

As the reference list for this chapter should show, there is a broad range of sources for PA economics. Some key sources of new developments are discussed in the following sections.

6.1 International Conference for Precision Agriculture (ICPA) and the European Conference for Precision Agriculture (ECPA)

There are numerous local and regional PA conferences, but the ICPA and ECPA usually provide the most comprehensive overview of PA research, including economic research. The proceedings of these conferences becomes a key record of the 'state of the art' in PA. The ICPA and ECPA are normally held in alternate years. Information about the next conferences can be found on the International Society of Precision Agriculture (ISPA) website: https://www.ispag.org/.

6.2 Refereed scientific journals

PA economics rarely captures the attention of academic economists and so it is rarely a topic in disciplinary economics journals. Advances in PA economics are more likely to be published in more applied subject matter journals like *Precision Agriculture*, https://link.springer.com/journal/11119; *Computers and Electronics in Agriculture*, https://www.journals.elsevier.com/computers-and-electronics-in-agriculture; or *Agricultural Systems*, https://www.journals.elsevier.com/agricultural-systems.

6.3 Agribusiness

Equipment manufacturers and software developers are eager to communicate benefits of their most recent products and services, but for a third-party view of the costs, benefits and potential profits, it is usually better to go to the farm media. It is their job to sift through the daily deluge of claims and identify those most worthy of attention. There are many to choose from worldwide, but in the United States they include:

- *Farm Journal AgTech* – https://www.farmjournalagtech.com/news
- *Farm Industry News* – http://www.farmindustrynews.com/
- *CropLife* magazine – focused on the agribusiness perspective – https://www.meistermedia.com/agribusiness-market/precision-farming/

7 References

Armstrong, D. V. and Daugherty, L. S. 1997. Milking robots in large dairy farms. *Computers and Electronics in Agriculture* 17(1): 123–8.

Auernhammer, H., Muhr, T. and Demmel, M. 1995. GPS and DGPS as a challenge for environmentally-friendly agriculture. *Journal of Navigation* 48(2): 268–78.

Awan, S. 2016. A review of the past, present and future of precision agriculture in the UK. Research Report No. 87. Kenilworth, UK: Agriculture and Horticulture Development Board (AHDB). https://cereals.ahdb.org.uk/media/1126475/rr87.pdf (Last accessed 1 May 2018).

Balafoutis, A. T., Beck, A., Fountas, S., Tsiropoulos, Z., Vangeyte, J., van der Wal, T., Soto-Embodas, I., Gómez-Barbero, M. and Pedersen, S. M. 2017. Smart farming technologies – Description, taxonomy and economic impact. In: Pedersen, S. and Lind, K. (Eds), *Precision Agriculture: Technology and Economic Perspectives. Progress in Precision Agriculture*. Dordrecht, the Netherlands: Springer.

Batte, M. and Ehsani, M. R. 2006. The economics of precision guidance with autoboom control for farmer-owned agricultural sprayers. *Computers and Electronics in Agriculture* 53: 28–44.

Bergtold, J. S., Raper, R. L. and Schwab, E. B. 2009. The economic benefit of improving the proximity of tillage and planting operations in cotton production with automatic steering. *Applied Engineering in Agriculture* 25(2): 133–43.

Biermacher, J., Brorsen, W. B., Epplin, F., Solie, J. and Raun, W. 2009. The economic potential of precision nitrogen application with wheat based on plant sensing. *Agricultural Economics* 40: 397–407.

Bijl, R., Kooistra, S. R. and Hogeveen, H. 2007. The profitability of automatic milking on Dutch dairy farms. *Journal of Dairy Science* 90: 239–48.

Blackmore, B. S., Fountas, S., Tang, L. and Have, H. 2004. System requirements for a small autonomous tractor. *Agricultural Engineering International: The CIGR Journal of Scientific Research and Development*. Manuscript PM 04 001. http://www.cigrjournal.org/index.php/Ejounral/article/view/525 (Last accessed 1 May 2018).

Boyer, C., Brorsen, B. W., Solie, J. B. and Raun, W. R. 2011. Profitability of variable rate nitrogen application in wheat production. *Precision Agriculture* 12: 473–87.

Castle, M., Lubben, B., Luck, J. and Mieno, T. 2017. Precision agriculture adoption and profitability. *Cornhusker Economics*. https://agecon.unl.edu/cornhusker-economics/2017/precision-agriculture-adoption-profitability (Last accessed 1 May 2018).

Coble, K., Mishra, A., Ferrell, S. L. and Griffin, T. W. 2017. Big data in agriculture: A challenge for the future. *Applied Economics Perspectives and Policy* 40(1): 79–96.

Cook, S. E., Adams, M. L. and Bramley, R. G. V. 2000. What is obstructing the wider adoption of precision agriculture technology? In: Robert, P. C., Rust, R.H and Larson, W. E. (Eds), *Proceedings of the 5th International Conference on Precision Agriculture*. Madison, WI: ASA-CSSA-SSSA (CDROM).

Department for Environment, Food and Rural Affairs (DEFRA). 2013. Farm practices survey autumn 2012 – England. https://www.gov.uk/government/uploads/system/uploads/attachment_data/file/181719/defra-stats-foodfarm-environ-fps-statsrelease-autumn2012edition-130328.pdf (Last accessed 1 May 2018).

Diacono, M., Rubino, P. and Montemurro, F. 2013. Precision nitrogen management of wheat: A Review. *Agronomy for Sustainable Development* 33: 219–41.

Erickson, B., Lowenberg-DeBoer, J. and Bradford, J. 2018. 2017 Precision agriculture dealership survey. West Lafayette, IN: Departments of Agricultural Economics and Agronomy, Purdue University. http://agribusiness.purdue.edu/precision-ag-survey (Last accessed 1 May 2018).

Franzen, D. W. 2017. Profitable use of site-specific nutrient management technologies. American Society of Agronomy. Annual Meeting Presentation 223-1. https://scisoc.confex.com/scisoc/2017am/videogateway.cgi/id/31003?recordingid=31003 (Last assessed 1 May 2018).

Gardener, D. 2011. Robotics and smart engineering. *Journal of Farm Management* 14(3): 249–70.

Griffin, T. and Lowenberg-DeBoer, J. 2005. Worldwide adoption and profitability of precision agriculture: Implications for Brazil. *Revista de Politica Agricola* 14(4): 20–38. https://seer.sede.embrapa.br/index.php/RPA/article/view/549/498 (Last accessed 1 May 2018).

Griffin, T., Lowenberg-DeBoer, J., Lambert, D., Peone, J., Payne, T. and Daberkow, S. 2004. Precision farming: Adoption, profitability and making better use of data. *2004 Triennial North Central Farm Management Conference*, Lexington, KY. https://www.agriculture.purdue.edu/ssmc/ (Last accessed 1 May 2018).

Griffin, T. W., Lambert, D. M. and Lowenberg-DeBoer, J. 2005. Economics of GPS lightbar navigation and auto-guidance technologies. In: Stafford, J. (Ed.), *Precision Agriculture-2005 Proceedings of the 5th European Conference on Precision Agriculture*. Wageningen, the Netherlands: Wageningen Academic Publishers, pp. 581–7.

Griffin, T. W., Shockley, J. M. and Mark, T. B. 2017a. Economics of precision farming. In: Shannon, D. K. and Clay, D. (Eds), *Precision Agriculture Basics*. Madison, WI: American Society of Agronomy, Crop Science Society of America and Soil Science Society of America.

Griffin, T. W., Miller, N. J., Bergtold, J., Shanoyan, A., Sharda, A. and Ciampitti, I. A. 2017b. Farm's sequence of adoption of information intensive precision agriculture technology. *Applied Engineering in Agriculture* 33: 1–7.

Hansen, B. G. 2015. Robotic milking – farmer experiences and adoption rate in Jaeren, Norway. *Journal of Rural Studies* 41: 109–17.

Kitchen, N. 2008. Emerging technologies for real-time and integrated agriculture decisions. *Computers and Electronics in Agriculture* 61: 1–3.

Kuang, B., Tekin, Y., Waine, T. and Mouazen, A. M. 2014. Variable rate lime application based on on-line visible and near infrared (vis-NIR) spectroscopy measurement of soil properties in a Danish field. *Proceedings of the AgEng Conference*, Zurich, Switzerland, 6–10 July 2014. http://www.geyseco.es/geystiona/adjs/comunicaciones/304/C03770001.pdf (Last accessed 30 April 2018).

Lambert, D. and Lowenberg-DeBoer, J. 2000. Precision farming profitability review. www.agriculture.purdue.edu/ssmc (Last accessed 1 May 2018).

Lawson, L. G., Pedersen, S. M., Sørensen, C. G., Pesonen, L., Fountas, S., Werner, A., Oudshoorn, F. W., Herold, L., Chatzinikos, T., Kirketerp, I. M. and Blackmore, S. 2011. A four nation survey of farm information management and advanced farming systems: A descriptive analysis of survey responses. *Computers and Electronics in Agriculture* 77: 7–20.

Li, M. and Yost, R. S. 2000. Management-oriented modeling: Optimizing Nitrogen management with artificial intelligence. *Agricultural Systems* 65: 1–27.

Llewellyn, R. and Ousman, J. 2014. Adoption of precision agriculture-related practices: Status, opportunities and the role of farm advisors. CSIRO. https://grdc.com.au/resources-and-publications/all-publications/publications/2014/12/adoption-of-precision-agriculture-related-practices (Last accessed 1 May 2018).

Lowenberg-DeBoer, J. 1998. Adoption patterns for precision agriculture. Agricultural Machine Systems SP-1383. Warrendale, PA: Society of Automotive Engineers. Also available as SAE Technical paper 982041. https://www.sae.org/publications/technical-papers/content/982041/ (Last accessed 1 May 2018).

Lowenberg-DeBoer, J. 1999. GPS based guidance systems for agriculture. Purdue Agricultural Economics Report. Purdue University, December 1999, pp. 8–9. https://ag.purdue.edu/agecon/Documents/PAER_December%201999.pdf (Last accessed 1 May 2018).

Lowenberg-DeBoer, J. 2002. Liability problems for autonomous equipment? Site-Specific Management Center Newsletter (February 2002). www.agriculture.purdue.edu/ssmc (Last accessed 1 May 2018).

Lowenberg-DeBoer, J. 2018. The future of agriculture: Challenges and opportunities of precision agriculture for medium and small-scale farms. InterAmerican Development Bank. https://blogs.iadb.org/sostenibilidad/en/2018/01/10/the-future-of-agriculture/ (Last accessed 1 May 2018).

Lowenberg-DeBoer, J. and Swinton, S. 1997. Economics of site-specific management of agronomic crops. In: Pierce, F. J., Robert, P. C. and Sadler, J. D. (Eds), *The State of Site-Specific Management in Agriculture*. Madison, WI: ASA-CSSA-SSSA.

McCarl, B., Candler, W., Doster, D. H. and Robbins, P. 1977. Experiences with farmer oriented linear programing for crop planning. *Canadian Journal of Agricultural Economics* 25(1): 17–30.

McFadden, B. B., Brorsen, W. and Raun, W. 2018. Nitrogen fertilizer recommendations based on plant sensing and bayesian updating. *Precision Agriculture* 19: 79–92.

Méndez, A., Vélez, J., Villarroel, D. and Scaramuzza, F. 2014. Evolución de la Agricultura de Precisión en Argentina en los últimos 15 años (Evolution of Precision Agriculture in Argentina in the last 15 Years). *13th International Precision Agriculture Course*, 24–25 September. Manfredi, Argentina: INTA. http://www.agriculturadeprecision.org/13roCursoAgPrec/Libro/g4_Libro13CursoAP2014.pdf (Last accessed 1 May 2018).

Miller, N., Griffin, T., Bergtold, J., Sharda, A. and Ciampitti, I. 2017. Adoption of precision agriculture technology bundles on Kansas farms. Selected Paper. Mobile, AL: Southern Agricultural Economics Association, 4–7 February 2017.

Mitchell, S., Weersink, A. and Erickson, B. 2017. Precision agriculture in Ontario: 2017 precision agriculture services dealership survey. Working Paper Series – WP 17-01. Canada: Institute for the Advanced Study of Food and Agricultural Policy, Department of Food, Agriculture and Resource Economics, University of Guelph.

Molin, J. 2016. Precision agriculture in Latin America. *INFO AG Conference*. https://www.infoag.org/presentations/2378.pdf.

National Research Council. 1997. *Precision Agriculture in the 21st Century: Geospatial and Information Technologies in Crop Management*. Washington DC: National Academy Press.

OECD. 2016. Farm management practices to foster green growth. OECD Publishing. http://dx.doi.org/10.1787/9789264238657-en

Ortiz, B. V., Balkcom, K. B., Duzy, L., van Santen, E. and Hartzog, D. L. 2013. Evaluation of agronomic and economic benefits of using RTK-GPS-based auto-steer guidance systems for peanut digging operations. *Precision Agriculture* 14: 357–75.

Ortiz-Monasterio, J. I. and Raun, W. 2007. Reduced nitrogen and improved farm income for irrigated spring wheat in the Yaqui Valley, Mexico, using sensor based nitrogen management. *Journal of Agricultural Science* 145: 215–22. doi:10.1017/S002185960700699.

Pedersen, S. M., Fountas, S., Have, H. and Blackmore, B. S. 2006. Agricultural robots – System analysis and economic feasibility. *Precision Agriculture* 7: 295–308.

Pedersen, S. M., Fountas, S., Sørensen, C. G., Van Evert, F. K. and Blackmore, B. S. 2017. Chapter 8: Robotic seeding: Economic perspectives. In: Pedersen, S. M. and Lind, K. M. (Eds), *Precision Agriculture: Technology and Economic Perspectives*. Heidelberg, Germany: Springer International Publishing.

Pickthall, T. and Trivett, E. 2017. An investigation into the barriers that prevent the adoption of precision farming technologies in combinable cropping in the UK. *Aspects of Applied Biology 135: Precision Systems in Agricultural and Horticultural Production*, pp. 29–37.

Pierpaoli, E., Carli, G., Pignatti, E. and Canavari, M. 2013. Drivers of precision agriculture technologies adoption: A literature review. *Procedia Technology* 8: 61–9.

Robert, P., Rust, R. and Larson, W. 1994. Site-specific management for agricultural systems. In: *Proceedings of the 2nd International Conference on Precision Agriculture*. Madison, WI: ASA-CSSA-SSSA.

Salfer, J., Endres, M., Lazarus, W., Minegishi, K. and Berning, B. 2017. Dairy robotic milking systems – What are the economics? University of Minnesota Extension. http://articles.extension.org/pages/73995/dairy-robotic-milking-systems-what-are-the-economics (Last accessed 1 May 2018).

Scaramuzza, F., Villarroel, D. and Vélez, J. P. 2017. Incremento de las herramientas de agricultura de precisión. Mayor automatismo y precision para la sustentabilidad del ambiente productivo (Increase in Precision Agriculture Tools. Greater Automation and Precision for the Sustainability of the Productive Environment). *16th International Course in Precision Agriculture and Livestock*, 29 September 2017. Manfredi, Argentina: INTA E.E.A. http://www.agriculturadeprecision.org/16toCursoAgPrec/Libro/03_Scaramuzza_F_et_al_Incremento-de-las-Herramientas-de-Agricultura-de-Precision.pdf (Last accessed 1 May 2018).

Scharf, P., Shannon, D. K., Palm, H. L., Sudduth, K. A., Drummond, S. T., Kitchen, N. R., Mueller, L. J., Hubbard, V. C. and Oliveira, L. F. 2011. Sensor-based nitrogen applications out-performed producers-chosen rates for corn in on-farm demonstrations. *Agronomy Journal* 103(6): 1683–91. http://extension.missouri.edu/sare/documents/ProducerStudies2012.pdf (Last accessed 1 May 2018).

Schimmelpfennig, D. 2016. Farm profits and the adoption of precision agriculture. Economic Research Report 217. USDA Economic Research Service.

Stefanini, M., Larson, J., Boyer, C., Cho, S-H., Lambert, D. and Yin, X. 2015. Profitability of variable rate technology in cotton production. Selected Paper. *Southern Agricultural Economics Association Annual Meeting*, Atlanta, GA. https://ageconsearch.umn.edu/bitstream/196995/2/SAEA_CottonVRT.pdf (Last accessed 1 May 2018).

Swinton, S. M. and Lowenberg-DeBoer, J. 1998. Evaluating the profitability of site-specific farming. *Journal of Production Agriculture* 11: 439–46.

Teboh, J., Tubaña, B. Udeigwe, T., Emenback, Y. Y. and Lofton, J. 2012. Applicability of ground-based remote sensors for crop N management in Sub Saharan Africa. *Journal of Agricultural Science* 4: 175–88. http://www.ccsenet.org/journal/index.php/jas/article/view/11757 (Last accessed 1 May 2018).

Tenkorang, F. and Lowenberg-DeBoer, J. 2008. On-farm profitability of remote sensing in agriculture. *Journal of Terrestrial Observation* 1(1): 50–9. https://docs.lib.purdue.edu/jto/vol1/iss1/art6/ (Last accessed 1 May 2018).

Tey, Y. S. and Brindal, M. 2013. Factors influencing the adoption of precision agricultural technologies: A review for policy implications. *Precision Agriculture* 13: 713–30.

Thorp, K. 2014. Precision agriculture. In: Njoku, E. G. (Ed.), *Encyclopedia of Remote Sensing. Encyclopedia of Earth Sciences Series*. New York, NY: Springer.

USDA. 2018. Tailored reports: Crop production practices. https://data.ers.usda.gov/reports.aspx?ID=17883 (Last accessed 1 May 2018).

Watson, M. and Lowenberg-DeBoer, J. 2004. Who will benefit from GPS auto guidance in the Corn Belt? Purdue Agricultural Economics Report. http://www.agecon.purdue.edu/extension/pubs/paer/2004/paer0204.pdf (Last accessed 1 May 2018).

Whelan, B. M. and McBratney, A. B. 2000. The 'Null Hypothesis' of precision agriculture management. *Precision Agriculture* 2(3): 265–79.

Wolfert, S., Ge, L., Verdouw, C. and Bogaardt, M.-J. 2017. Big data in smart farming – A review. *Agricultural Systems* 153:69–80.

World Bank. 2017. *ICT in Agriculture (Updated Edition): Connecting Smallholders to Knowledge, Networks, and Institutions*. Washington DC: World Bank. https://openknowledge.worldbank.org/handle/10986/27526 (Last accessed on 1 May 2018).

Zarco-Tejada, P., Hubbard, N. and Loudjani, P. 2014. Precision agriculture: An opportunity for EU farmers – Potential support with the CAP 2014–2020. Directorate General for Internal Policies, Policy Department B: Structural and Cohesion Policies, European Parliament Committee on Agriculture and Rural Development. http://www.europarl.europa.eu/RegData/etudes/note/join/2014/529049/IPOL-AGRI_NT%282014%29529049_EN.pdf (Last accessed on 1 May 2018).

Zhang, N., Wang, M. and Wang, N. 2002. Precision agriculture – A worldwide overview. *Computers and Electronics in Agriculture* 36: 113–32.

Index